VOLUME ONE HUNDRED AND FIFTY THREE

ADVANCES IN
CANCER RESEARCH

RAS: Past, Present, and Future

VOLUME ONE HUNDRED AND FIFTY THREE

ADVANCES IN CANCER RESEARCH

RAS: Past, Present, and Future

Edited by

JOHN P. O'BRYAN
Department of Cell and Molecular Pharmacology and Experimental Therapeutics, Hollings Cancer Center, Medical University of South Carolina;
Ralph H. Johnson VA Medical Center, Charleston, SC, United States

GARY A. PIAZZA
Department of Drug Discovery and Development, Harrison School of Pharmacy, Auburn University, Auburn, AL, United States

ELSEVIER

ACADEMIC PRESS
An imprint of Elsevier

Academic Press is an imprint of Elsevier
50 Hampshire Street, 5th Floor, Cambridge, MA 02139, United States
525 B Street, Suite 1650, San Diego, CA 92101, United States
The Boulevard, Langford Lane, Kidlington, Oxford OX5 1GB, United Kingdom
125 London Wall, London, EC2Y 5AS, United Kingdom

First edition 2022

ISBN: 978-0-12-824485-2
ISSN: 0065-230X

For information on all Academic Press publications
visit our website at https://www.elsevier.com/books-and-journals

Publisher: Zoe Kruze
Developmental Editor: Naiza Ermin Mendoza
Production Project Manager: James Selvam
Cover Designer: Greg Harris

Typeset by STRAIVE, India

Contents

Contributors ix
Preface xiii

1. A brief history of RAS and the RAS Initiative **1**
Frank McCormick
1. Discovery of v-Ha-RAS, v-Ki-RAS and N-RAS 1
2. Discovery of GAPs 5
3. Discovery of RAS GEFs 6
4. Discovery of RAF kinases and their relationship to RAS 11
5. The RAS Initiative at Frederick National Laboratories 16
References 18

2. Not all RAS mutations are equal: A detailed review of the functional diversity of RAS hot spot mutations **29**
Rachel A. Burge and G. Aaron Hobbs
1. Introduction 30
2. RAS mutational frequencies in cancer 32
3. *RAS* mutational frequencies across cancer 35
4. All RAS mutant proteins are not equal 38
5. RAS effector interactions and mutational frequencies 40
6. Intrinsic nucleotide (GDP and GTP) dissociation and GTP hydrolysis 44
7. GEFs and GAPs 51
8. Conclusions 53
Acknowledgments 54
References 54

3. Drug targeting opportunities en route to Ras nanoclusters **63**
Karolina Pavic, Rohan Chippalkatti, and Daniel Abankwa
1. Introduction 64
2. Ras processing and trafficking 69
3. Trafficking-factors and -chaperones of Ras 73
4. Organization of Ras in specific plasma membrane lipid domains and nanocluster 79
5. Lipid domain and Ras nanocluster modulating factors 82
6. Concluding remarks 88
References 89

4. Targeting the ERK mitogen-activated protein kinase cascade for the treatment of KRAS-mutant pancreatic cancer **101**

J. Nathaniel Diehl, Priya S. Hibshman, Irem Ozkan-Dagliyan, Craig M. Goodwin, Sarah V. Howard, Adrienne D. Cox, and Channing J. Der

1. Introduction 102
2. *KRAS*: The predominant RAS gene mutated in human cancer 103
3. Targeting KRAS for cancer treatment: Direct inhibition of RAS 104
4. Targeting KRAS for cancer treatment: A focus on RAS effector signaling 105
5. The ERK MAPK signaling network 107
6. ERK feedback inhibition and the Goldilock Principle: A "just right" level of ERK signaling supports cancer growth 109
7. Inhibition of the RAF-MEK-ERK cascade: Promises and challenges 110
8. ERK regulation of a diverse phosphoproteome and transcriptome 113
9. MYC: The critical determinant for ERK-dependent cancer growth? 117
10. KRAS, ERK, MYC and metabolism: An Achilles heel for KRAS-mutant PDAC? 119

Acknowledgments 121
Disclosure of potential conflicts of interest 121
References 121

5. Pan-RAS inhibitors: Hitting multiple RAS isozymes with one stone **131**

Alexander B. Coley, Antonio Ward, Adam B. Keeton, Xi Chen, Yulia Maxuitenko, Aishwarya Prakash, Feng Li, Jeremy B. Foote, Donald J. Buchsbaum, and Gary A. Piazza

1. Introduction 132
2. PAN-RAS inhibitors 138
3. Discussion 156
4. Conclusion 159

Grant support 159
Disclosure of potential conflicts of interest 159
References 159

6. Targeting RAS oncogenesis with SOS1 inhibitors **169**

Roman Christian Hillig and Benjamin Bader

1. Introduction: RAS proteins and their Guanine Nucleotide Exchange Factors (GEFs) 170
2. SOS1 as a molecular target for cancer therapy 174
3. Discovery of the hydrophobic pocket on SOS1 175
4. More binders for the hydrophobic pocket 183

5. The long way from activators to inhibitors 185
6. Molecular basis for SOS1 activation versus inhibition 189
7. Selectivity against SOS2 and other RasGEFs 190
8. Activity of SOS1 inhibitors in cellular assays 191
9. SOS1 inhibition affects RAS mutations with residual intrinsic GTPase activity 193
10. SOS1 inhibitors in RASopathies 193
11. SOS1 inhibitors in combination therapy 194
12. Clinical studies 196
13. Conclusions 196
References 197

7. Inhibition of SHP2 as an approach to block RAS-driven cancers 205

Yu-Ting Chou and Trever G. Bivona

1. Introduction 206
2. Biology of RTK/RAS signaling 207
3. General understanding of RAS-driven cancer and treatment 209
4. The biological and oncogenic role of SHP2 211
5. SHP2 inhibitors as potential therapies in RAS-driven cancers 220
6. Conclusions and perspectives 226
Acknowledgments 227
Potential competing interests 228
References 228

8. Targeting the "undruggable" RAS with biologics 237

Michael Whaby, Imran Khan, and John P. O'Bryan

1. Introduction 237
2. Anti-RAS biologics 240
3. Targeting RAS using peptide-PROTACs: A new frontier for RAS inhibition 255
4. Conclusion and future outlook 259
Acknowledgments 259
Disclosures 260
References 260

9. Unraveling and targeting RAS-driven metabolic signaling for therapeutic gain 267

Jonathan M. DeLiberty, Ryan Robb, Claire E. Gates, and Kirsten L. Bryant

1. Introduction 268
2. Glycolysis 274
3. Mitochondrial signaling and dynamics 277

4. Autophagy 280
5. Lipid scavenging and metabolism 289
6. Macropinocytosis 290
7. Concluding remarks 291
Acknowledgments 292
References 292

10. The RASopathies: Biology, genetics and therapeutic options 305
Jody Fromm Longo and Steven L. Carroll

1. Introduction 306
2. Neurofibromatosis type 1 (NF1) and related syndromes 306
3. Legius syndrome 317
4. Noonan syndrome and related syndromes 319
5. Noonan-related syndromes 322
6. Costello syndrome 324
7. Cardiofaciocutaneous syndrome 325
8. Gingival fibromatosis 326
9. Capillary malformation-arteriovenous malformation syndrome 326
10. Future directions 327
References 328

Contributors

Daniel Abankwa
Cancer Cell Biology and Drug Discovery Group, Department of Life Sciences and Medicine, University of Luxembourg, Esch-sur-Alzette, Luxembourg

Benjamin Bader
Screening, Lead Discovery, Nuvisan ICB GmbH, Berlin, Germany

Trever G. Bivona
Department of Medicine, Division of Hematology and Oncology, and The Helen Diller Comprehensive Cancer Center, University of California, San Francisco, CA, United States

Kirsten L. Bryant
Department of Pharmacology; Lineberger Comprehensive Cancer Center, University of North Carolina at Chapel Hill, Chapel Hill, NC, United States

Donald J. Buchsbaum
Department of Radiation Oncology, University of Alabama at Birmingham, Birmingham, AL, United States

Rachel A. Burge
Department of Cell and Molecular Pharmacology and Experimental Therapeutics, Medical University of South Carolina, Charleston, SC, United States

Steven L. Carroll
Department of Pathology and Laboratory Medicine, Medical University of South Carolina, Charleston, SC, United States

Xi Chen
Department of Drug Discovery and Development, Harrison School of Pharmacy, Auburn University, Auburn, AL, United States

Rohan Chippalkatti
Cancer Cell Biology and Drug Discovery Group, Department of Life Sciences and Medicine, University of Luxembourg, Esch-sur-Alzette, Luxembourg

Yu-Ting Chou
Department of Medicine, Division of Hematology and Oncology, and The Helen Diller Comprehensive Cancer Center, University of California, San Francisco, CA, United States

Alexander B. Coley
Department of Pharmacology, University of South Alabama; Mitchell Cancer Institute, Mobile, AL, United States

Adrienne D. Cox
Cell Biology and Physiology Curriculum; Department of Pharmacology; Lineberger Comprehensive Cancer Center; Department of Radiation Oncology, University of North Carolina at Chapel Hill, Chapel Hill, NC, United States

Jonathan M. DeLiberty
Department of Pharmacology, University of North Carolina at Chapel Hill, Chapel Hill, NC, United States

Channing J. Der
Curriculum in Genetics and Molecular Biology; Cell Biology and Physiology Curriculum; Department of Pharmacology; Lineberger Comprehensive Cancer Center, University of North Carolina at Chapel Hill, Chapel Hill, NC, United States

J. Nathaniel Diehl
Curriculum in Genetics and Molecular Biology, University of North Carolina at Chapel Hill, Chapel Hill, NC, United States

Jeremy B. Foote
Department of Microbiology, University of Alabama at Birmingham, Birmingham, AL, United States

Claire E. Gates
Department of Pharmacology, University of North Carolina at Chapel Hill, Chapel Hill, NC, United States

Craig M. Goodwin
Lineberger Comprehensive Cancer Center, University of North Carolina at Chapel Hill, Chapel Hill, NC, United States

Priya S. Hibshman
Cell Biology and Physiology Curriculum, University of North Carolina at Chapel Hill, Chapel Hill, NC, United States

Roman Christian Hillig
Structural Biology, Lead Discovery, Nuvisan ICB GmbH, Berlin, Germany

G. Aaron Hobbs
Department of Cell and Molecular Pharmacology and Experimental Therapeutics; Hollings Cancer Center, Medical University of South Carolina, Charleston, SC, United States

Sarah V. Howard
Lineberger Comprehensive Cancer Center, University of North Carolina at Chapel Hill, Chapel Hill, NC, United States

Adam B. Keeton
Department of Drug Discovery and Development, Harrison School of Pharmacy, Auburn University, Auburn, AL, United States

Imran Khan
Department of Cell and Molecular Pharmacology and Experimental Therapeutics, Hollings Cancer Center, Medical University of South Carolina; Ralph H. Johnson VA Medical Center, Charleston, SC, United States

Feng Li
Department of Drug Discovery and Development, Harrison School of Pharmacy, Auburn University, Auburn, AL, United States

Jody Fromm Longo
Department of Pathology and Laboratory Medicine, Medical University of South Carolina, Charleston, SC, United States

Yulia Maxuitenko
Department of Drug Discovery and Development, Harrison School of Pharmacy, Auburn University, Auburn, AL, United States

Frank McCormick
Helen Diller Family Comprehensive Cancer Center, University of California, San Francisco, San Francisco, CA; Frederick National Laboratory for Cancer Research, Frederick, MD, United States

John P. O'Bryan
Department of Cell and Molecular Pharmacology and Experimental Therapeutics, Hollings Cancer Center, Medical University of South Carolina; Ralph H. Johnson VA Medical Center, Charleston, SC, United States

Irem Ozkan-Dagliyan
Department of Pharmacology, University of North Carolina at Chapel Hill, Chapel Hill, NC, United States

Karolina Pavic
Cancer Cell Biology and Drug Discovery Group, Department of Life Sciences and Medicine, University of Luxembourg, Esch-sur-Alzette, Luxembourg

Gary A. Piazza
Department of Drug Discovery and Development, Harrison School of Pharmacy, Auburn University, Auburn, AL, United States

Aishwarya Prakash
Mitchell Cancer Institute; Department of Biochemistry & Molecular Biology, University of South Alabama, Mobile, AL, United States

Ryan Robb
Department of Pharmacology, University of North Carolina at Chapel Hill, Chapel Hill, NC, United States

Antonio Ward
Department of Pharmacology, University of South Alabama; Mitchell Cancer Institute, Mobile, AL, United States

Michael Whaby
Department of Cell and Molecular Pharmacology and Experimental Therapeutics, Hollings Cancer Center, Medical University of South Carolina, Charleston, SC, United States

Preface

Cancer represents the second leading cause of death in the United States and is one of the leading causes of death worldwide. The *RAS* family of oncogenes is frequently mutated in cancer and is an established driver of tumor development and progression. Indeed, three of the four most deadly cancers (pancreatic, lung, and colorectal) in the United States harbor the highest incidence of activating *RAS* mutations. Since the discovery of *RAS* as the transforming gene in both Harvey and Kirsten rat sarcoma viruses, coupled with the identification of mutated *RAS* alleles in human tumor samples in the early 1980s, there has been an explosion of studies to develop anti-RAS drugs. This year witnessed a historic achievement in this effort with the FDA approval of the first-in-class direct RAS inhibitor, sotorasib (LUMAKRAS™), to treat KRAS(G12C) mutant nonsmall cell lung cancers. This crowning achievement marks the culmination of decades of research to study various facets of RAS vulnerabilities and develop RAS-directed anticancer drugs. This success illustrates that the once "undruggable" RAS is indeed pharmacologically tractable. Although sotorasib provides a new weapon in our arsenal to treat cancer patients, its applicability is unfortunately limited to the subset of *KRAS(G12C)*-mutant cancers. In addition, reports describing various mechanisms of resistance have already appeared in the literature. This limitation highlights the remaining urgent need to develop more broadly acting anti-RAS drugs.

This volume of *Advances in Cancer Research* brings together a collection of reviews from experts in the field that provides an up-to-date overview of RAS biology, its role in cancer, and the various efforts to inhibit its activity. Chapter 1 provides a historical overview of the seminal discoveries in the RAS field combined with a discussion of the National Cancer Institute's RAS Initiative, which was established in 2013 to accelerate the development of new therapies for *RAS*-mutant cancers. Chapter 2 provides a detailed discussion of the differences in *RAS* mutation frequency between the three RAS isoforms, along with a description of the unique biochemical and biological properties of various mutant RAS proteins. Although many have viewed the different mutant RAS proteins as functionally equivalent, it is becoming clear that not all RAS mutants are created equal. While the biological and clinical importance of these differences remains an area of fundamental interest, the success of sotorasib provides a clear example of how the unique chemical properties of a specific RAS mutant can be exploited for the development of new anti-RAS therapies.

Chapters 3–7 provide in-depth overviews of various strategies to pharmacologically inhibit RAS using small molecule-based strategies. These include agents that interfere with membrane localization of RAS (Chapter 3), inhibitors of the highly conserved ERK signaling cascade (Chapter 4), pan-RAS inhibitors (Chapter 5), inhibitors of upstream regulators of RAS such as the guanine nucleotide exchange factors that promote loading of RAS with GTP (Chapter 6), and inhibitors of the protein tyrosine phosphatase SHP2 that has emerged as an important regulator of RAS activation (Chapter 7). Chapter 8 discusses the use of biologics as a strategy to uncover vulnerabilities in RAS mutants that can be exploited to interfere with its oncogenic activity in cells.

Although Otto Warburg observed in the 1920s that cancer cells have a higher metabolic consumption of glucose compared to normal cells, it was not until many decades later that researchers began to focus on the metabolic differences of tumor cells vs normal tissue as a potential Achilles' heel of cancer. Chapter 9 provides an overview of our current understanding of the metabolic vulnerabilities of KRAS-mutant cancers and how these properties are being exploited for the development of new cancer therapies. Finally, the volume concludes with a discussion of the genetics, biology, and therapeutic approaches to treat a class of inherited disorders referred to as RASopathies due to the presence of germline mutations in various components of the RAS signaling pathway.

Given the importance of RAS in human cancer and the recent clinical success of KRAS(G12C) inhibitors, we hope this volume provides the reader with a greater appreciation of the approaches being used to pharmacologically inhibit RAS. It is clear that much progress has been made in successfully drugging the once "undruggable" RAS, but more research is needed to develop additional RAS inhibitors. We look forward to future discoveries and successes with new RAS-specific pharmacologics.

John P. O'Bryan, PhD
Gary A. Piazza, PhD

CHAPTER ONE

A brief history of RAS and the RAS Initiative

Frank McCormick*
Helen Diller Family Comprehensive Cancer Center, University of California, San Francisco, San Francisco, CA, United States
Frederick National Laboratory for Cancer Research, Frederick, MD, United States
*Corresponding author: e-mail address: frank.mccormick@ucsf.edu

Contents

1. Discovery of v-Ha-RAS, v-Ki-RAS and N-RAS 1
2. Discovery of GAPs 5
3. Discovery of RAS GEFs 6
   3.1 RAS processing 7
   3.2 RAS structures 8
   3.3 The RAS superfamily 9
4. Discovery of RAF kinases and their relationship to RAS 11
   4.1 RAS and PI3′ kinase 13
   4.2 Which effectors drive RAS cancers? 13
   4.3 Therapeutic strategies 15
5. The RAS Initiative at Frederick National Laboratories 16
References 18

Abstract

In this review, I provide a brief history of the discovery of RAS and the GAPs and GEFs that regulate its activity from a personal perspective. Much of this history has been driven by technological breakthroughs that occurred concurrently, such as molecular cloning, cDNA expression to analyze RAS proteins and their structures, and application of PCR to detect mutations. I discuss the RAS superfamily and RAS proteins as therapeutic targets, including recent advances in developing RAS inhibitors. I also describe the role of the RAS Initiative at Frederick National Laboratory for Cancer Research in advancing development of RAS inhibitors and providing new insights into signaling complexes and interaction of RAS proteins with the plasma membrane.

1. Discovery of v-Ha-RAS, v-Ki-RAS and N-RAS

RAS genes were first identified in 1964 by Jennifer Harvey (Harvey, 1964). While passaging Moloney Leukemia Virus, Harvey noticed that

Advances in Cancer Research, Volume 153
ISSN 0065-230X
https://doi.org/10.1016/bs.acr.2021.07.003

plasma collected from leukemic rats acquired the ability to generate aggressive and deadly tumors. Recombination had occurred between the helper independent mouse C-type virus and novel sequences from the rats in which they were grown, generating Harvey sarcoma virus (Ha-SV). Kirsten sarcoma virus (Ki-SV) was identified by a similar protocol (Kirsten & Mayer, 1967).

Both Ha-SV and Ki-SV initiate tumors in vivo with remarkable efficiency. The oncogenes of these viruses each contain activating mutations at codon 12: G12R (Ha-SV) and G12S (Ki-SV) and both contain a second mutation, A59T. This latter mutation is rare in human cancers (0.08% prevalence), and confers unusual biochemical features that are only now being explored (Lou et al., 2021). The combination of two activating mutations in the same RAS gene has never been reported in human tumors. Nevertheless, this threonine played an important role in developing our understanding of RAS protein function. Metabolic labeling revealed that Ha-SV and Ki-SV p21 proteins are phosphoproteins, in common with many oncoproteins under intense scrutiny at that time, but phosphorylation derived from GTP not ATP (Shih, Papageorge, Stokes, Weeks, & Scolnick, 1980). This led to the major insight that these proteins may be related to heterotrimeric G-protein and were not protein kinases like many other oncoproteins under analysis at that time. Armed with these potent retroviruses, cell lines that they had transformed and monoclonal antibodies for their detection and purification (Furth, Davis, Fleurdelys, & Scolnick, 1982), considerable progress was made in characterizing the biology and biochemistry of Ha-RAS and Ki-RAS oncogenes.

1979 marked the beginning of a seismic shift in molecular oncology: Dr. Bob Weinberg showed that genomic DNA from a tumor cell line derived from chemical carcinogens could transform recipient normal cells in culture (Shih, Shilo, Goldfarb, Dannenberg, & Weinberg, 1979). The field then moved quickly from analysis of genes encoded by retroviruses, to genes encoded by tumor cell genomes. The series of landmark experiments from the laboratories of Drs. Cooper, Weinberg, Wigler, Aaronson and Barbacid that led to the identification HRAS G12V in human tumor-derived DNA has been told many times (Malumbres & Barbacid, 2003). A detailed history of the discovery of the *KRAS* gene has been published recently (Tsuchida, Murugan, & Grieco, 2016). Essential contributions included development of simple assays for transformation in cell culture, development of methods for transferring genomic DNA into cells, techniques for recognizing the human DNA in the background of recipient mouse DNA, and newly developed techniques for cloning and sequencing DNA.

Once HRAS was identified, cDNA cloning and expression in *E. coli* followed soon after. This was not a trivial exercise at that time, as techniques for manipulating cDNAs and producing recombinant proteins were still being developed. It is notable that the first description of recombinant HRAS proteins came from Genentech (McGrath, Capon, Goeddel, & Levinson, 1984), Merck (Gibbs, Sigal, Poe, & Scolnick, 1984) and Smith Kline and French (Sweet et al., 1984), who were already heavily invested in this technology for producing recombinant proteins in bacteria for therapeutic purposes. As expected, these proteins bound GTP and GDP, but a critical difference emerged: the G12V mutant had less GTPase activity than wild type, by a factor of almost 10-fold. This suggested that oncogenic mutations arise because RAS proteins cannot hydrolyze GTP and therefore remain in their GTP state persistently.

The technique of transfecting DNA derived from human tumors onto NIH/3T3 cells, followed by analysis of transformed foci, led to the discovery of many other oncogenes. DNA from the neuroblastoma cell line SK-N-SH, as well as the HL60 leukemia cells, and HT1080 and RD sarcomas, expressed activated versions of a new member of the RAS family designated NRAS (Hall, Marshall, Spurr, & Weiss, 1983; Taparowsky, Shimizu, Goldfarb, & Wigler, 1983).

Molecular analysis of KRAS (McGrath et al., 1983; Shimizu et al., 1983) revealed the presence of alternative spliced forms, KRAS4A and KRAS4B. The v-Ki-RAS p21 protein is KRAS4A, clearly a powerful oncogene when activated by mutation. However, KRAS4B is more highly expressed in most tissues and cancer cells, though KRAS4A expression is comparable to KRAS4B in some cases (Tsai et al., 2015). Fig. 1 shows that the KRAS4A and 4B proteins differ at their HVR sequences and indeed have unique functions. KRAS4A binds to hexokinase 1 (Amendola et al., 2019) and to SIN1 (Castel et al., 2021). KRAS4B binds to calmodulin (Villalonga et al., 2001) and is phosphorylated by PKC (Bivona et al., 2006). KRAS4B has a unique role in conferring stem-like properties to cancer cells (Quinlan, Quatela, Philips, & Settleman, 2008; Wang et al., 2015). Obviously, many interesting questions remain relating to these variants. From a drug discovery point of view, it seems likely that strategies for targeting mutant KRAS proteins will need to block functions of both proteins to be effective.

Forty years after their discovery in human tumor DNA, RAS genes have emerged as the most frequently mutated oncogene in human cancer. One giant step in the process of cataloging RAS' role in human cancer was the application the newly invented Polymerase Chain Reaction to amplify RAS genes and facilitate their sequencing. Dr. Hans Bos, was one of the first

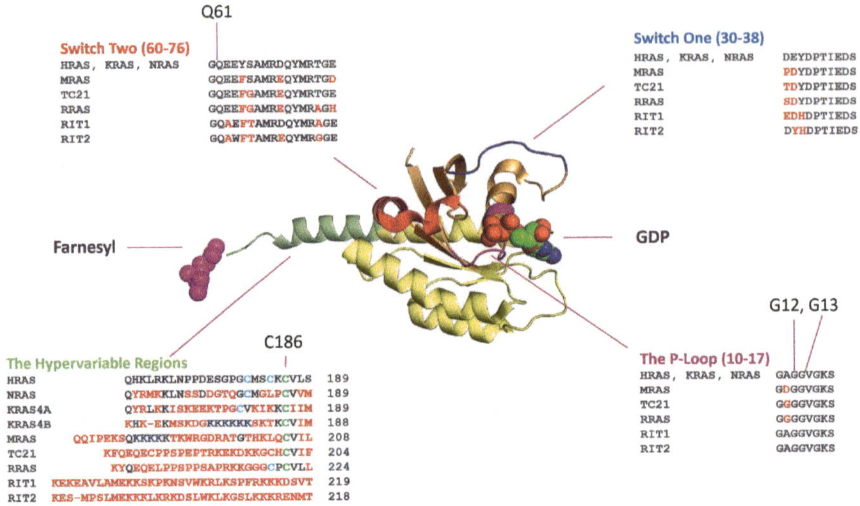

Fig. 1 Important features of KRAS4B, and related RAS proteins. The structure of full length, processed KRAS 4B is from the complex of this protein with PDEδ (Dharmaiah et al., 2016). Red letters represent differences in sequence relative to HRAS as reference. Polybasic sequences in KRAS 4B and MRAS are blue. Cysteines modified by palmitoylation are light blue, cysteines modified by farnesylation are green.

to apply this technique outside of Cetus Corporation where it was invented. In collaboration with Dr. Bert Vogelstein and colleagues, he showed frequent mutation of KRAS in colorectal cancers and established a protocol for identifying RAS mutations more efficiently and rapidly than previously imaginable (Bos et al., 1987).

RAS mutations are associated with approximately 19% of cancers diagnosed every year in the United States, and about 3.4 million globally. Seventy-five percent of these are KRAS (Prior, Hood, & Hartley, 2020). While these numbers are, by now, painfully familiar, it is noteworthy that we still do not understand why KRAS is much more frequent in human cancer than NRAS or HRAS, even though each of these are powerful transforming genes in model systems and are expressed widely in adult tissues. Furthermore, different mutant forms of KRAS may lead to different clinical outcomes, again for reasons we don't understand (Hobbs & Der, 2019; Ihle et al., 2012). NRAS mutations in myeloid leukemias are usually at codon 12, but in melanoma, they are usually at codon 61 (Burd et al., 2014). These anecdotes simply remind us we still have a lot to learn about these proteins. For a further discussion of these differences, please see the accompanying review by Burge and Hobbs.

2. Discovery of GAPs

Soon after Drs. Marshall, Hall and colleagues identified and cloned the NRAS gene, the author's group at Cetus Corporation collaborated with this group to express and characterize the NRAS protein, in its wild-type and G12D mutant forms, as well as the G12V mutant for comparison with HRAS G12V. To our surprise, the biochemical properties of NRAS were similar to its wild type counterpart: its intrinsic GTPase activity was only two- to threefold lower (Trahey et al., 1987). We tested the biological activity of NRAS, G12D, using an elegant system described by Drs. Carmen Birchmeier, Mike Wigler and colleagues, in which recombinant RAS proteins are injected into Xenopus oocytes (Birchmeier, Broek, & Wigler, 1985). In their active state, RAS proteins induce germinal vesicle breakdown, which is easily scored quantitatively. In this assay, NRAS G12D and NRAS G12V were equally active, both hundreds times more active than wild type NRAS. We concluded that the differences in nucleotide binding and hydrolysis could not account for this striking difference in biological activity and deduced that a protein in the cytoplasm of Xenopus oocytes inactivated wild type NRAS by stimulating GTP hydrolysis, but did not affect oncogenic mutants. We named this protein GTPase Activating Protein (GAP) (Trahey & McCormick, 1987). Purification of this activity from bovine brain (Vogel et al., 1988) or human placenta (Trahey et al., 1988) revealed GAP to be a 120K protein with SH2 and SH3 domains, signaling motifs that had recently been identified by Dr. Tony Pawson and colleagues (Pawson, 1988). This led to a model in which receptor tyrosine kinases recruit p120 RASGAP to terminate RAS signaling by converting RAS.GTP to RAS.GDP. The discovery that p120 RASGAP itself becomes phosphorylated during this process and binds to a RHO GAP, p190 (Settleman, Albright, Foster, & Weinberg, 1992), added a degree of complexity to this model and suggested that RAS and RAC/RHO pathways are closely coordinated.

In 1990, the gene responsible for the hereditary disease neurofibromatosis type 1 (NF1) was cloned, by classical positional cloning from families in which the disease is transmitted (Cawthon et al., 1990; Viskochil et al., 1990; Wallace et al., 1990). The gene contained a region of homology to p120 RAS GAP and to the RAS inhibitor genes from *S. cerevisiae* IRA1 and IRA2 (Ballester et al., 1990). The protein product, neurofibromin, was quickly shown to be a GAP for RAS proteins (Martin et al., 1990), immediately suggesting that many of the multiple phenotypes associated with NF1 could be due to increased levels of RAS GTP through loss of neurofibromin GAP activity. Proof that the NF1 gene was indeed a tumor suppressor came from Kevin Shannon (Shannon et al., 1994). Furthermore, the high degree of conservation of this gene, and its lack of recognizable signaling domains, suggested a function distinct from p120 RASGAP and receptor tyrosine kinase signaling. More than 30 years later, the functional significance of the majority of the neurofibromin protein remains a mystery. During this time frame, several other GAPs have been identified (Hennig, Markwart, Esparza-Franco, Ladds, & Rubio, 2015). In some cases, the importance of these proteins has been revealed by analysis of loss-of-function mutations in specific human diseases. For example, loss of p120 RASGAP (RASA1) leads to Capillary Malformation-Arteriovenous Malformations (Eerola et al., 2003), loss of RASA2 contributes to melanoma (Arafeh et al., 2015) and loss of SynGAP is a major cause of autism (Jeyabalan & Clement, 2016). In breast cancer, loss of expression of RASAL2 contributes to metastatic progression (McLaughlin et al., 2013).

3. Discovery of RAS GEFs

Analysis of RAS signaling pathways in *S. cerevisiae*, *C. elegans* and *D. melanogaster* played important roles in identifying and ordering critical components of the pathway. CDC25 was the first guanine nucleotide exchange factor (GEF) identified for proteins of the RAS family, although its precise biochemical function was not revealed until it had been placed upstream of RAS through genetic analysis of RAS and adenylyl cyclase signaling in *S. cerevisiae* (Broek et al., 1987) and the Sevenless signaling pathway in *D. melanogaster* (Simon, Bowtell, Dodson, Laverty, & Rubin, 1991). Son of Sevenless (SOS) was shown to be upstream of RAS and essential for its activation. The mammalian ortholog was shown to promote GTP exchange on RAS in mammalian cells in response to EGF receptor activation and Grb2 engagement (Buday & Downward, 1993). We now recognize multiple GEFs for RAS proteins (Bos, Rehmann, & Wittinghofer, 2007; Hennig et al., 2015). They activate RAS in response to different upstream signals and function in specific tissues. SOS (Rojas, Oliva, & Santos, 2011) has been the best studied as it is downstream of receptors such as EGFR, which are most frequently studied in cells in culture, especially cancer cell lines. Like GAPs, GEFs are large multi-domain proteins whose overall functions are not well understood. However, the presence of a DBL/RHO GEF domain in SOS proteins underscores the close parallel between RAS and Rac/Rho signaling mentioned above.

3.1 RAS processing

The discovery that RAS proteins are modified by farnesyl transferase, and that farnesylation is essential to RAS function, initiated a race to develop inhibitors (reviewed in Cox, Der, & Philips, 2015). Although many high quality compounds were developed and tested, they failed to provide any clinical benefit. This is because KRAS and NRAS can also be modified by geranylgeranyl transferase. HRAS, however, is not a substrate for alternative processing and, indeed, the farnesyl transferase inhibitor tipifarnib, is now in late-stage clinical trials for HRAS mutant squamous cell carcinoma of the head and neck (Ho et al., 2021), and other indications. Many proteins are farnesylated, and it is surprising that these drugs are relatively well tolerated. Indeed, another farnesyl transferase inhibitor, lonafarnib was recently approved for Hutchinson-Gilford Progeria Syndrome and for the treatment of processing-deficient progeroid laminopathies (Dhillon, 2021).

Following farnesylation, the C-terminal CAAX motif is further processed by the enzymes RCE1 and ICMT. Whether these represent suitable targets for blocking RAS function is not clear (Cox et al., 2015). However, it is interesting that an elegant screen for genes that are essential for mutant KRAS cancer cells, but not wild type KRAS cancer cells revealed that RCE1 and ICMT are among top hits, along with RAF1 and SHOC2 as well PREX1, a GEF for RAC proteins (Wang et al., 2017). The latter again underscores a close relationship between RAS and RAC signaling. RAC1 is essential for RAS transformation in culture (Qiu, Chen, Kirn, McCormick, & Symons, 1995), and for KRAS tumor formation in vivo (Kissil et al., 2007).

The strong dependency on RCE1 and ICMT suggest that inhibitors of RAS processing may yet have a role in treating KRAS cancers, though the precise context has yet to be defined. Recently, a small molecule inhibitor of ICMT1 was shown to prevent NRAS trafficking to the plasma membrane, suggesting a strategy for targeting NRAS mutant cancers (Ahearn, Court, Siddiqui, Abankwa, & Philips, 2021). In addition to the CAAX motif, Drs. Hancock and Marshall discovered that RAS proteins require a second signal to confer plasma membrane binding: for KRAS4B this is a polybasic sequence, for KRAS4A, HRAS and NRAS, palmitoylation is essential (Hancock, Magee, Childs, & Marshall, 1989). This process is dynamic, and facilitates cycling on and off the plasma membrane, for reasons that are not clear. Enzymes involved in palmitoylation and de-palmitoylation present potential therapeutic targets that have yet to be fully explored (Cox et al., 2015).

3.2 RAS structures

Dr. Ed Scolnick's discovery that RAS proteins bind and hydrolyze guanine nucleotides, was the first step toward understanding how these proteins function at the molecular level. Dr. Al Gilman and others showed that the RAS proteins may be a new branch of the G-Protein family, but the most direct prediction of the similarity between RAS and another GDP/GTP binding protein, the bacterial elongation factor EF-Tu, was published by Dr. Karen Halliday in 1983 (Halliday, 1983). Based on her alignments and on early crystal structures of EF-Tu, we presented a model for the tertiary structure of RAS (McCormick et al., 1985). The actual structure of human H-RAS in its active state was solved in 1989 by Dr. Fred Wittinghofer and colleagues (Pai et al., 1989), whose structural and biochemical analysis of RAS and its many cousins and regulators set the framework for RAS analysis for many years.

Until the last decade, most of the focus of RAS biology and biochemistry, including 90% of the structures in public databases, centered on HRAS. However, KRAS is now the must studied RAS protein because of its dominant role in human cancer. There are over 150 structures of KRAS proteins available, mostly of the G-domain complexed with small molecules (Pantsar, 2020). The only structure of fully processed RAS protein was solved by Dhirendra Simanshu and colleagues at the Frederick National Lab, who obtained high resolution structures of KRAS 4B complexed with PDEδ. This structure revealed the entire HVR, and also explained how PDEδ could bind either farnesylated or geranylgeranylated KRAS 4B despite the different lengths of these entities (Dharmaiah et al., 2016).

Many other important complexes have been solved by X-Ray crystallography and, more recently, by Cryo-Electron Microscopy. The structure of HRAS bound to the GAP domain of RASA1 yielded important insights into the mechanism of GAP-mediated GTP hydrolysis. In this structure, HRAS was bound to GDP and Al3—the transition state of GTP hydrolysis (Scheffzek et al., 1997). Many years later, the structure of KRAS in its ground state complexed to the GAP-domain of neurofibromin was solved

(Rabara et al., 2019). In 1998, resolution of the structure of nucleotide-free HRAS bound SOS1 led to a clear understanding of the mechanism of nucleotide exchange (Boriack-Sjodin, Margarit, Bar-Sagi, & Kuriyan, 1998). Multiple structures of RAS proteins and their relatives bound to their effectors have been solved and analyzed. The first of these effector structures was solved in 1995, again from Dr. Wittinghofer's group: this structure consisted of the RAS-binding domain of RAF-1 bound to RAP1a in its GTP-bound state (Nassar et al., 1995). This was followed in 1998 by the structure of the RAS-RalGDS complex the structure from Dr. Sung Ho Kim's group (Huang, Hofer, Martin, & Kim, 1998) and, in 2000, the structure of HRAS bound to a mutant form of PI3′ Kinase γ, this time from Dr. David Williams and colleagues (Pacold et al., 2000). In each of these structures, the effector interface consists of a ubiquitin fold that binds to Switch 1 of the RAS GTPase. The RAS-PI3′ kinase structure was the first complex that included the entire effector protein rather than the RAS binding domain: the footprint of this protein extended to Switch 2 of the RAS protein. Recently, Simanshu and colleagues solved the structure of the RAF-1 RBD-CRD complex bound to KRAS: this larger fragment of RAF-1 extended the footprint on KRAS even more extensively to include the inter-switch regions as well as points of contact at other sites in the G-domain (Tran et al., 2021). Interestingly, these regions of KRAS are not GDP/GTP dependent, but are essential for efficient activation of RAF-1 kinase activity. Structures of multiple RAS family effectors have been extensively reviewed by Drs. Mott and Owen (Mott & Owen, 2015).

3.3 The RAS superfamily

While the biology and biochemical properties of RAS proteins were being explored, it was clear that these proteins are members of a superfamily of small GTPases that includes over 150 proteins in eukaryotes, including yeast (Rojas, Fuentes, Rausell, & Valencia, 2012). At least 13 related members have been identified in eukaryotes as well (Dong, Wen, & Tian, 2007). In eukaryotes, the superfamily is divided into five major families, RAS, RHO, ARF, RAN and RAB. For the most part, these families share highly conserved GDP/GTP binding regions and are regulated by GEFs and GAPs. They serve as binary switches regulating diverse biological functions. Within the RAS superfamily, the RHO family is most closely entangled with RAS family. They often share regulatory proteins and, indeed, RAS transformation depends on RAC and RHO proteins (Qiu, Chen, Kirn, et al., 1995;

Qiu, Chen, McCormick, & Symons, 1995). Mutations in RAC and RHO regulatory proteins, and in RAC and RHO genes themselves occur frequently in cancer (Cardama, Gonzalez, Maggio, Menna, & Gomez, 2017; Kazanietz & Caloca, 2017).

RRAS proteins share many regulators and effectors with classical RAS proteins, including their ability to activate RAF and PI3′ kinase as seen in Fig. 2. Activating mutations in RRAS, RRAS2 (TC21) and MRAS occur in cancer at low frequency, and cause Noonan Syndrome (Rauen, 2013). They appear to be weak activators of the MAPK pathway in this context, although the degree to which wild type RRAS proteins participate in regulating the MAPK pathway in normal cells is unclear: more likely they perform specific biological functions that are still being characterized (Colicelli, 2004). MRAS, however, plays an important role in classical RAS signal transduction: it recruits the SHOC2 protein and the catalytic subunit of protein phosphatase 1 to the plasma membrane, facilitating de-phosphorylation of inhibitory sites on RAS-associated RAF kinases (Rodriguez-Viciana, Oses-Prieto, Burlingame, Fried, & McCormick, 2006). SHOC2 recruitment also regulates the polarity protein SCRIB and potentially other proteins (Young & Rodriguez-Viciana, 2018).

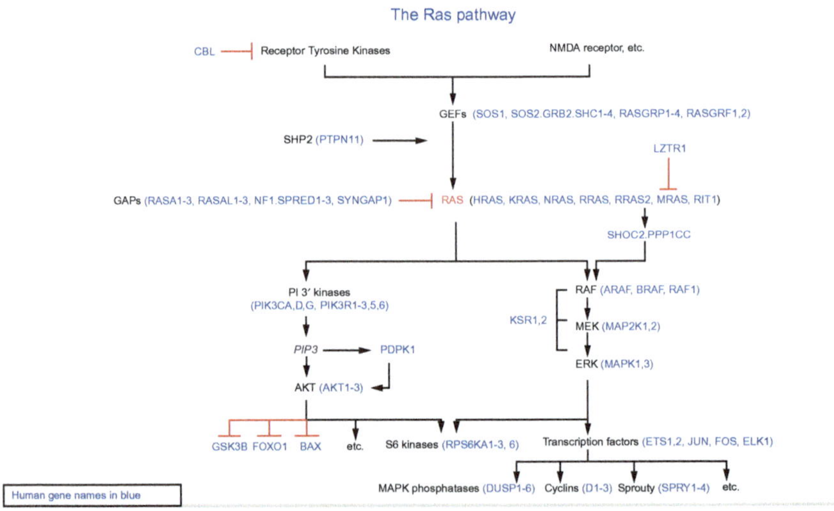

Fig. 2 Simplified RAS pathway. In this diagram, RAS proteins activate PI3′ kinases and RAF: other effector interactions are not shown for clarity. Black arrows represent activating interactions, red bars represent inhibitory interactions. This pathway is based on a previous, more comprehensive version developed at the RAS Initiative (The RAS pathway 2.0 RAS Pathway v2.0—National Cancer Institute) and has been modified to include LZTR1 and RIT1.

Like RRAS, the RAP family of RAS-like proteins (RAP1A and B, RAP2A,B and C) share very similar effector regions to classical RAS proteins but engage very different downstream effectors and, for the most part, are regulated by distinct GEFs and GAPs (Bos, 2018). RAP proteins regulate endothelial barrier functions and cell adhesion in mammalian cells. In lower organisms, their functions are distinct from RAS: in budding yeast, for example, the RAP ortholog RSR1 and its GEFs and GAPs are essential for bud site selection (reviewed in Kang, Beven, Hariharan, & Park, 2010).

RIT1 has received less attention than other members of the RAS superfamily, until recently. Mutant forms of RIT1, like RRAS, RRAS2 and MRAS, can cause Noonan Syndrome and cause malignant transformation (Van, Cuevas-Navarro, Castel, & McCormick, 2020). However, RIT1 has distinctive properties that sets it apart from other members of the RAS superfamily: first, it lacks a CAAX box. In spite of this, it localizes to the plasma membrane through its hypervariable region (Heo et al., 2006) and, second, it is regulated at the level of protein degradation rather than GDP/GTP loading (Castel et al., 2019). Mutations that activate RIT1 render the protein resistant to degradation by LZTR1, which binds wild type RIT1 and promotes ubiquitylation. Homozygous loss of LZTR1 also causes Noonan Syndrome, through accumulation of RIT1. Thirdly, RIT1 associates with MAD2 and regulates the mitotic spindle assembly checkpoint. Remarkably, the ability of RIT1 to function in the plasma membrane or at the spindle assembly checkpoint is mediated by phosphorylation of the hypervariable region by CDK1 (Cuevas-Navarro et al., 2021).

4. Discovery of RAF kinases and their relationship to RAS

The RAF1 gene was identified in 1983 in the mouse retrovirus called 3611 (Rapp et al., 1983). The name RAF derived from its ability to Rapidly Accelerate Fibrosarcoma. An avian virus, MH2, was found to contain a related gene, designated v-Mil (Sutrave et al., 1984). A-RAF was identified by screening a mouse cDNA library with a RAF-1 probe (Huleihel et al., 1986), whereas B-RAF was identified by transfecting DNA from Ewing's sarcoma cells onto NIH/3T3 cells and analyzing transformed foci (Ikawa et al., 1988). Following identification of individual transforming genes, whether from acutely transforming retroviruses, DNA transfection experiments or homology screening, efforts were made to place these genes into pathways and hierarchies. Parallel analysis of conserved pathways in model organisms played a major role in this process, as described above, but functional analysis in mammalian cells was equally powerful. The functional relationship between RAS and RAF was first revealed by microinjection of neutralizing antibodies in a remarkably informative series of experiments performed by Dr. Dennis Stacey and colleagues (Smith, DeGudicibus, & Stacey, 1986).

Direct binding of RAS to RAF was suggested by affinity purification protocols by Drs. Alan Wolfman and Shonna Moodie (Moodie, Willumsen, Weber, & Wolfman, 1993) and by a two-hybrid screen conducted by Drs. Linda van Aelst and Mike Wigler (Van Aelst, Barr, Marcus, Polverino, & Wigler, 1993). Interaction using pure recombinant proteins proved beyond doubt that this binding was a direct protein:protein interaction, and the RAF-binding domain of RAF-1 (the RBD) was clearly defined (Matallanas et al., 2011; Vojtek, Hollenberg, & Cooper, 1993; Warne, Viciana, & Downward, 1993; Zhang et al., 1993). Despite the advances, and others that now include cryo-EM structures (Kondo et al., 2019; Park et al., 2019) and extensive biochemical and genetic analysis of RAF complexes and associated proteins, the complicated process of RAS-dependent RAF activation is still not fully understood (Terrell & Morrison, 2019).

One critical step in this process involves de-phosphorylation of a RAF proteins at serine-259 in RAF-1, through recruitment of phosphatase PP1c in a complex with SHOC2 and MRAS (Rodriguez-Viciana et al., 2006). Removal of P-259 (in RAF-1) allows 14-3-3 to switch from stabilizing the inactive form of RAF to stabilizing active RAF dimers. These molecular gymnastics are facilitated by interacts of the cysteine-rich domain of RAF with both RAS.GTP and the plasma membrane (Tran et al., 2021), and underscore the complexity of signal transduction processes that are often represented as simple linear relationships.

My group at ONYX, in collaboration with Bayer, developed sorafenib as a RAF kinase inhibitor. This we did using traditional screening and medicinal chemistry: there were no structures available to guide us. Indeed, the first structure of a RAF kinase domain (B-RAF) was solved in a complex with sorafenib (Wan et al., 2004). While sorafenib was approved for several indications, it failed to show benefit to patients suffering from RAS-driven cancers. Its clinical activity has been attributed to anti-VEGFR activity, and possibly other kinases. Sorafenib, and subsequent more specific and potent RAF inhibitors, failed in RAS-driven cancers because of paradoxical activation of RAF kinase in these cells (reviewed in Karoulia, Gavathiotis, & Poulikakos, 2017). MEK inhibitors failed for a different reason: persistent activation of the MAPK pathway leads to feedback inhibition upstream (Mirzoeva et al., 2009; Pratilas et al., 2009), through phosphorylation of SOS and other mechanisms. Treatment with MEK inhibitors relieves feedback inhibition, resulting in a rebound of upstream signaling which offsets the effects of MEK inhibition (Lito, Rosen, & Solit, 2013).

4.1 RAS and PI3′ kinase

A strong correlation between activation of RAS and activation of PI3′ kinase had been noted by several investigators, but the first evidence of a more direct relationship was reported in 1991 by Dr. Eduardo Lapetina and colleagues, (Sjolander, Yamamoto, Huber, & Lapetina, 1991), who were able to detect PI3′ kinase activity in immune-precipitates using the monoclonal antibody Y13-259 developed by Dr. Mark Furth and colleagues referred to above. However, direct binding of PI3′ kinase to RAS.GTP was first demonstrated by Drs. Pablo Rodriguez-Viciana and Julian Downward (Rodriguez-Viciana et al., 1994). Oncogenic alleles of HRAS, NRAS, KRAS, RRAS and RRAS2 can bind and activate PI3′ kinases α and γ, whereas only RRAS and RRAS2 activate the δ isoform (Rodriguez-Viciana, Sabatier, & McCormick, 2004). The promiscuity of these interactions has made their significance more complicated to evaluate than the RAS-RAF interaction. Furthermore, PI3′ kinase can be activated by RAS-independent mechanisms (Fritsch & Downward, 2013). In addition, the role of PI3′ kinase activity in cancer biology is less clearly defined than that of the RAF MAPK pathway and likely contributes to multiple aspects of malignant transformation. Nevertheless, elegant experiments from Dr. Julian Downward have clearly validated the RAS-PI3′ kinase interaction as an attractive therapeutic target. In these experiments, the interaction was disrupted by mutagenesis at the binding interface. In mouse models, disruption prevented initiation of RAS driven cancers (Gupta et al., 2007) and caused partial regressions of established tumors, without significant effects on normal tissue (Castellano et al., 2013).

4.2 Which effectors drive RAS cancers?

Dr. Chris Marshall was first to propose that RAS proteins activate more than one effector pathway to induce DNA synthesis (Marshall, 1991) and many subsequent studies have shown cooperativity between effector pathways in

RAS transformation as seen in Fig. 2. Multiple proteins can bind to RAS in a GTP-dependent manner, thus qualifying as potential effectors (Kelly et al., 2020). Confidence of the importance of RAF and PI3′ kinase as RAS effectors comes from their frequent mutations in cancer and their relevance has been confirmed by elegant mouse model experiments discussed above. However, human cancers and mouse cancers may be different in this respect, and, indeed, RAS mutations may engage different effectors in different cell types and at different stages of tumor progression. Therefore, we need to acknowledge that other effectors may be critical in some contexts. For example, RALGDS proteins bind to activated RAS with high affinity and their potential roles in cancer has been strongly suggested in cells. Furthermore, RALGDS proteins may have different roles in mouse and human cells (Lim et al., 2005). However, analysis of combinations of potential effector pathways suggests that RAF-1, in combination with one of several partners, is an optimal strategy (Engelman et al., 2008; Yuan et al., 2014).

Analysis of RAS dependencies in panels of cancer cell lines shows that varying degrees of dependencies among RAF, PI3′ kinase, RAC, RSK and other pathways, suggesting that perhaps even within a clonal cell line, different subpopulations might depend on different signaling pathways (Yuan et al., 2018). Systematic elimination of all genes by CRISPR or shRNA reveals that loss of RAF1 most closely correlates with loss of KRAS, followed by loss of the RAC/RHO GEF DOCK5. Similar analysis of NRAS reveals SHOC2 and RAF1, whereas genes that correlate with HRAS are not obviously involved in RAF or PI3′ kinase signaling at all DepMap: The Cancer Dependency Map Project at Broad Institute.

Mouse models have played an extremely important role in understanding RAS functions in normal and cancer cells. One of the earlier surprises was that HRAS and NRAS are dispensable in mice, whereas KRAS is essential (Johnson et al., 1997), at least during development. Another was that wild type HRAS genes suppress oncogenic HRAS (To, Rosario, Westcott, Banta, & Balmain, 2013) and, likewise, wild type KRAS suppresses mutant KRAS (Zhou, Der, & Cox, 2016). These discoveries were among the first to suggest that RAS proteins may form oligomeric structures, a suggestion that has been borne out by multiple imaging and biophysical analyses.

Drs. Tyler Jacks and David Tuveson created mouse models that recapitulate KRAS-driven cancer initiation and progression (Tuveson & Jacks, 1999). Contributions of more than a dozen genes toward KRAS tumor formation have been analyzed in this, and related systems (summarized in

Drosten, Guerra, & Barbacid, 2018). Dr. Mariano Barbacid and colleagues have performed the most systematic analysis of RAS effectors in vivo to date. Using cleverly engineered mice, they have been able to assess the contribution of candidate effectors, and the RAS genes themselves, on tumor initiation and maintenance. Furthermore, ablation of these genes in adult mice predicts the therapeutic index we can expect with drugs targeting these proteins (Drosten et al., 2018). One of the most striking results from these experiments is that KRAS depends on RAF-1 but does not require RAF kinase activity (Sanclemente et al., 2021). Rather, the RAF-1 protein serves to protect from apoptosis, via a mechanism that is not yet understood. Ablation of RAF promotes apoptotic death in KRAS cancer cells but has little if any effect in normal tissues. For these reasons, strategies that ablate RAF-1 by targeted degradation, for example, appear very attractive opportunities.

4.3 Therapeutic strategies

Multiple strategies have been employed to target RAS cancers, including targeting downstream pathways (discussed above), synthetic lethal targets (Downward, 2015), immunological approaches (Wang et al., 2016), anti-sense approaches (Ross et al., 2017) and others. These have been summarized recently and will not be discussed here (Khan, Rhett, & O'Bryan, 2020; Moore, Rosenberg, McCormick, & Malek, 2021; Papke & Der, 2017; Stalnecker & Der, 2020; Zuberi, Khan, & O'Bryan, 2020). Please refer to the accompanying reviews in this volume for further discussion of approaches to inhibit RAS-mutant cancers. Since RAS proteins were first identified, efforts have been made to find drugs that inactivate its function directly, including efforts to restore GTP hydrolysis to oncogenic mutants (Ahmadian et al., 1999). Compounds that bind directly have indeed been identified (Maurer et al., 2012; Sun et al., 2012) but most of these have not progressed toward clinical evaluation because of low affinity or specificity (McCormick, 2019). However, In 2013, Dr. Kevan Shokat and his graduate student Jonathon Ostrem and colleagues published a paper that took the RAS world by surprise: they showed that covalent attack on the G12C allele of KRAS shut down signaling from this mutant protein by trapping it in the inactivate GDP-bound state (Ostrem, Peters, Sos, Wells, & Shokat, 2013). This brilliant discovery set off a race to develop G12C inhibitors. Amgen was first to enter the clinic, with AMG-510, with Mirati close behind. On May 28, 2021, AMG-510, sotorasib, was approved by the FDA for treatment of KRAS G12C lung adenocarcinoma. Three additional G12C compounds are currently in clinical trials, from Mirati, Genentech, and Novartis.

Analysis of crystal structures of KRAS complexed with G12C inhibitors has facilitated discovery of compounds that bind to other alleles, albeit non-covalently. We can expect a number of these approaches to generate clinical candidates for targeting G12D, G12V and other alleles in the next few years, including compounds identified at the Frederick National Laboratories as part of the RAS Initiative.

5. The RAS Initiative at Frederick National Laboratories

In 2013, Dr. Harold Varmus announced the launch of the RAS Initiative, an NCI-funded effort to target KRAS-driven cancers About the RAS Initiative—National Cancer Institute. This initiative was launched, in part, to address the lack of progress that had been made in developing drugs against KRAS-driven cancers, most notably pancreatic cancer. It also reflected advances in technology in attacking protein targets that had often been relegated to the "undruggable" category. These advances, among many others, prompted Dr. Varmus and colleagues to re-visit RAS as a drug target.

Dr. Varmus assembled a group of experts in the RAS field and a strategy began to develop. The author was invited to serve as Program Advisor to the initiative, which was based at the newly opened Advance Technology Research Facility at National Laboratories at Frederick, MD, just a few miles from the NCI in which many of the early discoveries in the oncogene field had been made, and close to Hood College, home to many memorable Oncogene Meetings. The initiative was supported through the NCI as part of its contract with what was then SAIC-Frederick (now Leidos Biomedical), led by CEO Dr. David Heimbrook. Coincidentally, Dr. Heimbrook had previously been part of the team at Merck Laboratories evaluating farnesyltransferase inhibitors for treating RAS cancers.

Under the leadership of Dr. Dwight Nissley, Director of the Cancer Research Technology Program, a group of about 50 individuals were assigned to the effort, including Dr. Dom Esposito, head of the Protein Engineering Laboratory, Dr. Andy Stephen, head of Biophysics and Dr. Tommy Turbyville, head of Cell Imaging. Several key investigators were recruited to Frederick to build the leadership group: Dr. Dhirendra

Simanshu (Structural Biology), Dr. Anna Maciag (Biochemistry) and Dr. David Turner (Chemistry). An ad hoc working group currently chaired by Dr. David Tuveson, helps oversee and guide the direction of the initiative and coordinates team interactions with Drs. Doug Lowy, Sara Hook and Ned Sharpless at the NCI.

When the initiative was launched in 2013, there were no structures of mutant KRAS proteins in their active states, free or bound to effectors. Since a major goal of the initiative was to target KRAS directly, this was an obvious knowledge gap. Resolution of many of these structures is on-going, but several have already been published, including the structure of fully processed KRAS4B complexed with PDEδ (Dharmaiah et al., 2016), the structure of a ternary complex of KRAS, the GAP domain of neurofibromin and the EVH1 domain of SPRED1 (Yan et al., 2020), and structure of KRAS4B bound to the RAF1 RBD.CRD domain (Tran et al., 2021). Dr. Simanshu's group has also collaborated with several groups in academia, including Dr. Channing Der and colleagues (Hobbs et al., 2020), and Dr. Bill Hahn and Andy Aguirre at the Broad Institute.

In 2013 it was unclear which cancers remain KRAS dependent, and to what extent different KRAS alleles depended on different effectors. The release of DepMap from the Broad Institute has made a tremendous contribution to this problem DepMap: The Cancer Dependency Map Project at Broad Institute. One limitation of this analysis is that effects of single genes can be masked by redundant genes that perform similar functions. To address this, a collaborative effort between scientists at MGH, Cold Spring Harbor, the Broad Institute, Memorial Sloan Kettering, NCI, UCSF and the RAS Initiative interrogated signaling nodes, rather than individual genes (Yuan et al., 2018). Other limitations to these approaches, including analysis of dependencies in 2D, have been addressed through collaboration with the RAS Synthetic Lethal Network led by NCI Program Director Dr. Joanna Watson.

Another early goal of the RAS Initiative was to develop new ways of analyzing KRAS behavior in the plasma membrane using advanced imaging technology (Goswami et al., 2020) and biophysical methods that detect highly specific interactions with specific membrane components (Van, Lopez, et al., 2020). This goal has evolved into a highly productive collaboration with the Department of Energy that includes the computational and analytic power of four National Labs under the leadership of Dr. Fred Streitz. This group has developed methods for modeling interactions of

KRAS with the plasma membrane using molecular dynamic simulations and computing methods that have not been possible previously (Bhattacharya et al., 2019).

A key strategy of the RAS Initiative has been to collaborate with colleagues in academia and the biopharma industry to accelerate development of drugs targeting KRAS cancers. Collaborations are currently in place with the Weizmann Institute, Sanofi, TheRAS Inc., Eli Lilly and Company and Evotec, each involving unique approaches to discovering KRAS inhibitors. Drug discovery with these partners has become the major priority within the RAS Initiative, and we expect clinical candidates from these collaborations to emerge in the coming years. We also believe that a more precise description of how RAS proteins interact with the plasma membrane and activate effectors, derived though iterative biophysical analysis, imaging and computational modeling, will lead to more drugs targets in the future.

The RAS Initiative has tried to encourage dialog between members of the RAS research community, through regular international meetings and through a lively blog, RAS Lab, hosted on Basecamp RAS Lab (basecamp.com). This community has over 1100 members and has hosted almost 400 discussions. We welcome readers to join and look forward to your contributions to this exciting and interactive field! In many ways, we feel that after many exhilarating years of research, the best may still be ahead of us.

References

Ahearn, I. M., Court, H. R., Siddiqui, F., Abankwa, D., & Philips, M. R. (2021). NRAS is unique among RAS proteins in requiring ICMT for trafficking to the plasma membrane. *Life Science Alliance*, *4*(5). https://doi.org/10.26508/lsa.202000972.

Ahmadian, M. R., Zor, T., Vogt, D., Kabsch, W., Selinger, Z., Wittinghofer, A., et al. (1999). Guanosine triphosphatase stimulation of oncogenic Ras mutants. *Proceedings of the National Academy of Sciences of the United States of America*, *96*(12), 7065–7070. https://www.ncbi.nlm.nih.gov/pubmed/10359839.

Amendola, C. R., Mahaffey, J. P., Parker, S. J., Ahearn, I. M., Chen, W. C., Zhou, M., et al. (2019). KRAS4A directly regulates hexokinase 1. *Nature*, *576*(7787), 482–486. https://doi.org/10.1038/s41586-019-1832-9.

Arafeh, R., Qutob, N., Emmanuel, R., Keren-Paz, A., Madore, J., Elkahloun, A., et al. (2015). Recurrent inactivating RASA2 mutations in melanoma. *Nature Genetics*, *47*(12), 1408–1410. https://doi.org/10.1038/ng.3427.

Ballester, R., Marchuk, D., Boguski, M., Saulino, A., Letcher, R., Wigler, M., et al. (1990). The NF1 locus encodes a protein functionally related to mammalian GAP and yeast IRA proteins. *Cell*, *63*(4), 851–859. https://www.ncbi.nlm.nih.gov/pubmed/2121371.

Bhattacharya, T., Brettin, T., Doroshow, J. H., Evrard, Y. A., Greenspan, E. J., Gryshuk, A. L., et al. (2019). AI meets exascale computing: Advancing cancer research with large-scale high performance computing. *Frontiers in Oncology*, *9*, 984. https://doi.org/10.3389/fonc.2019.00984.

Birchmeier, C., Broek, D., & Wigler, M. (1985). ras proteins can induce meiosis in Xenopus oocytes. Cell, 43(3 Pt. 2), 615–621. https://doi.org/10.1016/0092-8674(85)90233-8.

Bivona, T. G., Quatela, S. E., Bodemann, B. O., Ahearn, I. M., Soskis, M. J., Mor, A., et al. (2006). PKC regulates a farnesyl-electrostatic switch on K-Ras that promotes its association with Bcl-XL on mitochondria and induces apoptosis. *Molecular Cell, 21*(4), 481–493. https://doi.org/10.1016/j.molcel.2006.01.012.

Boriack-Sjodin, P. A., Margarit, S. M., Bar-Sagi, D., & Kuriyan, J. (1998). The structural basis of the activation of Ras by Sos. *Nature, 394*(6691), 337–343. https://doi.org/10.1038/28548.

Bos, J. L. (2018). From Ras to Rap and back, a journey of 35 years. *Cold Spring Harbor Perspectives in Medicine, 8*(2). https://doi.org/10.1101/cshperspect.a031468.

Bos, J. L., Fearon, E. R., Hamilton, S. R., Verlaan-de Vries, M., van Boom, J. H., van der Eb, A. J., et al. (1987). Prevalence of ras gene mutations in human colorectal cancers. *Nature, 327*(6120), 293–297. https://doi.org/10.1038/327293a0.

Bos, J. L., Rehmann, H., & Wittinghofer, A. (2007). GEFs and GAPs: Critical elements in the control of small G proteins. *Cell, 129*(5), 865–877. https://doi.org/10.1016/j.cell.2007.05.018.

Broek, D., Toda, T., Michaeli, T., Levin, L., Birchmeier, C., Zoller, M., Powers, S., & Wigler, M. (1987). The S. cerevisiae CDC25 gene product regulates the RAS/adenylate cyclase pathway. Cell, 48(5), 789–799. https://doi.org/10.1016/0092-8674(87)90076-6.

Buday, L., & Downward, J. (1993). Epidermal growth factor regulates p21ras through the formation of a complex of receptor, Grb2 adapter protein, and Sos nucleotide exchange factor. Cell, 73(3), 611–620. https://doi.org/10.1016/0092-8674(93)90146-h.

Burd, C. E., Liu, W., Huynh, M. V., Waqas, M. A., Gillahan, J. E., Clark, K. S., et al. (2014). Mutation-specific RAS oncogenicity explains NRAS codon 61 selection in melanoma. *Cancer Discovery, 4*(12), 1418–1429. https://doi.org/10.1158/2159-8290.CD-14-0729.

Cardama, G. A., Gonzalez, N., Maggio, J., Menna, P. L., & Gomez, D. E. (2017). Rho GTPases as therapeutic targets in cancer (review). *International Journal of Oncology, 51*(4), 1025–1034. https://doi.org/10.3892/ijo.2017.4093.

Castel, P., Cheng, A., Cuevas-Navarro, A., Everman, D. B., Papageorge, A. G., Simanshu, D. K., et al. (2019). RIT1 oncoproteins escape LZTR1-mediated proteolysis. *Science, 363*(6432), 1226–1230. https://doi.org/10.1126/science.aav1444.

Castel, P., Dharmaiah, S., Sale, M., Messing, S., Rizzuto, G., Cuevas-Navarro, A., et al. (2021). RAS interaction with Sin1 is dispensable for mTORC2 assembly and activity. *Proceedings of the National Academy of Sciences of the United States of America, 118*, 1–11. https://doi.org/10.1073/pnas.2103261118.

Castellano, E., Sheridan, C., Thin, M. Z., Nye, E., Spencer-Dene, B., Diefenbacher, M. E., et al. (2013). Requirement for interaction of PI3-kinase p110alpha with RAS in lung tumor maintenance. *Cancer Cell, 24*(5), 617–630. https://doi.org/10.1016/j.ccr.2013.09.012.

Cawthon, R. M., Weiss, R., Xu, G. F., Viskochil, D., Culver, M., Stevens, J., Robertson, M., Dunn, D., Gesteland, R., O'Connell, P. (1990). A major segment of the neurofibromatosis type 1 gene: cDNA sequence, genomic structure, and point mutations. Cell, 62(1), 193–201. https://doi.org/10.1016/0092-8674(90)90253-b.

Colicelli, J. (2004). Human RAS superfamily proteins and related GTPases. *Science's STKE, 2004*(250), RE13. https://doi.org/10.1126/stke.2502004re13.

Cox, A. D., Der, C. J., & Philips, M. R. (2015). Targeting RAS membrane association: Back to the future for anti-RAS drug discovery? *Clinical Cancer Research, 21*(8), 1819–1827. https://doi.org/10.1158/1078-0432.CCR-14-3214.

Cuevas-Navarro, A., Van, R., Cheng, A., Urisman, A., Castel, P., McCormick, F., et al. (2021). The RAS GTPase RIT1 compromises mitotic fidelity through spindle assembly checkpoint suppression. *Current Biology, 31*, 1–10. https://doi.org/10.1016/j.cub.2021.06.030.

Dharmaiah, S., Bindu, L., Tran, T. H., Gillette, W. K., Frank, P. H., Ghirlando, R., et al. (2016). Structural basis of recognition of farnesylated and methylated KRAS4b by PDEdelta. *Proceedings of the National Academy of Sciences of the United States of America, 113*(44), E6766–E6775. https://doi.org/10.1073/pnas.1615316113.
Dhillon, S. (2021). Lonafarnib: First approval. *Drugs, 81*(2), 283–289. https://doi.org/10.1007/s40265-020-01464-z.
Dong, J. H., Wen, J. F., & Tian, H. F. (2007). Homologs of eukaryotic Ras superfamily proteins in prokaryotes and their novel phylogenetic correlation with their eukaryotic analogs. *Gene, 396*(1), 116–124. https://doi.org/10.1016/j.gene.2007.03.001.
Downward, J. (2015). RAS synthetic lethal screens revisited: Still seeking the elusive prize? *Clinical Cancer Research, 21*(8), 1802–1809. https://doi.org/10.1158/1078-0432.CCR-14-2180.
Drosten, M., Guerra, C., & Barbacid, M. (2018). Genetically engineered mouse models of K-Ras-driven lung and pancreatic tumors: Validation of therapeutic targets. *Cold Spring Harbor Perspectives in Medicine, 8*(5). https://doi.org/10.1101/cshperspect.a031542.
Eerola, I., Boon, L. M., Mulliken, J. B., Burrows, P. E., Dompmartin, A., Watanabe, S., et al. (2003). Capillary malformation-arteriovenous malformation, a new clinical and genetic disorder caused by RASA1 mutations. *American Journal of Human Genetics, 73*(6), 1240–1249. https://doi.org/10.1086/379793.
Engelman, J. A., Chen, L., Tan, X., Crosby, K., Guimaraes, A. R., Upadhyay, R., et al. (2008). Effective use of PI3K and MEK inhibitors to treat mutant Kras G12D and PIK3CA H1047R murine lung cancers. *Nature Medicine, 14*(12), 1351–1356. https://doi.org/10.1038/nm.1890.
Fritsch, R., & Downward, J. (2013). SnapShot: Class I PI3K isoform signaling. *Cell, 154*(4), 940–940.e1. https://doi.org/10.1016/j.cell.2013.07.045.
Furth, M. E., Davis, L. J., Fleurdelys, B., & Scolnick, E. M. (1982). Monoclonal antibodies to the p21 products of the transforming gene of Harvey murine sarcoma virus and of the cellular ras gene family. *Journal of Virology, 43*(1), 294–304. https://doi.org/10.1128/JVI.43.1.294-304.1982.
Gibbs, J. B., Sigal, I. S., Poe, M., & Scolnick, E. M. (1984). Intrinsic GTPase activity distinguishes normal and oncogenic ras p21 molecules. *Proceedings of the National Academy of Sciences of the United States of America, 81*(18), 5704–5708. https://doi.org/10.1073/pnas.81.18.5704.
Goswami, D., Chen, D., Yang, Y., Gudla, P. R., Columbus, J., Worthy, K., et al. (2020). Membrane interactions of the globular domain and the hypervariable region of KRAS4b define its unique diffusion behavior. *eLife, 9*. https://doi.org/10.7554/eLife.47654.
Gupta, S., Ramjaun, A. R., Haiko, P., Wang, Y., Warne, P. H., Nicke, B., et al. (2007). Binding of ras to phosphoinositide 3-kinase p110alpha is required for ras-driven tumorigenesis in mice. *Cell, 129*(5), 957–968. http://www.ncbi.nlm.nih.gov/entrez/query.fcgi?cmd=Retrieve&db=PubMed&dopt=Citation&list_uids=17540175.
Hall, A., Marshall, C. J., Spurr, N. K., & Weiss, R. A. (1983). Identification of transforming gene in two human sarcoma cell lines as a new member of the ras gene family located on chromosome 1. *Nature, 303*(5916), 396–400. http://www.ncbi.nlm.nih.gov/pubmed/6304521.
Halliday, K. R. (1983). Regional homology in GTP-binding proto-oncogene products and elongation factors. *Journal of Cyclic Nucleotide and Protein Phosphorylation Research, 9*(6), 435–448. https://www.ncbi.nlm.nih.gov/pubmed/6396323.
Hancock, J. F., Magee, A. I., Childs, J. E., & Marshall, C. J. (1989). All ras proteins are polyisoprenylated but only some are palmitoylated. *Cell, 57*(7), 1167–1177. https://doi.org/10.1016/0092-8674(89)90054-8.
Harvey, J. J. (1964). An unidentified virus which causes the rapid production of tumours in mice. *Nature, 204*, 1104–1105. https://doi.org/10.1038/2041104b0.

Hennig, A., Markwart, R., Esparza-Franco, M. A., Ladds, G., & Rubio, I. (2015). Ras activation revisited: Role of GEF and GAP systems. *Biological Chemistry, 396*(8), 831–848. https://doi.org/10.1515/hsz-2014-0257.

Heo, W. D., Inoue, T., Park, W. S., Kim, M. L., Park, B. O., Wandless, T. J., et al. (2006). PI(3,4,5)P3 and PI(4,5)P2 lipids target proteins with polybasic clusters to the plasma membrane. *Science, 314*(5804), 1458–1461. https://doi.org/10.1126/science.1134389.

Ho, A. L., Brana, I., Haddad, R., Bauman, J., Bible, K., Oosting, S., et al. (2021). Tipifarnib in head and neck squamous cell carcinoma with HRAS mutations. *Journal of Clinical Oncology, 39*(17), 1856–1864. https://doi.org/10.1200/JCO.20.02903.

Hobbs, G. A., Baker, N. M., Miermont, A. M., Thurman, R. D., Pierobon, M., Tran, T. H., et al. (2020). Atypical KRAS(G12R) mutant is impaired in PI3K signaling and macropinocytosis in pancreatic cancer. *Cancer Discovery, 10*(1), 104–123. https://doi.org/10.1158/2159-8290.CD-19-1006.

Hobbs, G. A., & Der, C. J. (2019). RAS mutations are not created equal. *Cancer Discovery, 9*(6), 696–698. https://doi.org/10.1158/2159-8290.CD-19-0406.

Huang, L., Hofer, F., Martin, G. S., & Kim, S. H. (1998). Structural basis for the interaction of Ras with RalGDS. *Nature Structural Biology, 5*(6), 422–426. https://doi.org/10.1038/nsb0698-422.

Huleihel, M., Goldsborough, M., Cleveland, J., Gunnell, M., Bonner, T., & Rapp, U. R. (1986). Characterization of murine A-raf, a new oncogene related to the v-raf oncogene. *Molecular and Cellular Biology, 6*(7), 2655–2662. https://doi.org/10.1128/mcb.6.7.2655-2662.1986.

Ihle, N. T., Byers, L. A., Kim, E. S., Saintigny, P., Lee, J. J., Blumenschein, G. R., et al. (2012). Effect of KRAS oncogene substitutions on protein behavior: Implications for signaling and clinical outcome. *Journal of the National Cancer Institute, 104*(3), 228–239. https://doi.org/10.1093/jnci/djr523.

Ikawa, S., Fukui, M., Ueyama, Y., Tamaoki, N., Yamamoto, T., & Toyoshima, K. (1988). B-raf, a new member of the raf family, is activated by DNA rearrangement. *Molecular and Cellular Biology, 8*(6), 2651–2654. https://doi.org/10.1128/mcb.8.6.2651-2654.1988.

Jeyabalan, N., & Clement, J. P. (2016). SYNGAP1: Mind the gap. *Frontiers in Cellular Neuroscience, 10*, 32. https://doi.org/10.3389/fncel.2016.00032.

Johnson, L., Greenbaum, D., Cichowski, K., Mercer, K., Murphy, E., Schmitt, E., et al. (1997). K-ras is an essential gene in the mouse with partial functional overlap with N-ras. *Genes & Development, 11*(19), 2468–2481. https://www.ncbi.nlm.nih.gov/pubmed/9334313.

Kang, P. J., Beven, L., Hariharan, S., & Park, H. O. (2010). The Rsr1/Bud1 GTPase interacts with itself and the Cdc42 GTPase during bud-site selection and polarity establishment in budding yeast. *Molecular Biology of the Cell, 21*(17), 3007–3016. https://doi.org/10.1091/mbc.E10-03-0232.

Karoulia, Z., Gavathiotis, E., & Poulikakos, P. I. (2017). New perspectives for targeting RAF kinase in human cancer. *Nature Reviews. Cancer, 17*(11), 676–691. https://doi.org/10.1038/nrc.2017.79.

Kazanietz, M. G., & Caloca, M. J. (2017). The Rac GTPase in cancer: From old concepts to new paradigms. *Cancer Research, 77*(20), 5445–5451. https://doi.org/10.1158/0008-5472.CAN-17-1456.

Kelly, M. R., Kostyrko, K., Han, K., Mooney, N. A., Jeng, E. E., Spees, K., et al. (2020). Combined proteomic and genetic interaction mapping reveals new RAS effector pathways and susceptibilities. *Cancer Discovery, 10*(12), 1950–1967. https://doi.org/10.1158/2159-8290.CD-19-1274.

Khan, I., Rhett, J. M., & O'Bryan, J. P. (2020). Therapeutic targeting of RAS: New hope for drugging the "undruggable". *Biochimica et Biophysica Acta. Molecular Cell Research, 1867*(2), 118570. https://doi.org/10.1016/j.bbamcr.2019.118570.

Kirsten, W. H., & Mayer, L. A. (1967). Morphologic responses to a murine erythroblastosis virus. *Journal of the National Cancer Institute, 39*(2), 311–335. https://www.ncbi.nlm.nih.gov/pubmed/18623947.

Kissil, J. L., Walmsley, M. J., Hanlon, L., Haigis, K. M., Bender Kim, C. F., Sweet-Cordero, A., et al. (2007). Requirement for Rac1 in a K-ras induced lung cancer in the mouse. *Cancer Research, 67*(17), 8089–8094. https://doi.org/10.1158/0008-5472.CAN-07-2300.

Kondo, Y., Ognjenovic, J., Banerjee, S., Karandur, D., Merk, A., Kulhanek, K., et al. (2019). Cryo-EM structure of a dimeric B-Raf:14-3-3 complex reveals asymmetry in the active sites of B-Raf kinases. *Science, 366*(6461), 109–115. https://doi.org/10.1126/science.aay0543.

Lim, K. H., Baines, A. T., Fiordalisi, J. J., Shipitsin, M., Feig, L. A., Cox, A. D., et al. (2005). Activation of RalA is critical for Ras-induced tumorigenesis of human cells. *Cancer Cell,* 7(6), 533–545. https://doi.org/10.1016/j.ccr.2005.04.030.

Lito, P., Rosen, N., & Solit, D. B. (2013). Tumor adaptation and resistance to RAF inhibitors. *Nature Medicine, 19*(11), 1401–1409. https://doi.org/10.1038/nm.3392.

Lou, E., Xiu, J., Baca, Y., Nelson, A. C., Weinberg, B. A., Beg, M. S., et al. (2021). Expression of immuno-oncologic biomarkers is enriched in colorectal cancers and other solid tumors harboring the A59T variant of KRAS. *Cell, 10*(6). https://doi.org/10.3390/cells10061275.

Malumbres, M., & Barbacid, M. (2003). RAS oncogenes: The first 30 years. *Nature Reviews. Cancer, 3*(6), 459–465. https://doi.org/10.1038/nrc1097.

Marshall, C. J. (1991). How does p21ras transform cells? *Trends in Genetics,* 7(3), 91–95. http://www.ncbi.nlm.nih.gov/pubmed/2031288.

Martin, G. A., Viskochil, D., Bollag, G., McCabe, P. C., Crosier, W. J., Haubruck, H., et al. (1990). The GAP-related domain of the neurofibromatosis type 1 gene product interacts with ras p21. *Cell, 63*(4), 843–849. https://doi.org/10.1016/0092-8674(90)90150-d.

Matallanas, D., Birtwistle, M., Romano, D., Zebisch, A., Rauch, J., von Kriegsheim, A., et al. (2011). Raf family kinases: Old dogs have learned new tricks. *Genes & Cancer, 2*(3), 232–260. https://doi.org/10.1177/1947601911407323.

Maurer, T., Garrenton, L. S., Oh, A., Pitts, K., Anderson, D. J., Skelton, N. J., et al. (2012). Small-molecule ligands bind to a distinct pocket in Ras and inhibit SOS-mediated nucleotide exchange activity. *Proceedings of the National Academy of Sciences of the United States of America, 109*(14), 5299–5304. https://doi.org/10.1073/pnas.1116510109.

McCormick, F. (2019). Progress in targeting RAS with small molecule drugs. *The Biochemical Journal, 476*(2), 365–374. https://doi.org/10.1042/BCJ20170441.

McCormick, F., Clark, B. F., la Cour, T. F., Kjeldgaard, M., Norskov-Lauritsen, L., & Nyborg, J. (1985). A model for the tertiary structure of p21, the product of the ras oncogene. *Science, 230*(4721), 78–82. http://www.ncbi.nlm.nih.gov/pubmed/3898366.

McGrath, J. P., Capon, D. J., Goeddel, D. V., & Levinson, A. D. (1984). Comparative biochemical properties of normal and activated human ras p21 protein. *Nature, 310*(5979), 644–649. https://doi.org/10.1038/310644a0.

McGrath, J. P., Capon, D. J., Smith, D. H., Chen, E. Y., Seeburg, P. H., Goeddel, D. V., et al. (1983). Structure and organization of the human Ki-ras proto-oncogene and a related processed pseudogene. *Nature, 304*(5926), 501–506. https://doi.org/10.1038/304501a0.

McLaughlin, S. K., Olsen, S. N., Dake, B., De Raedt, T., Lim, E., Bronson, R. T., et al. (2013). The RasGAP gene, RASAL2, is a tumor and metastasis suppressor. *Cancer Cell, 24*(3), 365–378. https://doi.org/10.1016/j.ccr.2013.08.004.

Mirzoeva, O. K., Das, D., Heiser, L. M., Bhattacharya, S., Siwak, D., Gendelman, R., et al. (2009). Basal subtype and MAPK/ERK kinase (MEK)-phosphoinositide 3-kinase feedback signaling determine susceptibility of breast cancer cells to MEK inhibition. *Cancer Research, 69*(2), 565–572. https://doi.org/10.1158/0008-5472.CAN-08-3389.

Moodie, S., Willumsen, B., Weber, M., & Wolfman, A. (1993). Complexes of Ras.GTP withRaf-1 and mitogen-activated protein kinase kinase. *Science, 260*, 1658–1661. https://doi.org/10.1126/science.8503013.

Moore, A. R., Rosenberg, S. C., McCormick, F., & Malek, S. (2021). RAS-targeted therapies. *Nature Reviews. Drug Discovery*. https://doi.org/10.1038/s41573-021-00220-6.

Mott, H. R., & Owen, D. (2015). Structures of Ras superfamily effector complexes: What have we learnt in two decades? *Critical Reviews in Biochemistry and Molecular Biology, 50*(2), 85–133. https://doi.org/10.3109/10409238.2014.999191.

Nassar, N., Horn, G., Herrmann, C., Scherer, A., McCormick, F., & Wittinghofer, A. (1995). The 2.2 A crystal structure of the Ras-binding domain of the serine/threonine kinase c-Raf1 in complex with Rap1A and a GTP analogue. *Nature, 375*(6532), 554–560. http://www.ncbi.nlm.nih.gov/entrez/query.fcgi?cmd=Retrieve&db=PubMed&dopt=Citation&list_uids=7791872.

Ostrem, J. M., Peters, U., Sos, M. L., Wells, J. A., & Shokat, K. M. (2013). K-Ras(G12C) inhibitors allosterically control GTP affinity and effector interactions. *Nature, 503*(7477), 548–551. https://doi.org/10.1038/nature12796.

Pacold, M. E., Suire, S., Perisic, O., Lara-Gonzalez, S., Davis, C. T., Walker, E. H., Hawkins, P. T., Stephens, L., Eccleston, J. F., & Williams, R. L. (2000). Crystal structure and functional analysis of Ras binding to its effector phosphoinositide 3-kinase gamma. Cell, 103(6), 931–943. https://doi.org/10.1016/s0092-8674(00)00196-3.

Pai, E. F., Kabsch, W., Krengel, U., Holmes, K. C., John, J., & Wittinghofer, A. (1989). Structure of the guanine-nucleotide-binding domain of the Ha-ras oncogene product p21 in the triphosphate conformation. *Nature, 341*(6239), 209–214. https://doi.org/10.1038/341209a0.

Pantsar, T. (2020). The current understanding of KRAS protein structure and dynamics. *Computational and Structural Biotechnology Journal, 18*, 189–198. https://doi.org/10.1016/j.csbj.2019.12.004.

Papke, B., & Der, C. J. (2017). Drugging RAS: Know the enemy. *Science, 355*(6330), 1158–1163. https://doi.org/10.1126/science.aam7622.

Park, E., Rawson, S., Li, K., Kim, B. W., Ficarro, S. B., Pino, G. G., et al. (2019). Architecture of autoinhibited and active BRAF-MEK1-14-3-3 complexes. *Nature, 575*(7783), 545–550. https://doi.org/10.1038/s41586-019-1660-y.

Pawson, T. (1988). Non-catalytic domains of cytoplasmic protein-tyrosine kinases: Regulatory elements in signal transduction. *Oncogene, 3*(5), 491–495. https://www.ncbi.nlm.nih.gov/pubmed/3078956.

Pratilas, C. A., Taylor, B. S., Ye, Q., Viale, A., Sander, C., Solit, D. B., et al. (2009). (V600E) BRAF is associated with disabled feedback inhibition of RAF-MEK signaling and elevated transcriptional output of the pathway. *Proceedings of the National Academy of Sciences of the United States of America, 106*(11), 4519–4524. https://doi.org/10.1073/pnas.0900780106.

Prior, I. A., Hood, F. E., & Hartley, J. L. (2020). The frequency of Ras mutations in cancer. *Cancer Research, 80*(14), 2969–2974. https://doi.org/10.1158/0008-5472.CAN-19-3682.

Qiu, R. G., Chen, J., Kirn, D., McCormick, F., & Symons, M. (1995). An essential role for Rac in Ras transformation. *Nature, 374*(6521), 457–459. http://www.ncbi.nlm.nih.gov/entrez/query.fcgi?cmd=Retrieve&db=PubMed&dopt=Citation&list_uids=7700355.

Qiu, R. G., Chen, J., McCormick, F., & Symons, M. (1995). A role for Rho in Ras transformation. *Proceedings of the National Academy of Sciences of the United States of America, 92*(25), 11781–11785. http://www.ncbi.nlm.nih.gov/entrez/query.fcgi?cmd=Retrieve&db=PubMed&dopt=Citation&list_uids=8524848.

Quinlan, M. P., Quatela, S. E., Philips, M. R., & Settleman, J. (2008). Activated Kras, but not Hras or Nras, may initiate tumors of endodermal origin via stem cell expansion. *Molecular and Cellular Biology, 28*(8), 2659–2674. https://doi.org/10.1128/MCB.01661-07.

Rabara, D., Tran, T. H., Dharmaiah, S., Stephens, R. M., McCormick, F., Simanshu, D. K., et al. (2019). KRAS G13D sensitivity to neurofibromin-mediated GTP hydrolysis. *Proceedings of the National Academy of Sciences of the United States of America*, *116*(44), 22122–22131. https://doi.org/10.1073/pnas.1908353116.

Rapp, U. R., Goldsborough, M. D., Mark, G. E., Bonner, T. I., Groffen, J., Reynolds, F. H., Jr., et al. (1983). Structure and biological activity of v-raf, a unique oncogene transduced by a retrovirus. *Proceedings of the National Academy of Sciences of the United States of America*, *80*(14), 4218–4222. https://doi.org/10.1073/pnas.80.14.4218.

Rauen, K. A. (2013). The RASopathies. *Annual Review of Genomics and Human Genetics*, *14*, 355–369. https://doi.org/10.1146/annurev-genom-091212-153523.

Rodriguez-Viciana, P., Oses-Prieto, J., Burlingame, A., Fried, M., & McCormick, F. (2006). A phosphatase holoenzyme comprised of Shoc2/Sur8 and the catalytic subunit of PP1 functions as an M-Ras effector to modulate Raf activity. *Molecular Cell*, *22*(2), 217–230. http://www.ncbi.nlm.nih.gov/entrez/query.fcgi?cmd=Retrieve&db=PubMed&dopt=Citation&list_uids=16630891.

Rodriguez-Viciana, P., Sabatier, C., & McCormick, F. (2004). Signaling specificity by Ras family GTPases is determined by the full spectrum of effectors they regulate. *Molecular and Cellular Biology*, *24*(11), 4943–4954. http://www.ncbi.nlm.nih.gov/entrez/query.fcgi?cmd=Retrieve&db=PubMed&dopt=Citation&list_uids=15143186.

Rodriguez-Viciana, P., Warne, P. H., Dhand, R., Vanhaesebroeck, B., Gout, I., Fry, M. J., et al. (1994). Phosphatidylinositol-3-OH kinase as a direct target of Ras. *Nature*, *370*(6490), 527–532. http://www.ncbi.nlm.nih.gov/entrez/query.fcgi?cmd=Retrieve&db=PubMed&dopt=Citation&list_uids=8052307.

Rojas, A. M., Fuentes, G., Rausell, A., & Valencia, A. (2012). The Ras protein superfamily: Evolutionary tree and role of conserved amino acids. *The Journal of Cell Biology*, *196*(2), 189–201. https://doi.org/10.1083/jcb.201103008.

Rojas, J. M., Oliva, J. L., & Santos, E. (2011). Mammalian son of sevenless Guanine nucleotide exchange factors: Old concepts and new perspectives. *Genes & Cancer*, *2*(3), 298–305. https://doi.org/10.1177/1947601911408078.

Ross, S. J., Revenko, A. S., Hanson, L. L., Ellston, R., Staniszewska, A., Whalley, N., et al. (2017). Targeting KRAS-dependent tumors with AZD4785, a high-affinity therapeutic antisense oligonucleotide inhibitor of KRAS. *Science Translational Medicine*, *9*(394). https://doi.org/10.1126/scitranslmed.aal5253.

Sanclemente, M., Nieto, P., Garcia-Alonso, S., Fernandez-Garcia, F., Esteban-Burgos, L., Guerra, C., et al. (2021). RAF1 kinase activity is dispensable for KRAS/p53 mutant lung tumor progression. *Cancer Cell*, *39*(3), 294–296. https://doi.org/10.1016/j.ccell.2021.01.008.

Scheffzek, K., Ahmadian, M. R., Kabsch, W., Wiesmuller, L., Lautwein, A., Schmitz, F., et al. (1997). The Ras-RasGAP complex: Structural basis for GTPase activation and its loss in oncogenic Ras mutants. *Science*, *277*(5324), 333–338. http://www.ncbi.nlm.nih.gov/entrez/query.fcgi?cmd=Retrieve&db=PubMed&dopt=Citation&list_uids=9219684.

Settleman, J., Albright, C. F., Foster, L. C., & Weinberg, R. A. (1992). Association between GTPase activators for Rho and Ras families. *Nature*, *359*(6391), 153–154. https://doi.org/10.1038/359153a0.

Shannon, K. M., O'Connell, P., Martin, G. A., Paderanga, D., Olson, K., Dinndorf, P., et al. (1994). Loss of the normal NF1 allele from the bone marrow of children with type 1 neurofibromatosis and malignant myeloid disorders. *The New England Journal of Medicine*, *330*(9), 597–601. https://doi.org/10.1056/NEJM199403033300903.

Shih, T. Y., Papageorge, A. G., Stokes, P. E., Weeks, M. O., & Scolnick, E. M. (1980). Guanine nucleotide-binding and autophosphorylating activities associated with the p21src protein of Harvey murine sarcoma virus. *Nature*, *287*(5784), 686–691. https://doi.org/10.1038/287686a0.

Shih, C., Shilo, B. Z., Goldfarb, M. P., Dannenberg, A., & Weinberg, R. A. (1979). Passage of phenotypes of chemically transformed cells via transfection of DNA and chromatin. *Proceedings of the National Academy of Sciences of the United States of America, 76*(11), 5714–5718. https://www.ncbi.nlm.nih.gov/pubmed/230490.

Shimizu, K., Birnbaum, D., Ruley, M. A., Fasano, O., Suard, Y., Edlund, L., et al. (1983). Structure of the Ki-ras gene of the human lung carcinoma cell line Calu-1. *Nature, 304*(5926), 497–500. https://doi.org/10.1038/304497a0.

Simon, M. A., Bowtell, D. D., Dodson, G. S., Laverty, T. R., & Rubin, G. M. (1991). Ras1 and a putative guanine nucleotide exchange factor perform crucial steps in signaling by the sevenless protein tyrosine kinase. Cell, 67(4), 701–716. https://doi.org/10.1016/0092-8674(91)90065-7.

Sjolander, A., Yamamoto, K., Huber, B. E., & Lapetina, E. G. (1991). Association of p21ras with phosphatidylinositol 3-kinase. *Proceedings of the National Academy of Sciences of the United States of America, 88*(18), 7908–7912. https://doi.org/10.1073/pnas.88.18.7908.

Smith, M. R., DeGudicibus, S. J., & Stacey, D. W. (1986). Requirement for c-ras proteins during viral oncogene transformation. *Nature, 320*(6062), 540–543. https://doi.org/10.1038/320540a0.

Stalnecker, C. A., & Der, C. J. (2020). RAS, wanted dead or alive: Advances in targeting RAS mutant cancers. *Science Signaling, 13*(624). https://doi.org/10.1126/scisignal.aay6013.

Sun, Q., Burke, J. P., Phan, J., Burns, M. C., Olejniczak, E. T., Waterson, A. G., et al. (2012). Discovery of small molecules that bind to K-Ras and inhibit Sos-mediated activation. *Angewandte Chemie (International Ed. in English), 51*(25), 6140–6143. https://doi.org/10.1002/anie.201201358.

Sutrave, P., Bonner, T. I., Rapp, U. R., Jansen, H. W., Patschinsky, T., & Bister, K. (1984). Nucleotide sequence of avian retroviral oncogene v-mil: Homologue of murine retroviral oncogene v-raf. *Nature, 309*(5963), 85–88. https://doi.org/10.1038/309085a0.

Sweet, R. W., Yokoyama, S., Kamata, T., Feramisco, J. R., Rosenberg, M., & Gross, M. (1984). The product of ras is a GTPase and the T24 oncogenic mutant is deficient in this activity. *Nature, 311*(5983), 273–275. https://doi.org/10.1038/311273a0.

Taparowsky, E., Shimizu, K., Goldfarb, M., & Wigler, M. (1983). Structure and activation of the human N-ras gene. Cell, 34(2), 581–586. https://doi.org/10.1016/0092-8674(83)90390-2.

Terrell, E. M., & Morrison, D. K. (2019). Ras-mediated activation of the Raf family kinases. *Cold Spring Harbor Perspectives in Medicine, 9*(1). https://doi.org/10.1101/cshperspect.a033746.

To, M. D., Rosario, R. D., Westcott, P. M., Banta, K. L., & Balmain, A. (2013). Interactions between wild-type and mutant Ras genes in lung and skin carcinogenesis. *Oncogene, 32*(34), 4028–4033. https://doi.org/10.1038/onc.2012.404.

Trahey, M., & McCormick, F. (1987). A cytoplasmic protein stimulates normal N-ras p21 GTPase, but does not affect oncogenic mutants. *Science, 238*(4826), 542–545. http://www.ncbi.nlm.nih.gov/pubmed/2821624.

Trahey, M., Milley, R. J., Cole, G. E., Innis, M., Paterson, H., Marshall, C. J., et al. (1987). Biochemical and biological properties of the human N-ras p21 protein. *Molecular and Cellular Biology*, 7(1), 541–544. http://www.ncbi.nlm.nih.gov/pubmed/3550423.

Trahey, M., Wong, G., Halenbeck, R., Rubinfeld, B., Martin, G. A., Ladner, M., et al. (1988). Molecular cloning of two types of GAP complementary DNA from human placenta. *Science, 242*(4886), 1697–1700. https://www.ncbi.nlm.nih.gov/pubmed/3201259.

Tran, T. H., Chan, A. H., Young, L. C., Bindu, L., Neale, C., Messing, S., et al. (2021). KRAS interaction with RAF1 RAS-binding domain and cysteine-rich domain provides insights into RAS-mediated RAF activation. *Nature Communications, 12*(1), 1176. https://doi.org/10.1038/s41467-021-21422-x.

Tsai, F. D., Lopes, M. S., Zhou, M., Court, H., Ponce, O., Fiordalisi, J. J., et al. (2015). K-Ras4A splice variant is widely expressed in cancer and uses a hybrid membrane-targeting motif. *Proceedings of the National Academy of Sciences of the United States of America*, *112*(3), 779–784. https://doi.org/10.1073/pnas.1412811112.

Tsuchida, N., Murugan, A. K., & Grieco, M. (2016). Kirsten Ras* oncogene: Significance of its discovery in human cancer research. *Oncotarget*, 7(29), 46717–46733. https://doi.org/10.18632/oncotarget.8773.

Tuveson, D. A., & Jacks, T. (1999). Modeling human lung cancer in mice: Similarities and shortcomings. *Oncogene*, *18*(38), 5318–5324. https://doi.org/10.1038/sj.onc.1203107.

Van Aelst, L., Barr, M., Marcus, S., Polverino, A., & Wigler, M. (1993). Complex formation between RAS and RAF and other protein kinases. *Proceedings of the National Academy of Sciences of the United States of America*, *90*(13), 6213–6217. https://doi.org/10.1073/pnas.90.13.6213.

Van, R., Cuevas-Navarro, A., Castel, P., & McCormick, F. (2020). The molecular functions of RIT1 and its contribution to human disease. *The Biochemical Journal*, *477*(15), 2755–2770. https://doi.org/10.1042/BCJ20200442.

Van, Q. N., Lopez, C. A., Tonelli, M., Taylor, T., Niu, B., Stanley, C. B., et al. (2020). Uncovering a membrane-distal conformation of KRAS available to recruit RAF to the plasma membrane. *Proceedings of the National Academy of Sciences of the United States of America*, *117*(39), 24258–24268. https://doi.org/10.1073/pnas.2006504117.

Villalonga, P., Lopez-Alcala, C., Bosch, M., Chiloeches, A., Rocamora, N., Gil, J., et al. (2001). Calmodulin binds to K-Ras, but not to H- or N-Ras, and modulates its downstream signaling. *Molecular and Cellular Biology*, *21*(21), 7345–7354. http://www.ncbi.nlm.nih.gov/entrez/query.fcgi?cmd=Retrieve&db=PubMed&dopt=Citation&list_uids=11585916.

Viskochil, D., Buchberg, A. M., Xu, G., Cawthon, R. M., Stevens, J., Wolff, R. K., Culver, M., Carey, J. C., Copeland, N. G., Jenkins, N. A. (1990). Deletions and a translocation interrupt a cloned gene at the neurofibromatosis type 1 locus. Cell, 62(1), 187–192. https://doi.org/10.1016/0092-8674(90)90252-a.

Vogel, U. S., Dixon, R. A., Schaber, M. D., Diehl, R. E., Marshall, M. S., Scolnick, E. M., et al. (1988). Cloning of bovine GAP and its interaction with oncogenic ras p21. *Nature*, *335*(6185), 90–93. https://doi.org/10.1038/335090a0.

Vojtek, A. B., Hollenberg, S. M., & Cooper, J. A. (1993). Mammalian Ras interacts directly with the serine/threonine kinase Raf. Cell, 74(1), 205–214. https://doi.org/10.1016/0092-8674(93)90307-c.

Wallace, M. R., Marchuk, D. A., Andersen, L. B., Letcher, R., Odeh, H. M., Saulino, A. M., et al. (1990). Type 1 neurofibromatosis gene: Identification of a large transcript disrupted in three NF1 patients. *Science*, *249*(4965), 181–186. https://doi.org/10.1126/science.2134734.

Wan, P. T., Garnett, M. J., Roe, S. M., Lee, S., Niculescu-Duvaz, D., Good, V. M., Jones, C. M., Marshall, C. J., Springer, C. J., Barford, D., Marais, R., & Cancer Genome, P. (2004). Mechanism of activation of the RAF-ERK signaling pathway by oncogenic mutations of B-RAF. Cell, 116(6), 855–867. https://doi.org/10.1016/s0092-8674(04)00215-6.

Wang, M. T., Holderfield, M., Galeas, J., Delrosario, R., To, M. D., Balmain, A., et al. (2015). K-Ras promotes tumorigenicity through suppression of non-canonical Wnt signaling. *Cell*, *163*(5), 1237–1251. https://doi.org/10.1016/j.cell.2015.10.041.

Wang, Q. J., Yu, Z., Griffith, K., Hanada, K., Restifo, N. P., & Yang, J. C. (2016). Identification of T-cell receptors targeting KRAS-mutated human tumors. *Cancer Immunology Research*, *4*(3), 204–214. https://doi.org/10.1158/2326-6066.CIR-15-0188.

Wang, T., Yu, H., Hughes, N. W., Liu, B., Kendirli, A., Klein, K., et al. (2017). Gene essentiality profiling reveals gene networks and synthetic lethal interactions with oncogenic Ras. *Cell, 168*(5). 890–903.e15 https://doi.org/10.1016/j.cell.2017.01.013.

Warne, P. H., Viciana, P. R., & Downward, J. (1993). Direct interaction of Ras and the amino-terminal region of Raf-1 in vitro. *Nature, 364*(6435), 352–355. https://doi.org/10.1038/364352a0.

Yan, W., Markegard, E., Dharmaiah, S., Urisman, A., Drew, M., Esposito, D., et al. (2020). Structural insights into the SPRED1-neurofibromin-KRAS complex and disruption of SPRED1-neurofibromin interaction by oncogenic EGFR. *Cell Reports, 32*(3), 107909. https://doi.org/10.1016/j.celrep.2020.107909.

Young, L. C., & Rodriguez-Viciana, P. (2018). MRAS: A close but understudied member of the RAS family. *Cold Spring Harbor Perspectives in Medicine, 8*(12). https://doi.org/10.1101/cshperspect.a033621.

Yuan, T. L., Amzallag, A., Bagni, R., Yi, M., Afghani, S., Burgan, W., et al. (2018). Differential effector engagement by oncogenic KRAS. *Cell Reports, 22*(7), 1889–1902. https://doi.org/10.1016/j.celrep.2018.01.051.

Yuan, T. L., Fellmann, C., Lee, C. S., Ritchie, C. D., Thapar, V., Lee, L. C., et al. (2014). Development of siRNA payloads to target KRAS-mutant cancer. *Cancer Discovery, 4*(10), 1182–1197. https://doi.org/10.1158/2159-8290.CD-13-0900.

Zhang, X. F., Settleman, J., Kyriakis, J. M., Takeuchi-Suzuki, E., Elledge, S. J., Marshall, M. S., et al. (1993). Normal and oncogenic p21ras proteins bind to the amino-terminal regulatory domain of c-Raf-1. *Nature, 364*(6435), 308–313. https://doi.org/10.1038/364308a0.

Zhou, B., Der, C. J., & Cox, A. D. (2016). The role of wild type RAS isoforms in cancer. *Seminars in Cell & Developmental Biology, 58*, 60–69. https://doi.org/10.1016/j.semcdb.2016.07.012.

Zuberi, M., Khan, I., & O'Bryan, J. P. (2020). Inhibition of RAS: Proven and potential vulnerabilities. *Biochemical Society Transactions, 48*(5), 1831–1841. https://doi.org/10.1042/BST20190023.

CHAPTER TWO

Not all RAS mutations are equal: A detailed review of the functional diversity of RAS hot spot mutations

Rachel A. Burge[a] and G. Aaron Hobbs[a,b,*]

[a]Department of Cell and Molecular Pharmacology and Experimental Therapeutics, Medical University of South Carolina, Charleston, SC, United States
[b]Hollings Cancer Center, Medical University of South Carolina, Charleston, SC, United States
*Corresponding author: e-mail address: hobbsg@musc.edu

Contents

1. Introduction 30
2. RAS mutational frequencies in cancer 32
3. *RAS* mutational frequencies across cancer 35
4. All RAS mutant proteins are not equal 38
5. RAS effector interactions and mutational frequencies 40
6. Intrinsic nucleotide (GDP and GTP) dissociation and GTP hydrolysis 44
7. GEFs and GAPs 51
8. Conclusions 53
Acknowledgments 54
References 54

Abstract

The RAS family of small GTPases are among the most frequently mutated oncogenes in human cancer. Approximately 20% of cancers harbor a RAS mutation, and >150 different missense mutations have been detected. Many of these mutations have mutant-specific biochemical defects that alter nucleotide binding and hydrolysis, effector interactions and cell signaling, prompting renewed efforts in the development of anti-RAS therapies, including the mutation-specific strategies. Previously viewed as undruggable, the recent FDA approval of a $KRAS^{G12C}$-selective inhibitor has offered real promise to the development of allele-specific RAS therapies. A broader understanding of the mutational consequences on RAS function must be developed to exploit additional allele-specific vulnerabilities. Approximately 94% of RAS mutations occur at one of three mutational "hot spots" at Gly^{12}, Gly^{13} and Gln^{61}. Further, the single-nucleotide substitutions represent >99% of these mutations. Within this scope, we discuss the mutational frequencies of RAS isoforms in cancer, mutant-specific effector interactions and biochemical properties. By limiting our analysis to this mutational subset, we

Advances in Cancer Research, Volume 153
ISSN 0065-230X
https://doi.org/10.1016/bs.acr.2021.07.004

simplify the analysis while only excluding a small percentage of total mutations. Combined, these data suggest that the presence or absence of select RAS mutations in human cancers can be linked to their biochemical properties. Continuing to examine the biochemical differences in each RAS-mutant protein will continue to provide additional breakthroughs in allele-specific therapeutic strategies.

1. Introduction

The RAS proto-oncogene family of small GTPases is comprised of HRAS, NRAS and the two splice variants, KRAS4A and KRAS4B (Wennerberg, Rossman, & Der, 2005). *RAS* is the most frequently mutated oncogene, and approximately 20% of cancer patients have a *RAS* mutation, with *KRAS* being the most commonly mutated isoform (Prior, Hood, & Hartley, 2020). The isoform and mutational frequencies in the *RAS* genes observed in human cancers vary depending on the tissue type. *KRAS* mutations are common in pancreatic, colorectal and lung cancers, *NRAS* mutations frequently appear in skin, blood and thyroid cancers and *HRAS* mutations are infrequent overall but are detected in approximately 5%–10% of thyroid and urinary cancers (COSMIC v94).

RAS GTPases cycle between the GTP-bound active and GDP-bound inactive states. RAS is activated by guanine nucleotide exchange factors (GEFs), which catalyze GDP release, resulting in GTP binding due to the greater relative concentration of GTP in the cell (Boriack-Sjodin, Margarit, Bar-Sagi, & Kuriyan, 1998). As RAS GTPases have low intrinsic GTPase activity, GTP hydrolysis is stimulated by GTPase activating proteins (GAPs), which promote GTP hydrolysis (Scheffzek et al., 1997). Active RAS regulates numerous cellular signaling pathways (Vigil, Cherfils, Rossman, & Der, 2010), such as the RAF-MEK-ERK MAPK pathway, to promote proliferation and differentiation (Vojtek & Der, 1998). RAS isoforms are highly homologous aside from the carboxy-terminal hypervariable regions (HVRs), which regulate membrane binding and intracellular localization *via* posttranslational modifications and lipidation of the CAAX motif (Prior & Hancock, 2012). The RAS isoform-specific lipid post-translational modifications contributes to the cellular localization patterns and has been implicated in isoform-specific effector activation in cells (Cox, Der, & Philips, 2015; Jang et al., 2015; Lynch et al., 2015).

Hot spot mutations in RAS proteins are generally characterized by their decreased sensitivity to GAPs, leading to an accumulation of active mutant RAS (Eccleston, Moore, Morgan, Skinner, & Lowe, 1993; Scheffzek et al., 1998). Further, mutations at Gly^{13} and Gln^{61} also have increased intrinsic nucleotide dissociation rates (Mistou et al., 1992; Smith, Neel, & Ikura, 2013). Gly^{12} and Gly^{13} reside within the phosphoryl binding loop of RAS, and mutation of these residues primarily blocks the activity of GAPs by restricting the insertion of the catalytic arginine, termed the arginine finger, from promoting GAP-mediated catalysis (Scheffzek, Lautwein, Kabsch, Ahmadian, & Wittinghofer, 1996). Gln^{61}, however, plays a direct role in catalyzing the cleavage of the terminal phosphate of GTP (Rudack, Xia, Schlitter, Kotting, & Gerwert, 2012), and mutation at this site, aside from possibly the RAS^{Q61E} mutant (Frech et al., 1994), severely reduces intrinsic and GAP-mediated GTP hydrolysis (Novelli, First, & Webb, 2018).

Due to subtle differences in protein structure, each mutant protein is characterized by unique features in effector binding, regulator interactions and intrinsic rates of hydrolysis and exchange (Munoz-Maldonado, Zimmer, & Medova, 2019). To date, no effective pan-RAS-mutant therapies are available in the clinic. However, the recent approval of a novel $KRAS^{G12C}$-selective inhibitor has provided renewed hope for the development of RAS allele-specific therapies (Mullard, 2021). Whether the mutational frequencies for RAS-mutant cancers are due to mutant-specific biochemical features remains to be fully understood. Additionally, it is not clear if all RAS mutant proteins detected in human cancers are acting as oncogenic drivers. Thus, determining the allele-specific signaling and biochemical properties of each mutant protein will provide evidence for additional mutation-specific therapeutic vulnerabilities and remains a priority in the study of RAS-mutant cancers (Moore, Rosenberg, McCormick, & Malek, 2020).

Here, we highlight the mutational frequencies of RAS isoforms in select cancers before describing the known biochemical parameters of each RAS mutant protein. We present the mutational frequency data in combination with the biochemical properties to examine the relationship between mutational frequency and biochemical function. While initially a novel concept, the RAS field has pivoted to accept that all RAS mutations are not created equal (Hobbs & Der, 2019; Miller & Miller, 2011). This concept was supported in a study that introduced a library of Gly^{12} and Gly^{13} KRAS mutations into somatic mouse lung and pancreas tissues, which demonstrated that the tissue-of-origin as well as co-mutations in other tumor suppressor genes influenced which KRAS mutations could promote tumorigenesis (Winters et al., 2017). Additionally, not all KRAS mutants were able to promote pancreatic intraepithelial neoplasia (PanIN) formation and/or acinar-to-ductal metaplasia (ADM), precursor stages to PDAC tumorigenesis, in a mouse pancreatic cancer model system (Zafra et al., 2020). Combined, these studies highlight that not all RAS mutant proteins can promote tumorigenesis in all contexts. As these studies utilized only KRAS mutations, which should have similar expression and subcellular localization patterns, these results imply that the mutant-specific biochemical features dictate tumorigenic fitness. Understanding the role of these biochemical features will provide additional rationale and strategies for developing allele-specific therapeutic strategies in the future.

2. RAS mutational frequencies in cancer

More than 150 different mutations have been detected in the *RAS* genes, with mutations found at nearly every amino acid (Prior et al., 2020). Despite the overwhelming number of mutations in the *RAS* genes, the majority of mutations occur at one of three "hot spot" positions at Gly^{12}, Gly^{13} and Gln^{61}. Mutations at these sites account for ~97% of all *NRAS* and *KRAS* mutations and nearly 90% of *HRAS* mutations (Prior et al., 2020). The observed mutational frequency at these hot-spot mutations varies between RAS isoform and tissue type, although the biological reason for this variability is unclear (Cox, Fesik, Kimmelman, Luo, & Der, 2014). Because nearly all RAS-mutant cancers can be represented by focusing on the single-nucleotide hot-spot mutations, we have chosen to focus on these mutations to simplify our figures and analysis.

The absolute and relative mutational frequencies of each RAS isoform vary depending on cancer type. Of the *RAS* genes, *KRAS* is the most frequently mutated isoform at Gly^{12} and Gly^{13}, and *NRAS* is the most frequently mutated isoform at Gln^{61} (Fig. 1A–C). Despite the similarities in DNA sequence, each isoform maintains a distinct mutational frequency at each codon. In Fig. 1A–C, the bullseye charts highlight the relative frequencies of RAS mutations across all cancers, which are strikingly distinct (data compiled from COSMIC v94). For example, the G12S mutation is not frequently observed in *KRAS* (5%) relative to HRAS (22%) and NRAS (16%) while $NRAS^{G12V}$ (8%) is uncommon despite being the second most common mutation in *KRAS* (28%; Fig. 1A). At Gly^{13}, $KRAS^{G13D}$ represents 90% of the *KRAS* mutations at this codon, yet $HRAS^{G13R}$ accounts for 76% of *HRAS* mutations and $HRAS^{G13A}$ accounts for nearly 10% (Fig. 1B). $NRAS^{G13V}$ represents 13% of *NRAS* Gly^{13} mutations, yet this mutation is nearly undetectable in the other isoforms. Finally, at Gln^{61}, $KRAS^{Q61H}$ represents 44% of KRAS mutations overall at this codon, yet it is rare in the

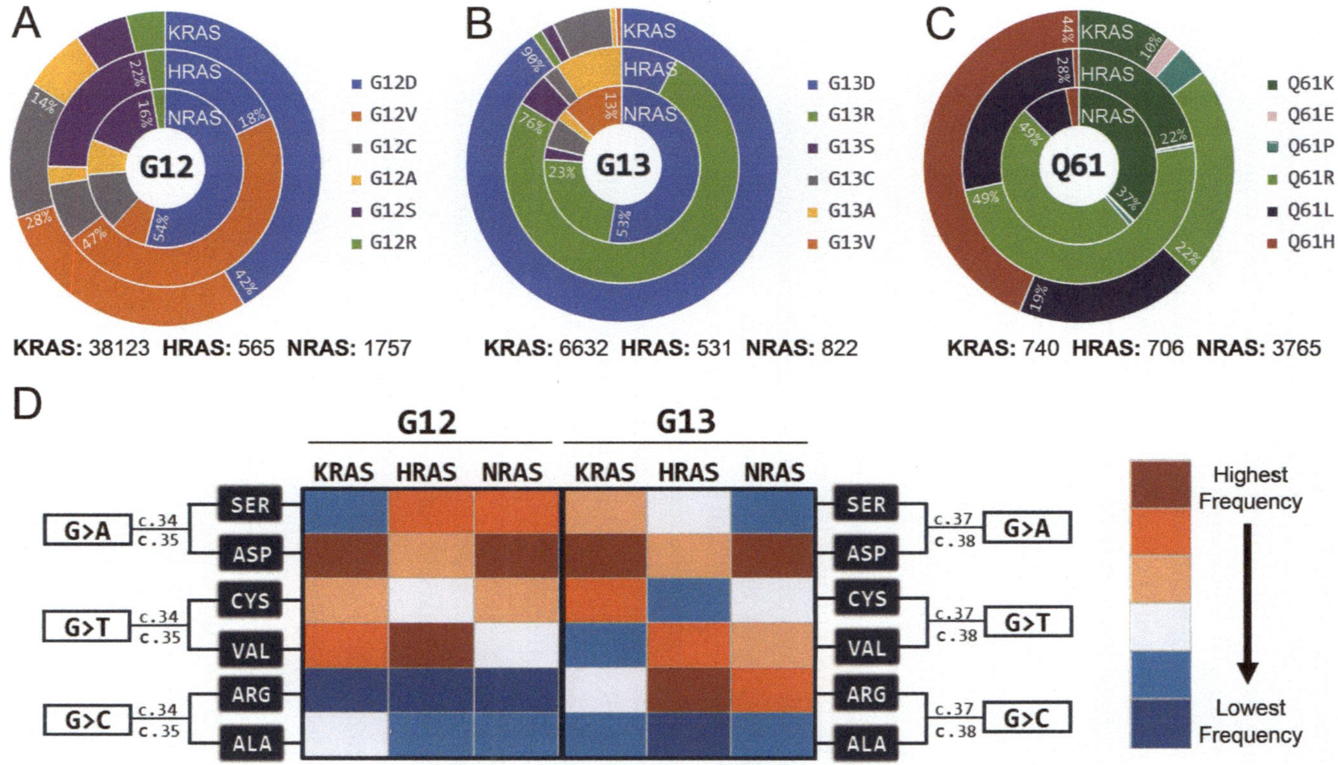

Fig. 1 Common RAS isoform-specific mutations in human cancers. (A) The six most commonly occurring mutations at Gly[12]. (B) The six most commonly occurring mutations at Gly[13]. (C) The six most commonly occurring mutations at Gln[61]. Mutations at Gly[12], Gly[13] and Gln[61] combine to represent 99.5% (KRAS), 86% (HRAS) and 97% (NRAS) of the total RAS isoform mutations. The mutations and relative percent values represent only the single-base nucleotide substitutions. Two-base substitutions, which account for <0.1% for KRAS, are omitted from the chart for simplicity. The total number of cases for each isoform are indicated below the chart. Frequency values are listed if >10%. (D) Relative rank-order of the isoform-specific frequency of Gly[12] and Gly[13] single-base mutations. The single-nucleotide substitutions at positions c. 34, c. 35, c. 37 and c. 38 result in the indicated amino acids. Standard nomenclature for nucleotide numbering begins at c. 1 of the start methionine codon; (*c, coding nucleotide strand numbers*). Mutations are ranked numerically from most frequent (*red/darkest*) to the least (*blue/hatched*) for each isoform. Mutations at c. 36 or c. 39 did not alter the amino acid and were not included. *Mutational data compiled from COSMIC v.94.*

other RAS isoforms (Fig. 1C). Another striking feature of Gln^{61} mutations is that some single-base mutations are rare in all three isoforms, such as RAS^{Q61E} and RAS^{Q61P}. While one explanation for the observed mutational frequencies is simply due to observational bias and the limited sequencing data available for all three RAS isoforms at each codon, other possibilities include carcinogen promoting only a subset of potential mutations and/or allele-specific signaling that drives the observed mutational frequencies, as has been previously suggested (Li, Balmain, & Counter, 2018).

The possibility that carcinogen exposure drives RAS mutational frequencies has strong evidence specifically in lung cancers. Smoking has been linked to the increased prevalence of the $KRAS^{G12C}$ mutation (Riely et al., 2008). In a study examining carcinogen and mutational burden, exposure of mice to urethane has been shown to promote multiple unique RAS allele mutations early after urethane exposure, although only a few mutations were detected in established tumors, suggesting that the tumorigenic fitness of each RAS mutant drives the observed mutational frequencies (Li, MacAlpine, & Counter, 2020). However, increasing the protein expression of KRAS altered the observed mutational frequencies after urethane exposure (Pershing et al., 2015). In this model, the frequency of $KRAS^{G12D}$ mutations was greatly increased despite the use of the same carcinogen, suggesting that additional factors aside from carcinogen dictate the tumorigenic fitness of RAS mutants. While there are several possibilities for these observations, these results imply that both carcinogen and biology influence the mutational frequency profiles observed in human tumors. In agreement with this possibility, a study analyzing 13,492 KRAS samples from colorectal, pancreatic, lung and blood cancers demonstrated that the frequency of the hotspot mutations are inconsistent with the distribution of genetic mutations in each cancer subtype (Cook, Melloni, Gulhan, Park, & Haigis, 2021). For example, $KRAS^{G12R}$ comprises 16% of the KRAS mutations in pancreatic ductal adenocarcinoma (PDAC) despite having a predicted frequency of approximately 5%. In contrast, $KRAS^{Q61L}$ has a predicted frequency of 1.5% in pancreatic cancer but is observed at a frequency of 0.18%. These data imply that carcinogen is not the only factor in promoting the KRAS mutational frequency in PDAC. Additionally, a mathematical evaluation of cancer risk has suggested that approximately two-thirds of all cancers can be ascribed to "bad luck", or mistakes by DNA polymerases during cell division (Tomasetti & Vogelstein, 2015). Taken together, these data suggest that the tumorigenic fitness of each RAS mutant protein is not simply an outcome of tissue-specific carcinogen exposure, but due to a host of properties, including protein expression levels, the biochemical properties of the RAS mutants and tissue-specific cellular signaling.

KRAS and *NRAS* share the same genetic codon [GGT] at Gly^{12}, while *HRAS* has [GGC] at Gly^{12} (Fig. 1D). For Gly^{13}, *HRAS* and *NRAS* share the same codon [GGT] and *KRAS* has a unique codon [GGC]. Coincidentally, any mutation in the wobble base generates a silent mutation in either codon in all three isoforms. Thus, three distinct mutations can occur at the two guanine bases to generate a missense mutation, G > A, G > T and G > C (Fig. 1D). While DNA mutation rates are influenced by DNA structure and the surrounding nucleotides, if random DNA replication errors or carcinogen were solely responsible for the observed RAS mutational frequencies, then one might expect an even frequency of mutations in comparison to the mutation patterns within the same cancer type (Cook et al., 2021). However, there are clear differences in the observed mutational frequencies in the RAS isoforms at these positions. In *NRAS*, the G > A mutation represents the two most common mutations, generating either $NRAS^{G12D}$ or $NRAS^{G12S}$, followed by the G > T and G > C mutations. In *HRAS*, the c. 35G > T mutation represents the most common mutation, while c. 34G > A has the second highest frequency. In KRAS, c. 35G > A represents the most common mutation yet the c. 34G > A (G12S) mutation is infrequent. Interestingly, mutations at c. 35 are the most common at Gly^{12}, yet the c. 35G > C (G12A) occurs at a much lower frequency, and the G > C transition is infrequent in all three isoforms at Gly^{12}. However, Gly^{13}, encoded by a [GG(T/C)] codon, the G > C transition is common in both HRAS and NRAS. Due to the proximity of codons for Gly^{12} and Gly^{13}, it is surprising that there are such dramatic differences in the DNA mutation frequencies between similar codons, although other factors, such as the surrounding nucleotides as well as DNA accessibility can also dictate the overall mutation rates within a codon (Makova & Hardison, 2015).

3. *RAS* mutational frequencies across cancer

Despite the similarity in the codons for each RAS isoform, there are clear distinctions between the mutated codon and isoform in different cancer types. In Fig. 2, we highlight the three cancers with the greatest number of RAS mutational data available for each isoform, detailing the RAS mutations in cancers that have been highly studied and have a high frequency

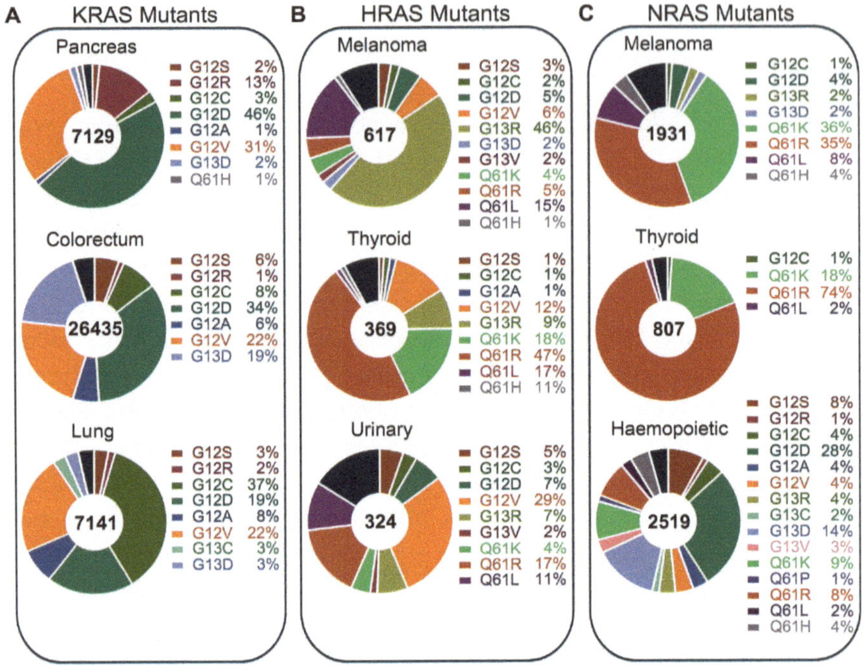

Fig. 2 RAS isoform mutations in common cancers. (A) KRAS mutation frequencies in cancers with the greatest number of KRAS mutations. (B) HRAS mutation frequencies in cancers with the greatest number of HRAS mutations. Coloring the same as in (A). (C) NRAS mutation frequencies in cancers with the greatest number of NRAS mutations. Coloring the same as in (A). Mutations that appear at a frequency >1% are listed for each cancer. Mutations that appear with <1% are grouped and shown in *black*. Total RAS-mutant tumors for each cancer type are in the center of each graph. Graphs begin with the first mutations at 12 o'clock and continue clockwise. *Data compiled from COSMIC v.94.*

of RAS mutations. In contrast, some cancers have a high prevalence of RAS mutations but have limited sequencing data, such as peritoneum cancer. This cancer has the second highest mutation rate of KRAS mutations (166/371) of all data available on COSMIC. However, with only 371 total cases, these data are limited compared to the total mutational data available for colorectal cancer (CRC) and lung cancer.

KRAS is the most commonly mutated RAS isoform (Simanshu, Nissley, & McCormick, 2017). Comparing the KRAS mutant cancers (Fig. 2A), there are three tissue-specific mutations detected. (1) $KRAS^{G12R}$ is the third most frequent mutation in PDAC and is rarely detected in CRC or lung cancer. (2) $KRAS^{G13D}$ is frequent in CRC and (3) $KRAS^{G12C}$ is present at an outsized frequency in LAC. Additionally, $KRAS^{G12C}$ mutants were mutually exclusive with other oncogenic driver mutations that are commonly associated with RAS mutations in LAC (Araujo et al., 2021), highlighting the unique tumorigenesis of KRAS^{G12C}-mutant lung cancers. Because KRAS^{G12C} mutations are associated with smoking (Alexandrov et al., 2013; Le Calvez et al., 2005), it is tempting to conclude that the KRAS mutations that occur at a higher frequency in a tissue-specific manner are due to tissue-specific carcinogens.

Interestingly, CRC is associated with fast-cycling RAS mutants, including KRAS^{G13D} and KRAS^{A146T} (Haigis, 2017). KRAS^{A146T} accounts for approximately 1% of all *KRAS*-mutant CRC cases but is nearly undetectable in other cancers. KRAS^{A146T} has an elevated intrinsic nucleotide exchange compared to KRAS^{WT} and provides another example of a cancer-type specific KRAS mutation with unique structural and biochemical defects (Poulin et al., 2019). Finally, KRAS^{G12R} is unable to bind to PI3Kα (Hobbs et al., 2020), yet PI3Kα is necessary for PDAC and lung tumorigenesis in mouse models of disease (Gupta et al., 2007; Murillo et al., 2014; Wu et al., 2014). One possibility is that insulin signaling relieves the dependence on KRAS-mediated PI3K activation, which allows for KRAS^{G12R} to promote tumorigenesis in the pancreas. Another possibility is that upregulation of additional PI3K isoforms in PDAC tumor tissue, specifically PIK3CD and PIK3CG (Baer et al., 2014), allows for KRAS^{G12R}-mediated PDAC tumorigenesis. However, as RAS mutations are considered the initiating event in PDAC (Iacobuzio-Donahue, Velculescu, Wolfgang, & Hruban, 2012), and the KRAS^{G12R} mutation cannot promote PanIN formation in mice models of PDAC (Zafra et al., 2020), it is unlikely that the upregulation of the additional PI3K isoforms allow for KRAS^{G12R}-mediated tumor initiation. Finally, KRAS is not the initiating mutation in CRC (Fearon & Vogelstein, 1990; Waliszewski, 1995) and lung cancers (Chung et al., 1995). Thus, these data support the possibility that co-mutations and tissue-specific signaling promotes the observed RAS mutational frequencies, selecting some mutations in certain contexts while limiting the same mutations in others.

While KRAS is the most commonly mutated isoform, *NRAS* and *HRAS* have unique cancer-specific frequencies as well. Melanoma exhibits a higher frequency of *NRAS* and *HRAS* mutations (15% vs 10% of tested cases; COSMIC v94). In melanoma, *NRAS*^{Q61K} and *NRAS*^{Q61R} represent ~70% of all *NRAS*-mutant melanoma cases while *HRAS*^{G13R} accounts for 46% of *HRAS*-mutant melanoma cases. (Fig. 2B and C). Despite this variability in melanoma, *NRAS* and *HRAS*-mutant thyroid cancers show a higher degree of similarity for mutation frequencies, although *HRAS* is somewhat more variable (Fig. 2B and C). Finally, hemopoietic cancers show almost no mutational preference in *NRAS* and is one of the most diverse RAS-mutant cancers, suggesting that perhaps hemopoietic cancers are able to tolerate a wide array of NRAS activity.

Given the lack of a clear pattern of mutational frequencies at the hot spot codons in the RAS isoforms despite their sequence similarity, these observations suggest that many factors simultaneously contribute to the observed mutation frequencies. While many factors can dictate tissue-specific RAS-mutant tumorigenic fitness, the biochemical properties of each mutant can also promote or limit tumorigenic fitness. However, aside from a few limited studies (Hobbs et al., 2020; Hunter et al., 2015; Poulin et al., 2019), the unique properties of each RAS mutant protein are often ignored when evaluating tumorigenic fitness and therapeutic strategies of RAS-mutant cancers.

4. All RAS mutant proteins are not equal

The mutational frequency data clearly demonstrate tissue-specific mutational patterns in the RAS isoforms. RAS isoforms are highly similar in the first 166 amino acids, termed the G-domain and is separated into two lobes, the effector lobe, residues 1–86, and the allosteric lobe, residues 87–166 (Buhrman, Kumar, Cirit, Haugh, & Mattos, 2011). The effector lobe shares 100% sequence identity between the RAS isoforms and the allosteric lobe has approximately 80% sequence identity. The allosteric lobe of RAS has been linked to RAS dimerization (Ambrogio et al., 2018; Guldenhaupt et al., 2012; Spencer-Smith et al., 2017) as well as isoform-specific differences in RAF1 binding and GTP hydrolysis (Johnson et al., 2017). A cartoon diagram is depicted in Fig. 3A, which highlights the important structural regions in RAS. The proximity of the Gly^{12}, Gly^{13} and Gln^{61} mutations to the nucleotide are presented. Finally, the c-terminal hyper-variable domain shares minimal sequence identity and dictates differential localization within the cell (Abankwa, Gorfe, Inder, & Hancock, 2010), which can dictate effector recognition (Amendola et al., 2019). RAS post-translational modifications also dictate membrane localization (Ahearn, Zhou, & Philips, 2018). In the cell, HRAS and NRAS traffic through the Golgi on their way to the plasma membrane (Choy et al., 1999). HRAS is distributed throughout the Golgi, while NRAS localizes toward one side of the Golgi stack (Lynch et al., 2015). Both HRAS and NRAS interact with lipid rafts, but NRAS palmitoylation enables it to preferentially interact with these microdomains (Roy et al., 2005). Prior et al. showed the interaction of HRAS with lipid rafts is a dynamic process, with GTP-loading

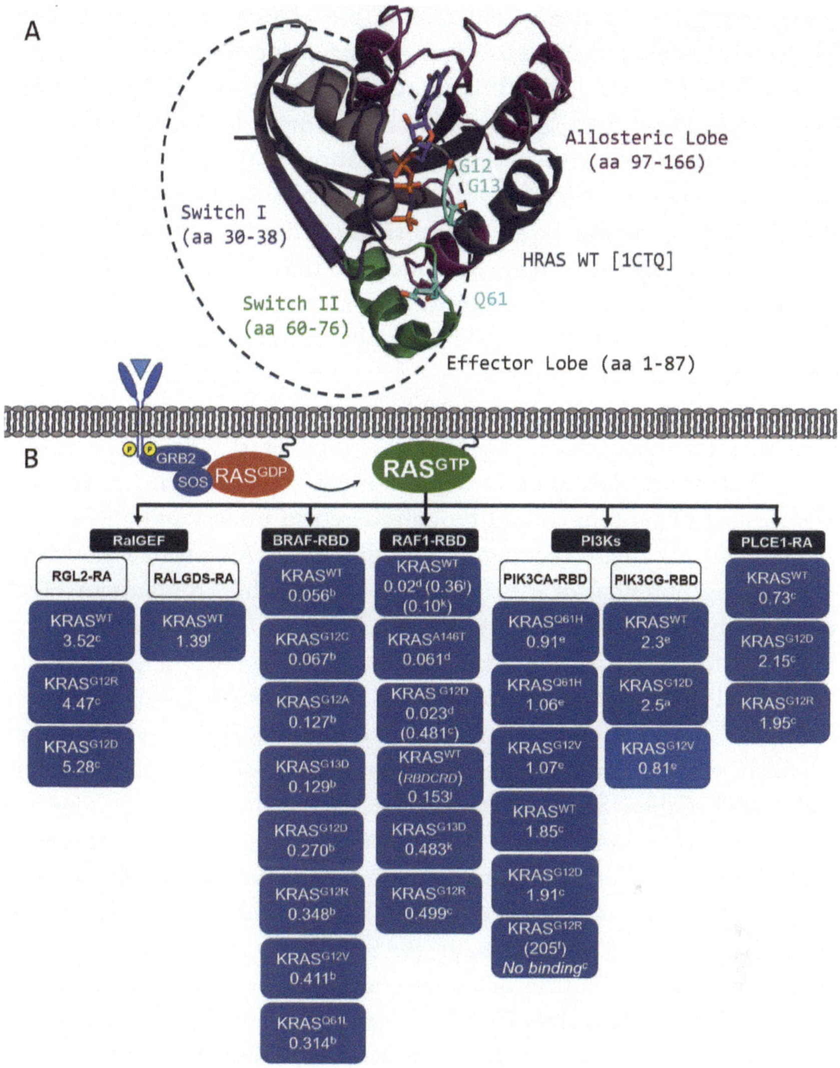

Fig. 3 RAS protein structure and effector binding. (A) Cartoon model of $HRAS^{WT}$ depicting the 3D locations of RAS hot-spot mutations. Other regions important for RAS structure are labeled. *Cyan*, Gly^{12}, Gly^{13} and Gln^{61}; *purple*, allosteric lobe; *black circle*, effector lobe; *blue*, switch I; *green*, switch II; GTP analog is shown as sticks (C, *purple*; N, *blue*; O, *red*; P, *orange*); Mg^{2+}, gray. (B) KRAS is active at the plasma membrane, where GTP loading promotes binding to effector proteins. Determination of the binding constants between KRAS mutant proteins and several effectors have been determined using three common methods of detection: isothermal titration calorimetry, fluorescence polarization and inhibition of guanine nucleotide dissociation. Binding constants are all presented in μM. Superscripts reference original publications. [a]Burd et al. (2014); [b]Hunter et al. (2015); [c]Hobbs et al. (2020); [d]Poulin et al. (2019); [e]Martinez et al. (2021); [f]Nakhaeizadeh, Amin, Nakhaei-Rad, Dvorsky, and Ahmadian (2016); [g]Linnemann, Kiel, Herter, and Herrmann (2002); [h]Herrmann, Martin, and Wittinghofer (1995); [i]Pacold et al. (2000); [j]Tran et al. (2021); [k]Johnson et al. (2019).

redistributing HRAS from lipid rafts in the plasma membrane, which is necessary to activate RAF. However, KRAS is localized outside lipid rafts regardless of its nucleotide bound state (Prior et al., 2001). KRAS is phosphorylated on Ser^{181} in the HVR by Protein Kinase C, localizing it to internal membranes, including translocation to the ER, Golgi, and the outer mitochondrial membrane, where it can associate with BcL-XL (BCL2L1), to function as an apoptosis-inducing protein (Bivona et al., 2006). Taken together, these data indicate that the RAS isoforms have isoform-specific subcellular localization, which contributes to the modulation of signaling pathways and potentially accounts for isoform-specific signaling in the cellular context.

It is widely accepted many RAS mutant proteins share distinct differences in their biochemical properties (Hunter et al., 2015; Smith et al., 2013). Therefore, a detailed understanding of the biochemical properties of each mutant protein should be considered to understand allele-specific signaling. The earliest biochemical studies on RAS probed the cellular transforming potential of a panel of Gly^{12} and Gln^{61} mutants in HRAS. Interestingly, all Gly^{12} mutants except $HRAS^{G12P}$ induced transformation in a focus formation assay (Seeburg, Colby, Capon, Goeddel, & Levinson, 1984). It was later shown that the intrinsic GTP hydrolysis rates of $HRAS^{G12P}$ and $KRAS^{G12P}$ were nearly double RAS^{WT}, which was used to explain why this mutant was unable to induce transformation (Eccleston et al., 1993; Gibbs, Schaber, Allard, Sigal, & Scolnick, 1988). The transforming properties of HRAS Gln^{61} mutants were similarly effective and 15 of 17 mutants promoted transformation, albeit some were weakly transforming (Der, Finkel, & Cooper, 1986). Consolidating the known biochemical features of each mutant RAS protein will provide additional insight into the distribution of mutational frequencies of the various RAS isoforms. The following biochemical properties should be considered; effector binding, intrinsic nucleotide exchange, which results in RAS activation, GTP hydrolysis, which promotes RAS inactivation, and the regulatory GAPs and GEFs that regulate RAS function.

5. RAS effector interactions and mutational frequencies

RAS has been shown to bind to numerous effector proteins with diverse functions (Kiel, Matallanas, & Kolch, 2021). RAS effectors are characterized by a commonly conserved RAS binding domain (RBD) or RAS-association (RA) domain. RBDs consist of a ubiquitin fold that forms an inter-protein β-sheet (Wohlgemuth et al., 2005). RA domains have a similar structure and binding mode, although they form different intermolecular interactions. The first RAS effector identified was the RAF1 kinase (Morrison, Kaplan, Rapp, & Roberts, 1988; Rapp et al., 1983), which binds RAS *via* an RBD and initiates a signaling cascade that activates RAF/MEK/ERK MAPK signaling. Despite the diverse array of RAS effectors detailed to date, a large siRNA screen targeting 41 KRAS effector nodes across 92 cell lines determined that RAF and PI3K signaling were the most critical nodes for RAS-mediated proliferation, thereby representing the targets with the greatest therapeutic potential (Yuan et al., 2018).

Despite our detailed understanding of RAS-effector interactions (Gupta et al., 2007; Kiel, Serrano, & Herrmann, 2004; Nakhaeizadeh et al., 2016; Rodriguez-Viciana, Warne, Vanhaesebroeck, Waterfield, & Downward, 1996), studies comparing the binding affinities between effectors and commonly occurring RAS-mutant proteins are limited. We have consolidated many of the effector binding studies that have compared the binding affinities between KRAS mutants and select RBD or RA domains of KRAS effectors (Fig. 3B). While there is considerable variability between mutants and effector proteins, the most striking consistency is that RAF-RBD binding is the strongest binding RAS effector; RAF binding is consistently measured in the hundreds of nanomolar while most other effector binding affinities are in the micromolar range (Burd et al., 2014; Herrmann et al., 1995; Hobbs et al., 2020; Hunter et al., 2015; Johnson et al., 2017, 2019; Linnemann et al., 2002; Martinez et al., 2021; Nakhaeizadeh et al., 2016; Pacold et al., 2000; Poulin et al., 2019; Thurman, Siraliev-Perez, & Campbell, 2020; Tran et al., 2021). Due to these binding data, RAF has been considered to be the most critical RAS effector for RAS-mediated tumorigenesis (Klomp, Klomp, & Der, 2021) and has been the target of numerous chemical inhibitors and clinical trials (Cox et al., 2014; Moore et al., 2020).

However, the biochemical data on RAS and RAF binding are highly variable, due in part to the method of data collection and experimental conditions. In an early study measuring the interaction between CRAF and each RAS isoform, NRAS was found to preferentially bind the RAF1-RBD while KRAS bound three-fold worse, albeit still with a tight interaction (Nakhaeizadeh et al., 2016). In contrast, a more recent study found that NRAS bound two-fold worse than KRAS and HRAS (Johnson et al., 2017). In addition, other studies have found that RAS-RAF binding is exceptionally tight, but these data do not show a clear RAS isoform specificity (Burd et al., 2014; Hobbs et al., 2020; Poulin et al., 2019; Thurman et al., 2020). However, the biochemical binding affinities ignore the possible effects of subcellular localization and membrane binding on RAF activity. Whether the differences in binding affinity result in isoform-specific RAF activation in cells is not clear from just the biochemical data, although BRAF and RAF1 are the strongest effector complexes when compared to all RAS effectors (Fig. 3). Importantly, common cancer-associated RAS mutations do not show binding affinities that match their overall frequencies (Hunter et al., 2015), such as $KRAS^{G12V}$ having one of the poorest RAF binding affinities despite being the second most common *KRAS* mutation overall.

Recent crystal structures have shown that the cysteine-rich domain (CRD) of RAF1 interacts with KRAS, dramatically increasing the RAF-RAS binding affinity (Tran et al., 2021), in line with previous predictions (Mott et al., 1996). Thus, it is unclear how relevant the subtle differences in binding affinities between the different RAS isoforms and mutant proteins are in the absence of the CRD. Further complicating the in vitro biochemical binding data, mutations generated to disrupt the KRAS-CRD interface only minimally decreased the biochemical binding affinity yet significantly reduced RAF1 kinase activity in cells. In agreement with this observation, ERK signaling in cells was shown to be similar for a number of oncogenic KRAS mutants (Gillies et al., 2020), suggesting that biochemical properties other than the binding affinity of the RBD domain drive ERK MAPK signaling. To add further complexity, RAS:BRAF binding in cells has been shown to be dependent on the c-terminal HVR of KRAS. Utilizing chimeric RAS proteins, the HVR of KRAS dictated an increased specificity for the KRAS:BRAF interaction over HRAS and NRAS while CRAF showed no isoform selectivity regardless of HVR (Terrell et al., 2019). Minor changes in binding affinity are not physiologically relevant if the competition for forming a complex is determined by concentrations of RAS effectors (Kiel, Verschueren, Yang, & Serrano, 2013). Additionally, proteomic/computational analysis shows that tissue-specific dependencies are likely to affect the effector binding profile of RAS in cancers (Catozzi, Halasz, & Kiel, 2021). Taken together, these data indicate that the differences in the biochemically measured binding affinities of RAF do not correlate with RAS oncogenicity in vivo. Additionally, there are important structural components of the RAS-effector interactions that have been missed by focusing only on the limited interactions of the RBDs, including the HVR, membrane association and components of

RAF such as the CRD. Thus, while in vitro and in vivo data all point to the critical role of RAF on RAS-mediated tumorigenesis, the current biochemical data are insufficient to rationalize RAF binding affinity to the observed mutational frequencies of RAS isoforms.

Regarded as a critical RAS effector, phosphatidylinositol-3-kinases (PI3Ks) are a class of lipid kinases that are activated by RAS to stimulate signaling pathways that control differentiation and survival (Fruman et al., 2017). However, RAS binds to PI3K isoforms with significantly weaker affinity compared to RAF (Pacold et al., 2000). Martinez et al. demonstrated that PI3K RBDs show differences in structure and stability when comparing PIK3CA and PIK3CG isoforms, and RAS mutations discriminate between these two PI3K isoforms (Martinez et al., 2021). However, the role of PI3K activated by RAS in tumorigenesis is unclear. While the necessity for the RAS PI3K-RBD interaction has been shown in lung cancer models, and loss of PI3Kα reduces tumors in mouse pancreatic cancer models (Castellano et al., 2013; Murillo et al., 2014; Wu et al., 2014), $KRAS^{G12R}$ cannot bind to PI3Kα yet is able to efficiently promote pancreatic cancer (Hobbs et al., 2020). However, sequencing data from a small cohort of PDAC patients indicated that $KRAS^{G12R}$ PDAC is associated with increased PI3K pathway mutations, implying a need for this kinase in PDAC tumorigenesis (Diehl, Hannan, & Chiorean, 2021). As $KRAS^{WT}$ and KRAS mutants have a relatively poor binding affinity to PI3Kα, it is not surprising that a computational model of KRAS signaling suggested that the RAS:PI3K interactions do not represent a dominant effector interaction (Catozzi et al., 2021). One hypothesis is that PI3K signaling is necessary in cancer, but direct RAS activation of PI3K is not necessary in all cancer types or can be easily overcome by additional co-mutations.

Several recent studies have identified additional RAS effectors that may play a role in tumorigenesis. PIP5K1A was identified as a potential target for KRAS-mutant cancers, and genetic ablation of PIP5K1A reduced KRAS signaling as well as synergistically reduced cell viability when combined with a MAPK inhibitor in many PDAC cell lines (Adhikari & Counter, 2018). Proximity ligation proteomics methods have also identified mTORC2 as a novel RAS-related effector. mTORC2 interaction *via* the effector binding domain was demonstrated across all RAS isoforms (Kovalski et al., 2019). In another study, a direct, GTP-dependent interaction between the KRAS4A isoform and hexokinase 1 (HK1) demonstrated specificity between the KRAS splice variants, which was linked to isoform-specific localization

(Amendola et al., 2019). This study suggested that mutant KRAS4A activation of HK1 promotes metabolic rewiring. Two additional KRAS effectors were identified in RAS-mutant lung cancer cell lines: RADIL (Rap-associating and Dilute Domain-Containing Protein), a cell adhesion regulator, and RIN1 (RAS and RAB Interactor 1), an endocytosis regulator that was implicated in KRAS-dependent macropinocytosis in lung cancer cell lines (Kelly et al., 2020). RIN1 had been previously identified as an HRAS effector (Han & Colicelli, 1995; Tall, Barbieri, Stahl, & Horazdovsky, 2001). These newly discovered RAS effector proteins provide new avenues for targeting additional chemical inhibitor combination therapies. However, whether these effector interactions are differentially activated by allele-specific signaling or sensitive to therapeutic intervention are unclear.

Ultimately, the differences in binding affinity between RAS mutants and effector proteins when using only RBD fragments remain inconsistent, which makes it difficult to draw any significant conclusions. The results with RAF binding and ERK signaling highlight how the binding affinities can be disconnected from the cellular signaling. However, RAS mutants are characterized by increased GTP levels, so the relative binding affinity may not be relevant simply due to the increased levels of mutant RAS in cancers. Therefore, the increased levels of GTP-loading may be more critical in dictating tumorigenic potential rather than absolute binding affinities.

6. Intrinsic nucleotide (GDP and GTP) dissociation and GTP hydrolysis

The intrinsic nucleotide dissociation and GTP hydrolysis rates have been measured for many mutant RAS proteins (Table 1). The intrinsic nucleotide dissociation and hydrolysis rates are key features of RAS mutant proteins, dictating the relative GTP:GDP ratios in cells. While mutant RAS proteins are generally regarded as having decreased GAP-mediated hydrolysis, the intrinsic biochemical properties further modulate the activity of each mutant. By collating the observed biochemical rates, there are several interesting trends that emerge. Foremost, while there is a fair magnitude of variability in the biochemical rates across isoforms, there is just as much variability between studies of the same isoform. Using $KRAS^{G12D}$ as an example, hydrolysis was measured to be $3.98 \times 10^{-3}\ min^{-1}$ (Escobar-Hoyos et al., 2020), $5.6 \times 10^{-3}\ min^{-1}$ (Bollag et al., 1996), $7.2 \times 10^{-3}\ min^{-1}$ (Eccleston et al., 1993) and $11.4 \times 10^{-3}\ min^{-1}$ (Hunter et al., 2015;

Table 1 Biochemical rates of mutant RAS proteins.

Mutant	WT	G12S	G12D	G12C	G12V	G12R	G12A	G13S	G13D	G13C	G13V	G13R	G13A	Q61H	Q61R	Q61L	Q61K	Q61P	Q61E
Units ($x10^{-3} min^{-1}$)																			
GDP dissociation	13[m] (3.8[g]) (7.9[a]) (6.5[h])	28.2[f]	10[m] (9.6[f]) (13[e])	13.2[f]	2.3[a] (2.0[g])	3.7[a]	28.8[f]	23[d]	9.6[f] (56.7[g])	15[f]	630[d] (20.4[f])	No Data	No Data	7.2[a] (9.6[h])	11[e]	16[h] (9[g])	9[h]	6.8[h]	130[b]
GTP dissociation	14[m] (5.0[a]) (15[h])	1.2[f]	36[m] (57[f]) (16[e])	3.6[f]	4.7[a]	1.1[a]	5.4[f]	5[d]	37.8[f]	4.8[f]	30[d] (7.2[f])			3.7[a] (9.2[h])	4[e]	31.0[c]	9.1[h]	19[h]	240[b]
GTP hydrolysis	13[m] (2.42[i]) (28.0[a]) (8.8[g]) (40.8[j]) (20.9[h])	7.2[f]	5.6[m] (11.4[j]) (8.4[f]) (23[e])	29.4[j] (10.8[f]) (0.68[i])	2.0[a] (0.9[g]) (2.5[j])	1.4[a] (1.1[j])	3[f] (0.8[j])	32[d] (19.2[f])	11.4[f] (3.1[g]) (0.92[i]) (5.8[j])	6.6[f]	13[d] (12[f])			1.9[a] (1.63[h]) (0.25[i]) (0.8[j])	0.02[e]	1.3[c] (0.11[g]) (0.5[j])	0.348[h]	1.63[h]	570[b]
RASA1 K_D (K_M) units in μM	4.8[d] (5.0[h]) (8.9[f])	(15.8[f])	5[l] (4.8[f])	12.3[f]	20[d] (>100[f])	35[d]	46.8[f]	(19.6[f])	(5.2[f])	(26.6[f])	(87.0[f])			2.0[h] (2.0[d])	8.0[h]	0.1[d] (0.5[h])	6.0[h]	1.2[h]	
NF1 units in μM	1.2[k] (0.035[h]) (0.16[l])		8.1[k] (0.16[l])						6.4[k]	11.0[k]				0.025[h]	18.4[k] 1.5[h]	0.002[h]	2[h]	0.05[h]	

[a]John, Frech, and Wittinghofer (1988), John, Schlichting, Schiltz, Rosch, and Wittinghofer (1989), HRAS, 37 °C.
[b]Frech et al. (1994), HRAS, 37 °C.
[c]Krengel et al. (1990), HRAS, 37 °C.
[d]Gideon et al. (1992), HRAS 30 °C.
[e]Burd et al. (2014), NRAS, 25 °C.
[f]Wey, Lee, Jeong, Kim, and Heo (2013), HRAS.
[g]Smith et al. (2013), HRAS, 25 °C.
[h]Donovan, Shannon, and Bollag (2002), HRAS.
[i]Moghadamchargari et al. (2019), KRAS, 25 °C.
[j]Hunter et al. (2015), KRAS, 25 °C.
[k]Rabara et al. (2019), KRAS, 25 °C.
[l]Bollag and McCormick (1991), KRAS, 25 °C.
[m]Bollag et al. (1996), KRAS, 37 °C.
Units for dissociation and hydrolysis are ($x10^{-3} min^{-1}$), the bracketed values indicate the referenced studies; isoform as indicated.

Poulin et al., 2019). Despite the variability, these rates are within a factor of three of each other, and they are all slower than RAS^{WT} (28×10^{-3} min^{-1}) (John et al., 1989). While some of the variability can be inferred from the reaction conditions, construct and phosphate detection reagents, the biochemical rates can provide information on relative rates and trends. This table does not represent all biochemical data collected on RAS proteins. It is, however, designed to highlight several key points using rates for 16 of the 18 mutants represented by single-base nucleotide substitutions at the three hot spot mutation sites. First, RAS^{WT} is defined by a fast intrinsic GTP hydrolysis rate relative to nucleotide dissociation (Gideon et al., 1992), which maintains RAS^{WT} in the inactive state and under tight control by its regulators (GEFs, GAPs). Second, aside from RAS^{G13S} (Gideon et al., 1992; Wey et al., 2013) and RAS^{Q61E} (Frech et al., 1994), all RAS mutants have decreased intrinsic GTP hydrolysis rates. Third, RAS mutants show variable rates for GDP and GTP dissociation (Frech et al., 1994; Hunter et al., 2015; Mistou et al., 1992), highlighting that each mutant has unique binding kinetics for the nucleotides (Lenzen, Cool, Prinz, Kuhlmann, & Wittinghofer, 1998). Finally, considering the intrinsic GTP dissociation rate, which is often overlooked in RAS biochemical studies, is necessary to describe the RAS mutant proteins. For example, RAS^{G12D} is not considered to have a fast nucleotide dissociation rate, yet the GTP dissociation rate for this mutant is significantly elevated (Table 1). Thus, given the restricted GTP hydrolysis rates for this mutant, it is possible that RAS^{G12D} mutants will cycle between the GTP-bound form and nucleotide free form, thereby limiting the occupancy of this mutant in the GTP-bound state.

Several additional limitations must be considered when interpreting the biochemical data. First, due to the challenges of generating full-length prenylated protein, most recent RAS biochemical studies utilize a truncated construct containing the G-domain, residues 1–166 in HRAS, 1–169 in KRAS and 1–172 in NRAS. Direct comparison between full length and truncated RAS has been performed for HRAS^{WT} and HRAS^{G12V} (John et al., 1989), in which the truncated forms had variable effects on exchange and hydrolysis. Second, the seminal studies on RAS kinetics utilized radioactive nucleotides to measure nucleotide dissociation and GTP hydrolysis. More recent approaches have utilized fluorescent nucleotides and nucleotide analogs to measure dissociation and hydrolysis rates. While the fluorescent-GDP/GTP analogs have similar kinetics (Eccleston et al., 1993), GTP analogs such as GmPPNP and GmPPCP, which are commonly used with the fluorescent approaches, have faster exchange kinetics than

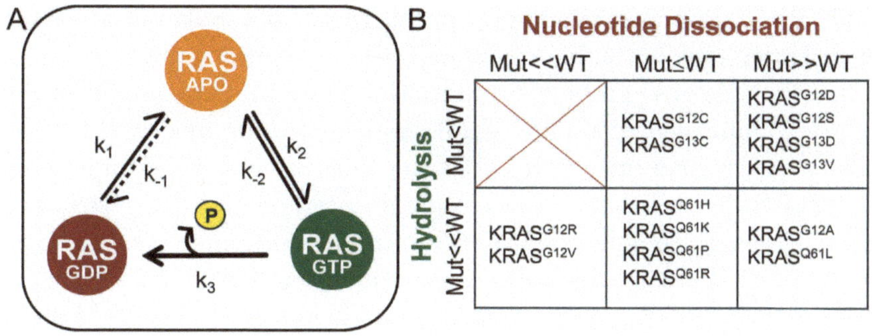

Fig. 4 RAS mutant proteins can be grouped based on intrinsic rates of dissociation and hydrolysis. (A) Simplified diagram of the RAS GTPase cycle. RAS cycles between the GTP, GDP and nucleotide-free states. Although assisted by GAPs and GEFs, mutant RAS proteins are also regulated by the intrinsic nucleotide dissociation and hydrolysis rates. The GDP dissociation rate is denoted by k_1, GTP dissociation by k_{-2} and GTP hydrolysis by k_3. The rates k_{-1} and k_2 denote nucleotide association. k_{-1} is not predicted to occur in cells due to the GTP:GDP ratio in cells. (B) RAS mutants are categorized into groups by their intrinsic dissociation and hydrolysis rates found in Table 1. For hydrolysis, Mut<<WT have hydrolysis rates under 3×10^{-3} min^{-1}. For dissociation, Mut<<WT, $k_1 < 4 \times 10^{-3}$ min^{-1} or $k_{-2} < 2.5 \times 10^{-3}$ min^{-1}; Mut~WT, $k_1 < 20 \times 10^{-3}$ min^{-1} or $k_{-2} < 15 \times 10^{-3}$ min^{-1}; Mut>>WT, $k_1 > 20 \times 10^{-3}$ min^{-1} or $k_{-2} > 15 \times 10^{-3}$ min^{-1}. G13S and Q61E were excluded.

GTP (John et al., 1989). Further, the nitrocellulose binding assay, which uses radiolabeled nucleotides, cannot be performed on the truncated RAS isoforms (Gideon et al., 1992; John et al., 1989), making direct comparisons between the radiolabeled data and the more recent fluorescent data difficult. Finally, only a few mutants have been fully characterized in multiple RAS isoforms. However, the wild-type isoforms have similar intrinsic nucleotide exchange and hydrolysis kinetics but varying sensitivity to GEFs and GAPs (Ahmadian, Hoffmann, Goody, & Wittinghofer, 1997; Lenzen et al., 1998). For clarification, nucleotide exchange represents the combined rate of dissociation and association of nucleotide (Fig. 4A). Due to the picomolar affinity of GDP/GTP to RAS, nucleotide association is rapid (John et al., 1990). Most biochemical studies measure nucleotide dissociation due to the technical limitations of performing association assays, as it is a second order kinetic rate that is dependent on the concentrations of substrates (John et al., 1990). Combined, these technical limitations should be considered when comparing the biochemical properties of each RAS mutant protein. Table 1 aggregates the biochemical data and displays the KRAS protein kinetics when available. Further, the

GTP dissociation rate, in addition to the more commonly measured GDP dissociation rate, must be considered when discussing the biochemical properties (Fig. 4A) as the GTP dissociation rate provides insight into the relative abundance of GDP-, GTP- and nucleotide-free levels. Finally, despite careful consideration of all the kinetic rates, predicting relative GTP/GDP levels based on nucleotide dissociation and hydrolysis rates is often markedly higher compared to in cell measurements (Donovan et al., 2002). These observations suggest numerous possibilities, specifically that GAPs, known or unknown, may facilitate mutant RAS GTP hydrolysis (Moghadamchargari et al., 2019; Zhao et al., 2020) or that RAS effector binding may facilitate GTP hydrolysis in some instances (Johnson et al., 2017, 2019; Zhao et al., 2020). Alternatively, the inability to accurately predict GTP/GDP levels may be due to the potential limitations of model cell systems to compared to *bona fide* cancer cells.

Comparing the hydrolysis and dissociation rates leads to several possible outcomes: (1) If GTP hydrolysis is the fastest step, then RAS will remain mostly inactive. (2) Conversely, if GDP dissociation is the fastest step, then RAS will populate the GTP-bound state yet continue to hydrolyze GTP to GDP at a mutation-dependent rate constant. (3) Finally, if both GDP and GTP nucleotide dissociation rates are fast and hydrolysis is the slowest step, RAS will likely rarely visit the GDP-bound state but cycle between a GTP-bound and nucleotide free state (Fig. 4A). Importantly, the nucleotide free state has been proposed to promote RAS interaction with PIK3C2B (Wong et al., 2012). In some mutants, such as RAS^{G12A}, GDP dissociation is accelerated (Xu et al., 2017). However, in other mutants, such as RAS^{G12D} and RAS^{G13D}, GTP dissociation is accelerated relative to WT (Table 1) (Smith et al., 2013; Wey et al., 2013). For $\text{HRAS}^{\text{G12V}}$ and $\text{HRAS}^{\text{G12R}}$, GTP hydrolysis and GTP dissociation are dramatically slower relative to other Gly^{12} mutants (John et al., 1988). For $\text{HRAS}^{\text{G12R}}$, the fastest rate is GDP dissociation followed by GTP hydrolysis (Krengel et al., 1990), suggesting that $\text{HRAS}^{\text{G12R}}$ can slowly cycle between the GDP and GTP forms while $\text{HRAS}^{\text{G12V}}$ is more likely to remain GTP loaded. Consequently, measurement of the GTP bound state of $\text{NRAS}^{\text{G12V}}$ in cells shows near 100% GTP loading in cells (Trahey et al., 1987), which is significantly higher than the 35% GTP loaded levels of $\text{KRAS}^{\text{G12D}}$ (Bollag et al., 1996). While these data imply that GTP hydrolysis rates may dictate total GTP levels in cells, this was not the case for Gln^{61} mutants (Donovan et al., 2002). Thus, intracellular GTP levels are likely dictated by the aggregate effect of the biochemical rates, making their prediction difficult.

Accordingly, we grouped the RAS mutant proteins according to their kinetics of nucleotide dissociation and hydrolysis (Fig. 4B). By considering both the GTP and GDP dissociation rates, most mutations display elevated rates of nucleotide dissociation of either GDP, GTP or both. However, nearly all mutations show decreased hydrolysis rates. Importantly, the biochemical features do not group the mutations according to overall frequency, implying that the biochemical kinetics are not a singular factor in the overall mutational frequencies observed in human cancers.

While many mutant biochemical rates were similar between RAS isoforms, (N/K)RAS^{G12D} has been reported to have isoform-specific differences in the intrinsic kinetics that may translate to tumorigenic potential. The reported intrinsic rate of NRAS^{G12D} GTP hydrolysis is faster than intrinsic GDP or GTP dissociation rates (Burd et al., 2014). Based on these data, NRAS^{G12D} kinetics would favor an increased population of the inactive GDP-bound state. KRAS^{G12D}, surprisingly, has the fastest GTP exchange rate of all Gly^{12} and Gly^{13} mutants (Bollag & McCormick, 1991) (Table 1), which could limit the downstream signaling efficiency of this mutant. HRAS^{G12D} has similar exchange kinetics (Wey et al., 2013). Functionally, NRAS^{G12D} could not promote tumorigenesis in a mouse melanoma model while KRAS^{G12D} promoted tumorigenesis in the same melanoma model (Burd et al., 2014). However, NRAS^{Q61R} was able to promote tumorigenesis, indicating that the NRAS isoform was capable of promoting tumor growth in this system. Perhaps these subtle differences between RAS isoforms dictate isoform-specific tumorigenic potential. However, NRAS^{G12D} is still detected in hematopoietic cancers, implying that it may be tumorigenic in some contexts. These subtle shifts in relative intrinsic rates provide a potential rational for why NRAS^{G12D} cannot promote tumorigenesis in melanoma models. However, other factors such as the overexpression of GEFs or mutations in GAPs, can also alter the tumorigenic potential of RAS mutants (Vigil et al., 2010). Overall, the lack of complete kinetic information from multiple isoforms limits the depth of this type of potential analysis.

The most comprehensive assessment of the intrinsic biochemical properties of KRAS mutants measured intrinsic GTP hydrolysis, GAP-mediated hydrolysis and nucleotide association kinetics (Hunter et al., 2015), which are not directly comparable to dissociation assays. Here, Hunter et al. reported an accelerated intrinsic hydrolysis for KRAS^{G12C} relative to other Gly^{12} mutants, and these results served as the rationale for the success of the GDP-specific KRAS^{G12C} inhibitors (Lito, Solomon, Li,

Hansen, & Rosen, 2016; Ostrem, Peters, Sos, Wells, & Shokat, 2013; Patricelli et al., 2016). In agreement with these results, HRAS^{G12C} was found to have a similar elevated rate of intrinsic GTP hydrolysis (Wey et al., 2013). In contrast, a recent study comparing KRAS hydrolysis rates using a mass-spectrometry-based technique found the intrinsic rate of KRAS^{G12C} GTP hydrolysis to be considerably slower relative to other common mutant-proteins in these studies (Moghadamchargari et al., 2019). However, the RAS^{G12C} GDP dissociation rate is faster than the rate of GTP hydrolysis, thus this mutant should largely populate the GTP-bound state (Table 1), which has led to the suggestion that other cell intrinsic factors or unknown GAPs could regulate RAS mutant GTP levels in cells (Zhao et al., 2020). While the KRAS^{G12C} intrinsic hydrolysis rates have been reported to be somewhat variable, they are not dramatically different from RAS^{G12D} (Table 1), a mutant that is only 35% GTP-bound when over-expressed in COS cells (Bollag et al., 1996). These results suggest the possibility that KRAS^{G12D} mutants may be susceptible to the same GDP-trapping mechanism originally presented by Kevan Shokat's group (Ostrem et al., 2013) if a molecule can be targeted to the carboxylic acid of RAS^{G12D}.

The biochemical properties may also explain the low frequency of some single-base mutants in human cancers. HRAS^{G13S} has an elevated GDP dissociation rate but the intrinsic hydrolysis rate is similar to HRAS^{WT} (Gideon et al., 1992; Wey et al., 2013). HRAS^{G13S} has also been reported to be partially sensitive to RASA1 (p120RASGAP)-catalyzed GTP hydrolysis (Gideon et al., 1992) and consequently, RAS^{G13S} is not a common oncogenic mutation. In contrast, KRAS^{G13D} is responsive to the RASGAP NF1-catalyzed GTP hydrolysis yet can still promote tumorigenesis, most frequently in CRC (Sherekar et al., 2020). As the intrinsic rate of GTP hydrolysis for KRAS^{G13D} has been reported to be considerably reduced (Hunter et al., 2015; Moghadamchargari et al., 2019), these data suggest that the intrinsic biochemical rates may play a determining role in the tumorigenic fitness of the RAS mutants. As such, HRAS^{G13V} has dramatically accelerated GDP and GTP dissociation rates (Table 1), which likely limit its ability to maintain nucleotide binding and could affect protein stability. Consequently, the RAS^{G13V} mutation is uncommon in cancer overall (Gideon et al., 1992). In yet another example, HRAS^{Q61E} has accelerated GTP hydrolysis coupled with rapid nucleotide dissociation and is considered thermodynamically unstable (Frech et al., 1994). The fast intrinsic hydrolysis and instability likely contribute to the infrequent detection of this mutation in human tumors. Consequently, RAS^{Q61E} was unable to induce transformation in the pioneering transformation assays characterizing Gln^{61} mutants (Der et al., 1986). However, $HRAS^{Q61H}$ was shown to be marginally sensitive to NF1-mediated GTP hydrolysis (Gideon et al., 1992; Mistou et al., 1992) yet $KRAS^{Q61H}$ is fully GAP insensitive (Hunter et al., 2015). Further, $KRAS^{Q61H}$ is a common mutation while $HRAS^{Q61H}$ is rarely observed, suggesting that this isoform-specific GAP sensitivity may dictate the oncogenic potential of this mutant. Thus, the mutants with biochemical rates that represent outliers are generally not detected with high frequency in human cancers. Therefore, it remains a possibility that some RAS mutant proteins may not serve the same oncogenic driver function that is commonly associated with mutant RAS cancers.

Many of the mutants discussed in this section represent the outliers, or the most extreme examples of intrinsic nucleotide dissociation and hydrolysis rates. While it is tempting to assign these features as the singular rationale behind their mutational frequency, such assumptions may be overstated. Overall, the intrinsic biochemical data between the various RAS mutants point to several conclusions. First, the intrinsic rates are generally similar between isoforms, although a few examples of isoform-specific differences exist, most notably $KRAS^{G12D}$ and $NRAS^{G12D}$. Second, the biochemical data on Gly^{12} and Gln^{61} mutants match the transforming abilities of over-expressed RAS in early studies, namely that RAS^{G12P} and RAS^{Q61E} should be rare. However, it should be pointed out that RAS^{Q61P} is rare in cancer overall yet there are no clear distinctions in its biochemical data (Donovan et al., 2002), suggesting other factors are at play with this mutant. Finally, these data provide hints that regulator interactions also play more significant role in defining the tumorigenic fitness and the observed mutation frequencies of cancer, specifically the isoform-specific GAP sensitivity of $HRAS^{Q61H}$. In this next section, we look at regulator interactions and their role in promoting mutant- RAS activity.

7. GEFs and GAPs

While nearly all RAS hotspot mutations decrease GTP hydrolysis, it is not universal. $HRAS^{G13S}$ and $HRAS^{Q61E}$ are two examples where the intrinsic hydrolysis rate is similar or greater than RAS^{WT} rates. Additionally, $HRAS^{G13S}$, but not $HRAS^{G13V}$, is sensitive to RASA1-mediated hydrolysis (Gideon et al., 1992). Recently, $KRAS^{G13D}$ and $KRAS^{G13C}$ were shown to be sensitive to NF1-mediated hydrolysis despite being resistant to

RASA1 (Rabara et al., 2019). Despite the similar NF1-mediated hydrolysis rates, the KRAS^{G13C} mutant is considerably less common than KRAS^{G13D} (Fig. 1).

Another defining feature of Gln^{61} mutants is increased binding affinity to GAPs (Gideon et al., 1992; Krengel et al., 1990). Surprisingly, most RAS Gln^{61} mutants have a tighter binding affinity to NF1 than CRAF (Fig. 3; Table 1). It is possible that Gln^{61} mutants are less common because they nonproductively bind to GAPs, reducing the levels of free RAS to interact with effectors. Further, HRAS^{Q61H} has a limited but measurable response to NF1, although the biological significance of the response is unclear. While there is some variability in the hydrolysis rates at the Gly^{13} and Gln^{61} positions, all RAS Gly^{12} mutants are GAP resistant, potentially contributing to the prevalence of Gly^{12} mutations in all RAS-mutant cancers.

While GTP hydrolysis receives more attention in RAS mutant proteins, GEF activity also plays a role. The first study to probe the role of GEF function utilized the yeast RAS-related GEF SDC25 on a panel of biochemically purified HRAS mutants (Mistou et al., 1992). In this study, HRAS^{G12D} and HRAS^{G12V} had impaired intrinsic and GEF-mediated exchange rates, an observation that has been corroborated with KRAS mutants in the RAS GEF Son of Sevenless (SOS) (Hobbs et al., 2020). HRAS^{G13V} and HRAS^{G13S} were both characterized as having fast intrinsic and GEF-mediated dissociation, similar to observations with HRAS^{G13D} (Smith et al., 2013). Finally, HRAS^{Q61H} and HRAS^{Q61L} were both characterized with an elevated intrinsic exchange, yet HRAS^{Q61H} was resistant to SDC25-mediated exchange and HRAS^{Q61L} was sensitive to SOS-mediated exchange (Smith et al., 2013). Interestingly, NRAS^{Q61R} is resistant to SOS-mediated exchange (Burd et al., 2014). While these data provide potential insight into mutation-specific GEF function, isoform-specific effects on binding cannot be ruled out. Further complicating the analysis is that GEF-mediated exchange on KRAS mutant proteins has been rather limited, which limits the ability to draw any large conclusions.

While KRAS^{G12R} is not sensitive to SOS stimulation due to the inability to bind to the allosteric site, KRAS^{G12R} is responsive to GRP1-mediated exchange (Hobbs et al., 2020). Despite the extensive characterization of other RAS GEFs, such as GRP1 and GRF1 (Iwig et al., 2013), the selectivity of RAS hot-spot mutants to the various RAS GEFs has not been studied in detail. Importantly, other RAS GEFs have tissue specific expression and localization (Clyde-Smith et al., 2000; Jaumot, Yan, Clyde-Smith, Sluimer, & Hancock, 2002; Matallanas et al., 2003). Perhaps there will be a correlation between the RAS cancer-associated mutational frequencies and GEF sensitivity when accounting for the tissue-specific expression of each GEF protein and the RAS-mutant sensitivities toward each GEF protein.

8. Conclusions

Despite the extensive study of RAS mutant proteins, the connection between the overall mutational frequencies in cancer and biochemical properties remain unclear. This limitation is due to the challenges in studying the various mutants and isoforms as well as the limitations from using minimal interacting domains to measure protein binding constants. New studies that are utilizing cell-based probes to determine the relative interactions between mutants are providing novel details that will increase our understanding of RAS signaling and function.

The aggregation and analysis of the biochemical properties demonstrates two main features of RAS proteins. First, the intrinsic biochemical rates of RAS isoforms can be used to bin the RAS mutants (Fig. 4B). Second, the biochemical kinetics provide a strong rationale for the limited detection of some mutations, although the current biochemical data are not fully predictive. Thus, not all RAS mutations are created equal. While these analyses cannot wholly explain the observed mutational frequencies in human cancers, they do provide increased clarity. The biochemical data provide a potential rationale for why some mutants are present at low frequencies, in agreement with the observed focus formation assays that originally characterized these mutant proteins. However, the rationale for the tissue-specific prevalence of some mutations remains unknown. The biochemical data suggest that RAS mutations and their frequencies in human cancers can be linked to their biochemical properties, although one cannot rule out other context-specific factors, such as DNA mutation rates, GEF expression and other co-mutations. Therefore, careful examination of the biochemistry of RAS mutants should continue to provide additional insight into the biology of RAS-mutant cancers. Studies that have examined atypical mutations at greater detail have highlighted unexpected differences in RAS mutant proteins and have proven instrumental in developing our understanding of the unique biochemical and cellular features of RAS cancers (Hobbs et al., 2020; Poulin et al., 2019). A deeper understanding of these differences, through continued investigation of tissue-specific mutants, will uncover distinct therapeutic vulnerabilities, leading to more effective RAS anti-cancer therapies in the future.

Acknowledgments

We would like to thank Dr. Antje Schaefer for editing and critical evaluation of the manuscript. GAH acknowledges funding from American Cancer Society IRG-19-137-20. The authors report no conflicts of interest.

References

Abankwa, D., Gorfe, A. A., Inder, K., & Hancock, J. F. (2010). Ras membrane orientation and nanodomain localization generate isoform diversity. *Proceedings of the National Academy of Sciences of the United States of America, 107*(3), 1130–1135. https://doi.org/10.1073/pnas.0903907107.

Adhikari, H., & Counter, C. M. (2018). Interrogating the protein interactomes of RAS isoforms identifies PIP5K1A as a KRAS-specific vulnerability. *Nature Communications, 9*(1), 3646. https://doi.org/10.1038/s41467-018-05692-6.

Ahearn, I., Zhou, M., & Philips, M. R. (2018). Posttranslational modifications of RAS proteins. *Cold Spring Harbor Perspectives in Medicine, 8*(11), a031484. https://doi.org/10.1101/cshperspect.a031484.

Ahmadian, M. R., Hoffmann, U., Goody, R. S., & Wittinghofer, A. (1997). Individual rate constants for the interaction of Ras proteins with GTPase-activating proteins determined by fluorescence spectroscopy. *Biochemistry, 36*(15), 4535–4541. https://doi.org/10.1021/bi962556y.

Alexandrov, L. B., Nik-Zainal, S., Wedge, D. C., Aparicio, S. A., Behjati, S., Biankin, A. V., et al. (2013). Signatures of mutational processes in human cancer. *Nature, 500*(7463), 415–421. https://doi.org/10.1038/nature12477.

Ambrogio, C., Kohler, J., Zhou, Z. W., Wang, H., Paranal, R., Li, J., et al. (2018). KRAS dimerization impacts MEK inhibitor sensitivity and oncogenic activity of mutant KRAS. *Cell, 172*(4), 857–868 e815. https://doi.org/10.1016/j.cell.2017.12.020.

Amendola, C. R., Mahaffey, J. P., Parker, S. J., Ahearn, I. M., Chen, W. C., Zhou, M., et al. (2019). KRAS4A directly regulates hexokinase 1. *Nature, 576*(7787), 482–486. https://doi.org/10.1038/s41586-019-1832-9.

Araujo, L. H., Souza, B. M., Leite, L. R., Parma, S. A. F., Lopes, N. P., Malta, F. S. V., et al. (2021). Molecular profile of KRAS G12C-mutant colorectal and non-small-cell lung cancer. *BMC Cancer, 21*(1), 193. https://doi.org/10.1186/s12885-021-07884-8.

Baer, R., Cintas, C., Dufresne, M., Cassant-Sourdy, S., Schonhuber, N., Planque, L., et al. (2014). Pancreatic cell plasticity and cancer initiation induced by oncogenic Kras is completely dependent on wild-type PI 3-kinase p110alpha. *Genes & Development, 28*(23), 2621–2635. https://doi.org/10.1101/gad.249409.114.

Bivona, T. G., Quatela, S. E., Bodemann, B. O., Ahearn, I. M., Soskis, M. J., Mor, A., et al. (2006). PKC regulates a farnesyl-electrostatic switch on K-Ras that promotes its association with Bcl-XL on mitochondria and induces apoptosis. *Molecular Cell, 21*(4), 481–493. https://doi.org/10.1016/j.molcel.2006.01.012.

Bollag, G., Adler, F., elMasry, N., McCabe, P. C., Conner, E., Jr., Thompson, P., et al. (1996). Biochemical characterization of a novel KRAS insertion mutation from a human leukemia. *The Journal of Biological Chemistry, 271*(51), 32491–32494. https://doi.org/10.1074/jbc.271.51.32491.

Bollag, G., & McCormick, F. (1991). Differential regulation of rasGAP and neurofibromatosis gene product activities. *Nature, 351*(6327), 576–579. https://doi.org/10.1038/351576a0.

Boriack-Sjodin, P. A., Margarit, S. M., Bar-Sagi, D., & Kuriyan, J. (1998). The structural basis of the activation of Ras by Sos. *Nature, 394*(6691), 337–343. https://doi.org/10.1038/28548.

Buhrman, G., Kumar, V. S., Cirit, M., Haugh, J. M., & Mattos, C. (2011). Allosteric modulation of Ras-GTP is linked to signal transduction through RAF kinase. *The Journal of Biological Chemistry*, *286*(5), 3323–3331. https://doi.org/10.1074/jbc.M110.193854.

Burd, C. E., Liu, W., Huynh, M. V., Waqas, M. A., Gillahan, J. E., Clark, K. S., et al. (2014). Mutation-specific RAS oncogenicity explains NRAS codon 61 selection in melanoma. *Cancer Discovery*, *4*(12), 1418–1429. https://doi.org/10.1158/2159-8290.CD-14-0729.

Castellano, E., Sheridan, C., Thin, M. Z., Nye, E., Spencer-Dene, B., Diefenbacher, M. E., et al. (2013). Requirement for interaction of PI3-kinase p110alpha with RAS in lung tumor maintenance. *Cancer Cell*, *24*(5), 617–630. https://doi.org/10.1016/j.ccr.2013.09.012.

Catozzi, S., Halasz, M., & Kiel, C. (2021). Predicted 'wiring landscape' of Ras-effector interactions in 29 human tissues. *NPJ Systems Biology and Applications*, 7(1), 10. https://doi.org/10.1038/s41540-021-00170-0.

Choy, E., Chiu, V. K., Silletti, J., Feoktistov, M., Morimoto, T., Michaelson, D., et al. (1999). Endomembrane trafficking of ras: The CAAX motif targets proteins to the ER and Golgi. *Cell*, *98*(1), 69–80. https://doi.org/10.1016/S0092-8674(00)80607-8.

Chung, G. T., Sundaresan, V., Hasleton, P., Rudd, R., Taylor, R., & Rabbitts, P. H. (1995). Sequential molecular genetic changes in lung cancer development. *Oncogene*, *11*(12), 2591–2598. Retrieved from https://www.ncbi.nlm.nih.gov/pubmed/8545116.

Clyde-Smith, J., Silins, G., Gartside, M., Grimmond, S., Etheridge, M., Apolloni, A., et al. (2000). Characterization of RasGRP2, a plasma membrane-targeted, dual specificity Ras/Rap exchange factor. *The Journal of Biological Chemistry*, *275*(41), 32260–32267. https://doi.org/10.1074/jbc.M006087200.

Cook, J. H., Melloni, G. E. M., Gulhan, D. C., Park, P. J., & Haigis, K. M. (2021). The origins and genetic interactions of KRAS mutations are allele- and tissue-specific. *Nature Communications*, *12*(1), 1808. https://doi.org/10.1038/s41467-021-22125-z.

Cox, A. D., Der, C. J., & Philips, M. R. (2015). Targeting RAS membrane association: Back to the future for anti-RAS drug discovery? *Clinical Cancer Research*, *21*(8), 1819–1827. https://doi.org/10.1158/1078-0432.CCR-14-3214.

Cox, A. D., Fesik, S. W., Kimmelman, A. C., Luo, J., & Der, C. J. (2014). Drugging the undruggable RAS: Mission possible? *Nature Reviews. Drug Discovery*, *13*(11), 828–851. https://doi.org/10.1038/nrd4389.

Der, C. J., Finkel, T., & Cooper, G. M. (1986). Biological and biochemical properties of human rasH genes mutated at codon 61. *Cell*, *44*(1), 167–176. https://doi.org/10.1016/0092-8674(86)90495-2.

Diehl, A., Hannan, L. M., & Chiorean, E. G. (2021). Prognostic value of KRAS and PI3K pathway mutations for advanced pancreatic ductal adenocarcinoma (PDAC) patients (pts). *Journal of Clinical Oncology*, *39*(3_suppl), 424. https://doi.org/10.1200/JCO.2021.39.3_suppl.424.

Donovan, S., Shannon, K. M., & Bollag, G. (2002). GTPase activating proteins: Critical regulators of intracellular signaling. *Biochimica et Biophysica Acta*, *1602*(1), 23–45. https://doi.org/10.1016/s0304-419x(01)00041-5.

Eccleston, J. F., Moore, K. J., Morgan, L., Skinner, R. H., & Lowe, P. N. (1993). Kinetics of interaction between normal and proline 12 Ras and the GTPase-activating proteins, p120-GAP and neurofibromin. The significance of the intrinsic GTPase rate in determining the transforming ability of ras. *The Journal of Biological Chemistry*, *268*(36), 27012–27019. Retrieved from https://www.ncbi.nlm.nih.gov/pubmed/8262937.

Escobar-Hoyos, L. F., Penson, A., Kannan, R., Cho, H., Pan, C. H., Singh, R. K., et al. (2020). Altered RNA splicing by mutant p53 activates oncogenic RAS signaling in pancreatic cancer. *Cancer Cell*, *38*(2), 198–211 e198. https://doi.org/10.1016/j.ccell.2020.05.010.

Fearon, E. R., & Vogelstein, B. (1990). A genetic model for colorectal tumorigenesis. *Cell, 61*(5), 759–767. https://doi.org/10.1016/0092-8674(90)90186-i.

Frech, M., Darden, T. A., Pedersen, L. G., Foley, C. K., Charifson, P. S., Anderson, M. W., et al. (1994). Role of glutamine-61 in the hydrolysis of GTP by p21H-ras: An experimental and theoretical study. *Biochemistry, 33*(11), 3237–3244. https://doi.org/10.1021/bi00177a014.

Fruman, D. A., Chiu, H., Hopkins, B. D., Bagrodia, S., Cantley, L. C., & Abraham, R. T. (2017). The PI3K pathway in human disease. *Cell, 170*(4), 605–635. https://doi.org/10.1016/j.cell.2017.07.029.

Gibbs, J. B., Schaber, M. D., Allard, W. J., Sigal, I. S., & Scolnick, E. M. (1988). Purification of ras GTPase activating protein from bovine brain. *Proceedings of the National Academy of Sciences of the United States of America, 85*(14), 5026–5030. https://doi.org/10.1073/pnas.85.14.5026.

Gideon, P., John, J., Frech, M., Lautwein, A., Clark, R., Scheffler, J. E., et al. (1992). Mutational and kinetic analyses of the GTPase-activating protein (GAP)-p21 interaction: The C-terminal domain of GAP is not sufficient for full activity. *Molecular and Cellular Biology, 12*(5), 2050–2056. https://doi.org/10.1128/mcb.12.5.2050-2056.1992.

Gillies, T. E., Pargett, M., Silva, J. M., Teragawa, C. K., McCormick, F., & Albeck, J. G. (2020). Oncogenic mutant RAS signaling activity is rescaled by the ERK/MAPK pathway. *Molecular Systems Biology, 16*(10), e9518. https://doi.org/10.15252/msb.20209518.

Guldenhaupt, J., Rudack, T., Bachler, P., Mann, D., Triola, G., Waldmann, H., et al. (2012). N-Ras forms dimers at POPC membranes. *Biophysical Journal, 103*(7), 1585–1593. https://doi.org/10.1016/j.bpj.2012.08.043.

Gupta, S., Ramjaun, A. R., Haiko, P., Wang, Y., Warne, P. H., Nicke, B., et al. (2007). Binding of ras to phosphoinositide 3-kinase p110alpha is required for ras-driven tumorigenesis in mice. *Cell, 129*(5), 957–968. https://doi.org/10.1016/j.cell.2007.03.051.

Haigis, K. M. (2017). KRAS alleles: The devil is in the detail. *Trends Cancer, 3*(10), 686–697. https://doi.org/10.1016/j.trecan.2017.08.006.

Han, L., & Colicelli, J. (1995). A human protein selected for interference with Ras function interacts directly with Ras and competes with Raf1. *Molecular and Cellular Biology, 15*(3), 1318–1323. https://doi.org/10.1128/MCB.15.3.1318.

Herrmann, C., Martin, G. A., & Wittinghofer, A. (1995). Quantitative analysis of the complex between p21ras and the Ras-binding domain of the human Raf-1 protein kinase. *The Journal of Biological Chemistry, 270*(7), 2901–2905. https://doi.org/10.1074/jbc.270.7.2901.

Hobbs, G. A., Baker, N. M., Miermont, A. M., Thurman, R. D., Pierobon, M., Tran, T. H., et al. (2020). Atypical KRAS(G12R) mutant is impaired in PI3K signaling and macropinocytosis in pancreatic cancer. *Cancer Discovery, 10*(1), 104–123. https://doi.org/10.1158/2159-8290.CD-19-1006.

Hobbs, G. A., & Der, C. J. (2019). RAS mutations are not created equal. *Cancer Discovery, 9*(6), 696–698. https://doi.org/10.1158/2159-8290.CD-19-0406.

Hunter, J. C., Manandhar, A., Carrasco, M. A., Gurbani, D., Gondi, S., & Westover, K. D. (2015). Biochemical and structural analysis of common cancer-associated KRAS mutations. *Molecular Cancer Research, 13*(9), 1325–1335. https://doi.org/10.1158/1541-7786.MCR-15-0203.

Iacobuzio-Donahue, C. A., Velculescu, V. E., Wolfgang, C. L., & Hruban, R. H. (2012). Genetic basis of pancreas cancer development and progression: Insights from whole-exome and whole-genome sequencing. *Clinical Cancer Research, 18*(16), 4257–4265. https://doi.org/10.1158/1078-0432.CCR-12-0315.

Iwig, J. S., Vercoulen, Y., Das, R., Barros, T., Limnander, A., Che, Y., et al. (2013). Structural analysis of autoinhibition in the Ras-specific exchange factor RasGRP1. *eLife, 2*, e00813. https://doi.org/10.7554/eLife.00813.

Jang, H., Abraham, S. J., Chavan, T. S., Hitchinson, B., Khavrutskii, L., Tarasova, N. I., et al. (2015). Mechanisms of membrane binding of small GTPase K-Ras4B farnesylated hypervariable region. *The Journal of Biological Chemistry*, *290*(15), 9465–9477. https://doi.org/10.1074/jbc.M114.620724.

Jaumot, M., Yan, J., Clyde-Smith, J., Sluimer, J., & Hancock, J. F. (2002). The linker domain of the Ha-Ras hypervariable region regulates interactions with exchange factors, Raf-1 and phosphoinositide 3-kinase. *The Journal of Biological Chemistry*, *277*(1), 272–278. https://doi.org/10.1074/jbc.M108423200.

John, J., Frech, M., & Wittinghofer, A. (1988). Biochemical properties of Ha-ras encoded p21 mutants and mechanism of the autophosphorylation reaction. *The Journal of Biological Chemistry*, *263*(24), 11792–11799. Retrieved from https://www.ncbi.nlm.nih.gov/pubmed/3042780.

John, J., Schlichting, I., Schiltz, E., Rosch, P., & Wittinghofer, A. (1989). C-terminal truncation of p21H preserves crucial kinetic and structural properties. *The Journal of Biological Chemistry*, *264*(22), 13086–13092. Retrieved from https://www.ncbi.nlm.nih.gov/pubmed/2502546.

John, J., Sohmen, R., Feuerstein, J., Linke, R., Wittinghofer, A., & Goody, R. S. (1990). Kinetics of interaction of nucleotides with nucleotide-free H-ras p21. *Biochemistry*, *29*(25), 6058–6065. https://doi.org/10.1021/bi00477a025.

Johnson, C. W., Lin, Y. J., Reid, D., Parker, J., Pavlopoulos, S., Dischinger, P., et al. (2019). Isoform-specific destabilization of the active site reveals a molecular mechanism of intrinsic activation of KRas G13D. *Cell Reports*, *28*(6), 1538–1550 e1537. https://doi.org/10.1016/j.celrep.2019.07.026.

Johnson, C. W., Reid, D., Parker, J. A., Salter, S., Knihtila, R., Kuzmic, P., et al. (2017). The small GTPases K-Ras, N-Ras, and H-Ras have distinct biochemical properties determined by allosteric effects. *The Journal of Biological Chemistry*, *292*(31), 12981–12993. https://doi.org/10.1074/jbc.M117.778886.

Kelly, M. R., Kostyrko, K., Han, K., Mooney, N. A., Jeng, E. E., Spees, K., et al. (2020). Combined proteomic and genetic interaction mapping reveals new RAS effector pathways and susceptibilities. *Cancer Discovery*, *10*(12), 1950–1967. https://doi.org/10.1158/2159-8290.CD-19-1274.

Kiel, C., Matallanas, D., & Kolch, W. (2021). The ins and outs of RAS effector complexes. *Biomolecules*, *11*(2), 236. https://doi.org/10.3390/biom11020236.

Kiel, C., Serrano, L., & Herrmann, C. (2004). A detailed thermodynamic analysis of ras/effector complex interfaces. *Journal of Molecular Biology*, *340*(5), 1039–1058. https://doi.org/10.1016/j.jmb.2004.05.050.

Kiel, C., Verschueren, E., Yang, J. S., & Serrano, L. (2013). Integration of protein abundance and structure data reveals competition in the ErbB signaling network. *Science Signaling*, *6*(306), ra109. https://doi.org/10.1126/scisignal.2004560.

Klomp, J. E., Klomp, J. A., & Der, C. J. (2021). The ERK mitogen-activated protein kinase signaling network: The final frontier in RAS signal transduction. *Biochemical Society Transactions*, *49*(1), 253–267. https://doi.org/10.1042/BST20200507.

Kovalski, J. R., Bhaduri, A., Zehnder, A. M., Neela, P. H., Che, Y., Wozniak, G. G., et al. (2019). The functional proximal proteome of oncogenic Ras includes mTORC2. *Molecular Cell*, *73*(4), 830–844 e812. https://doi.org/10.1016/j.molcel.2018.12.001.

Krengel, U., Schlichting, I., Scherer, A., Schumann, R., Frech, M., John, J., et al. (1990). Three-dimensional structures of H-ras p21 mutants: Molecular basis for their inability to function as signal switch molecules. *Cell*, *62*(3), 539–548. https://doi.org/10.1016/0092-8674(90)90018-a.

Le Calvez, F., Mukeria, A., Hunt, J. D., Kelm, O., Hung, R. J., Taniere, P., et al. (2005). TP53 and KRAS mutation load and types in lung cancers in relation to tobacco smoke: Distinct patterns in never, former, and current smokers. *Cancer Research*, *65*(12), 5076–5083. https://doi.org/10.1158/0008-5472.CAN-05-0551.

Lenzen, C., Cool, R. H., Prinz, H., Kuhlmann, J., & Wittinghofer, A. (1998). Kinetic analysis by fluorescence of the interaction between Ras and the catalytic domain of the guanine nucleotide exchange factor Cdc25Mm. *Biochemistry, 37*(20), 7420–7430. https://doi.org/10.1021/bi972621j.

Li, S., Balmain, A., & Counter, C. M. (2018). A model for RAS mutation patterns in cancers: Finding the sweet spot. *Nature Reviews. Cancer, 18*(12), 767–777. https://doi.org/10.1038/s41568-018-0076-6.

Li, S., MacAlpine, D. M., & Counter, C. M. (2020). Capturing the primordial Kras mutation initiating urethane carcinogenesis. *Nature Communications, 11*(1), 1800. https://doi.org/10.1038/s41467-020-15660-8.

Linnemann, T., Kiel, C., Herter, P., & Herrmann, C. (2002). The activation of RalGDS can be achieved independently of its Ras binding domain. Implications of an activation mechanism in Ras effector specificity and signal distribution. *The Journal of Biological Chemistry, 277*(10), 7831–7837. https://doi.org/10.1074/jbc.M110800200.

Lito, P., Solomon, M., Li, L. S., Hansen, R., & Rosen, N. (2016). Allele-specific inhibitors inactivate mutant KRAS G12C by a trapping mechanism. *Science, 351*(6273), 604–608. https://doi.org/10.1126/science.aad6204.

Lynch, S. J., Snitkin, H., Gumper, I., Philips, M. R., Sabatini, D., & Pellicer, A. (2015). The differential palmitoylation states of N-Ras and H-Ras determine their distinct Golgi subcompartment localizations. *Journal of Cellular Physiology, 230*(3), 610–619. https://doi.org/10.1002/jcp.24779.

Makova, K. D., & Hardison, R. C. (2015). The effects of chromatin organization on variation in mutation rates in the genome. *Nature Reviews. Genetics, 16*(4), 213–223. https://doi.org/10.1038/nrg3890.

Martinez, N. G., Thieker, D. F., Carey, L. M., Rasquinha, J. A., Kistler, S. K., Kuhlman, B. A., et al. (2021). Biophysical and structural characterization of novel RAS-binding domains (RBDs) of PI3Kalpha and PI3Kgamma. *Journal of Molecular Biology, 433*(8), 166838. https://doi.org/10.1016/j.jmb.2021.166838.

Matallanas, D., Arozarena, I., Berciano, M. T., Aaronson, D. S., Pellicer, A., Lafarga, M., et al. (2003). Differences on the inhibitory specificities of H-Ras, K-Ras, and N-Ras (N17) dominant negative mutants are related to their membrane microlocalization. *The Journal of Biological Chemistry, 278*(7), 4572–4581. https://doi.org/10.1074/jbc.M209807200.

Miller, M. S., & Miller, L. D. (2011). RAS mutations and oncogenesis: Not all RAS mutations are created equally. *Frontiers in Genetics, 2*, 100. https://doi.org/10.3389/fgene.2011.00100.

Mistou, M. Y., Jacquet, E., Poullet, P., Rensland, H., Gideon, P., Schlichting, I., et al. (1992). Mutations of ha-ras p21 that define important regions for the molecular mechanism of the SDC25 C-domain, a guanine nucleotide dissociation stimulator. *The EMBO Journal, 11*(7), 2391–2397. Retrieved from https://www.ncbi.nlm.nih.gov/pubmed/1628612.

Moghadamchargari, Z., Huddleston, J., Shirzadeh, M., Zheng, X., Clemmer, D. E., F, M. R., et al. (2019). Intrinsic GTPase activity of K-RAS monitored by native mass spectrometry. *Biochemistry, 58*(31), 3396–3405. https://doi.org/10.1021/acs.biochem.9b00532.

Moore, A. R., Rosenberg, S. C., McCormick, F., & Malek, S. (2020). RAS-targeted therapies: Is the undruggable drugged? *Nature Reviews. Drug Discovery, 19*(8), 533–552. https://doi.org/10.1038/s41573-020-0068-6.

Morrison, D. K., Kaplan, D. R., Rapp, U., & Roberts, T. M. (1988). Signal transduction from membrane to cytoplasm: Growth factors and membrane-bound oncogene products increase Raf-1 phosphorylation and associated protein kinase activity. *Proceedings of the National Academy of Sciences of the United States of America, 85*(23), 8855–8859. https://doi.org/10.1073/pnas.85.23.8855.

Mott, H. R., Carpenter, J. W., Zhong, S., Ghosh, S., Bell, R. M., & Campbell, S. L. (1996). The solution structure of the Raf-1 cysteine-rich domain: A novel ras and phospholipid binding site. *Proceedings of the National Academy of Sciences of the United States of America*, *93*(16), 8312–8317. https://doi.org/10.1073/pnas.93.16.8312.

Mullard, A. (2021). Amgen overcomes historically undruggable target, with FDA nod for first KRAS inhibitor. *Nature Reviews. Drug Discovery*, *20*(7), 496. https://doi.org/10.1038/d41573-021-00098-4.

Munoz-Maldonado, C., Zimmer, Y., & Medova, M. (2019). A comparative analysis of individual RAS mutations in cancer biology. *Frontiers in Oncology*, *9*, 1088. https://doi.org/10.3389/fonc.2019.01088.

Murillo, M. M., Zelenay, S., Nye, E., Castellano, E., Lassailly, F., Stamp, G., et al. (2014). RAS interaction with PI3K p110alpha is required for tumor-induced angiogenesis. *The Journal of Clinical Investigation*, *124*(8), 3601–3611. https://doi.org/10.1172/JCI74134.

Nakhaeizadeh, H., Amin, E., Nakhaei-Rad, S., Dvorsky, R., & Ahmadian, M. R. (2016). The RAS-effector interface: Isoform-specific differences in the effector binding regions. *PLoS One*, *11*(12). https://doi.org/10.1371/journal.pone.0167145, e0167145.

Novelli, E. T., First, J. T., & Webb, L. J. (2018). Quantitative measurement of intrinsic GTP hydrolysis for carcinogenic glutamine 61 mutants in H-Ras. *Biochemistry*, *57*(44), 6356–6366. https://doi.org/10.1021/acs.biochem.8b00878.

Ostrem, J. M., Peters, U., Sos, M. L., Wells, J. A., & Shokat, K. M. (2013). K-Ras(G12C) inhibitors allosterically control GTP affinity and effector interactions. *Nature*, *503*(7477), 548–551. https://doi.org/10.1038/nature12796.

Pacold, M. E., Suire, S., Perisic, O., Lara-Gonzalez, S., Davis, C. T., Walker, E. H., et al. (2000). Crystal structure and functional analysis of Ras binding to its effector phosphoinositide 3-kinase gamma. *Cell*, *103*(6), 931–943. https://doi.org/10.1016/s0092-8674(00)00196-3.

Patricelli, M. P., Janes, M. R., Li, L. S., Hansen, R., Peters, U., Kessler, L. V., et al. (2016). Selective inhibition of oncogenic KRAS output with small molecules targeting the inactive state. *Cancer Discovery*, *6*(3), 316–329. https://doi.org/10.1158/2159-8290.CD-15-1105.

Pershing, N. L., Lampson, B. L., Belsky, J. A., Kaltenbrun, E., MacAlpine, D. M., & Counter, C. M. (2015). Rare codons capacitate Kras-driven de novo tumorigenesis. *The Journal of Clinical Investigation*, *125*(1), 222–233. https://doi.org/10.1172/JCI77627.

Poulin, E. J., Bera, A. K., Lu, J., Lin, Y. J., Strasser, S. D., Paulo, J. A., et al. (2019). Tissue-specific oncogenic activity of KRAS(A146T). *Cancer Discovery*, *9*(6), 738–755. https://doi.org/10.1158/2159-8290.CD-18-1220.

Prior, I. A., & Hancock, J. F. (2012). Ras trafficking, localization and compartmentalized signalling. *Seminars in Cell & Developmental Biology*, *23*(2), 145–153. https://doi.org/10.1016/j.semcdb.2011.09.002.

Prior, I. A., Harding, A., Yan, J., Sluimer, J., Parton, R. G., & Hancock, J. F. (2001). GTP-dependent segregation of H-ras from lipid rafts is required for biological activity. *Nature Cell Biology*, *3*(4), 368–375. https://doi.org/10.1038/35070050.

Prior, I. A., Hood, F. E., & Hartley, J. L. (2020). The frequency of Ras mutations in cancer. *Cancer Research*, *80*(14), 2969–2974. https://doi.org/10.1158/0008-5472.CAN-19-3682.

Rabara, D., Tran, T. H., Dharmaiah, S., Stephens, R. M., McCormick, F., Simanshu, D. K., et al. (2019). KRAS G13D sensitivity to neurofibromin-mediated GTP hydrolysis. *Proceedings of the National Academy of Sciences of the United States of America*, *116*(44), 22122–22131. https://doi.org/10.1073/pnas.1908353116.

Rapp, U. R., Goldsborough, M. D., Mark, G. E., Bonner, T. I., Groffen, J., Reynolds, F. H., Jr., et al. (1983). Structure and biological activity of v-raf, a unique oncogene transduced by a retrovirus. *Proceedings of the National Academy of Sciences of the United States of America*, *80*(14), 4218–4222. https://doi.org/10.1073/pnas.80.14.4218.

Riely, G. J., Kris, M. G., Rosenbaum, D., Marks, J., Li, A., Chitale, D. A., et al. (2008). Frequency and distinctive spectrum of KRAS mutations in never smokers with lung adenocarcinoma. *Clinical Cancer Research*, *14*(18), 5731–5734. https://doi.org/10.1158/1078-0432.CCR-08-0646.

Rodriguez-Viciana, P., Warne, P. H., Vanhaesebroeck, B., Waterfield, M. D., & Downward, J. (1996). Activation of phosphoinositide 3-kinase by interaction with Ras and by point mutation. *The EMBO Journal*, *15*(10), 2442–2451. Retrieved from https://www.ncbi.nlm.nih.gov/pubmed/8665852.

Roy, S., Plowman, S., Rotblat, B., Prior, I. A., Muncke, C., Grainger, S., et al. (2005). Individual palmitoyl residues serve distinct roles in H-ras trafficking, microlocalization, and signaling. *Molecular and Cellular Biology*, *25*(15), 6722–6733. https://doi.org/10.1128/MCB.25.15.6722-6733.2005.

Rudack, T., Xia, F., Schlitter, J., Kotting, C., & Gerwert, K. (2012). Ras and GTPase-activating protein (GAP) drive GTP into a precatalytic state as revealed by combining FTIR and biomolecular simulations. *Proceedings of the National Academy of Sciences of the United States of America*, *109*(38), 15295–15300. https://doi.org/10.1073/pnas.1204333109.

Scheffzek, K., Ahmadian, M. R., Kabsch, W., Wiesmuller, L., Lautwein, A., Schmitz, F., et al. (1997). The Ras-RasGAP complex: Structural basis for GTPase activation and its loss in oncogenic Ras mutants. *Science*, *277*(5324), 333–338. https://doi.org/10.1126/science.277.5324.333.

Scheffzek, K., Ahmadian, M. R., Wiesmuller, L., Kabsch, W., Stege, P., Schmitz, F., et al. (1998). Structural analysis of the GAP-related domain from neurofibromin and its implications. *The EMBO Journal*, *17*(15), 4313–4327. https://doi.org/10.1093/emboj/17.15.4313.

Scheffzek, K., Lautwein, A., Kabsch, W., Ahmadian, M. R., & Wittinghofer, A. (1996). Crystal structure of the GTPase-activating domain of human p120GAP and implications for the interaction with Ras. *Nature*, *384*(6609), 591–596. https://doi.org/10.1038/384591a0.

Seeburg, P. H., Colby, W. W., Capon, D. J., Goeddel, D. V., & Levinson, A. D. (1984). Biological properties of human c-Ha-ras1 genes mutated at codon 12. *Nature*, *312*(5989), 71–75. https://doi.org/10.1038/312071a0.

Sherekar, M., Han, S. W., Ghirlando, R., Messing, S., Drew, M., Rabara, D., et al. (2020). Biochemical and structural analyses reveal that the tumor suppressor neurofibromin (NF1) forms a high-affinity dimer. *The Journal of Biological Chemistry*, *295*(4), 1105–1119. https://doi.org/10.1074/jbc.RA119.010934.

Simanshu, D. K., Nissley, D. V., & McCormick, F. (2017). RAS proteins and their regulators in human disease. *Cell*, *170*(1), 17–33. https://doi.org/10.1016/j.cell.2017.06.009.

Smith, M. J., Neel, B. G., & Ikura, M. (2013). NMR-based functional profiling of RASopathies and oncogenic RAS mutations. *Proceedings of the National Academy of Sciences of the United States of America*, *110*(12), 4574–4579. https://doi.org/10.1073/pnas.1218173110.

Spencer-Smith, R., Koide, A., Zhou, Y., Eguchi, R. R., Sha, F., Gajwani, P., et al. (2017). Inhibition of RAS function through targeting an allosteric regulatory site. *Nature Chemical Biology*, *13*(1), 62–68. https://doi.org/10.1038/nchembio.2231.

Tall, G. G., Barbieri, M. A., Stahl, P. D., & Horazdovsky, B. F. (2001). Ras-activated endocytosis is mediated by the Rab5 guanine nucleotide exchange activity of RIN1. *Developmental Cell*, *1*(1), 73–82. https://doi.org/10.1016/s1534-5807(01)00008-9.

Terrell, E. M., Durrant, D. E., Ritt, D. A., Sealover, N. E., Sheffels, E., Spencer-Smith, R., et al. (2019). Distinct binding preferences between Ras and Raf family members and the impact on oncogenic Ras signaling. *Molecular Cell*, *76*(6), 872–884 e875. https://doi.org/10.1016/j.molcel.2019.09.004.

Thurman, R., Siraliev-Perez, E., & Campbell, S. L. (2020). RAS ubiquitylation modulates effector interactions. *Small GTPases*, *11*(3), 180–185. https://doi.org/10.1080/21541248.2017.1371267.

Tomasetti, C., & Vogelstein, B. (2015). Cancer etiology. Variation in cancer risk among tissues can be explained by the number of stem cell divisions. *Science*, *347*(6217), 78–81. https://doi.org/10.1126/science.1260825.

Trahey, M., Milley, R. J., Cole, G. E., Innis, M., Paterson, H., Marshall, C. J., et al. (1987). Biochemical and biological properties of the human N-ras p21 protein. *Molecular and Cellular Biology*, 7(1), 541–544. https://doi.org/10.1128/mcb.7.1.541.

Tran, T. H., Chan, A. H., Young, L. C., Bindu, L., Neale, C., Messing, S., et al. (2021). KRAS interaction with RAF1 RAS-binding domain and cysteine-rich domain provides insights into RAS-mediated RAF activation. *Nature Communications*, *12*(1), 1176. https://doi.org/10.1038/s41467-021-21422-x.

Vigil, D., Cherfils, J., Rossman, K. L., & Der, C. J. (2010). Ras superfamily GEFs and GAPs: Validated and tractable targets for cancer therapy? *Nature Reviews. Cancer*, *10*(12), 842–857. https://doi.org/10.1038/nrc2960.

Vojtek, A. B., & Der, C. J. (1998). Increasing complexity of the Ras signaling pathway. *The Journal of Biological Chemistry*, *273*(32), 19925–19928. https://doi.org/10.1074/jbc.273.32.19925.

Waliszewski, P. (1995). Controversies about the genetic model of colorectal tumorigenesis. *Polish Journal of Pathology*, *46*(4), 239–243. Retrieved from https://www.ncbi.nlm.nih.gov/pubmed/8713291.

Wennerberg, K., Rossman, K. L., & Der, C. J. (2005). The Ras superfamily at a glance. *Journal of Cell Science*, *118*(Pt. 5), 843–846. https://doi.org/10.1242/jcs.01660.

Wey, M., Lee, J., Jeong, S. S., Kim, J., & Heo, J. (2013). Kinetic mechanisms of mutation-dependent Harvey Ras activation and their relevance for the development of Costello syndrome. *Biochemistry*, *52*(47), 8465–8479. https://doi.org/10.1021/bi400679q.

Winters, I. P., Chiou, S. H., Paulk, N. K., McFarland, C. D., Lalgudi, P. V., Ma, R. K., et al. (2017). Multiplexed in vivo homology-directed repair and tumor barcoding enables parallel quantification of Kras variant oncogenicity. *Nature Communications*, *8*(1), 2053. https://doi.org/10.1038/s41467-017-01519-y.

Wohlgemuth, S., Kiel, C., Kramer, A., Serrano, L., Wittinghofer, F., & Herrmann, C. (2005). Recognizing and defining true Ras binding domains I: Biochemical analysis. *Journal of Molecular Biology*, *348*(3), 741–758. https://doi.org/10.1016/j.jmb.2005.02.048.

Wong, K. A., Russo, A., Wang, X., Chen, Y. J., Lavie, A., & O'Bryan, J. P. (2012). A new dimension to Ras function: A novel role for nucleotide-free Ras in class II phosphatidylinositol 3-kinase beta (PI3KC2beta) regulation. *PLoS One*, 7(9), e45360. https://doi.org/10.1371/journal.pone.0045360.

Wu, C. Y., Carpenter, E. S., Takeuchi, K. K., Halbrook, C. J., Peverley, L. V., Bien, H., et al. (2014). PI3K regulation of RAC1 is required for KRAS-induced pancreatic tumorigenesis in mice. *Gastroenterology*, *147*(6), 1405–1416 e1407. https://doi.org/10.1053/j.gastro.2014.08.032.

Xu, S., Long, B. N., Boris, G. H., Chen, A., Ni, S., & Kennedy, M. A. (2017). Structural insight into the rearrangement of the switch I region in GTP-bound G12A K-Ras. *Acta Crystallographica Section D: Structural Biology*, *73*(Pt. 12), 970–984. https://doi.org/10.1107/S2059798317015418.

Yuan, T. L., Amzallag, A., Bagni, R., Yi, M., Afghani, S., Burgan, W., et al. (2018). Differential effector engagement by oncogenic KRAS. *Cell Reports*, *22*(7), 1889–1902. https://doi.org/10.1016/j.celrep.2018.01.051.

Zafra, M. P., Parsons, M. J., Kim, J., Alonso-Curbelo, D., Goswami, S., Schatoff, E. M., et al. (2020). An in vivo Kras allelic series reveals distinct phenotypes of common oncogenic variants. *Cancer Discovery*, *10*(11), 1654–1671. https://doi.org/10.1158/2159-8290.CD-20-0442.

Zhao, Q., Fujimiya, R., Kubo, S., Marshall, C. B., Ikura, M., Shimada, I., et al. (2020). Real-time in-cell NMR reveals the intracellular modulation of GTP-bound levels of RAS. *Cell Reports*, *32*(8). https://doi.org/10.1016/j.celrep.2020.108074, 108074.

CHAPTER THREE

Drug targeting opportunities en route to Ras nanoclusters

Karolina Pavic[†], Rohan Chippalkatti[†], and Daniel Abankwa*
Cancer Cell Biology and Drug Discovery Group, Department of Life Sciences and Medicine, University of Luxembourg, Esch-sur-Alzette, Luxembourg
*Corresponding author: e-mail address: daniel.abankwa@uni.lu

Contents

1. Introduction 64
2. Ras processing and trafficking 69
  2.1 Ras prenylation 69
  2.2 CAAX processing by Ras converting enzyme 1 (RCE1) 70
  2.3 Carboxymethylation 71
  2.4 Palmitoylation 72
3. Trafficking-factors and -chaperones of Ras 73
  3.1 Phosphodiesterase 6 delta subunit (PDEδ, PDE6D) 73
  3.2 Calmodulin 75
  3.3 VPS35 76
  3.4 Nogo-B receptor (NgBR) 76
  3.5 GPR31 77
  3.6 SNARE proteins SNAP23, SNAP29 and VAMP3 78
4. Organization of Ras in specific plasma membrane lipid domains and nanocluster 79
  4.1 The H-Ras and N-Ras lipid domains 79
  4.2 The K-Ras lipid domain 80
5. Lipid domain and Ras nanocluster modulating factors 82
  5.1 ORP5 and ORP8 82
  5.2 TRPML1 (gene *MCOLN1*) 83
  5.3 Prohibitin-1 83
  5.4 Galectin-1 84
  5.5 Galectin-3 85
  5.6 ASPP2 86
  5.7 Nucleophosmin 1 (NPM1) 87
  5.8 DIRAS3 88
6. Concluding remarks 88
References 89

[†] Contributed equally.

Advances in Cancer Research, Volume 153
ISSN 0065-230X
https://doi.org/10.1016/bs.acr.2021.07.005

Abstract

Disruption of the native membrane organization of Ras by the farnesyltransferase inhibitor tipifarnib in the late 1990s constituted the first indirect approach to drug target Ras. Since then, our understanding of how dynamically Ras shuttles between subcellular locations has changed significantly.

Ras proteins have to arrive at the plasma membrane for efficient MAPK-signal propagation. On the plasma membrane Ras proteins are organized into isoform specific proteo-lipid assemblies called nanocluster. Recent evidence suggests that Ras nanocluster have a specific lipid composition, which supports the recruitment of effectors such as Raf. Conversely, effectors possess lipid-recognition motifs, which appear to serve as co-incidence detectors for the lipid domain of a given Ras isoform. Evidence suggests that dimeric Raf proteins then co-assemble dimeric Ras in an immobile complex, thus forming the minimal unit of an active nanocluster.

Here we review established and novel trafficking chaperones and trafficking factors of Ras, along with the set of lipid and protein modulators of Ras nanoclustering. We highlight drug targeting approaches and opportunities against these determinants of functional Ras membrane organization. Finally, we reflect on implications for Ras signaling in polarized cells, such as epithelia, which are a common origin of tumorigenesis.

1. Introduction

Ras proteins are membrane bound small GTPases, which relay in particular mitogenic, extracellular stimuli to intracellular signaling pathways, such as the mitogen-activated protein kinase (MAPK) pathway (Simanshu, Nissley, & McCormick, 2017). Their single globular domain operates as a molecular switch, which undergoes conformational changes upon GTP-binding that allows them to interact with downstream effectors, such as Raf or phosphoinositide-3-kinase (PI3K), to propagate signaling. In their active conformation Ras recognizes Ras binding domains (RBD) or structurally related Ras association (RA) domains of effectors proteins, which then further transmit signaling, ultimately altering gene expression, proliferation, differentiation, invasion or apoptosis (Kiel, Matallanas, & Kolch, 2021; Simanshu et al., 2017).

The four cancer associated Ras proteins, H-Ras, N-Ras, K-Ras4A and K-Ras4B, show the greatest structural divergence in their carboxy-terminal hypervariable region (HVR, residues 166-188/189), with less than 20% sequence homology (Simanshu et al., 2017). While there are several other Ras subfamily members, we here refer to these four isoforms, when referring to Ras (Wennerberg, Rossman, & Der, 2005). All Ras proteins are prenylated on a conserved cysteine within the HVR. In addition, K-Ras4A is palmitoylated on Cys180, N-Ras on Cys181, and H-Ras is dually palmitoylated on Cys181 and Cys184. Importantly, the lipid modified HVR determines

Ras isoform specific properties, including their cellular trafficking itineraries (Mo, Coulson, & Prior, 2018). Various proteins have been identified, which facilitate Ras trafficking through the cytoplasm (Schmick, Kraemer, & Bastiaens, 2015). Both trafficking chaperones and other trafficking factors impact on Ras trafficking between subcellular membrane compartments and by shifting the subcellular distribution or organization of lipids that affect the binding and organization of Ras in membranes.

The farnesyl transferase inhibitor (FTI) tipifarnib essentially targets this trafficking machinery and despite initial clinical failure of this compound, inhibition of Ras trafficking to the plasma membrane (PM) still holds today as a valid approach to target H-Ras (Cox, Der, & Philips, 2015). It was recognized early that Ras transforming activity requires its localization to the PM (Willumsen, Christensen, Hubbert, Papageorge, & Lowy, 1984). Within the PM, Ras is laterally segregated into specific lipid-domains (Zhou & Hancock, 2018). Intriguingly, the specific lipid environment of each Ras isoform is shaped by the HVR and accessory proteins of Ras, thus establishing specific proteo-lipid signaling complexes of Ras (Yurugi et al., 2020; Zhou & Hancock, 2018). Effectors have typically evolved membrane association motifs or specific domains for co-incidence detection of lipids, which can read-out the specific lipid content around Ras and potentially also the nature of the HVR (Kiel et al., 2021; Terrell et al., 2019). This might explain how effector paralogs prefer one or the other Ras isoform in the context of the intact cell membrane (Plowman, Ariotti, Goodall, Parton, & Hancock, 2008). Hence pharmacological disruption of plasma membrane lipid distribution can effectively perturb PM localization of Ras and attenuate downstream signaling (Zhou & Hancock, 2018) (Fig. 1).

Recent evidence further suggests that the Ras-effector complexes impact on how Ras engages with the membrane (Lakshman et al., 2019; Travers et al., 2018). For instance, binding of the extended RBD that contains also the cysteine rich domain of Raf, appears to impose a preferred orientation of Ras relative to the membrane and increases the residence of the whole complex in the membrane (Fang et al., 2020; Lakshman et al., 2019). Ultimately, the oligomeric complex of lipids, Ras, and effector Raf organizes into nanoclusters (Abankwa & Gorfe, 2020; Van et al., 2021). Whether other effectors than Raf are in nanoclusters is not known.

Nanoclusters were first identified as "observables," i.e. electron micrographs of the PM showed Ras in non-random collections (Prior, Muncke, Parton, & Hancock, 2003). A number of cellular and in vitro biophysical methods helped to describe nanoclusters further. We employed in particular fluorescence lifetime imaging microscopy (FLIM) to measure Förster resonance energy transfer (FRET) of cells expressing FRET-fluorophore tagged

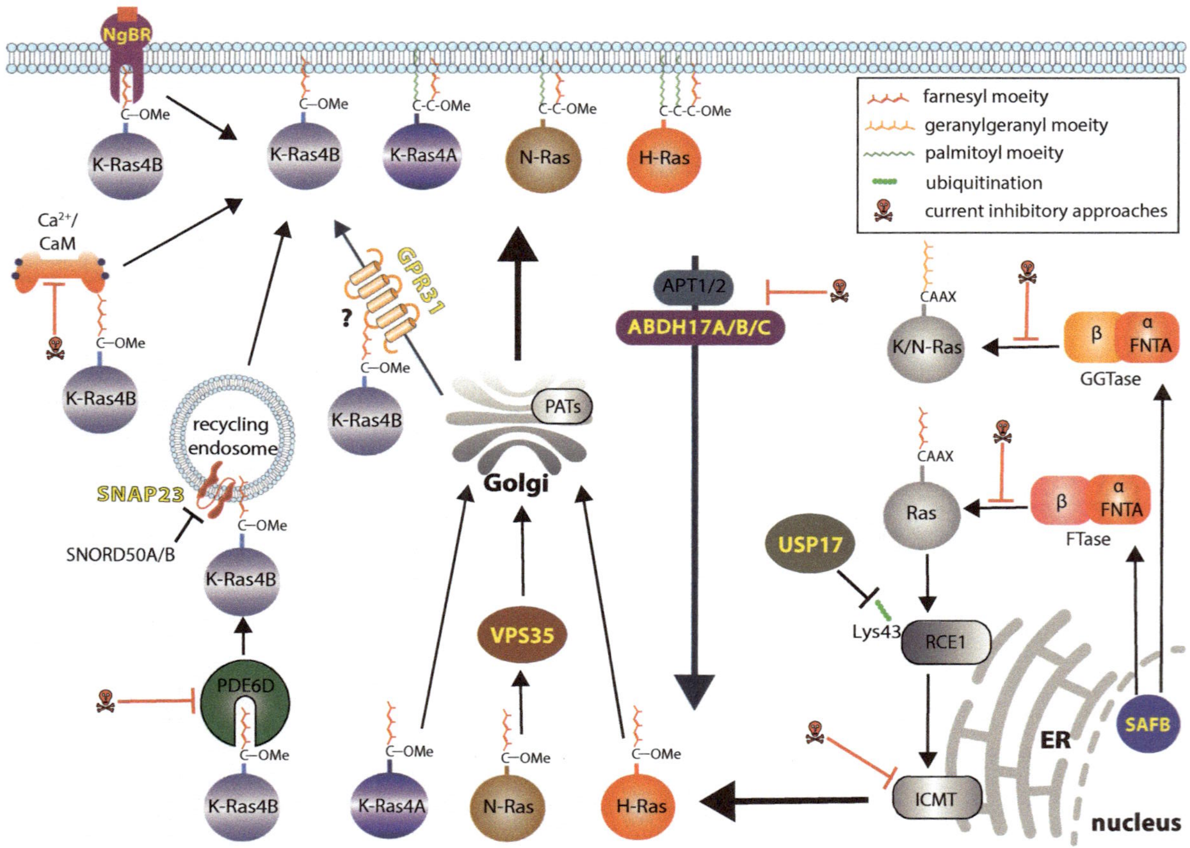

Fig. 1 See figure legend on opposite page.

Ras-constructs (Guzman, Oetken-Lindholm, & Abankwa, 2016). Dense packing of Ras or other signaling proteins in nanocluster leads to average fluorophore distances in the nanometer range allowing for FRET to occur. Importantly, loss of FRET reports on loss of nanoclustering and any upstream process such as mislocalization to the cytoplasm or disturbed lipid modification (Najumudeen, Kohnke, Solman, Alexandrov, & Abankwa, 2013).

These data have provided us with the following, emergent hallmarks for Ras-nanocluster:

- Driven by their HVR, Ras isoforms laterally segregate into specific proteolipid membrane domains. The HVR alone can instruct non-random assemblies of Ras, which can be regarded as neutral, membrane anchor-dependent nanoclusters (Li, Janosi, & Gorfe, 2012; Plowman, Muncke, Parton, & Hancock, 2005).
- In the case of K-Ras4B it was established that interactions between lysines of the HVR and membrane phospholipids, encode a specific nanoscale lipid environment of Ras (Zhou et al., 2017; Zhou, Prakash, Liang, Gorfe, & Hancock, 2021). Similar mechanisms for the other Ras isoforms and other peripheral membrane proteins are plausible.
- Most Ras nanocluster contain 2–3 Ras proteins, with the majority present as dimers (Sarkar-Banerjee et al., 2017). However, K-Ras4B does not dimerize significantly on its own even in the membrane (Chung et al., 2018; Van et al., 2021).
- Importantly, active, GTP-Ras in complex with dimeric Raf-effectors, appears to form active, transiently immobilized nanoclusters (Abankwa

Fig. 1 Trafficking pathways and points of drug intervention of K-Ras4B, K-Ras4A, N-Ras and H-Ras: All Ras isoforms undergo farnesylation on the "CAAX" motif but K-Ras and N-Ras can also be alternatively geranylgeranylated (*right part*). SAFB regulates the expression of *FNTA,* the α subunit of both FTase and GGTase-I and therefore supports Ras processing. All Ras isoforms are then processed further by RCE1, which removes the terminal "AAX" and ICMT which carboxymethylates the now terminal cysteine. USP17 can inhibit the activity of RCE1 isoform 2 by de-ubiquitinating it, thus impeding mostly N-Ras and H-Ras processing. Farnesylated K-Ras4A, N-Ras and H-Ras are trapped by palmitoylation on the Golgi and then shuttled to the PM via the secretory pathway (*middle part*). Palmitoylation can be reversed by depalmitoylases APT1/2 and ABDH17A/B/C. VPS35 aids N-Ras trafficking by promoting its association with the Golgi. GPR31 facilitates K-Ras4B, K-Ras4A and N-Ras PM localization through the secretory pathway. K-Ras4B and other farnesylated, but not palmitoylated proteins, can be solubilized from cellular membranes by PDE6D or CaM (*left*). In addition, NgBR is capable of binding farnesylated proteins and aids PM localization of K-Ras4B and H-Ras. Forward trafficking of K-Ras4B from the recycling endosome to the PM is supported by SNARE proteins and SNAP23 directly binds to K-Ras4B. This interaction is typically disrupted by a small nucleolar RNA, SNORD50A/B, which therefore promotes K-Ras4B retention on the recycling endosome. Interestingly, this RNA is frequently deleted in common cancers. Yellow highlighted proteins were only recently established as potential drug targets.

Table 1 Summary of reviewed modulators of Ras nanoclustering.

Protein	H-Ras	N-Ras	K-Ras4B	K-Ras4A	Mechanism
ORP5 and ORP8					Increasing PS content on the PM by exchanging PI_4P.
TRPML1					Stimulates increase of cholesterol on the PM by increasing its processing in lysosomes.
Prohibitin-1					Associates with PS, PA and C-Raf.
Galectin-1					Acts as a dimer and binds to the RBD, suggesting stabilization of Raf dimers on Ras.
Galectin-3					Binds to C-Raf and PI3Kα RBDs, suggesting similar mechanism as Gal-1.
ASPP2					Increases heterodimerization of B/C-Raf, outcompetes Gal-1.
Nucleophosmin-1					Associates with HVR region of K-Ras4B.
DIRAS3					Disrupts Ras-Raf interaction and Ras-GTP loading; binds to K-Ras4B.

Modulators of the plasma membrane lipid content (*top, light gray*) and protein modulators (*bottom, dark gray*) are listed with their mechanism of action and Ras isoform specificity. *Red*, Ras nanoclustering increasing; *white*, neutral effect; *blue*, Ras nanoclustering decreasing. *Gray* fields indicate unknown effects. In addition, other proteins (H-RasG12V) and structures (caveolae) are described in the text.

& Gorfe, 2020). This implies dimeric Ras-proteins, however, in a complex not driven by Ras dimerization, but Raf dimerization and allosteric coupling between Raf and Ras (Jin et al., 2017).

Just about half a dozen modulators of Ras nanoclusters are known, and this ensemble suggests that both certain lipids as well as structurally unrelated proteins can impact on the "observable" nanocluster (Table 1). Given that nanocluster-modulator or -scaffold proteins are structurally quite diverse, it is likely that the portfolio of such nanocluster-modulators is still incomplete and that the mechanisms by which nanoclusters can be regulated are also quite distinct. Due to the implied requirement of active nanoclusters for Ras-signaling and the isoform specificity of nanoclusters, these complexes have more recently been nominated as novel intervention sites for Ras targeted drugs (Cho & Hancock, 2013; Van et al., 2021).

In this chapter, we will review essential steps of Ras lipid modification, trafficking and plasma membrane organization into nanocluster. We will discuss modulators of trafficking and nanoclustering with a focus on emerging drug targeting opportunities.

2. Ras processing and trafficking

2.1 Ras prenylation

The C-terminal CAAX-box of Ras is a tetrapeptide motif that is recognized by prenyltransferases (Nguyen, Goodall, Alexandrov, & Abankwa, 2010). The cysteine residue (C in the tetrapeptide) is irreversibly modified via a thioether linkage with either farnesyl- (15-carbon lipid chain) or geranylgeranyl- (20-carbon lipid chain), by farnesyltransferase (FTase) or geranylgeranyl transferase I (GGTase-I), respectively. These hetero-dimeric enzymes share a common α-subunit, while their β-subunit is specific. The first aliphatic residue (A) of the CAAX-box is flexible, as it is solvent exposed when bound to the transferases. For the second "A," amino acids with larger aliphatic side chains are preferred, consistent with the hydrophobic binding site on the prenyltransferases. The terminal amino acid, X, then significantly impacts on the specific recognition by FTase or GGTase-I. However, several amino acids, such as Met and Phe, are tolerated by both prenyltransferases. Therefore, K-Ras4B (CVIM) and N-Ras (CVVM) can be alternatively prenylated by GGTase-I (Lerner et al., 1997; Whyte et al., 1997). While the functional significance of their alternative prenylation under normal conditions is not well understood, it is the dire hallmark of the clinical failure of the FTase inhibitor (FTI) program (Berndt, Hamilton, & Sebti, 2011). This event has largely halted most of the Ras drug development efforts at around the turn of the millennium. Regardless, utilizing CAAX-peptidomimetic derived FTIs to target H-Ras membrane localization has gained track again. FTase inhibitors such as tipifarnib shown promise in clinical trials for the treatment of

HRAS-mutant head and neck squamous cell carcinoma (HNSCC) and thyroid cancer (Moore, Rosenberg, McCormick, & Malek, 2020).

Additional approaches targeting the prenylation step are emerging. A recent genetic screen identified scaffold association factor B (SAFB), a nuclear DNA/RNA binding protein, as a regulator of Ras and Rap1a PM localization (Zhou et al., 2020). SAFB is necessary for the expression of *FNTA*, which encodes the α-subunit of both FTase and GGTase-I (Zhou et al., 2020). Therefore, knockdown of SAFB and simultaneous treatment with FTIs prevented alternative prenylation of K-Ras4B and reduced its PM association. SAFB thus represents a novel drug target that could synergize with FTIs, such as the approved Ras inhibitor tipifarnib.

An innovative strategy to target K-Ras4B farnesylation was demonstrated by Novotny and colleagues (Novotny, Hamilton, McCormick, & Shokat, 2017). Compounds with a chemically reactive group that could covalently react with the CAAX-box cysteine were developed. A farnesylpyrophosphate (i.e. the substrate of FTase) mimetic FTI served as a starting point, thus leading to a neosubstrate for FTase that is carrying a reactive group. FTase then delivers this neosubstrate to the CAAX-box for reaction with the cysteine and irreversible inactivation of Ras. Several such conjugates were tested and were found to mis-localize K-Ras4B-G12C from the PM. The authors speculated that the non-natural attachment of the farnesyl-like neosubstrate on K-Ras4B would prevent subsequent processing and trafficking steps. Surprisingly, a selectivity for K-Ras4B over H-Ras was found, however, further investigation of the target selectivity seems warranted.

2.2 CAAX processing by Ras converting enzyme 1 (RCE1)

Prenylated Ras proteins are processed further by the endoprotease Ras converting enzyme 1 (RCE1), an integral membrane protein of the endoplasmic reticulum (ER) that removes the terminal "AAX" peptide (Ahearn, Zhou, & Philips, 2018).

USP17 is a ubiquitin-specific protease (USP), i.e. a de-ubiquitinating enzyme, which co-localizes with RCE1 on the ER (Burrows et al., 2009). USP17 de-ubiquitinates and blocks RCE1 activity which disrupts PM localization of H-Ras, N-Ras but not K-Ras4B (Burrows et al., 2009; de la Vega et al., 2010). This finding agrees with previous observations, where K-Ras4B PM localization seemed unaffected in RCE1 knockout cells (Roberts et al., 2008). Alternative splicing creates a novel isoform of RCE1 (isoform 2) which is N-terminally truncated and thus distinct from the full length RCE1 isoform 1 (Jaworski et al., 2014). USP17 has a differential effect on both isoforms as it downregulated the protein levels of isoform 2 but not isoform 1. Interestingly, ubiquitination at Lys43 is also required for the ER localization of RCE1 isoform 2. The authors proposed that de-ubiquitination by USP17 may prevent ER association of RCE isoform 2 and would therefore affect CAAX-processing. Nonetheless, activity of both isoforms was required to reconstitute PM localization of H-Ras (Jaworski et al., 2014).

While investigations are still ongoing to identify selective and potent RCE1 inhibitors, the regulation of RCE1 ubiquitination may present an additional opportunity for drug development (Hampton, Dore, & Schmidt, 2018).

2.3 Carboxymethylation

Isoprenylcysteine carboxymethyltransferase (ICMT) methyl-esterifies the α-carboxyl group of the C-terminal prenyl-cysteine and thus neutralizes the negatively charged carboxyl group. Like RCE1, ICMT localizes to the ER with its active sites facing the cytoplasm, allowing access to Ras proteins as they transit through the ER (Dai et al., 1998).

A recent study demonstrated that ICMT is necessary for N-Ras PM trafficking, as compared to H-Ras, K-Ras4A and K-Ras4B (Ahearn, Court, Siddiqui, Abankwa, & Philips, 2021). This was because loss of ICMT decreased both palmitoylation and binding to the trafficking chaperone PDE6D on N-Ras.

Because of the general relevance of ICMT for Ras processing, it is pursued as a drug target. Cysmethynil, the first small molecule inhibitor of ICMT, dislodges K-Ras4B from the membrane at micromolar concentrations (Winter-Vann et al., 2005).

A structurally distinct inhibitor, UCM-1336 ($IC_{50} = 2\,\mu M$), is superior to cysmethynil in reducing PM localization of all four Ras isoforms and reduces cell proliferation at low micromolar concentrations in *KRAS*- and *NRAS*-mutant cell lines, but not in non-tumor cells (Marin-Ramos et al., 2019). It is not clear, why genetic and inhibitor studies show a different Ras isoform selectivity pattern.

The most potent inhibitor so far, compound 75 (IC_{50} = 1.3 nM) with a distinct scaffold from cysmethynil, increases the cytosolic fraction as detected with a pan-Ras antibody at sub-micromolar concentrations, while inhibiting most *KRAS*- and *NRAS*-mutant cancer cell lines only at low micromolar concentrations (Judd et al., 2011). Of note, this compound showed efficacy in a Hutchinson-Gilford progeria syndrome mouse model, where ICMT inactivation reduces methylation of progerin, the causative aberrant protein of the disease (Chen et al., 2021).

2.4 Palmitoylation

Palmitoylation is mediated by the DHHC family of protein acyl-transferases (PATs) (Mitchell, Vasudevan, Linder, & Deschenes, 2006). While certain PATs for Ras, such as DHHC9 and GCP16, have been proposed (Swarthout et al., 2005), a specificity for Ras of the DHHC PAT activity was later questioned (Rocks et al., 2010). Instead, a broad set of substrates may be accepted by PATs, if they contain a free cysteine and transit through the Golgi. Given this functional redundancy among the 23 mammalian DHHC proteins and the inability to target the highly mutated, non-palmitoylated K-Ras4B, inhibitor development is stalling (Cox et al., 2015).

Farnesylated Ras proteins partition across all cellular membranes, with H-Ras and N-Ras then becoming trapped at the Golgi apparatus by repalmitoylation for transport via the secretory pathway back to the PM (Goodwin et al., 2005; Rocks et al., 2005; Rocks, Peyker, & Bastiaens, 2006). This route may at least partially apply to K-Ras4A, which could however in the absence of its palmitoylation site and mislocalization to internal membranes still be found at the PM (Tsai et al., 2015). Even a fully and stably lipid modified N-Ras (with hexadecyl-, instead of palmitoyl) redistributes across all cellular membranes, with a strong decoration of the ER due to its large surface area. However, spontaneous dissociation of H-Ras would be slower. Instead, ubiquitous depalmitoylation releases H-Ras and N-Ras from the membrane (Rocks et al., 2010). This activity is superseded on the Golgi by PAT activity, allowing for trapping of these Ras proteins. Because of the necessity of the palmitoylation-depalmitoylation cycle to refocus Ras on the Golgi for PM transport, interruption of this cycle at the point of depalmitoylation blocks PM localization of Ras.

The acyl protein thioesterase (APT) 1/2 was nominated as candidate depalmitoylase and a depalmitoylation inhibitor Palmostatin B was developed (Dekker et al., 2010). Treatment with this sub-micromolar active compound caused endomembrane accumulation of palmitoylated N-Ras and H-Ras and decreased the GTP-loaded pool of H-Ras at the Golgi. Ultimately, inhibitor treatment was sufficient to reverse the aberrant phenotype of H-Ras G12V transformed MDCK cells (Dekker et al., 2010).

However, in another study APT1/2 knockdown or specific inhibition did not affect N-Ras trafficking (Lin & Conibear, 2015). Instead, additional serine hydrolases, including members of the α/β hydrolase domain 17 (ABDH17) family were identified as targets of Palmostatin B. ABDH17 enzymes are themselves palmitoylated and localize to the PM and endosomal compartments (Lin & Conibear, 2015). Overexpression or knockdown of either ABDH17A, ABDH17B or ABDH17C perturbed depalmitoylation of N-Ras, while ABHD17A overexpression was shown to block its PM localization (Lin & Conibear, 2015). Of note, the triple knockdown of ABDH17A/B/C was only about half as efficient as Palmostatin B treatment in retaining palmitoylated N-Ras, suggesting additional depalmitoylases as targets of the drug.

3. Trafficking-factors and -chaperones of Ras

3.1 Phosphodiesterase 6 delta subunit (PDEδ, PDE6D)

Selectivity: K-Ras4B, N-Ras >> H-Ras; not K-Ras4A

PDE6D was identified as the non-catalytic δ-subunit of the membrane bound photoreceptor phosphodiesterase 6 (PDE6), however, it has a much broader cellular activity than suggested by its name, as it is a general trafficking chaperone for prenylated proteins (Chandra et al., 2011). Given its requirement for K-Ras4B PM localization and signaling, PDE6D has been nominated as a surrogate target for K-Ras4B (Zimmermann et al., 2013).

PDE6D sequesters the farnesyl moiety of K-Ras4B and prevents it from binding to cellular membranes, thereby effectively enhancing its diffusion through the cytoplasm. With the farnesyl moiety being the major binding determinant, the interaction is independent of whether K-Ras4B is GDP- or GTP-bound (Chandra et al., 2011). K-Ras4B is subsequently released in the perinuclear region by the action of release factors Arl2 and in the primary cilium by Arl3 and transported back to the PM by Rab11-positive recycling endosomes (Ismail et al., 2011; Schmick et al., 2015).

Structural studies have revealed two different binding modes of K-Ras4B with different insertion depths of the farnesyl tail into the hydrophobic pocket of PDE6D (Dharmaiah et al., 2016). While these structures suggest that also geranylgeranylated proteins are clients of PDE6D, this was not seen experimentally (Chandra et al., 2011). Of note, K-Ras4B without carboxymethylation has a more than 35-fold reduced affinity to PDE6D ($K_d = 82 \pm 9\,\mu M$ without vs $K_d = 2.3 \pm 0.3\,\mu M$ with methylation) (Dharmaiah et al., 2016). A similar significance of carboxymethylation in this context has recently been demonstrated for N-Ras (Ahearn et al., 2021).

Importantly, PDE6D has a different client spectrum than FTase. The ability to bind to PDE6D is abrogated by C-terminal palmitoylation or a second prenylation, and in addition only small side chain containing amino acids are tolerated at positions −1 and −2 relative to the farnesylated cysteine. Importantly, replacement of Ser181 by a phosphomimetic Glu reduced the affinity more than 6-fold, while that of the Ala-replacement had a comparable affinity (Dharmaiah et al., 2016). This contrasts with cell biological studies, where bryostatin-1 treatment, that was meant to stimulate Ser181-phosphorylation, enhanced the solubilization of K-Ras4B by PDE6D (Chandra et al., 2011).

Despite low nanomolar affinity, PDE6D inhibitors Deltarasin and Deltazinone suffered from the Arl2/3-associated ejection mechanism (Martin-Gago et al., 2017; Papke et al., 2016; Zimmermann et al., 2013). While this was overcome by Deltasonamides, which bound with subnanomolar affinity, low cell penetration remained an issue (Martin-Gago et al., 2017).

To overcome the ejection mechanism, Deltaflexins with a flexible middle part in the inhibitor molecule were developed. Structure-function data suggest that a hexamethylene-linker acts like a molecular spring enabling resistance to ejection (Siddiqui et al., 2020). Furthermore, the second generation of these compounds consisting of the flexible linker and more stably interacting moieties could be easily generated by molecular hybridization. As a result, low micromolar in vitro activity was matched by micromolar cellular activity, which was a significant improvement as compared to the almost 1000-fold activity difference seen with the first PDE6D inhibitors.

Another solution to the low target occupation due to the ejection mechanism was the diversification of the high affinity Deltasonamide into a PROTAC (proteolysis-targeting chimera), by linking it to pomalidomide for cereblon-associated E3-ligase engagement (Winzker et al., 2020). This powerful approach exploits the fact that PROTACs can be recycled after target destruction and hence do not need to permanently bind to their target.

Interestingly, PROTAC induced loss of PDE6D increased sterol regulatory element binding protein (SREBP)-associated gene expression, which may point to other PDE6D clients being more significantly impacted on than K-Ras4B.

Finally, stabilization of the K-Ras4B/PDE6D complex is being explored, as it may specifically take the K-Ras4B/PDE6D-associated activity out of action (Casique-Aguirre et al., 2018). However, current experimental evidence is limited for this concept.

3.2 Calmodulin

Selectivity: K-Ras4B/not H-Ras N-Ras; K-Ras4A unknown

Calmodulin (CaM) is a calcium sensor that binds to a diverse group of client proteins and regulates several important biological pathways (Tidow & Nissen, 2013). CaM is composed of C- and N- terminal Ca^{2+} binding globular lobes which are connected by a flexible linker. Each lobe possesses a hydrophobic pocket, which is exposed upon Ca^{2+} binding and mediates interactions with client proteins (Tidow & Nissen, 2013; Zhang, Tanaka, & Ikura, 1995).

One group of non-canonical client proteins features a lipid modified polybasic region, such as found in K-Ras4B (Grant et al., 2020). Probably for similar reasons as with PDE6D, namely interference of a second lipid modification with binding to the hydrophobic pocket, CaM interacts specifically with K-Ras4B but not H-Ras or N-Ras and for the same reason likely not with K-Ras4A (Fivaz & Meyer, 2005; Sperlich, Kapoor, Waldmann, Winter, & Weise, 2016; Villalonga et al., 2001). Of note, this interaction has been implicated in K-Ras4B-driven cancer cell stemness, which provides a novel rational to pursue CaM as a drug target (Manoharan, Kopra, Eskonen, Harma, & Abankwa, 2019; Najumudeen et al., 2016; Wang et al., 2015).

Ca^{2+}/CaM binds K-Ras4B by sequestering its farnesyl tail in its hydrophobic pockets, possibly with additional affinity contributions by some basic residues in the HVR (minimal CaM-binding motif KSKTKC-farnesyl/carboxymethyl) (Agamasu et al., 2019; Grant et al., 2020). Consequently, this interaction is independent of the activation state of K-Ras4B and can potentially involve two K-Ras4B molecules per CaM (Agamasu et al., 2019; Grant et al., 2020). Both in vitro, as well as in cells can Ca^{2+}/CaM solubilize membrane bound K-Ras4B in a Ca^{2+}-dependent manner (Fivaz & Meyer, 2005; Grant et al., 2020; Sperlich et al., 2016).

A large variety of CaM inhibitors is available, however, no dedicated CaM-targeting cancer therapy has been established, in part maybe because of the broad involvement of CaM in many biological processes (Berchtold & Villalobo, 2014). It will therefore be important to gain a better understanding of the CaM/K-Ras4B biology in cancer (stem) cells, to derive more focused interventions.

3.3 VPS35

Selectivity: N-Ras/not H-Ras, K-Ras4A and K-Ras4B

VPS35 was identified as a regulator for subcellular trafficking of N-Ras (Zhou et al., 2016). VPS35 is a component of the retromer complex which is well characterized for its roles in the endosome-Golgi trafficking pathway (Seaman, 2021). Non-palmitoylated, but farnesylated N-Ras and VPS35 are part of a cytosolic, high molecular weight complex that is mostly devoid of H-Ras or K-Ras4B (Zhou et al., 2016). Farnesylation of N-Ras was required for VPS35 interaction, while the nucleotide-bound state was not relevant. These constraints are reminiscent of those for PDE6D and CaM, yet a distinct Ras isoform selectivity is realized.

VPS35 knockdown increased N-Ras decoration on vesicular structures, while Golgi localization was still maintained. Consequently, this manipulation decreased MAPK signaling output in oncogenic *NRAS*-driven melanoma cells (Zhou et al., 2016). From these findings, the authors proposed that VPS35 sequesters N-Ras from the cytosol and promotes its association with the Golgi from where it is trafficked to the PM. As a direct interaction was not demonstrated, it remains to be shown whether the α-helical HEAT repeat protein VPS35, which typically scaffolds the interaction of VPS29 and VPS26, directly binds to Ras, or whether and intermediary is required for binding of Ras.

3.4 Nogo-B receptor (NgBR)

Selectivity: H-Ras, K-Ras4B; potentially other farnesylated or geranylgeranylated proteins

NgBR is known as a receptor for neurite growth inhibitor type B (Nogo-B) and participates in axon guidance and blood vessel development (Park, Grabinska, Guan, & Sessa, 2016). NgBR is also expressed in invasive ductal carcinoma and is known to promote epithelial to mesenchymal transition (Wang et al., 2013).

Overexpression of NgBR increased H-Ras PM localization, while knockdown significantly decreased PM levels, irrespective of whether wild type, G12V (active)- or S17N (inactive)-mutant H-Ras was studied. Similar observations were made with geranylgeranylated H-Ras mutants, mutants containing the HVR of K-Ras4B or with K-Ras4B itself. This suggests a broad chaperoning activity of NgBR for several farnesylated or geranylgeranylated proteins. The effect of NgBR is mediated via a direct interaction with Ras, as purified NgBR binds to farnesylated H-Ras, and mutations of several conserved hydrophobic residues in the C-terminal cis-isoprenyltransferase domain of NgBR reduces binding to H-Ras (Zhao et al., 2017). NgBR expression further augments downstream signaling of PI3K- or MAPK-pathways and colony formation. These activities are sensitive to mutations in the C-terminal cis-isoprenyltransferase domain that would abrogate farnesylated H-Ras interaction (Zhao et al., 2017). Intriguingly, overexpression of NgBR in otherwise non-transformed NIH3T3 cells grafted into nude mice induces tumors, while the inducible knockdown of NgBR in MDA-MB-231 xenografts decreases tumor growth. These data suggest an important contribution of NgBR in Ras-driven tumorigenesis and recommend this receptor as future drug-target.

3.5 GPR31

Selectivity: K-Ras4B, K-Ras4A, N-Ras >> H-Ras

In a genome-wide screen for regulators of K-Ras4B membrane trafficking 13 candidate genes remained after stringent filtering (Fehrenbacher et al., 2017). Among these hits were two G-protein coupled receptors (GPCRs), of which the orphan GPR31 was further investigated for reasons of potential drug-targetability of GPCRs. While a farnesylation-dependence of the interaction of GPR31 with K-Ras4B could be shown, it could not be determined whether the interaction is direct. Furthermore, GPR31 interacted with K-Ras4A, N-Ras, but less with H-Ras, yet no binding to geranylgeranylated Rap1b, a related Ras-like small GTPase, was observed. These findings were interpreted to reflect a potential involvement of the Ras globular domain for recognition.

GPR31 localizes to the PM and endomembranes and thus it partially overlaps with the K-Ras4B distribution. Silencing of GPR31 reduces the amount of basolaterally delivered K-Ras4B in Caco-2 cells, while the inverse is seen if GPR31 is overexpressed. Knockdown of GPR31 reduces growth of K-Ras4B dependent cancer cell lines, however, effects on MAPK-signaling are not detected (Fehrenbacher et al., 2017). As GPR31 is an integral membrane protein that traffics through the secretory pathway, it may assist Ras trafficking along that pathway.

3.6 SNARE proteins SNAP23, SNAP29 and VAMP3

Selectivity: K-Ras4B; potentially K-Ras4A/less relevant N-Ras, H-Ras

K-Ras4B is solubilized in the cytosol by PDE6D and upon unloading at the recycling endosome shuttled back to the PM (Schmick et al., 2015). Finer details of K-Ras4B delivery to the PM from RE have recently been shown to involve SNARE (soluble NSF attachment protein receptor) family proteins SNAP23, SNAP29 and VAMP3 (Che et al., 2019). Using a proximity proteomics approach, not only K-Ras4B, but also to a lesser extent H-Ras and N-Ras were found in proximity of these three SNAREs. SNARE proteins are best known for their role in neurons where they carry out an energy requiring zippering process of synaptic vesicle fusion with the target membrane (Goda, 1997). This enables either the release of vesicular cargo into the extracellular space or incorporation of cargo proteins in the target membrane.

Further investigations focused on K-Ras4B, which co-localizes partially with any of the three identified SNAREs on the recycling endosome, and upon triple knock-out of the above SNAREs K-Ras4B was retained on recycling endosomes.

The striking novelty is that the interaction between K-Ras4B and SNAREs is modulated by a small nucleolar RNA, named SNORD50A/B, which is deleted in 10–40% of 12 common cancers (Siprashvili et al., 2016). The snoRNAs are known to reside in the nucleolus to regulate modifications of small nuclear RNAs, tRNAs and rRNAs, thus aiding the synthesis of functional ribosomes (Falaleeva & Stamm, 2013). Both SNORD50A/B and SNARE compete for the same basic residues on K-Ras4B and probably also K-Ras4A, which SNORD50A/B likewise interacts with. Direct, submicromolar interactions could be demonstrated for K-Ras4B with SNAP23 and SNAP29, with the latter having affinities also for H-Ras and N-Ras (Che et al., 2019).

In agreement with this constellation, depletion of Ras trafficking promoting SNARE, decreases the interaction of Ras-GTP with up- and downstream interaction partners, such as EGFR and effectors, and attenuates both MAPK- and PI3K-signaling. Moreover, tumor growth of H23 pancreatic cancer cell xenografts is reduced with either SNAP23 or VAMP3 knockdown (Che et al., 2019). CRISPR-screening based co-essentiality analysis of 11 cancer cell lines then further confirmed a *KRAS*-specific role of SNAP23, as compared to *HRAS* or *NRAS*. Further analysis then supported the idea that SNAP23 is particularly important for forward trafficking of K-Ras4B from the recycling endosome to the PM, a step that is blocked by SNORD50A/B. Finally, SNAP23 targeting with an engineered SNARE-cleaving botulism toxin protease reduced *KRAS*-driven tumorigenesis in vivo, providing proof-of-concept for SNARE proteins as drug targets.

4. Organization of Ras in specific plasma membrane lipid domains and nanocluster

4.1 The H-Ras and N-Ras lipid domains

The major affinity providing interactions of H-Ras and N-Ras with PM lipids occur via the C-terminal farnesyl and palmitoyl modifications (Janosi, Li, Hancock, & Gorfe, 2012). In addition, membrane proximal residues of the HVR interact with a distinct set of PM lipids (Gorfe, Hanzal-Bayer, Abankwa, Hancock, & McCammon, 2007). Parts of the globular domain, notably within the so-called allosteric lobe (approximately residues 87 to 171) engage with the membrane, during activation-dependent membrane reorientation of Ras (Abankwa et al., 2008; Gorfe et al., 2007; Guzman et al., 2014). It is not clear, whether such contacts significantly alter the local lipid environment of Ras.

Reorientation alters the insertion depth of palmitates in the C-terminal region, which may diminish their interaction with cholesterol rich PM domains (Zhou & Hancock, 2015). This could explain why GTP-H-Ras assembles in cholesterol-independent nanoclusters, while the opposite is true for GDP-H-Ras. While the conformational transitions of N-Ras at the PM are less well understood (Abankwa, Gorfe, Inder, & Hancock, 2010), they may explain the inverse cholesterol-dependence of GTP- and GDP-N-Ras (Zhou & Hancock, 2015).

Among PM phospholipids, in particular phosphatidylinositol 4,5-bisphosphate (PIP_2), colocalized preferentially with the shortened HVR of H-Ras if expressed alone (Zhou et al., 2014). This construct often serves as a proxy for GDP-H-Ras distribution. Similarly, phosphatidylinositol (3,4,5)-trisphosphate (PIP_3), phosphatidylinositol 3-phosphate (PI_3P) and phosphatidylserine (PS) are enriched in this way, whereas phosphatidylinositol 4-phosphate (PI_4P) or phosphatidic acid (PA) are less co-clustering with H-Ras. Nevertheless, H-Ras nanoclustering, H-Ras-dependent signaling and 3D spheroid growth of *HRAS*-mutant cells is decreased by global phosphatidic acid (PA) level increases that are observed, if the activity of SREBP1 downstream in the mTORC1-pathway is reduced (Posada et al., 2017). Intriguingly, the opposite is true for K-Ras4B, therefore PA represents in addition to PS (see below) another lipid that regulates these two Ras-isoforms oppositely.

4.2 The K-Ras lipid domain

K-Ras4A interacts with the PM through its "hybrid" HVR, which contains several basic residues, in addition to palmitoylation- and farnesylation sites (Laude & Prior, 2008; Tsai et al., 2015). In a simulation of farnesylated K-Ras4A, additional contributions to the membrane interaction from the globular domain were predicted, which again stabilize a reoriented Ras at the membrane (Li & Buck, 2017). However, K-Ras4A is relatively under-studied, hence this section will focus on K-Ras4B.

While K-Ras4B co-clusters with phosphatidylinositides, its nanoclustering critically depends on phosphatidylserine (PS) (Zhou et al., 2014). In contrast to the long held believe that the stretch of lysine residues in the HVR (also known as polybasic region) of K-Ras4B serves as a simple electrostatic tether to the plasma membrane, the Hancock and Gorfe groups recently demonstrated that the HVR encodes a distinct PM lipid selectivity. This is enabled by well-defined conformational states of the HVR, which is stabilized by distinct interactions of certain basic residues with PS (Zhou et al., 2017). This specificity extends even to the acyl chain composition of PS species, whereby the mix of one saturated with one unsaturated fatty acid species is preferred in K-Ras4B nanocluster (Zhou et al., 2021). As a result, a distinct response of mixed chain PS to an increase in membrane curvature dictates a corresponding decrease in K-Ras4B nanoclustering and membrane anchorage.

Membrane potential depolarization (which renders cells more positive), is another physiological stimulus, that increases PS- and PIP_2-nanoclustering, with the former again augmenting K-Ras4B nanoclustering (Zhou et al., 2015). This mechanism in fact explains why membrane potential elevation leads to cell proliferation, as increased K-Ras4B nanoclustering results in higher MAPK-signaling output (Zhou et al., 2015).

Phosphorylation of K-Ras4B on Ser181 drives conformational changes in the lipidated HVR, which alter its relative interaction preference from PS to PIP_2 (Zhou et al., 2017). This phosphorylation was reported to be mediated by either protein kinase C (PKC) or cyclic GMP (cGMP) dependent protein kinase G 2 (PKG2) (Bivona et al., 2006; Cho et al., 2016). Acute stimulation of PKG by 8-Br-cGMP increases within 15 min overall PM association and nanoclustering, as well as pERK- and pS473-Akt-output of K-Ras4B. Similarly, FDA approved pharmacological agents sildenafil and metformin can elevate cellular cGMP levels and thus modulate K-Ras4B PM localization by stimulating Ser181-phosphorylation (Cho, Casteel, et al., 2016). Sildenafil inhibits phosphodiesterase 5, thereby blocking the hydrolysis of cGMP, while metformin triggers nitric oxide release, which augments cGMP levels through activation of guanyl cyclase (Davis, Xie, Viollet, & Zou, 2006; Turko, Ballard, Francis, & Corbin, 1999). Of note, computational simulations suggest that phosphorylation does not dislodge the K-Ras4B HVR from the membrane. Instead, K-Ras4B PM localization is lost in the long run probably during the recycling phase. The authors argue that Ser181-phosphorylation reduces the rebinding of "soluble" K-Ras4B-pSer181 to anionic, PS-rich recycling endosomes (Cho, Casteel, et al., 2016). In addition, reduced solubilization by PDE6D, which binds less to the phosphomimetic mutant K-Ras4B-Ser181E, could similarly impact on trafficking phosphorylated K-Ras4B back to the PM. These mechanisms may explain the reduced cancer cell stemness activity of Ser181-phosphomimetic mutants of K-Ras4B (Wang et al., 2015).

PS can be segregated by several proteins, such as cavins (Kovtun, Tillu, Ariotti, Parton, & Collins, 2015). Cavins together with caveolin-1 are the major building blocks of caveolae, membrane pits that can be found in many epithelial cell types (Parton, Tillu, McMahon, & Collins, 2021). Therefore, loss of caveolae, by loss of either of its major building blocks or increased membrane tension that leads to flattening of caveolae and cavin-coat disassembly, releases PS and increases K-Ras4B nanoclustering (Ariotti et al., 2014). Similarly, GTP-H-Ras (or its HVR alone) can sequester PS via non-specific hydrophobic interactions involving its palmitoyl moieties (Zhou & Hancock, 2018). This then likewise negatively modulates K-Ras4B nanoclustering, but not H-Ras nanoclustering (Zhou et al., 2014).

Several inhibitory approaches have targeted the metabolism of sphingomyelin, which regulates endosomal delivery of lipids, notably PS, to the PM. Fendiline, targets acid sphingomyelinase, which leads to accumulation of sphingomyelin and cholesterol in the lysosomal membrane, thus preventing lipid transport and eventually mis-localization of PS from the PM (Cho et al., 2016). Accordingly, fendiline treatment disrupts K-Ras4B PM-localization and -nanoclustering, thus abrogating downstream signaling (Cho, van der Hoeven, et al., 2016). Improved, nanomolar-active derivatives of fendiline showed potent inhibition of PM localization, and blocked oncogenic K-Ras4B transformed MEF cell proliferation and tumor growth of MIA PaCa-2 xenografts in mice (Wang et al., 2021). Staurosporine inhibits in addition to many kinases the protein ORMDL3, a modulator of sphingomyelin biosynthesis, thus reducing PS from the PM and mis-localizing K-Ras4B from the PM (Cho et al., 2012; Maekawa, Lee, Wei,

Ridgway, & Fairn, 2016). Similar effects were observed upon treatment with oxanthroquinone G0, which targets sphingomyelin metabolism and disrupted the localization of PM lipids (Tan, Cho, Neupane, Capon, & Hancock, 2018).

5. Lipid domain and Ras nanocluster modulating factors

5.1 ORP5 and ORP8

Nanoclustering effect: K-Ras4B increasing; K-Ras4A, N-Ras, H-Ras unknown

Given the importance of PS for K-Ras4B PM localization and nanoclustering, the oxysterol binding proteins (OSBP)-related proteins (ORP) ORP5 and ORP8 were nominated as K-Ras4B associated targets (Kattan et al., 2019). ORP5 and ORP8 are ER proteins tethered to PM phosphatidylinositol-4-phosphate (PI_4P) via their pleckstrin homology domains (Chung et al., 2015). They maintain cellular lipid homeostasis by serving as transport proteins exchanging PS from the ER with PI_4P of the PM thus increasing PM PS content.

Therefore, depletion of either ORP5 or ORP8 mis-localized K-Ras4B and PS to endomembranes and ORP8 knockdown further decreased K-Ras4B-G12V nanoclustering and PM abundance (Kattan et al., 2019). However, single- or double-knockdowns had variable effects on pancreatic cancer cell line proliferation and anchorage independent growth, with the strongest effect observed for *KRAS*-dependent MIA PaCa-2 and MOH cells. Surprisingly, the single and dual knockdowns were accompanied by paradoxical pERK- and pS473-Akt- level increases, which was interpreted as a release of negative feedback on upstream components of the MAPK-pathway.

Targeting of PI4IIa, which synthesizes PI_4P, with a specific small molecule inhibitor (compound 7) blocks ORP5/8 action (Kattan et al., 2019). Thus, treatment with this compound phenocopied the negative effect of the ORP5/8 depletion on K-Ras4B and PS PM localization and nanoclustering, as well as on proliferation of *KRAS*-dependent pancreatic cancer cell lines. Consistent with the observed cellular phenotype, ORP5 is significantly upregulated in *KRAS*-mutant pancreatic adenocarcinomas and non-small cell lung cancers. Furthermore, analysis of a pan-cancer cohort of 33 different cancer types revealed an inverse correlation between ORP5 levels and the one-year survival rate (Koga et al., 2008).

5.2 TRPML1 (gene *MCOLN1*)

Nanoclustering effect: H-Ras increasing; K-Ras4B, K-Ras4A, N-Ras, unknown

TRPML1 is one of three transient receptor potential (TRMP) mucolipin group endolysosomal ion channels, which have been implicated in cancer (Xu & Dong, 2021). It transports various cations from lysosomes or the extracellular space into the cytosol and is predominantly regulated by phosphatidylinositides, in particular PI(3,5)P_2. TRPML1 in late endosomes regulates endosomal fusion with lysosomes, thus increasing de-esterification of cholesterol esters, or fusion with the PM for cholesterol recycling back to the PM (Jung et al., 2019).

TRPML1 is upregulated by oncogenic H-Ras and then further stabilizes PM abundance and nanoclustering of both oncogenic and wt H-Ras (Jung et al., 2019). Blocking TRPML1 with a direct inhibitor or by depleting PI(3,5)P_2, selectively reduced the activity of oncogenic H-Ras. Downregulating the TRPML1-encoding gene, *MCOLN1,* reduced ERK1/2 phosphorylation in oncogenic H-Ras expressing cancer cells. Consistently, TRPML1 was essential for sustaining *HRAS*-mutant cancer cell growth. Tumors bearing oncogenic *HRAS* mutations displayed upregulated *MCOLN1*, and higher levels were correlated with poorer patient outcome (Jung et al., 2019).

5.3 Prohibitin-1

Nanoclustering effect: Probably K-Ras4B stabilizing/not N-Ras, H-Ras; K-Ras4A, unknown

Prohibitins (PHBs) are evolutionary highly conserved proteins characterized by a SPFH (stomatin/prohibitin/flotillin/HflK/HflC) domain, which is found in archaeal, bacterial as well as eukaryotic proteins (Wang et al., 2020). In humans the two isoforms, PHB1 and PHB2, act as scaffolding proteins involved in various protein interactions and cellular processes. They are subject to over 70 different post-translational modifications, which regulate their interactions and subcellular localization. PHBs can be found in virtually all cellular compartments, including the plasma membrane and mitochondria (Wang et al., 2020).

Flavaglines, also known as rocaglamides, are natural compounds that have a broad range of anti-cancer, but also neuroprotective and other biologically relevant activities (Wang et al., 2020). Rocaglamide robustly inhibits MAPK-signaling in cells and tumor growth in a *KRAS*-driven lung tumor model without severe side effects (Polier et al., 2012; Yurugi et al., 2020).

Already at nanomolar concentrations rocaglamide reduces Ras-GTP-loading. Consistently, K-Ras4B-G12V-, but not H-RasG12V- or N-RasG12V-nanoclustering is blocked by rocaglamide, which disrupts co-clustering of K-Ras4B with PS and PA (Yurugi et al., 2020). Given that PHB shows affinity for these lipids, as well as its interaction with C-Raf via its C-terminal region, it appears that PHB acts as an accessory protein to K-Ras4B- and C-Raf-containing active nanocluster (Rajalingam et al., 2005; Yurugi et al., 2020). Thus, PHB could further shape the lipid environment in favor of K-Ras4B activation and nanoclustering.

5.4 Galectin-1

Nanoclustering effect: H-RasG12V stabilizing/N-RasG12V neutral/ K-Ras4B-G12V destabilizing; K-Ras4A unknown

Galectin-1 (Gal-1) and galectin-3 (Gal-3) are to date the best profiled modulators of Ras nanoclustering. Galectins are a family of soluble proteins with at least one conserved carbohydrate recognition domain, which binds to β-galactoside containing glycoconjugates intracellularly, on the cell surface or in the extracellular matrix (Johannes, Jacob, & Leffler, 2018). They mediate cell-cell and cell-matrix interactions depending on the glycosylation status and may thereby reflect nutritional availability (Johannes et al., 2018). Galectins have been linked to different pathologies, including various cancers (Timoshenko, 2015).

Initially it was proposed that Gal-1 increases GTP-H-Ras nanocluster and downstream MAPK-, but not PI3K/Akt-signaling, by binding to the farnesyl-tail of H-Ras (Belanis, Plowman, Rotblat, Hancock, & Kloog, 2008; Elad-Sfadia, Haklai, Ballan, Gabius, & Kloog, 2002; Rotblat et al., 2010). However, as the examples of PDE6D and CaM show, such a protein would solubilize Ras independently of whether GTP or GDP is bound. In agreement with an indirect complexation, direct interactions of Gal-1 with farnesylated Ras-constructs could not be demonstrated by two independent groups (Blazevits et al., 2016; Mejuch, van Hattum, Triola, Jaiswal, & Waldmann, 2015).

Instead, Gal-1 binds the RBD of Ras effector proteins with sub-micromolar affinity and exhibits a further preference for full length B-Raf in cells (Blazevits et al., 2016). Given that Gal-1 enhances Ras signaling, this interaction does not interfere with Raf or RBD/GTP-Ras binding. Likewise, a cell permeable inhibitor of the carbohydrate recognition domain did not affect binding of Gal-1 to the RBD of C-Raf-RBD (Blazevits et al., 2016). Computational docking provided some insight into the Gal-1/RBD interface, which was validated by mutating interacting residues D113 and D117 on the RBD of C-Raf.

The prototypical Gal-1 dimerizes via an interface comprising both the N- and C-terminus ($K_d = 7\,\mu M$) (Cho & Cummings, 1995). Mutational disruption of this interface abolished the H-Ras nanoclustering and MAPK-activity (Blazevits et al., 2016), suggesting that dimeric Gal-1 scaffolds dimeric Raf-proteins, by binding to their RBD in order to augment H-Ras nanoclustering.

At the same time, Gal-1 has a negative effect on K-Ras4B nanoclustering (Blazevits et al., 2016; Posada et al., 2016). It is currently not understood, how this Ras isoform specific activity is realized. However, evidence suggests that ultimately distinct preferred conformations of Ras on the PM are involved (Abankwa et al., 2010).

Current galectin-directed drug development efforts focus on targeting its carbohydrate binding pocket (Blanchard, Bum-Erdene, Bohari, & Yu, 2016; Modenutti, Capurro, Di Lella, & Marti, 2019). Such compounds would not compete with the interaction surface of Gal-1 and the RBD, which would rather require a protein-protein interface inhibitor approach.

5.5 Galectin-3

Nanoclustering effect: K-Ras4B stabilizing; N-Ras, H-Ras, K-Ras4A unknown

Like Gal-1, Gal-3 can form dimers or multimers through its unique N-terminal 120 amino acid extension. Pentamers and hexamers of Gal-3, are suggested to be involved in the functioning of immune cells (Farhad, Rolig, & Redmond, 2018).

While Gal-1 increases H-RasG12V-nanoclustering, Gal-3 specifically increases K-Ras4B-G12V nanoclustering (Plowman et al., 2008). An analogous farnesyl-binding activity was proposed, but again no binding to farnesylated-Ras constructs could be detected (Mejuch et al., 2015; Shalom-Feuerstein et al., 2008). Instead, Gal-3 binds to the C-Raf-RBD and PI3Kα-RBD like Gal-1, suggesting that it may have a similar Raf-dependent mechanism of action as Gal-1 (Blazevits et al., 2016). Gal-3 increases signaling output with increased GTP-Ras levels in addition to C-Raf- and PI3K-activities (Elad-Sfadia, Haklai, Balan, & Kloog, 2004).

Additional galectins may impact on Ras-signaling, such as galectin-8 and galectin-7 that were recently identified as part of the K-Ras4B interactome using proximity-proteomics (Kovalski et al., 2019).

5.6 ASPP2

Nanoclustering effect: H-RasG12V, N-RasG12V, K-Ras4B-G12V stabilizing; K-Ras4A unknown

Apoptosis-stimulating of p53 protein (ASPP) family member 2 is a novel positive pan-RasG12V nanocluster modulator, hence the first nanocluster modulator identified for N-Ras (Posada et al., 2016). Like Gal-1, it increases engagement of the RBD of C-Raf by RasG12V and downstream pERK1/2 output, but different from Gal-1, it also increases pAkt levels.

ASPP2 contains a N-terminal ubiquitin like domain structurally similar to the RBD or RA domains of Ras effectors. Others have shown that only full length, but not N-terminally deleted ASPP2 coimmunoprecipitates with H-RasG12V, increases B/C-Raf-heterodimerization and -activation and increases pERK levels, thus potentiating H-RasG12V-induced senescence (Wang et al., 2013). Increased MAPK-activity is needed to phosphorylate ASPP2, which triggers its translocation to the nucleus where it associates with p53 via its C-terminal domain to promote apoptosis or senescence (Godin-Heymann, Wang, Slee, & Lu, 2013).

However, deletion of N- and C-terminal ends of ASPP2 had no effect on promoting Ras nanoclustering, therefore suggesting that the middle α- helical domain was needed to mediate Ras nanocluster formation (Posada et al., 2016). When Gal-1 and ASPP2 were co-expressed, the Gal-1 or ASPP2-mediated increase in nanoclustering-dependent FRET of oncogenic H-RasG12V was neutralized (Blazevits et al., 2016; Posada et al., 2016). Similarly, the inhibitory effect of Gal-1 on K-Ras4B-G12V nanoclustering was overridden by the nanocluster increasing effect of ASPP2. In addition, Posada et al. showed that tumorosphere formation that is promoted by Gal-1 in Ras wild-type and H-Ras mutant cancer cells, is almost completely abrogated by ASPP2 co-expression.

An interesting connection between the nanocluster-mediating helical domain of ASPP2 and RASSF proteins exists. RASSF are a large family of putative Ras effectors with an RA-domain. They mediate pro-apoptotic signaling via the Hippo pathway (Badouel & McNeill, 2011). While only RASSF5 strongly binds to active H-Ras and K-Ras4B as effector, three (RASSF8-10) of the four so called N-RASSF with N-terminal RA domain formed oligomers with ASPP2 and its sibling ASPP1 (Dhanaraman et al., 2020). At the same time, this subgroup did not bind to the Hippo kinase MST1, suggesting that their oncogenic Ras-induced pro-apoptotic activity is mediated via ASPP1/2. As both N-RASSF and ASPP1/2 possess a similar coiled-coil region, it is plausible to assume that they interact via this region to mediate between Ras-binding and p53-apoptosis signaling. However, it is currently unclear, whether the detected weak interaction of RASSF8 with active K-Ras would be sufficient to engage this pathway (Dhanaraman et al., 2020).

5.7 Nucleophosmin 1 (NPM1)

Nanoclustering effect: K-Ras4B stabilizing, no binding to H-Ras; N-Ras, K-Ras4A unknown

NPM1 is frequently mutated in acute myeloid leukemia (AML), with C-terminal mutations affecting its nuclear localization signal, leading to abnormal cytoplasmic distribution of the protein (Bolli et al., 2007). Thus, the normal interactions with nucleolin become perturbed (Sasinkova, Holoubek, Otevrelova, Kuzelova, & Brodska, 2018). Both NPM1 and nucleolin were identified as K-Ras4B interacting proteins in a proteomics approach that employed the HVR of K-Ras4B as bait (Inder et al., 2009).

Each of these proteins alone can increase PM localization and NPM1 in addition increases nanoclustering of wildtype and oncogenic K-Ras4B. It is not fully understood, how the natively nucleolar NPM1 affects K-Ras4B at the PM. However, in the instance where the nuclear envelope is resolved during mitosis, an interaction possibility between NPM1 and K-Ras4B would emerge (Chippalkatti & Abankwa, 2021). In this context it may be interesting to note that a non-farnesylated K-Ras4B mutant binds strongest to NPM1 in immunoprecipitation assays.

Given that the interaction between K-Ras4B and NPM1 or nucleolin increases upon EGF-stimulation, suggests some effector associated processes for their engagement. Consistently, the PM fraction of K-Ras4B-G12V contains in addition to Raf also NPM1. Moreover, the interaction between the HVR of K-Ras4B and NPM1 was mapped to the non-polar N-terminal 119 residues of NPM1 (Inder, Hill, & Hancock, 2010). This potential biochemical mismatch further suggests an indirect adaptor mediated interaction. Similar to what is observed for CaM, the phosphomimetic Ser181-mutants of K-Ras4B bind somewhat less to NPM1, which may relate to perturbed interaction or colocalization of these constructs.

The NPM1 N-terminus harbors a nuclear exclusion signal and mediates oligomerization and nuclear chaperone functions (Grisendi, Mecucci, Falini, & Pandolfi, 2006). This domain and the chaperone activity are shared by the family of nucloeplasmins, which prevent protein aggregation in the dense nucleolar environment. Given the interaction mapping data, it would be interesting to know, whether other family members affect K-Ras4B specifically.

5.8 DIRAS3

Nanoclustering effect: K-Ras4B-G12V perturbing, probably effect on H-RasG12V/probably not on N-RasG12V; K-Ras4A unknown

DIRAS3 is a small GTPase with over 50% homology to Ras differing mostly by a functionally necessary N-terminal extension (Sutton et al., 2019). Moreover, it contains sequence alterations that would be detrimental for GTPase activity and effector recruitment (Luo et al., 2003). It is a tumor suppressor encoded by the maternally imprinted *ARHI* gene, i.e., genetic or epigenetic events in the single active allele can trigger a complete loss of function.

Both K-Ras4B and DIRAS3 partially colocalize at the PM and within nanocluster. Thus, DIRAS3 suppresses both K-Ras4B-G12V and H-RasG12V transforming activities, unless its unique N-terminus is removed (Sutton et al., 2019). Furthermore, DIRAS3 blocks Ras-GTP-loading and pERK output.

DIRAS3 forms a high affinity complex with K-Ras4B (12.1 ± 0.3 nM) that is partially dependent on helix α5 of the latter, but largely independent of the unique N-terminus of DIRAS3. Yet, only full length DIRAS3 blocks K-Ras4B multimerization, native nanoclustering and pERK-signaling, while removal of this N-terminus abrogates these properties. Hence, K-Ras4B nanoclustering is not abolished but altered for smaller cluster radii with full length DIRAS3, but larger cluster radii with N-terminally truncated DIRAS3. Thus, the most consistent perturbation that explains the signaling effects appears to be the disruption of Ras-Raf interaction, which depends on the N-terminus of DIRAS3, as did transforming and tumor suppressive activities.

6. Concluding remarks

Most cancers emerge from epithelia, which are characterized by at least one layer of polarized cells, typically with a luminal apical, a basement membrane facing basal side and laterally connected by tight junctions (Martin-Belmonte & Perez-Moreno, 2011). This polarized phenotype is established and maintained by the integration of extracellular cues, such as growth factor input, and cell-attachment interactions. Intracellular polarized transport processes determine the often unequal apico-basal membrane lipid composition, for instance with $PI(4,5)P_2$ demarcating the apical and $PI(3,4,5)P_3$ the basolateral membrane (Gassama-Diagne et al., 2006; Martin-Belmonte et al., 2007). In addition, many proteins are asymmetrically distributed, notably caveolae on the basal side, and signaling proteins that are often instructing the plane of mitosis and cell division (Martin-Belmonte & Perez-Moreno, 2011). Therefore, the entire cellular machinery is adjusted for polarization and thus specializations that are critical during cell fate determination (Bonifacino, 2014; Chippalkatti & Abankwa, 2021).

It is not well understood whether Ras proteins are distributed or activated in a polarized fashion. However, in neurons with a somewhat extreme polarization of apical axon and basal soma and dendrites, H-Ras activity localizes predominantly to the axon, while that of K-Ras4B is found in soma and dendrites (Fivaz, Bandara, Inoue, & Meyer, 2008). The H-Ras activity is maintained via its axonal transport and a phosphatidylinositide-dependent positive feedback loop downstream of its effector PI3K. Conversely, SNAP23, which was implicated in K-Ras4B PM delivery is highly enriched in the soma and dendrites (Che et al., 2019; Suh et al., 2010). Moreover, PS clustering is modulated by curvature, which is for instance high in dendritic spines, and by other lipids as well as the membrane potential (Zhou et al., 2021), suggesting in particular in neurons several spatio-temporally fine-tuned instances of K-Ras4B nanocluster regulation by PS. The consequences of such fine regulation are however unclear in these postmitotic cells.

Moreover, certain organelles of polarized cells, such as the apical primary cilium, depend on specific trafficking processes. PDE6D, a trafficking chaperone of K-Ras4B, shuttles several proteins into the primary cilium (Fansa, Kosling, Zent, Wittinghofer, & Ismail, 2016). The fate of K-Ras4B in this context has not been explored.

A hallmark of oncogenic Ras activity is the loss of polarity in epithelial cells (Schoenenberger, Zuk, Kendall, & Matlin, 1991), which strongly suggests Ras as a master regulator of polarity. Therefore, it is plausible to assume that cell polarization associated processes add additional layers of complexity to Ras signaling that need further investigation in order to devise new targeted interventions against Ras in cancer.

References

Abankwa, D., & Gorfe, A. A. (2020). Mechanisms of Ras membrane organization and signaling: Ras rocks again. *Biomolecules*, *10*(11), 1522. https://doi.org/10.3390/biom10111522.

Abankwa, D., Gorfe, A. A., Inder, K., & Hancock, J. F. (2010). Ras membrane orientation and nanodomain localization generate isoform diversity. *Proceedings of the National Academy of Sciences of the United States of America*, *107*(3), 1130–1135. https://doi.org/10.1073/pnas.0903907107.

Abankwa, D., Hanzal-Bayer, M., Ariotti, N., Plowman, S. J., Gorfe, A. A., Parton, R. G., et al. (2008). A novel switch region regulates H-ras membrane orientation and signal output. *The EMBO Journal, 27*(5), 727–735. https://doi.org/10.1038/emboj.2008.10.

Agamasu, C., Ghirlando, R., Taylor, T., Messing, S., Tran, T. H., Bindu, L., et al. (2019). KRAS prenylation is required for bivalent binding with calmodulin in a nucleotide-independent manner. *Biophysical Journal, 116*(6), 1049–1063. https://doi.org/10.1016/j.bpj.2019.02.004.

Ahearn, I., Zhou, M., & Philips, M. R. (2018). Posttranslational modifications of RAS proteins. *Cold Spring Harbor Perspectives in Medicine, 8*(11), a031484. https://doi.org/10.1101/cshperspect.a031484.

Ahearn, I. M., Court, H. R., Siddiqui, F., Abankwa, D., & Philips, M. R. (2021). NRAS is unique among RAS proteins in requiring ICMT for trafficking to the plasma membrane. *Life Science Alliance, 4*(5), e202000972. https://doi.org/10.26508/lsa.202000972.

Ariotti, N., Fernandez-Rojo, M. A., Zhou, Y., Hill, M. M., Rodkey, T. L., Inder, K. L., et al. (2014). Caveolae regulate the nanoscale organization of the plasma membrane to remotely control Ras signaling. *The Journal of Cell Biology, 204*(5), 777–792. https://doi.org/10.1083/jcb.201307055.

Badouel, C., & McNeill, H. (2011). SnapShot: The hippo signaling pathway. *Cell, 145*(3), 484–484 e481. https://doi.org/10.1016/j.cell.2011.04.009.

Belanis, L., Plowman, S. J., Rotblat, B., Hancock, J. F., & Kloog, Y. (2008). Galectin-1 is a novel structural component and a major regulator of h-ras nanoclusters. *Molecular Biology of the Cell, 19*(4), 1404–1414. https://doi.org/10.1091/mbc.E07-10-1053.

Berchtold, M. W., & Villalobo, A. (2014). The many faces of calmodulin in cell proliferation, programmed cell death, autophagy, and cancer. *Biochimica et Biophysica Acta, 1843*(2), 398–435. https://doi.org/10.1016/j.bbamcr.2013.10.021.

Berndt, N., Hamilton, A. D., & Sebti, S. M. (2011). Targeting protein prenylation for cancer therapy. *Nature Reviews. Cancer, 11*(11), 775–791. https://doi.org/10.1038/nrc3151.

Bivona, T. G., Quatela, S. E., Bodemann, B. O., Ahearn, I. M., Soskis, M. J., Mor, A., et al. (2006). PKC regulates a farnesyl-electrostatic switch on K-Ras that promotes its association with Bcl-XL on mitochondria and induces apoptosis. *Molecular Cell, 21*(4), 481–493. https://doi.org/10.1016/j.molcel.2006.01.012.

Blanchard, H., Bum-Erdene, K., Bohari, M. H., & Yu, X. (2016). Galectin-1 inhibitors and their potential therapeutic applications: A patent review. *Expert Opinion on Therapeutic Patents, 26*(5), 537–554. https://doi.org/10.1517/13543776.2016.1163338.

Blazevits, O., Mideksa, Y. G., Solman, M., Ligabue, A., Ariotti, N., Nakhaeizadeh, H., et al. (2016). Galectin-1 dimers can scaffold Raf-effectors to increase H-ras nanoclustering. *Scientific Reports, 6*, 24165. https://doi.org/10.1038/srep24165.

Bolli, N., Nicoletti, I., De Marco, M. F., Bigerna, B., Pucciarini, A., Mannucci, R., et al. (2007). Born to be exported: COOH-terminal nuclear export signals of different strength ensure cytoplasmic accumulation of nucleophosmin leukemic mutants. *Cancer Research, 67*(13), 6230–6237. https://doi.org/10.1158/0008-5472.CAN-07-0273.

Bonifacino, J. S. (2014). Adaptor proteins involved in polarized sorting. *The Journal of Cell Biology, 204*(1), 7–17. https://doi.org/10.1083/jcb.201310021.

Burrows, J. F., Kelvin, A. A., McFarlane, C., Burden, R. E., McGrattan, M. J., De la Vega, M., et al. (2009). USP17 regulates Ras activation and cell proliferation by blocking RCE1 activity. *The Journal of Biological Chemistry, 284*(14), 9587–9595. https://doi.org/10.1074/jbc.M807216200.

Casique-Aguirre, D., Briseno-Diaz, P., Garcia-Gutierrez, P., la Rosa, C. H. G., Quintero-Barceinas, R. S., Rojo-Dominguez, A., et al. (2018). KRas4B-PDE6delta complex stabilization by small molecules obtained by virtual screening affects Ras signaling in pancreatic cancer. *BMC Cancer, 18*(1), 1299. https://doi.org/10.1186/s12885-018-5142-7.

Chandra, A., Grecco, H. E., Pisupati, V., Perera, D., Cassidy, L., Skoulidis, F., et al. (2011). The GDI-like solubilizing factor PDEdelta sustains the spatial organization and signalling of Ras family proteins. *Nature Cell Biology*, *14*(2), 148–158. https://doi.org/10.1038/ncb2394.

Che, Y., Siprashvili, Z., Kovalski, J. R., Jiang, T., Wozniak, G., Elcavage, L., et al. (2019). KRAS regulation by small non-coding RNAs and SNARE proteins. *Nature Communications*, *10*(1), 5118. https://doi.org/10.1038/s41467-019-13106-4.

Chen, X., Yao, H., Kashif, M., Revechon, G., Eriksson, M., Hu, J., et al. (2021). A small-molecule ICMT inhibitor delays senescence of Hutchinson-Gilford progeria syndrome cells. *eLife*, *10*, e63284. https://doi.org/10.7554/eLife.63284.

Chippalkatti, R., & Abankwa, D. (2021). Promotion of cancer cell stemness by Ras. *Biochemical Society Transactions*, *49*(1), 467–476. https://doi.org/10.1042/BST20200964.

Cho, K. J., Casteel, D. E., Prakash, P., Tan, L., van der Hoeven, D., Salim, A. A., et al. (2016). AMPK and endothelial nitric oxide synthase signaling regulates K-Ras plasma membrane interactions via cyclic gmp-dependent protein kinase 2. *Molecular and Cellular Biology*, *36*(24), 3086–3099. https://doi.org/10.1128/MCB.00365-16.

Cho, M., & Cummings, R. D. (1995). Galectin-1, a beta-galactoside-binding lectin in Chinese hamster ovary cells. I. Physical and chemical characterization. *The Journal of Biological Chemistry*, *270*(10), 5198–5206. https://doi.org/10.1074/jbc.270.10.5198.

Cho, K. J., & Hancock, J. F. (2013). Ras nanoclusters: A new drug target? *Small GTPases*, *4*(1), 57–60. https://doi.org/10.4161/sgtp.23145.

Cho, K. J., Park, J. H., Piggott, A. M., Salim, A. A., Gorfe, A. A., Parton, R. G., et al. (2012). Staurosporines disrupt phosphatidylserine trafficking and mislocalize Ras proteins. *The Journal of Biological Chemistry*, *287*(52), 43573–43584. https://doi.org/10.1074/jbc.M112.424457.

Cho, K. J., van der Hoeven, D., Zhou, Y., Maekawa, M., Ma, X., Chen, W., et al. (2016). Inhibition of acid sphingomyelinase depletes cellular phosphatidylserine and mislocalizes K-Ras from the plasma membrane. *Molecular and Cellular Biology*, *36*(2), 363–374. https://doi.org/10.1128/MCB.00719-15.

Chung, J. K., Lee, Y. K., Denson, J. P., Gillette, W. K., Alvarez, S., Stephen, A. G., et al. (2018). K-Ras4B remains monomeric on membranes over a wide range of surface densities and lipid compositions. *Biophysical Journal*, *114*(1), 137–145. https://doi.org/10.1016/j.bpj.2017.10.042.

Chung, J., Torta, F., Masai, K., Lucast, L., Czapla, H., Tanner, L. B., et al. (2015). Intracellular transport. PI4P/phosphatidylserine countertransport at ORP5- and ORP8-mediated ER-plasma membrane contacts. *Science*, *349*(6246), 428–432. https://doi.org/10.1126/science.aab1370.

Cox, A. D., Der, C. J., & Philips, M. R. (2015). Targeting RAS membrane association: Back to the future for anti-RAS drug discovery? *Clinical Cancer Research*, *21*(8), 1819–1827. https://doi.org/10.1158/1078-0432.CCR-14-3214.

Dai, Q., Choy, E., Chiu, V., Romano, J., Slivka, S. R., Steitz, S. A., et al. (1998). Mammalian prenylcysteine carboxyl methyltransferase is in the endoplasmic reticulum. *The Journal of Biological Chemistry*, *273*(24), 15030–15034. https://doi.org/10.1074/jbc.273.24.15030.

Davis, B. J., Xie, Z., Viollet, B., & Zou, M. H. (2006). Activation of the AMP-activated kinase by antidiabetes drug metformin stimulates nitric oxide synthesis in vivo by promoting the association of heat shock protein 90 and endothelial nitric oxide synthase. *Diabetes*, *55*(2), 496–505. https://doi.org/10.2337/diabetes.55.02.06.db05-1064.

de la Vega, M., Burrows, J. F., McFarlane, C., Govender, U., Scott, C. J., & Johnston, J. A. (2010). The deubiquitinating enzyme USP17 blocks N-Ras membrane trafficking and activation but leaves K-Ras unaffected. *The Journal of Biological Chemistry*, *285*(16), 12028–12036. https://doi.org/10.1074/jbc.M109.081448.

Dekker, F. J., Rocks, O., Vartak, N., Menninger, S., Hedberg, C., Balamurugan, R., et al. (2010). Small-molecule inhibition of APT1 affects Ras localization and signaling. *Nature Chemical Biology, 6*(6), 449–456. https://doi.org/10.1038/nchembio.362.

Dhanaraman, T., Singh, S., Killoran, R. C., Singh, A., Xu, X., Shifman, J. M., et al. (2020). RASSF effectors couple diverse RAS subfamily GTPases to the Hippo pathway. *Science Signaling, 13*(653), eabb4778. https://doi.org/10.1126/scisignal.abb4778.

Dharmaiah, S., Bindu, L., Tran, T. H., Gillette, W. K., Frank, P. H., Ghirlando, R., et al. (2016). Structural basis of recognition of farnesylated and methylated KRAS4b by PDEdelta. *Proceedings of the National Academy of Sciences of the United States of America, 113*(44), E6766–E6775. https://doi.org/10.1073/pnas.1615316113.

Elad-Sfadia, G., Haklai, R., Balan, E., & Kloog, Y. (2004). Galectin-3 augments K-Ras activation and triggers a Ras signal that attenuates ERK but not phosphoinositide 3-kinase activity. *The Journal of Biological Chemistry, 279*(33), 34922–34930. https://doi.org/10.1074/jbc.M312697200.

Elad-Sfadia, G., Haklai, R., Ballan, E., Gabius, H. J., & Kloog, Y. (2002). Galectin-1 augments Ras activation and diverts Ras signals to Raf-1 at the expense of phosphoinositide 3-kinase. *The Journal of Biological Chemistry, 277*(40), 37169–37175. https://doi.org/10.1074/jbc.M205698200.

Falaleeva, M., & Stamm, S. (2013). Processing of snoRNAs as a new source of regulatory non-coding RNAs: snoRNA fragments form a new class of functional RNAs. *BioEssays, 35*(1), 46–54. https://doi.org/10.1002/bies.201200117.

Fang, Z., Lee, K. Y., Huo, K. G., Gasmi-Seabrook, G., Zheng, L., Moghal, N., et al. (2020). Multivalent assembly of KRAS with the RAS-binding and cysteine-rich domains of CRAF on the membrane. *Proceedings of the National Academy of Sciences of the United States of America, 117*(22), 12101–12108. https://doi.org/10.1073/pnas.1914076117.

Fansa, E. K., Kosling, S. K., Zent, E., Wittinghofer, A., & Ismail, S. (2016). PDE6delta-mediated sorting of INPP5E into the cilium is determined by cargo-carrier affinity. *Nature Communications*, 7, 11366. https://doi.org/10.1038/ncomms11366.

Farhad, M., Rolig, A. S., & Redmond, W. L. (2018). The role of Galectin-3 in modulating tumor growth and immunosuppression within the tumor microenvironment. *Oncoimmunology*, 7(6), e1434467. https://doi.org/10.1080/2162402X.2018.1434467.

Fehrenbacher, N., Tojal da Silva, I., Ramirez, C., Zhou, Y., Cho, K. J., Kuchay, S., et al. (2017). The G protein-coupled receptor GPR31 promotes membrane association of KRAS. *The Journal of Cell Biology, 216*(8), 2329–2338. https://doi.org/10.1083/jcb.201609096.

Fivaz, M., Bandara, S., Inoue, T., & Meyer, T. (2008). Robust neuronal symmetry breaking by Ras-triggered local positive feedback. *Current Biology, 18*(1), 44–50. https://doi.org/10.1016/j.cub.2007.11.051.

Fivaz, M., & Meyer, T. (2005). Reversible intracellular translocation of KRas but not HRas in hippocampal neurons regulated by Ca2+/calmodulin. *The Journal of Cell Biology, 170*(3), 429–441. https://doi.org/10.1083/jcb.200409157.

Gassama-Diagne, A., Yu, W., ter Beest, M., Martin-Belmonte, F., Kierbel, A., Engel, J., et al. (2006). Phosphatidylinositol-3,4,5-trisphosphate regulates the formation of the basolateral plasma membrane in epithelial cells. *Nature Cell Biology, 8*(9), 963–970. https://doi.org/10.1038/ncb1461.

Goda, Y. (1997). SNAREs and regulated vesicle exocytosis. *Proceedings of the National Academy of Sciences of the United States of America, 94*(3), 769–772. https://doi.org/10.1073/pnas.94.3.769.

Godin-Heymann, N., Wang, Y., Slee, E., & Lu, X. (2013). Phosphorylation of ASPP2 by RAS/MAPK pathway is critical for its full pro-apoptotic function. *PLoS One, 8*(12), e82022. https://doi.org/10.1371/journal.pone.0082022.

Goodwin, J. S., Drake, K. R., Rogers, C., Wright, L., Lippincott-Schwartz, J., Philips, M. R., et al. (2005). Depalmitoylated Ras traffics to and from the Golgi complex via a nonvesicular pathway. *The Journal of Cell Biology*, *170*(2), 261–272. https://doi.org/10.1083/jcb.200502063.

Gorfe, A. A., Hanzal-Bayer, M., Abankwa, D., Hancock, J. F., & McCammon, J. A. (2007). Structure and dynamics of the full-length lipid-modified H-Ras protein in a 1,2-dimyristoylglycero-3-phosphocholine bilayer. *Journal of Medicinal Chemistry*, *50*(4), 674–684. https://doi.org/10.1021/jm061053f.

Grant, B. M. M., Enomoto, M., Back, S. I., Lee, K. Y., Gebregiworgis, T., Ishiyama, N., et al. (2020). Calmodulin disrupts plasma membrane localization of farnesylated KRAS4b by sequestering its lipid moiety. *Science Signaling*, *13*(625), eaaz0344. https://doi.org/10.1126/scisignal.aaz0344.

Grisendi, S., Mecucci, C., Falini, B., & Pandolfi, P. P. (2006). Nucleophosmin and cancer. *Nature Reviews. Cancer*, *6*(7), 493–505. https://doi.org/10.1038/nrc1885.

Guzman, C., Oetken-Lindholm, C., & Abankwa, D. (2016). Automated high-throughput fluorescence lifetime imaging microscopy to detect protein-protein interactions. *Journal of Laboratory Automation*, *21*(2), 238–245. https://doi.org/10.1177/2211068215606048.

Guzman, C., Solman, M., Ligabue, A., Blazevits, O., Andrade, D. M., Reymond, L., et al. (2014). The efficacy of Raf kinase recruitment to the GTPase H-ras depends on H-ras membrane conformer-specific nanoclustering. *The Journal of Biological Chemistry*, *289*(14), 9519–9533. https://doi.org/10.1074/jbc.M113.537001.

Hampton, S. E., Dore, T. M., & Schmidt, W. K. (2018). Rce1: Mechanism and inhibition. *Critical Reviews in Biochemistry and Molecular Biology*, *53*(2), 157–174. https://doi.org/10.1080/10409238.2018.1431606.

Inder, K. L., Hill, M. M., & Hancock, J. F. (2010). Nucleophosmin and nucleolin regulate K-Ras signaling. *Communicative & Integrative Biology*, *3*(2), 188–190. https://doi.org/10.4161/cib.3.2.10923.

Inder, K. L., Lau, C., Loo, D., Chaudhary, N., Goodall, A., Martin, S., et al. (2009). Nucleophosmin and nucleolin regulate K-Ras plasma membrane interactions and MAPK signal transduction. *The Journal of Biological Chemistry*, *284*(41), 28410–28419. https://doi.org/10.1074/jbc.M109.001537.

Ismail, S. A., Chen, Y. X., Rusinova, A., Chandra, A., Bierbaum, M., Gremer, L., et al. (2011). Arl2-GTP and Arl3-GTP regulate a GDI-like transport system for farnesylated cargo. *Nature Chemical Biology*, 7(12), 942–949. https://doi.org/10.1038/nchembio.686.

Janosi, L., Li, Z., Hancock, J. F., & Gorfe, A. A. (2012). Organization, dynamics, and segregation of Ras nanoclusters in membrane domains. *Proceedings of the National Academy of Sciences of the United States of America*, *109*(21), 8097–8102. https://doi.org/10.1073/pnas.1200773109.

Jaworski, J., Govender, U., McFarlane, C., de la Vega, M., Greene, M. K., Rawlings, N. D., et al. (2014). A novel RCE1 isoform is required for H-Ras plasma membrane localization and is regulated by USP17. *The Biochemical Journal*, *457*(2), 289–300. https://doi.org/10.1042/BJ20131213.

Jin, T., Lavoie, H., Sahmi, M., David, M., Hilt, C., Hammell, A., et al. (2017). RAF inhibitors promote RAS-RAF interaction by allosterically disrupting RAF autoinhibition. *Nature Communications*, *8*(1), 1211. https://doi.org/10.1038/s41467-017-01274-0.

Johannes, L., Jacob, R., & Leffler, H. (2018). Galectins at a glance. *Journal of Cell Science*, *131*(9), jcs208884. https://doi.org/10.1242/jcs.208884.

Judd, W. R., Slattum, P. M., Hoang, K. C., Bhoite, L., Valppu, L., Alberts, G., et al. (2011). Discovery and SAR of methylated tetrahydropyranyl derivatives as inhibitors of isoprenylcysteine carboxyl methyltransferase (ICMT). *Journal of Medicinal Chemistry*, *54*(14), 5031–5047. https://doi.org/10.1021/jm200249a.

Jung, J., Cho, K. J., Naji, A. K., Clemons, K. N., Wong, C. O., Villanueva, M., et al. (2019). HRAS-driven cancer cells are vulnerable to TRPML1 inhibition. *EMBO Reports*, *20*(4). https://doi.org/10.15252/embr.201846685.

Kattan, W. E., Chen, W., Ma, X., Lan, T. H., van der Hoeven, D., van der Hoeven, R., et al. (2019). Targeting plasma membrane phosphatidylserine content to inhibit oncogenic KRAS function. *Life Science Alliance*, *2*(5), e201900431. https://doi.org/10.26508/lsa.201900431.

Kiel, C., Matallanas, D., & Kolch, W. (2021). The ins and outs of RAS effector complexes. *Biomolecules*, *11*(2), 236. https://doi.org/10.3390/biom11020236.

Koga, Y., Ishikawa, S., Nakamura, T., Masuda, T., Nagai, Y., Takamori, H., et al. (2008). Oxysterol binding protein-related protein-5 is related to invasion and poor prognosis in pancreatic cancer. *Cancer Science*, *99*(12), 2387–2394. https://doi.org/10.1111/j.1349-7006.2008.00987.x.

Kovalski, J. R., Bhaduri, A., Zehnder, A. M., Neela, P. H., Che, Y., Wozniak, G. G., et al. (2019). The functional proximal proteome of oncogenic Ras includes mTORC2. *Molecular Cell*, *73*(4), 830–844 e812. https://doi.org/10.1016/j.molcel.2018.12.001.

Kovtun, O., Tillu, V. A., Ariotti, N., Parton, R. G., & Collins, B. M. (2015). Cavin family proteins and the assembly of caveolae. *Journal of Cell Science*, *128*(7), 1269–1278. https://doi.org/10.1242/jcs.167866.

Lakshman, B., Messing, S., Schmid, E. M., Clogston, J. D., Gillette, W. K., Esposito, D., et al. (2019). Quantitative biophysical analysis defines key components modulating recruitment of the GTPase KRAS to the plasma membrane. *The Journal of Biological Chemistry*, *294*(6), 2193–2207. https://doi.org/10.1074/jbc.RA118.005669.

Laude, A. J., & Prior, I. A. (2008). Palmitoylation and localisation of RAS isoforms are modulated by the hypervariable linker domain. *Journal of Cell Science*, *121*(Pt. 4), 421–427. https://doi.org/10.1242/jcs.020107.

Lerner, E. C., Zhang, T. T., Knowles, D. B., Qian, Y., Hamilton, A. D., & Sebti, S. M. (1997). Inhibition of the prenylation of K-Ras, but not H- or N-Ras, is highly resistant to CAAX peptidomimetics and requires both a farnesyltransferase and a geranylgeranyltransferase I inhibitor in human tumor cell lines. *Oncogene*, *15*(11), 1283–1288. https://doi.org/10.1038/sj.onc.1201296.

Li, Z. L., & Buck, M. (2017). Computational modeling reveals that signaling lipids modulate the orientation of K-Ras4A at the membrane reflecting protein topology. *Structure*, *25*(4), 679–689 e672. https://doi.org/10.1016/j.str.2017.02.007.

Li, Z., Janosi, L., & Gorfe, A. A. (2012). Formation and domain partitioning of H-ras peptide nanoclusters: Effects of peptide concentration and lipid composition. *Journal of the American Chemical Society*, *134*(41), 17278–17285. https://doi.org/10.1021/ja307716z.

Lin, D. T., & Conibear, E. (2015). ABHD17 proteins are novel protein depalmitoylases that regulate N-Ras palmitate turnover and subcellular localization. *eLife*, *4*, e11306. https://doi.org/10.7554/eLife.11306.

Luo, R. Z., Fang, X., Marquez, R., Liu, S. Y., Mills, G. B., Liao, W. S., et al. (2003). ARHI is a Ras-related small G-protein with a novel N-terminal extension that inhibits growth of ovarian and breast cancers. *Oncogene*, *22*(19), 2897–2909. https://doi.org/10.1038/sj.onc.1206380.

Maekawa, M., Lee, M., Wei, K., Ridgway, N. D., & Fairn, G. D. (2016). Staurosporines decrease ORMDL proteins and enhance sphingomyelin synthesis resulting in depletion of plasmalemmal phosphatidylserine. *Scientific Reports*, *6*, 35762. https://doi.org/10.1038/srep35762.

Manoharan, G. B., Kopra, K., Eskonen, V., Harma, H., & Abankwa, D. (2019). High-throughput amenable fluorescence-assays to screen for calmodulin-inhibitors. *Analytical Biochemistry*, *572*, 25–32. https://doi.org/10.1016/j.ab.2019.02.015.

Marin-Ramos, N. I., Balabasquer, M., Ortega-Nogales, F. J., Torrecillas, I. R., Gil-Ordonez, A., Marcos-Ramiro, B., et al. (2019). A potent isoprenylcysteine carboxylmethyltransferase (ICMT) inhibitor improves survival in ras-driven acute myeloid leukemia. *Journal of Medicinal Chemistry*, *62*(13), 6035–6046. https://doi.org/10.1021/acs.jmedchem.9b00145.

Martin-Belmonte, F., Gassama, A., Datta, A., Yu, W., Rescher, U., Gerke, V., et al. (2007). PTEN-mediated apical segregation of phosphoinositides controls epithelial morphogenesis through Cdc42. *Cell*, *128*(2), 383–397. https://doi.org/10.1016/j.cell.2006.11.051.

Martin-Belmonte, F., & Perez-Moreno, M. (2011). Epithelial cell polarity, stem cells and cancer. *Nature Reviews. Cancer*, *12*(1), 23–38. https://doi.org/10.1038/nrc3169.

Martin-Gago, P., Fansa, E. K., Klein, C. H., Murarka, S., Janning, P., Schurmann, M., et al. (2017). A PDE6delta-KRas inhibitor chemotype with up to seven H-bonds and picomolar affinity that prevents efficient inhibitor release by Arl2. *Angewandte Chemie (International Ed. in English)*, *56*(9), 2423–2428. https://doi.org/10.1002/anie.201610957.

Mejuch, T., van Hattum, H., Triola, G., Jaiswal, M., & Waldmann, H. (2015). Specificity of lipoprotein chaperones for the characteristic lipidated structural motifs of their cognate lipoproteins. *Chembiochem*, *16*(17), 2460–2465. https://doi.org/10.1002/cbic.201500355.

Mitchell, D. A., Vasudevan, A., Linder, M. E., & Deschenes, R. J. (2006). Protein palmitoylation by a family of DHHC protein S-acyltransferases. *Journal of Lipid Research*, *47*(6), 1118–1127. https://doi.org/10.1194/jlr.R600007-JLR200.

Mo, S. P., Coulson, J. M., & Prior, I. A. (2018). RAS variant signalling. *Biochemical Society Transactions*, *46*(5), 1325–1332. https://doi.org/10.1042/BST20180173.

Modenutti, C. P., Capurro, J. I. B., Di Lella, S., & Marti, M. A. (2019). The structural biology of galectin-ligand recognition: Current advances in modeling tools, protein engineering, and inhibitor design. *Frontiers in Chemistry*, 7, 823. https://doi.org/10.3389/fchem.2019.00823.

Moore, A. R., Rosenberg, S. C., McCormick, F., & Malek, S. (2020). RAS-targeted therapies: Is the undruggable drugged? *Nature Reviews. Drug Discovery*, *19*(8), 533–552. https://doi.org/10.1038/s41573-020-0068-6.

Najumudeen, A. K., Jaiswal, A., Lectez, B., Oetken-Lindholm, C., Guzman, C., Siljamaki, E., et al. (2016). Cancer stem cell drugs target K-ras signaling in a stemness context. *Oncogene*, *35*(40), 5248–5262. https://doi.org/10.1038/onc.2016.59.

Najumudeen, A. K., Kohnke, M., Solman, M., Alexandrov, K., & Abankwa, D. (2013). Cellular FRET-biosensors to detect membrane targeting inhibitors of n-myristoylated proteins. *PLoS One*, *8*(6), e66425. https://doi.org/10.1371/journal.pone.0066425.

Nguyen, U. T., Goodall, A., Alexandrov, K., & Abankwa, D. (2010). Isoprenoid modifications. *Post-Translational Modifications in Health and Disease*, *13*, 486.

Novotny, C. J., Hamilton, G. L., McCormick, F., & Shokat, K. M. (2017). Farnesyltransferase-mediated delivery of a covalent inhibitor overcomes alternative prenylation to mislocalize K-Ras. *ACS Chemical Biology*, *12*(7), 1956–1962. https://doi.org/10.1021/acschembio.7b00374.

Papke, B., Murarka, S., Vogel, H. A., Martin-Gago, P., Kovacevic, M., Truxius, D. C., et al. (2016). Identification of pyrazolopyridazinones as PDEdelta inhibitors. *Nature Communications*, 7, 11360. https://doi.org/10.1038/ncomms11360.

Park, E. J., Grabinska, K. A., Guan, Z., & Sessa, W. C. (2016). NgBR is essential for endothelial cell glycosylation and vascular development. *EMBO Reports*, *17*(2), 167–177. https://doi.org/10.15252/embr.201540789.

Parton, R. G., Tillu, V., McMahon, K. A., & Collins, B. M. (2021). Key phases in the formation of caveolae. *Current Opinion in Cell Biology*, *71*, 7–14. https://doi.org/10.1016/j.ceb.2021.01.009.

Plowman, S. J., Ariotti, N., Goodall, A., Parton, R. G., & Hancock, J. F. (2008). Electrostatic interactions positively regulate K-Ras nanocluster formation and function. *Molecular and Cellular Biology*, *28*(13), 4377–4385. https://doi.org/10.1128/MCB.00050-08.

Plowman, S. J., Muncke, C., Parton, R. G., & Hancock, J. F. (2005). H-ras, K-ras, and inner plasma membrane raft proteins operate in nanoclusters with differential dependence on the actin cytoskeleton. *Proceedings of the National Academy of Sciences of the United States of America*, *102*(43), 15500–15505. https://doi.org/10.1073/pnas.0504114102.

Polier, G., Neumann, J., Thuaud, F., Ribeiro, N., Gelhaus, C., Schmidt, H., et al. (2012). The natural anticancer compounds rocaglamides inhibit the Raf-MEK-ERK pathway by targeting prohibitin 1 and 2. *Chemistry & Biology*, *19*(9), 1093–1104. https://doi.org/10.1016/j.chembiol.2012.07.012.

Posada, I. M. D., Lectez, B., Siddiqui, F. A., Oetken-Lindholm, C., Sharma, M., & Abankwa, D. (2017). Opposite feedback from mTORC1 to H-ras and K-ras4B downstream of SREBP1. *Scientific Reports*, 7(1), 8944. https://doi.org/10.1038/s41598-017-09387-8.

Posada, I. M., Serulla, M., Zhou, Y., Oetken-Lindholm, C., Abankwa, D., & Lectez, B. (2016). ASPP2 is a novel Pan-Ras nanocluster scaffold. *PLoS One*, *11*(7), e0159677. https://doi.org/10.1371/journal.pone.0159677.

Prior, I. A., Muncke, C., Parton, R. G., & Hancock, J. F. (2003). Direct visualization of Ras proteins in spatially distinct cell surface microdomains. *The Journal of Cell Biology*, *160*(2), 165–170. https://doi.org/10.1083/jcb.200209091.

Rajalingam, K., Wunder, C., Brinkmann, V., Churin, Y., Hekman, M., Sievers, C., et al. (2005). Prohibitin is required for Ras-induced Raf-MEK-ERK activation and epithelial cell migration. *Nature Cell Biology*, 7(8), 837–843. https://doi.org/10.1038/ncb1283.

Roberts, P. J., Mitin, N., Keller, P. J., Chenette, E. J., Madigan, J. P., Currin, R. O., et al. (2008). Rho Family GTPase modification and dependence on CAAX motif-signaled posttranslational modification. *The Journal of Biological Chemistry*, *283*(37), 25150–25163. https://doi.org/10.1074/jbc.M800882200.

Rocks, O., Gerauer, M., Vartak, N., Koch, S., Huang, Z. P., Pechlivanis, M., et al. (2010). The palmitoylation machinery is a spatially organizing system for peripheral membrane proteins. *Cell*, *141*(3), 458–471. https://doi.org/10.1016/j.cell.2010.04.007.

Rocks, O., Peyker, A., & Bastiaens, P. I. (2006). Spatio-temporal segregation of Ras signals: One ship, three anchors, many harbors. *Current Opinion in Cell Biology*, *18*(4), 351–357. https://doi.org/10.1016/j.ceb.2006.06.007.

Rocks, O., Peyker, A., Kahms, M., Verveer, P. J., Koerner, C., Lumbierres, M., et al. (2005). An acylation cycle regulates localization and activity of palmitoylated Ras isoforms. *Science*, *307*(5716), 1746–1752. https://doi.org/10.1126/science.1105654.

Rotblat, B., Belanis, L., Liang, H., Haklai, R., Elad-Zefadia, G., Hancock, J. F., et al. (2010). H-Ras nanocluster stability regulates the magnitude of MAPK signal output. *PLoS One*, *5*(8), e11991. https://doi.org/10.1371/journal.pone.0011991.

Sarkar-Banerjee, S., Sayyed-Ahmad, A., Prakash, P., Cho, K. J., Waxham, M. N., Hancock, J. F., et al. (2017). Spatiotemporal analysis of K-Ras plasma membrane interactions reveals multiple high order homo-oligomeric complexes. *Journal of the American Chemical Society*, *139*(38), 13466–13475. https://doi.org/10.1021/jacs.7b06292.

Sasinkova, M., Holoubek, A., Otevrelova, P., Kuzelova, K., & Brodska, B. (2018). AML-associated mutation of nucleophosmin compromises its interaction with nucleolin. *The International Journal of Biochemistry & Cell Biology*, *103*, 65–73. https://doi.org/10.1016/j.biocel.2018.08.008.

Schmick, M., Kraemer, A., & Bastiaens, P. I. (2015). Ras moves to stay in place. *Trends in Cell Biology*, *25*(4), 190–197. https://doi.org/10.1016/j.tcb.2015.02.004.

Schoenenberger, C. A., Zuk, A., Kendall, D., & Matlin, K. S. (1991). Multilayering and loss of apical polarity in MDCK cells transformed with viral K-ras. *The Journal of Cell Biology*, *112*(5), 873–889. https://doi.org/10.1083/jcb.112.5.873.

Seaman, M. N. J. (2021). The retromer complex: From genesis to revelations. *Trends in Biochemical Sciences*, *46*, 608–620. https://doi.org/10.1016/j.tibs.2020.12.009.

Shalom-Feuerstein, R., Plowman, S. J., Rotblat, B., Ariotti, N., Tian, T., Hancock, J. F., et al. (2008). K-ras nanoclustering is subverted by overexpression of the scaffold protein galectin-3. *Cancer Research*, *68*(16), 6608–6616. https://doi.org/10.1158/0008-5472.CAN-08-1117.

Siddiqui, F. A., Alam, C., Rosenqvist, P., Ora, M., Sabt, A., Manoharan, G. B., et al. (2020). PDE6D inhibitors with a new design principle selectively block K-Ras activity. *ACS Omega*, *5*(1), 832–842. https://doi.org/10.1021/acsomega.9b03639.

Simanshu, D. K., Nissley, D. V., & McCormick, F. (2017). RAS proteins and their regulators in human disease. *Cell*, *170*(1), 17–33. https://doi.org/10.1016/j.cell.2017.06.009.

Siprashvili, Z., Webster, D. E., Johnston, D., Shenoy, R. M., Ungewickell, A. J., Bhaduri, A., et al. (2016). The noncoding RNAs SNORD50A and SNORD50B bind K-Ras and are recurrently deleted in human cancer. *Nature Genetics*, *48*(1), 53–58. https://doi.org/10.1038/ng.3452.

Sperlich, B., Kapoor, S., Waldmann, H., Winter, R., & Weise, K. (2016). Regulation of K-Ras4B membrane binding by calmodulin. *Biophysical Journal*, *111*(1), 113–122. https://doi.org/10.1016/j.bpj.2016.05.042.

Suh, Y. H., Terashima, A., Petralia, R. S., Wenthold, R. J., Isaac, J. T., Roche, K. W., et al. (2010). A neuronal role for SNAP-23 in postsynaptic glutamate receptor trafficking. *Nature Neuroscience*, *13*(3), 338–343. https://doi.org/10.1038/nn.2488.

Sutton, M. N., Lu, Z., Li, Y. C., Zhou, Y., Huang, T., Reger, A. S., et al. (2019). DIRAS3 (ARHI) blocks RAS/MAPK signaling by binding directly to RAS and disrupting RAS clusters. *Cell Reports*, *29*(11), 3448–3459 e3446. https://doi.org/10.1016/j.celrep.2019.11.045.

Swarthout, J. T., Lobo, S., Farh, L., Croke, M. R., Greentree, W. K., Deschenes, R. J., et al. (2005). DHHC9 and GCP16 constitute a human protein fatty acyltransferase with specificity for H- and N-Ras. *The Journal of Biological Chemistry*, *280*(35), 31141–31148. https://doi.org/10.1074/jbc.M504113200.

Tan, L., Cho, K. J., Neupane, P., Capon, R. J., & Hancock, J. F. (2018). An oxanthroquinone derivative that disrupts RAS plasma membrane localization inhibits cancer cell growth. *The Journal of Biological Chemistry*, *293*(35), 13696–13706. https://doi.org/10.1074/jbc.RA118.003907.

Terrell, E. M., Durrant, D. E., Ritt, D. A., Sealover, N. E., Sheffels, E., Spencer-Smith, R., et al. (2019). Distinct binding preferences between Ras and Raf family members and the impact on oncogenic ras signaling. *Molecular Cell*, *76*(6), 872–884 e875. https://doi.org/10.1016/j.molcel.2019.09.004.

Tidow, H., & Nissen, P. (2013). Structural diversity of calmodulin binding to its target sites. *The FEBS Journal*, *280*(21), 5551–5565. https://doi.org/10.1111/febs.12296.

Timoshenko, A. V. (2015). Towards molecular mechanisms regulating the expression of galectins in cancer cells under microenvironmental stress conditions. *Cellular and Molecular Life Sciences*, *72*(22), 4327–4340. https://doi.org/10.1007/s00018-015-2008-x.

Travers, T., Lopez, C. A., Van, Q. N., Neale, C., Tonelli, M., Stephen, A. G., et al. (2018). Molecular recognition of RAS/RAF complex at the membrane: Role of RAF cysteine-rich domain. *Scientific Reports*, *8*(1), 8461. https://doi.org/10.1038/s41598-018-26832-4.

Tsai, F. D., Lopes, M. S., Zhou, M., Court, H., Ponce, O., Fiordalisi, J. J., et al. (2015). K-Ras4A splice variant is widely expressed in cancer and uses a hybrid membrane-targeting motif. *Proceedings of the National Academy of Sciences of the United States of America*, *112*(3), 779–784. https://doi.org/10.1073/pnas.1412811112.

Turko, I. V., Ballard, S. A., Francis, S. H., & Corbin, J. D. (1999). Inhibition of cyclic GMP-binding cyclic GMP-specific phosphodiesterase (Type 5) by sildenafil and related compounds. *Molecular Pharmacology*, *56*(1), 124–130. https://doi.org/10.1124/mol.56.1.124.

Van, Q. N., Prakash, P., Shrestha, R., Balius, T. E., Turbyville, T. J., & Stephen, A. G. (2021). RAS nanoclusters: Dynamic signaling platforms amenable to therapeutic intervention. *Biomolecules, 11*(3), 377. https://doi.org/10.3390/biom11030377.

Villalonga, P., Lopez-Alcala, C., Bosch, M., Chiloeches, A., Rocamora, N., Gil, J., et al. (2001). Calmodulin binds to K-Ras, but not to H- or N-Ras, and modulates its downstream signaling. *Molecular and Cellular Biology, 21*(21), 7345–7354. https://doi.org/10.1128/MCB.21.21.7345-7354.2001.

Wang, Y., Godin-Heymann, N., Dan Wang, X., Bergamaschi, D., Llanos, S., & Lu, X. (2013). ASPP1 and ASPP2 bind active RAS, potentiate RAS signalling and enhance p53 activity in cancer cells. *Cell Death and Differentiation, 20*(4), 525–534. https://doi.org/10.1038/cdd.2013.3.

Wang, M. T., Holderfield, M., Galeas, J., Delrosario, R., To, D., Balmain, A., et al. (2015). K-Ras promotes tumorigenicity through suppression of non-canonical Wnt signaling. *Cell, 163*(5), 1237–1251. https://doi.org/10.1016/j.cell.2015.10.041.

Wang, D., Tabti, R., Elderwish, S., Djehal, A., Chouha, N., Pinot, F., et al. (2020). SFPH proteins as therapeutic targets for a myriad of diseases. *Bioorganic & Medicinal Chemistry Letters, 30*(22), 127600. https://doi.org/10.1016/j.bmcl.2020.127600.

Wang, P., van der Hoeven, D., Ye, N., Chen, H., Liu, Z., Ma, X., et al. (2021). Scaffold repurposing of fendiline: Identification of potent KRAS plasma membrane localization inhibitors. *European Journal of Medicinal Chemistry, 217*, 113381. https://doi.org/10.1016/j.ejmech.2021.113381.

Wang, B., Zhao, B., North, P., Kong, A., Huang, J., & Miao, Q. R. (2013). Expression of NgBR is highly associated with estrogen receptor alpha and survivin in breast cancer. *PLoS One, 8*(11), e78083. https://doi.org/10.1371/journal.pone.0078083.

Wennerberg, K., Rossman, K. L., & Der, C. J. (2005). The Ras superfamily at a glance. *Journal of Cell Science, 118*(Pt. 5), 843–846. https://doi.org/10.1242/jcs.01660.

Whyte, D. B., Kirschmeier, P., Hockenberry, T. N., Nunez-Oliva, I., James, L., Catino, J. J., et al. (1997). K- and N-Ras are geranylgeranylated in cells treated with farnesyl protein transferase inhibitors. *The Journal of Biological Chemistry, 272*(22), 14459–14464. https://doi.org/10.1074/jbc.272.22.14459.

Willumsen, B. M., Christensen, A., Hubbert, N. L., Papageorge, A. G., & Lowy, D. R. (1984). The p21 ras C-terminus is required for transformation and membrane association. *Nature, 310*(5978), 583–586. https://doi.org/10.1038/310583a0.

Winter-Vann, A. M., Baron, R. A., Wong, W., dela Cruz, J., York, J. D., Gooden, D. M., et al. (2005). A small-molecule inhibitor of isoprenylcysteine carboxyl methyltransferase with antitumor activity in cancer cells. *Proceedings of the National Academy of Sciences of the United States of America, 102*(12), 4336–4341. https://doi.org/10.1073/pnas.0408107102.

Winzker, M., Friese, A., Koch, U., Janning, P., Ziegler, S., & Waldmann, H. (2020). Development of a PDEdelta-targeting PROTACs that impair lipid metabolism. *Angewandte Chemie (International Ed. in English), 59*(14), 5595–5601. https://doi.org/10.1002/anie.201913904.

Xu, M., & Dong, X. P. (2021). Endolysosomal TRPMLs in cancer. *Biomolecules, 11*(1), 65. https://doi.org/10.3390/biom11010065.

Yurugi, H., Zhuang, Y., Siddiqui, F. A., Liang, H., Rosigkeit, S., Zeng, Y., et al. (2020). A subset of flavaglines inhibits KRAS nanoclustering and activation. *Journal of Cell Science, 133*(12), jcs244111. https://doi.org/10.1242/jcs.244111.

Zhang, M., Tanaka, T., & Ikura, M. (1995). Calcium-induced conformational transition revealed by the solution structure of apo calmodulin. *Nature Structural Biology, 2*(9), 758–767. https://doi.org/10.1038/nsb0995-758.

Zhao, B., Hu, W., Kumar, S., Gonyo, P., Rana, U., Liu, Z., et al. (2017). The Nogo-B receptor promotes Ras plasma membrane localization and activation. *Oncogene, 36*(24), 3406–3416. https://doi.org/10.1038/onc.2016.484.

Zhou, Y., & Hancock, J. F. (2015). Ras nanoclusters: Versatile lipid-based signaling platforms. *Biochimica et Biophysica Acta, 1853*(4), 841–849. https://doi.org/10.1016/j.bbamcr.2014.09.008.

Zhou, Y., & Hancock, J. F. (2018). Deciphering lipid codes: K-Ras as a paradigm. *Traffic, 19*(3), 157–165. https://doi.org/10.1111/tra.12541.

Zhou, M., Kuruvilla, L., Shi, X., Viviano, S., Ahearn, I. M., Amendola, C. R., et al. (2020). Scaffold association factor B (SAFB) is required for expression of prenyltransferases and RAS membrane association. *Proceedings of the National Academy of Sciences of the United States of America, 117*(50), 31914–31922. https://doi.org/10.1073/pnas.2005712117.

Zhou, Y., Liang, H., Rodkey, T., Ariotti, N., Parton, R. G., & Hancock, J. F. (2014). Signal integration by lipid-mediated spatial cross talk between Ras nanoclusters. *Molecular and Cellular Biology, 34*(5), 862–876. https://doi.org/10.1128/MCB.01227-13.

Zhou, Y., Prakash, P., Liang, H., Cho, K. J., Gorfe, A. A., & Hancock, J. F. (2017). Lipid-sorting specificity encoded in k-ras membrane anchor regulates signal output. *Cell, 168*(1–2), 239–251 e216. https://doi.org/10.1016/j.cell.2016.11.059.

Zhou, M., Wiener, H., Su, W., Zhou, Y., Liot, C., Ahearn, I., et al. (2016). VPS35 binds farnesylated N-Ras in the cytosol to regulate N-Ras trafficking. *The Journal of Cell Biology, 214*(4), 445–458. https://doi.org/10.1083/jcb.201604061.

Zhou, Y., Prakash, P. S., Liang, H., Gorfe, A. A., & Hancock, J. F. (2021). The KRAS and other prenylated polybasic domain membrane anchors recognize phosphatidylserine acyl chain structure. *Proceedings of the National Academy of Sciences of the United States of America, 118*(6), e2014605118. https://doi.org/10.1073/pnas.2014605118.

Zhou, Y., Wong, C. O., Cho, K. J., van der Hoeven, D., Liang, H., Thakur, D. P., et al. (2015). Signal transduction. membrane potential modulates plasma membrane phospholipid dynamics and K-Ras signaling. *Science, 349*(6250), 873–876. https://doi.org/10.1126/science.aaa5619.

Zimmermann, G., Papke, B., Ismail, S., Vartak, N., Chandra, A., Hoffmann, M., et al. (2013). Small molecule inhibition of the KRAS-PDEdelta interaction impairs oncogenic KRAS signalling. *Nature, 497*(7451), 638–642. https://doi.org/10.1038/nature12205.

CHAPTER FOUR

Targeting the ERK mitogen-activated protein kinase cascade for the treatment of KRAS-mutant pancreatic cancer

J. Nathaniel Diehl[a,†], Priya S. Hibshman[b,†], Irem Ozkan-Dagliyan[c], Craig M. Goodwin[d], Sarah V. Howard[d], Adrienne D. Cox[b,c,d,e], and Channing J. Der[a,b,c,d,*]

[a]Curriculum in Genetics and Molecular Biology, University of North Carolina at Chapel Hill, Chapel Hill, NC, United States
[b]Cell Biology and Physiology Curriculum, University of North Carolina at Chapel Hill, Chapel Hill, NC, United States
[c]Department of Pharmacology, University of North Carolina at Chapel Hill, Chapel Hill, NC, United States
[d]Lineberger Comprehensive Cancer Center, University of North Carolina at Chapel Hill, Chapel Hill, NC, United States
[e]Department of Radiation Oncology, University of North Carolina at Chapel Hill, Chapel Hill, NC, United States
*Corresponding author: e-mail address: cjder@med.unc.edu

Contents

1. Introduction 102
2. *KRAS*: The predominant RAS gene mutated in human cancer 103
3. Targeting KRAS for cancer treatment: Direct inhibition of RAS 104
4. Targeting KRAS for cancer treatment: A focus on RAS effector signaling 105
5. The ERK MAPK signaling network 107
6. ERK feedback inhibition and the Goldilock Principle: A "just right" level of ERK signaling supports cancer growth 109
7. Inhibition of the RAF-MEK-ERK cascade: Promises and challenges 110
8. ERK regulation of a diverse phosphoproteome and transcriptome 113
9. MYC: The critical determinant for ERK-dependent cancer growth? 117
10. KRAS, ERK, MYC and metabolism: An Achilles heel for KRAS-mutant PDAC? 119
Acknowledgments 121
Disclosure of potential conflicts of interest 121
References 121

† These authors contributed equally.

Advances in Cancer Research, Volume 153
ISSN 0065-230X
https://doi.org/10.1016/bs.acr.2021.07.008

Abstract

Mutational activation of the *KRAS* oncogene is found in ~95% of pancreatic ductal adenocarcinoma (PDAC), the major form of pancreatic cancer. With substantial experimental evidence that continued aberrant KRAS function is essential for the maintenance of PDAC tumorigenic growth, the National Cancer Institute has identified the development of effective anti-KRAS therapies as one of four major initiatives for pancreatic cancer research. The recent clinical success in the development of an anti-KRAS therapy targeting one specific KRAS mutant (G12C) supports the significant potential impact of anti-KRAS therapies. However, $KRAS^{G12C}$ mutations comprise only 2% of *KRAS* mutations in PDAC. Thus, there remains a dire need for additional therapeutic approaches for targeting the majority of *KRAS*-mutant PDAC. Among the different directions currently being pursued for anti-KRAS drug development, one of the most promising involves inhibitors of the key KRAS effector pathway, the three-tiered RAF-MEK-ERK mitogen-activated protein kinase (MAPK) cascade. We address the promises and challenges of targeting ERK MAPK signaling as an anti-KRAS therapy for PDAC. In particular, we also summarize the key role of the MYC transcription factor and oncoprotein in supporting ERK-dependent growth of *KRAS*-mutant PDAC.

1. Introduction

Since their discovery in 1982 as the first human cancer genes, *RAS* genes (*HRAS*, *KRAS* and *NRAS*) have been the subject of intense research scrutiny (Cox & Der, 2010). With now nearly four decades of research, RAS oncoproteins have cemented their place among the most significant and key cancer-promoting proteins due to their capacity to initiate and maintain tumor growth in a diverse range of tissue types (Ledford, 2015). RAS mutations are found in 19% of all cancers and are key cancer drivers (Prior, Hood, & Hartley, 2020). Consequently, there have been intensive efforts by both academia and industry to develop anti-RAS therapies for cancer treatment. With the recent accelerated approval by the Food and Drug Administration of sotorasib/AMG 510, a direct inhibitor of the G12C missense mutation of RAS, and with other G12C inhibitors (e.g., adagrasib/MRTX849) following closely in the pipeline, a clinically effective therapy for RAS-mutant cancers has finally been achieved. However, as G12C represents only a fraction of RAS mutations, this milestone achievement marks only the beginning of a new chapter in anti-RAS drug discovery. Further, resistance to G12C inhibitor treatment has already arisen. Recent studies of these resistance mechanisms have repeatedly demonstrated genetic reactivation of the key KRAS effector signaling network, the ERK mitogen-activated protein kinase (MAPK) cascade (Awad et al., 2021;

Tanaka et al., 2021), emphasizing the critical role that ERK inhibition will play in the long-term success of essentially all anti-RAS therapies. We review the current status of therapeutic targeting of the ERK MAPK signaling network to facilitate the development of effective anti-KRAS therapies, with a focus on the most RAS-addicted cancer, pancreatic ductal adenocarcinoma, and on the importance of the MYC oncoprotein in modulating the response to ERK MAPK inhibition.

2. *KRAS*: The predominant RAS gene mutated in human cancer

Mutationally activated *RAS* genes are found in a diverse spectrum of human cancer types (Cox, Fesik, Kimmelman, Luo, & Der, 2014), with highest frequencies seen in cancers that comprise the top three causes of projected cancer deaths in the United States in 2021 (lung, colorectal and pancreatic cancer) (Siegel, Miller, Fuchs, & Jemal, 2021). Pancreatic ductal adenocarcinoma (PDAC), the predominant form of pancreatic cancer, is unique among cancer types with its near-universal occurrence of RAS mutations (~95%) (Waters & Der, 2018). The next most frequently RAS-mutant cancer types are colorectal (CRC; ~50%), multiple myeloma (~40%), and lung adenocarcinoma (LAC; ~30%), the major form of lung cancer. In contrast, *RAS* mutations are rarely seen in breast and prostate cancers, the fourth and fifth leading causes of cancer deaths in the United States (Siegel et al., 2021).

Among *RAS* family members, *KRAS* is the most frequently mutated in cancer, representing 83% of all *RAS* mutations in cancer overall (Cox et al., 2014). However, the distribution of mutations in *KRAS* versus *NRAS* versus *HRAS* differs widely among various cancer types (Haigis, 2017). PDAC is a *KRAS*-mutant disease, and *RAS* mutations in LAC and CRC are nearly exclusively in *KRAS* as well (Cox et al., 2014). In contrast, cutaneous melanoma and acute myelogenous leukemia are dominated by activating *NRAS* mutations, whereas bladder cancer and head and neck squamous cell carcinoma (HNSCC) are most frequently associated with *HRAS* mutations.

For many years following the initial discoveries of *RAS* mutations in cancer, research centered on *HRAS* (Cox & Der, 2010). This focus was driven primarily by the availability of reagents to study HRAS and by the assumption that the highly conserved RAS proteins (~85% identical) shared essentially identical structural, biochemical and oncogenic functions. However, the unexpected discovery that farnesyltransferase inhibitors

(FTIs) blocked membrane association and function of HRAS but not KRAS or NRAS shifted the focus of the field to study *KRAS*, the predominant *RAS* gene mutated in cancer (Berndt, Hamilton, & Sebti, 2011; Cox, Der, & Philips, 2015). *KRAS* encodes two transcriptional splice variants due to alternative exon four utilization (Tsai et al., 2015). *KRAS* 4A and 4B expression levels vary in a tissue-dependent manner, and *KRAS4B* is the most abundant splice variant in PDAC and other cancer types. The vast majority of experimental studies focus on KRAS4B rather than KRAS4A.

Mutational activation of the *KRAS* oncogene is found in the earliest preneoplastic stages of PDAC progression (Kanda et al., 2012). Genetic studies in mouse models of PDAC support *Kras* mutational activation as the initiating genetic event in PDAC development (Hingorani et al., 2003). Additionally, substantial experimental evidence supports the essential role of KRAS in the maintenance of PDAC growth (Collins et al., 2012; Ying et al., 2012). Consequently, the National Cancer Institute has identified the development of anti-KRAS therapeutic strategies as one of four major initiatives for the field (The Pancreatic Cancer Working Group, 2013). PDAC is arguably the most KRAS-addicted cancer type, and consequently, the most clinically relevant cancer model to study KRAS function.

3. Targeting KRAS for cancer treatment: Direct inhibition of RAS

Many different approaches have been pursued intensively in efforts to develop anti-RAS therapies (Papke & Der, 2017). Following repeated failed attempts to develop direct RAS inhibitors, RAS itself was branded as "undruggable". However, recent discoveries of small molecules that directly bind to RAS and disrupt its function have challenged the perception of RAS as undruggable, and direct inhibition now comprises one of the most exciting new directions for anti-RAS drug discovery (Dang, Reddy, Shokat, & Soucek, 2017; Uprety & Adjei, 2020). In particular, recent early-stage clinical findings with direct inhibitors of one KRAS mutant, G12C, support the potential clinical impact of an effective direct KRAS inhibitor (Canon et al., 2019; Hallin et al., 2020). Interestingly, there are cancer type-specific differences in clinical response. While the response in $KRAS^{G12C}$-mutant lung cancers is ~35%, the response in $KRAS^{G12C}$-mutant CRC is considerably lower (Hong et al., 2020). Promisingly, the earliest KRAS^{G12C} inhibitor into clinical trials, sotorasib/AMG 510, was recently approved for the treatment of KRAS^{G12C} mutant lung cancers.

4. Targeting KRAS for cancer treatment: A focus on RAS effector signaling

The majority of anti-RAS drug development efforts have involved indirect strategies, particularly pharmacological blockade of RAS effector signaling (Hymowitz & Malek, 2018; Klomp, Klomp, & Der, 2021). Of the multitude of KRAS effectors, at least four are validated drivers of KRAS-dependent PDAC growth (Waters & Der, 2018). The best characterized and potentially most crucial is the RAF-MEK-ERK mitogen-activated protein kinase (MAPK) cascade (Fig. 1A). Supporting the key role of the MAPK cascade in driving Kras-dependent PDAC growth, one group demonstrated that activated *Braf* can phenocopy mutant *Kras* and, together with *Tp53* inactivation, drive the development of metastatic PDAC in mouse models, (Collisson et al., 2012). Our studies showed that pharmacologic inhibition of ERK blocked the tumorigenic growth of KRAS-mutant PDAC, validating ERK as a significant therapeutic target (Hayes et al., 2016).

The second best characterized KRAS effectors are the class I phosphoinositide 3-kinases (PI3K) (Castellano & Downward, 2011). KRAS activation of PI3K promotes formation of phosphatidylinositol (3,4,5)-triphosphate, leading to activation of the AKT and mTOR serine/threonine kinases. Activation of PI3K-AKT-mTOR signaling has also been observed as a compensatory response to ERK inhibition, driving resistance to ERK MAPK inhibitors (Hayes et al., 2016; Villanueva et al., 2010). Thus, concurrent PI3K inhibition synergistically enhances their anti-tumor activity (Collisson et al., 2012). Other key KRAS effectors include the RAL and RAC small GTPases, which activate TBK1 and PAK serine/threonine kinases, respectively (Waters & Der, 2018). With their high tractability as therapeutic targets, inhibition of these and other protein kinase components of KRAS effector signaling have been pursued as indirect approaches to KRAS inhibition (Furuse & Nagashima, 2017). Of these, due to their well-validated driver roles in supporting RAS-driven oncogenesis, the development and evaluation of pharmacologic inhibitors of RAF-MEK-ERK signaling have been the most intensively pursued.

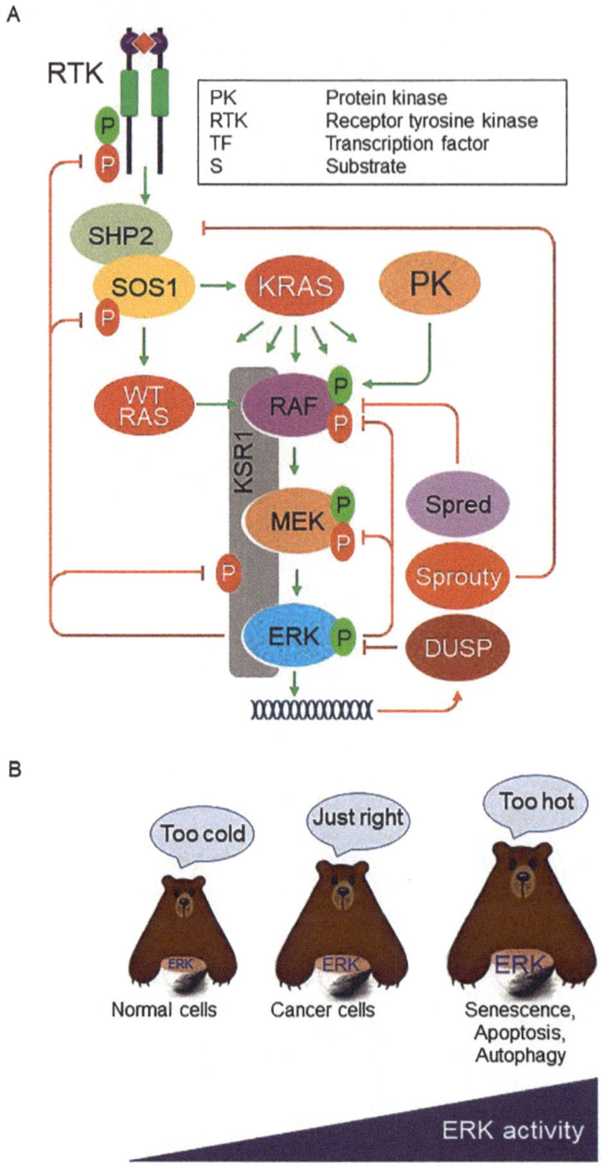

Fig. 1 KRAS regulation of the ERK MAPK effector signaling network. (A) Mutational activation of KRAS causes persistent and deregulated activation of the RAF-MEK-ERK protein kinase cascade. ERK activation then initiates multiple negative feedback mechanisms to prevent excessive ERK signaling. This includes ERK-dependent phosphorylation and inhibition of EGFR, SOS1, RAF and MEK kinases. Others involve transcriptional upregulation of genes encoding proteins that act as negative regulators of ERK signaling either upstream or downstream of ERK. Green lines indicate positive signaling mechanisms whereas red lines indicate negative feedback mechanisms. (B) The Goldilocks principle is a concept that originated from the fairy tale "Goldilocks and the Three Bears." A girl named Goldilocks entered a cottage in the woods and tasted three different bowls of porridge. She found one to be too hot, another to be too cold, and the third "just right." By analogy, in both normal and cancer cells, ERK-dependent feedback mechanisms dynamically maintain the level of ERK activity to be "just right." Mutant KRAS drives a level of elevated and persistent ERK activity that is necessary for tumorigenic growth. However, excessive ERK activity is deleterious to cancer cell growth, resulting in cellular senescence, autophagy, and apoptosis.

5. The ERK MAPK signaling network

The three-tier RAF-MEK-ERK MAPK cascade is a conserved protein kinase signaling pathway that regulates normal cell proliferation, differentiation, and survival (Klomp et al., 2021; Morrison, 2012). The regulation of ERK MAPK signaling has been studied extensively, as the components of this cascade are among the most deregulated proteins in cancer. Briefly, growth factor binding leads to phosphorylation of cell surface receptor tyrosine kinases (RTKs; e.g., EGFR), which in turn recruit the RAS guanine nucleotide exchange factor SOS1 to the plasma membrane where it can then convert inactive membrane-bound RAS-GDP to its active GTP-bound form, leading to RAS association with its effectors and to activation of RAS effector-mediated signaling (Fig. 1A).

Each level of the RAF-MEK-ERK cascade comprises multiple paralogs (Fig. 2A) (Klomp et al., 2021; Morrison, 2012). There are three highly related human RAF serine/threonine kinases that share 55–58% overall amino acid identity and 77–81% identity in their kinase domains (Terrell & Morrison, 2019). One key additional domain is a RAS-binding domain (RBD) that promotes RAF association with RAS-GTP. Upon association with RAS-GTP, RAF proteins translocate to the plasma membrane where a complex set of phosphorylation events catalyzed by multiple protein kinases including SRC leads to activation of RAF catalytic activity. An additional basis for RAS activation of RAF involves RAS-dependent formation of RAF homo- and hetero-dimers, most commonly BRAF-CRAF heterodimers (Durrant & Morrison, 2018).

Genetic ablation studies in mouse models of cancer have also addressed the importance of the three *RAF* genes in RAS-driven cancer development and growth. A deficiency in *Craf*, but not *Braf*, prevented *Kras*-driven lung cancer (Blasco et al., 2011; Karreth, Frese, DeNicola, Baccarini, & Tuveson, 2011). In contrast, *Craf* deficiency alone did not prevent *Kras*-driven pancreatic cancer growth (Eser et al., 2013). Thus, there may exist tissue-specific requirements for the different RAF genes for KRAS-driven cancers that arise in different tissues. These mouse studies also included ablation of the *Raf* gene concurrently with activation of the *Kras* gene, and thus addressed the role of Raf in cancer initiation rather than maintenance of cancer growth. In a subsequent study where *Craf* was ablated after *Kras*-driven lung cancer development, it was concluded that Craf loss alone was sufficient to block Kras-dependent lung cancer growth (Blasco et al., 2011).

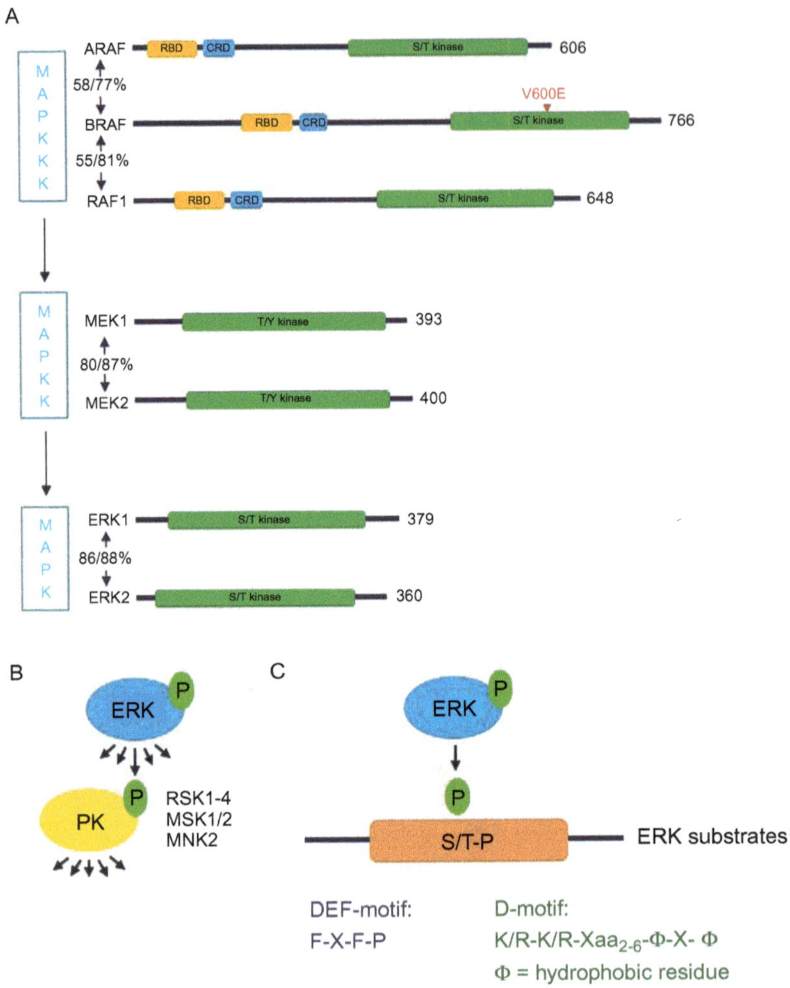

Fig. 2 Components of the ERK MAPK cascade. (A) The kinases that comprise the three-tiered ERK MAPK cascade. RAF kinases (MAPK kinase kinases; MAPKKK), in addition to the kinase domain, contain amino-terminal RAS-binding domains (RBDs), where activated RAS-GTP binds to initiate a series of protein kinase-mediated phosphorylation events that lead to RAF kinase activation. The adjacent cysteine-rich domain (CRD) facilitates RAF association with membranes. The MEK1 and MEK2 kinases (MAPK kinases; MAPKK) are comprised mainly of their dual specificity threonine/tyrosine (T/Y) kinase domains, while ERK1 and ERK2 kinases (MAPK) primarily consist of a serine/threonine (S/T) kinase domain. CLUSTALW multiple sequence alignment analysis was used to determine sequence identities of the total protein and kinase domains, indicated (%/%), respectively. (B) ERK substrates include protein kinases (PK) that further diversify the ERK-regulated phosphoproteome, including RSK1-4, MSK1/2, and MNK2. (C) ERK substrates are characterized by a S/T-P consensus phosphorylation site and DEF- and D-motif ERK binding sequences.

Further complexity was also indicated by recent studies that described kinase-independent functions of Raf in supporting Kras-dependent cancer growth (Sanclemente et al., 2018; Yen et al., 2021). While the kinase-independent functions remain to be fully elucidated, these findings suggest that inhibitors of RAF kinase activity may not be sufficient to fully ablate the functional role of RAF in oncogenesis.

The only well-characterized and -validated RAF substrates are the highly identical MEK1 and MEK2 threonine/tyrosine dual specificity kinases. MEK1 and MEK2 exhibit 80% overall amino acid identity, and 87% identity in their kinase domains. Following RAF phosphorylation of tandem serine residues, MEK1/2 catalytic function is activated. Similar to RAF, MEK1/2 substrate utilization is highly restricted: they phosphorylate a TEY motif in the highly related ERK1 and ERK2 serine/threonine kinases, that show 86% overall amino acid identity and 88% identity in their kinase domains. As discussed below, in striking contrast to the highly restricted substrates of RAF and MEK1/2, a complex spectrum of functionally diverse proteins has been identified as substrates of ERK1/2 (Fig. 2B and C). These include substrates that act in feedback inhibitory loops, additional protein kinases, and transcription factors, many of which are well-validated cancer drivers.

6. ERK feedback inhibition and the Goldilock Principle: A "just right" level of ERK signaling supports cancer growth

In addition to *RAS* or *BRAF* mutations, mutational activation or overexpression of RTKs (e.g., EGFR) can also cause aberrant activation of the ERK MAPK cascade. The RAS GTPase activating protein neurofibromin (encoded by the *NF1* tumor suppressor gene) is a negative RAS regulator that returns RAS-GTP to the resting, inactive state of RAS-GDP. Thus, loss of NF1 also causes hyperactivation of wild-type RAS. Within a specific cancer type, genetic mutations that cause ERK hyperactivation occur in a mutually exclusive pattern. For example, BxPC-3, the sole widely studied PDAC cell line that is *KRAS* wild-type, harbors mutationally activated BRAF (Chen et al., 2016; Foster et al., 2016). Similarly, *KRAS*, *NRAS* and *BRAF* mutations are mutually exclusive in CRC, whereas *EGFR* and *KRAS* mutations are mutually exclusive in lung cancer (cBioPortal).

While it is clear that an elevated level of ERK signaling is required for cancer growth, the mutually exclusive occurrence of mutations that drive

ERK hyperactivation supports another important premise of ERK signaling, that excessive ERK activation can also be detrimental (Fig. 1B). Experimental studies assessing the basis for this mutual exclusivity found that combining genetic alterations that drive ERK hyperactivation caused ERK-dependent tumor cell senescence, autophagy and/or apoptosis (Chi et al., 1999; Cisowski, Sayin, Liu, Karlsson, & Bergo, 2016; Unni et al., 2018; Unni, Lockwood, Zejnullahu, Lee-Lin, & Varmus, 2015). Consequently, there is a sweet spot where a "just right" level of ERK hyperactivation can drive cancer growth, and a "just right" level of pharmacologic inhibition of ERK suppresses cancer growth. To achieve the appropriate level of ERK signaling, the signaling intensity of the ERK MAPK network is under tight dynamic regulation in both normal and neoplastic cells (Klomp et al., 2021; Lake, Correa, & Muller, 2016).

ERK signaling intensity in response to ERK activation is normally downregulated by multiple negative feedback phosphorylation events and gene transcription (Fig. 1A) (Klomp et al., 2021; Lake et al., 2016). Upstream phosphorylation of EGFR, RAF and MEK by ERK can lead to inhibition of signaling through the entire cascade. Similarly, ERK activation stimulates the transcription of *SPRY*, *SPROUTY* and *DUSP* to dampen ERK signaling both upstream of ERK and at the level of ERK itself (Klomp et al., 2021). SPRY proteins block RTK activation of SOS1, preventing activation of wild-type RAS proteins and activation of RAF independent of mutant RAS. SPROUTY promotes activation of the RASGAP neurofibromin, which decreases RAS activation by promoting GTP hydrolysis (Lorenzo & McCormick, 2020). The dual specificity protein phosphatases (DUSPs) dephosphorylate ERK itself, at the MEK-phosphorylated TEY motif. Disruption of these negative modulators can therefore result in inappropriate persistence or reactivation of ERK signaling.

7. Inhibition of the RAF-MEK-ERK cascade: Promises and challenges

With the importance of aberrant RAF-MEK-ERK signaling in cancer, the pharmaceutical industry has comprehensively pursued and developed many inhibitors that can act at each level of the pathway (Fig. 3) (Moore, Rosenberg, McCormick, & Malek, 2020; Ryan & Corcoran, 2018). Initial efforts to target the ERK MAPK cascade focused on MEK1/2-selective inhibitors (Samatar & Poulikakos, 2014). Whereas these inhibitors have met with significant success in *BRAF*-mutant melanoma, to

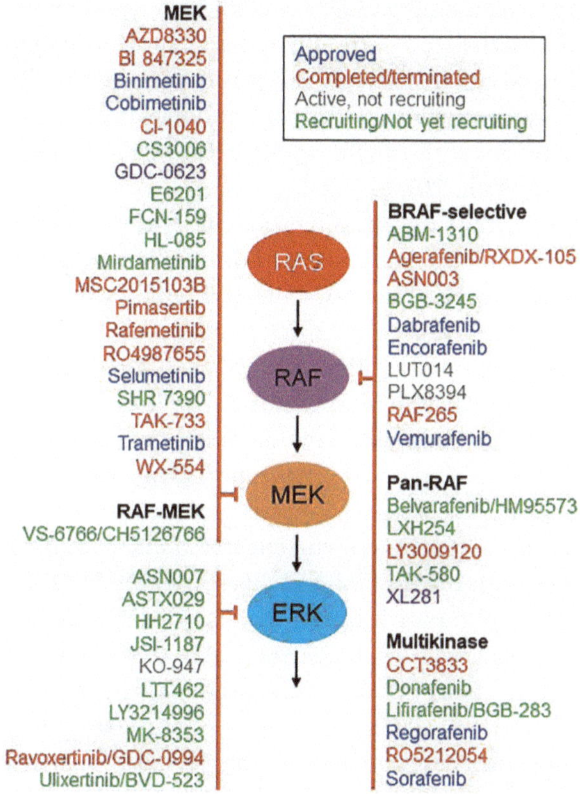

Fig. 3 Clinical evaluation of RAF-MEK-ERK inhibitors. BRAF-selective and MEK1/2 inhibitors are approved for the treatment of BRAF-mutant melanoma and other cancers. The MEK1/2 inhibitor selumetinib has been approved for NF1-deficient cancers. *Compiled from clinicaltrials.gov.*

date, they have not met with much success in RAS-mutant PDAC and other cancers, where toxicity has been a limiting factor (Blumenschein Jr. et al., 2015; Bodoky et al., 2012; Infante et al., 2014; Janne et al., 2013, 2017). A second limitation of MEK inhibitors, particularly when used as monotherapy, is the loss of ERK-dependent negative feedback signaling, resulting in ERK reactivation (Duncan et al., 2012).

Initially, with the misperception that the RAF-MEK-ERK cascade acts as a simple unidirectional linear pathway, and with successful development of potent and selective MEK inhibitors, there was limited interest in the development of RAF or ERK inhibitors for therapeutic inhibition of ERK-driven cancer growth (Samatar & Poulikakos, 2014). However, the appreciation that the RAF-MEK-ERK kinase cascade is a component of a complex and dynamic signaling network led to development of RAF and ERK inhibitors, with many currently under clinical evaluation (Fig. 3). Our studies evaluated both pan-RAF and ERK-selective inhibitors as monotherapy in KRAS-mutant PDAC cell models and we observed strong anti-tumor activity with single agent treatment (Hayes et al., 2016; Ozkan-Dagliyan et al., 2020).

More recently, the concept of vertical inhibition of ERK signaling has emerged, where combined inhibition of distinct components of the cascade can act synergistically by reducing vulnerability to loss of negative feedback inhibition (Fernandes Neto et al., 2020; Lamba et al., 2014; Merchant et al., 2017; Ozkan-Dagliyan et al., 2020; Yen et al., 2018). This premise was first demonstrated in BRAF-mutant melanomas, where combination of a MEK inhibitor and a BRAF inhibitor showed greater efficacy, and reduced toxicity, compared with single agent treatment (Flaherty et al., 2012; Larkin et al., 2014; Long et al., 2014). This concept was later extended to RAS-mutant cancers, where concurrent treatment with inhibitors that target two or more components of the EGFR-RAS-RAF-MEK-ERK cascade showed greater anti-tumor efficacy than any single agent. Initially, two groups showed that dual inhibition of RAF and MEK could overcome ERK reactivation that occurs with single agent MEK inhibition in KRAS-mutant cancer cells in vitro (Yen et al., 2018). Another study in KRAS-mutant cancers expanded upon this premise to show that inhibition of any two distinct nodes of the EGFR-RAF-MEK-ERK pathway was more potent than any single agent treatment (Ozkan-Dagliyan et al., 2020). Current ongoing clinical trials are evaluating vertical inhibition-based combinations.

As mentioned above, a second key RAS effector pathway involves PI3K-AKT-mTOR signaling. Additionally, inhibition of ERK MAPK is commonly associated with further upregulation of RTK activity and increased PI3K signaling (Chandarlapaty, 2012). Thus, it is not surprising that studies in preclinical cell culture and mouse models showed that concurrent inhibition of RAF and PI3K effector signaling exhibited greater anti-tumor efficacy than either alone (Collisson et al., 2012). However, the toxicity of this combination has prevented extension of this combination approach to clinical practice (Hanker, Kaklamani, & Arteaga, 2019).

In addition to combinations based on overcoming loss of ERK feedback inhibition, numerous unbiased system-wide screening studies have been done to identify additional combinations to enhance ERK MAPK anti-tumor activity. These include powerful genetic approaches such as

CRISPR- or RNAi-mediated loss-of-function or cDNA overexpression/activation gain-of-function screens (Wood, 2015). For example, a CRISPR-Cas9 screen demonstrated that ERK reactivation is a primary mechanism limiting the efficacy of $KRAS^{G12C}$ inhibitors (Lou et al., 2019). Indeed, two recent studies have now demonstrated that genetic alterations resulting in activated BRAF and MEK1 are associated with the onset of resistance to $KRAS^{G12C}$ inhibitors (Awad et al., 2021; Tanaka et al., 2021). Thus, reactivation of the ERK MAPK cascade is one significant mechanism of resistance to direct inhibitors of RAS. To overcome this, current clinical trials combine inhibitors of $KRAS^{G12C}$ with inhibitors that target either upstream (e.g., SOS1, SHP2, EGFR) or downstream (MEK) components of the ERK MAPK pathway (clinicaltrials.gov).

Similarly, Wood et al. applied an activated signaling expression library to identify both known and novel mechanisms that drive resistance to ERK MAPK inhibitors (Martz et al., 2014). In a complementary approach (Ozkan-Dagliyan et al., 2020), our chemical library screen showed that concurrent inhibition of ERK MAPK components, PI3K signaling, and other functionally diverse proteins (e.g., microtubules, HSP90 and histone deacetylases) can synergistically enhance the efficacy of RAF inhibitors in KRAS-mutant PDAC.

8. ERK regulation of a diverse phosphoproteome and transcriptome

While the substrates of RAF and MEK kinases are highly restricted, activation of ERK1/2 serine/threonine kinases can cause direct or indirect phosphorylation of a diverse spectrum of more than one thousand proteins (Unal, Uhlitz, & Bluthgen, 2017). Since ERK substrates include numerous other protein kinases (e.g., RSK1–4, MNK2 and MSK1/2) (Fig. 2B), ERK activation can regulate a highly diverse phosphoproteome (Cargnello & Roux, 2011). Additional ERK substrates include transcription factors (e.g., MYC), resulting in ERK regulation of a complex transcriptome.

Because the ERK consensus sequence (S/T-P) is shared by protein targets of other MAPK family members such as JNK and p38 (Fig. 2C), multiple substrate-interacting regions confer specificity to ERK-substrate binding (Tanoue, Adachi, Moriguchi, & Nishida, 2000). The D-site recruitment site (DRS) on ERK contains a negatively charged common docking domain and hydrophobic groove that, along with the F-site recruitment site (FRS) that is exclusively accessible when ERK is activated, enables favorable, specific interactions between phosphorylated ERK and its substrates (Burkhard, Chen, & Shapiro, 2011; Lee et al., 2004; Sheridan, Kong, Parker, Dalby, & Turk, 2008). Similarly, the majority of ERK target proteins contain a D-site and/or DEF (F-site) domain, which promote stable association with ERK (Roskoski, 2012). Mutating these domains reduces ERK-dependent phosphorylation and binding (Dimitri, Dowdle, MacKeigan, Blenis, & Murphy, 2005). Further, peptides engineered to express these ERK docking sites display dramatic increases in ERK binding affinity (Fernandes, Bailey, Vanvranken, & Allbritton, 2007). However, while most ERK substrates contain a D-site or DEF domain, these are not strictly required. Among the hundreds of currently identified direct ERK substrates, approximately 80% harbor a D-site or a DEF domain (Murphy, MacKeigan, & Blenis, 2004; Unal et al., 2017).

Many groups have undertaken ERK substrate identification, utilizing model systems ranging from untransformed rat cells to human cancer cell lines, and methods encompassing genetic, biochemical, and proteomic approaches. Together, these studies have revealed an extensive list of ERK targets, diverse in identity and function that are highly context-dependent. Carlson et al. used a chemical genetic screen to identify novel ERK substrates involved in mRNA processing and Rho GTPase signaling, containing both canonical and novel target motifs (Carlson et al., 2011). Another group used quantitative phosphoproteomics to identify 155 candidate ERK substrates associated with transcription, RNA splicing, and cytoskeleton dynamics (Courcelles et al., 2013). Still others have utilized pharmacologic ERK inhibition coupled with mass spectrometry to determine ERK target genes implicated in nuclear and cytoplasmic transport, nucleotide excision repair, nucleosome assembly, and membrane trafficking (Kosako et al., 2009; Lewis et al., 2000). SILAC (stable isotope labeling by amino acids in cell culture) identified hundreds of ERK substrates in cells treated with MAPK-stimulating growth factors both with and without ERK inhibition, expanding the repertoire of ERK-mediated oncogenicity to include chromatin remodeling processes (Pan, Olsen, Daub, & Mann, 2009).

A comprehensive survey of major ERK substrate-focused phosphoproteomic studies identified 659 "direct" ERK substrates, verified by in vitro kinase binding assays, and 1848 "indirect" or putative ERK substrates (Unal et al., 2017). Strikingly, only 30% of the direct targets and 10% of the indirect targets were observed in more than one study. These stark differences are likely due to distinct experimental techniques, model systems, or methods of ERK perturbation, in addition to intrinsic limitations of proteomics such as undersampling, low coverage, and target cut-off variability (Aslam, Basit, Nisar, Khurshid, & Rasool, 2017; Bensaddek, Nicolas, & Lamond, 2015). However, 90% of direct targets and 50% of indirect targets contained the canonical ERK phosphorylation motif (S/T-P), and approximately 80% of direct substrates contained either a D-site or a DEF domain, reinforcing known mechanisms of ERK signal transduction while also demonstrating our incomplete understanding of ERK signaling regulation (Unal et al., 2017). A searchable online database of all currently identified ERK substrates was created from this comprehensive meta-analysis and can be accessed at http://sys-bio.net/erk_targets.

These and other studies have identified transcription factors, large GTPases, and other protein kinases as direct ERK substrates, which together rapidly amplify and diversify MAPK signaling output (Fig. 4A). Upon ERK activation, the ETS transcription factor ELK1 is immediately phosphorylated; together with serum response factor (SRF), this stimulates promoters of immediate early genes (IEGs) (Marais, Wynne, & Treisman, 1993; Mylona et al., 2016). As ELK1 is constitutively expressed, transcription of such IEGs, most notably MYC, JUN, FOS, FOSL1, IER3, and EGR1, can occur within minutes of ERK phosphorylation (Gille, Sharrocks, & Shaw, 1992; Murphy et al., 2004; Murphy, Smith, Chen, Fingar, & Blenis, 2002; Sears et al., 2000). Additionally, the proteins encoded by many of these genes, including MYC, FOS, and FOSL1 (encoding Fra-1), are also direct ERK substrates (Basbous, Chalbos, Hipskind, Jariel-Encontre, & Piechaczyk, 2007; Murphy et al., 2002). They contain ERK docking sites and undergo stabilizing phosphorylation events as a result of sustained MAPK signaling, a signature characteristic of most RAS-driven cancers (Murphy et al., 2002, 2004). Unsurprisingly, KRAS-mutant tumors display heightened expression of many of these genes. Indeed, in mammalian cells MYC, JUN, and FOS have intrinsic transforming ability of their own (Belguise, Kersual, Galtier, & Chalbos, 2005; Land, Parada, & Weinberg, 1983; Okazaki & Sagata, 1995). Upon ERK activation, FOS and Fra-1 each form heterodimeric AP-1 transcriptional complexes with JUN, stimulating gene expression at promoter TPA response elements and driving cell growth, proliferation, and survival (Adiseshaiah, Peddakama, Zhang, Kalvakolanu, & Reddy, 2005; Basbous et al., 2007; Chinenov & Kerppola, 2001). These ERK-dependent complexes also repress transcription of growth-inhibitory genes such as JunD, Gadd45a, Sox6, Ddit3, and Tob1 (Eblen, 2018). Tob1, which suppresses cyclin D1 expression, is also a

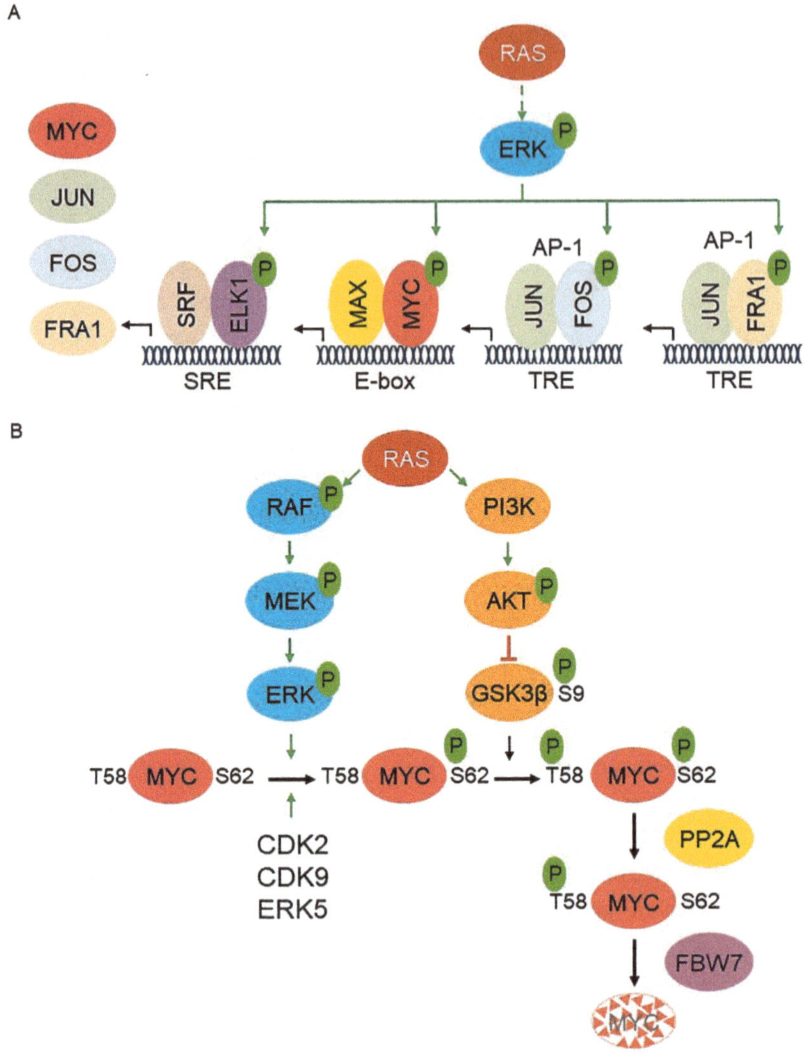

Fig. 4 ERK regulation of MYC gene transcription and protein stability. (A) RAS effector signaling regulates MYC protein stability in part through phosphorylation events that regulate FBXW7 E3 ligase-mediated protein degradation. Unphosphorylated wild-type MYC is unstable with a half-life of ~20 min in normal cells. ERK phosphorylation at S62 promotes MYC protein stability and transcriptional activity. Phospho-S62 additionally primes MYC for GSK3β – mediated phosphorylation at MYC residue T58. Protein phosphatase 2A (PP2A) catalyzes removal of phosphorylation at S62, leading to FBXW7-mediated polyubiquitination and MYC degradation. PI3K-AKT signaling leads to phosphorylation and inactivation of GSK3β, reducing phosphorylation at T58 and increasing MYC stability. (B) ERK regulates the expression of MYC and other oncoprotein transcription factors through both transcriptional and posttranslational mechanisms that promote protein stability. MYC forms heterodimers with the MAX transcription factor, which then recognizes and stimulates gene transcription from *E*-box DNA element-containing promoters. However, MYC overexpression leads to MYC-MAX stimulation of virtually all active promoters, thereby promoting cancer growth. ERK phosphorylation and activation of the ELK1 transcription factor together with serum response factor (SRF) stimulates promoters containing serum response elements (SRE) that are present in immediate early gene (IEG) promoters. FOS/FRA1 and JUN heterodimers form AP-1 complexes that further stimulate gene transcription from TPA response element (TRE) containing promoters.

direct ERK substrate. ERK phosphorylation of Tob1 inhibits its antiproliferative functions and contributes to G1 progression (Suzuki et al., 2002). ERK also downregulates repressive cell cycle genes through inhibitory phosphorylation of the FOXO3 transcription factor, promoting its ubiquitination and degradation (Eblen, 2018; Yang et al., 2008).

Protein kinase ERK targets, whose own substrates directly influence cell proliferation and survival, include p90RSK, MSK1/2, and MNK (Arthur, 2008; Doehn et al., 2009; Sturgill, Ray, Erikson, & Maller, 1988; Waskiewicz, Flynn, Proud, & Cooper, 1997). The kinase DYRK1B is also involved in cell cycle progression, and ERK-driven phosphorylation at its S421 site is sufficient to drive tumorigenesis (Ashford et al., 2016). The GTPase DRP1, a key regulator of mitochondrial morphology, is another major ERK target. Abrogation of DRP1 function or ERK signaling caused accumulation of fused mitochondria and decreased cell growth (Bryant et al., 2019; Nagdas et al., 2019). ERK also directly phosphorylates myosin light chain kinase as well as FAK, thereby integrating cell adhesion and migration signaling with cell survival pathways (Klemke et al., 1997). Further, ERK can regulate protein synthesis via phosphorylation of TSC1/2 (Ma, Chen, Erdjument-Bromage, Tempst, & Pandolfi, 2005), activate rRNA transcription and accelerate cell cycle progression via UBF phosphorylation (Stefanovsky et al., 2001), and regulate chromatin remodeling through phosphorylation of PARP-1 (Cohen-Armon et al., 2007).

9. MYC: The critical determinant for ERK-dependent cancer growth?

Despite the diversity of ERK targets, early seminal studies by Sears and colleagues showed that the MYC transcription factor and oncoprotein is a key ERK substrate critical for driving RAS-dependent growth (Fig. 4B) (Sears et al., 2000). MYC is a master transcriptional regulator, forming heterodimers with its binding partner MAX to drive transcription at *E*-box promoter sequences as well as to increase transcription at all active promoters (Dang, 2012; Grandori, Cowley, James, & Eisenman, 2000; Nie et al., 2020). RAS-driven activation of ERK rapidly activates MYC gene transcription through ELK1 IEG transcription. Importantly, ERK also directly stabilizes the MYC protein by catalyzing phosphorylation of MYC at its serine-62 site, dramatically increasing its half-life (Sears et al., 2000). Parallel activation of the PI3K pathway by RAS in many RAS-mutant cancers further supports ERK-driven MYC oncogenicity by inhibiting GSK3β activity. Under normal conditions, MYC phosphorylation at its S62 site induces GSK3β-mediated phosphorylation of its T58 residue, causing subsequent S62 dephosphorylation and labeling the protein for proteasomal degradation (Yeh et al., 2004). However, RAS activation of PI3K-AKT signaling inhibits GSK3β activity. Thus, MYC expression is normally tightly regulated and displays high rates of transcriptional and translational turnover. However, sustained ERK MAPK signaling, often coupled with aberrant activation of parallel RAS-driven pathways such as PI3K, can drive significant overexpression of MYC downstream of ERK in multiple malignancies, such as PDAC.

MYC expression is deregulated in over 70% of human cancer cases (Dang, 2012). While gene amplification is the most common mechanism of MYC oncogenicity, MYC is also overexpressed in many malignancies, including prostate cancer and triple negative breast cancer where RAS mutations are infrequent (Gurel et al., 2008; Palaskas et al., 2011). In many RAS-mutant cancers, such as PDAC, aberrant MYC expression as a result of RAS-ERK signaling correlates with disease severity (Vaseva et al., 2018). One study found that, in clinical samples from 69 patients, MYC overexpression occurred in 44% of primary tumors and MYC amplification in 32% of tumors, events that were largely mutually exclusive (Schleger, Verbeke, Hildenbrand, Zentgraf, & Bleyl, 2002). In a comprehensive analysis of 149 PDAC patient samples by The Cancer Genome Atlas (TCGA), MYC gene amplifications were observed in only 5% of PDAC patient cases, but MYC protein overexpression was not assessed (TCGA, 2017). Another group examined normal and malignant pancreatic tissue and concluded that while untransformed pancreas was negative for immunohistochemical (IHC) staining of MYC, 38% of PDAC samples exhibited positive staining (Lin et al., 2013).

Studies in mouse models have shown that Myc overexpression alone is sufficient to phenocopy mutant Kras and drive the development of metastatic PDAC (Lin et al., 2013). Conversely, direct suppression of MYC impairs the growth of KRAS-mutant human PDAC cell lines and causes regression of Kras-driven tumors in syngeneic orthotopic PDAC mouse models (Vaseva et al., 2018). Rapid regression of both precursor lesions and established tumors was similarly observed upon Myc suppression in Kras-driven lung tumors in mice (Soucek et al., 2013). Recently, the dominant negative omoMYC mini-protein not only caused rapid regression of murine lung, glioma, and PDAC tumors, but also demonstrated reversible off-target effects in normal proliferating tissues (Annibali et al., 2014;

Beaulieu et al., 2019; Savino et al., 2011; Sodir et al., 2020; Soucek et al., 2008). Importantly, KRAS-mutant PDAC cells rely on ERK-driven MYC stabilization more than on the MYC-protective functions of PI3K. Sensitivity or resistance to ERK inhibition in a panel of KRAS-mutant human PDAC cell lines correlated with loss or accumulation of MYC protein, respectively (Hayes et al., 2016), and pharmacologic inhibition of RAF or ERK significantly suppressed expression of MYC target genes as determined by RNA-sequencing analysis (Ozkan-Dagliyan et al., 2020). However, pharmacologic inhibition of PI3K or AKT did not alter MYC protein abundance or stabilization, suggesting that MYC overexpression in PDAC is solely due to ERK signaling (Vaseva et al., 2018). Taken together, these data indicate that ERK-driven MYC activity is necessary and sufficient to support KRAS-driven PDAC growth.

10. KRAS, ERK, MYC and metabolism: An Achilles heel for KRAS-mutant PDAC?

A primary role of MYC in supporting KRAS/ERK-driven oncogenicity in PDAC is regulation of key metabolic processes. RAS-mutant cancers are characterized by abnormal metabolic processes that include increased glycolysis, altered mitochondrial function, and increased activities of two lysosome-dependent processes, macropinocytosis and autophagy (Bryant, Mancias, Kimmelman, & Der, 2014). An elegant study showed that $KRAS^{G12D}$ stimulated glucose uptake and shunted glycolytic intermediates through the hexosamine biosynthesis (HBP) and pentose phosphate pathways (PPP) in murine pancreatic tumors (Ying et al., 2012). Inhibition of the MAPK pathway with the MEK inhibitor AZD8330 recapitulated these changes, similarly causing upregulation of glycolytic, HBP, and non-oxidative PPP genes. Pharmacologic ERK inhibition in human pancreatic cancer cell lines showed similar metabolic dependencies (Bryant et al., 2019), establishing the MAPK pathway as a critical mediator of KRAS-driven metabolic rewiring (Bryant et al., 2019). Importantly, shRNA knockdown of Myc significantly downregulated metabolite changes in identical pathways, while pharmacologic inhibition of PI3K did not recapitulate these results, indicating that MYC regulates glucose utilization in KRAS/ERK-driven PDAC largely independently of PI3K signaling (Ying et al., 2012). Another group has similarly shown that ERK MAPK signaling activates MYC-driven transcription of glycolytic genes and the non-oxidative PPP gene *RPIA*, ultimately driving nucleotide biosynthesis in response to the heightened energetic demands associated with mutant KRAS activity (Santana-Codina et al., 2018). PI3K was dispensable for this process, but was able to drive resistance to MAPK inhibition and replenish intracellular nucleotide pools.

RAS-driven ERK stabilization of MYC is also implicated in other aspects of metabolic regulation. For example, ERK regulates mitochondrial fragmentation directly via activating phosphorylation of its substrate DRP1 (Kashatus et al., 2015; Nagdas et al., 2019; Serasinghe et al., 2015) and through phosphorylation of MYC, which can bind the DRP1 promoter and initiate its transcription (Agarwal, Altman, Ho Seo, et al., 2019). MYC activity has also been linked to mitochondrial trafficking, reactive oxygen species regulation, and local metabolic alterations (Agarwal, Altman, Ho Seo, et al., 2019; Agarwal, Altman, Seo, et al., 2019).

To supplement their nutrient supply and satisfy stringent metabolic requirements, RAS-transformed cells internalize extracellular components in endocytic vesicles through the process of macropinocytosis, which also releases their dependence on glutamine (Commisso et al., 2013; Son et al., 2013). Suppression of either KRAS or MYC in KRAS-mutant PDAC cell lines impairs endocytic nutrient uptake, suggesting a role for MYC in facilitating yet another aspect of RAS-driven metabolic alterations (Hobbs et al., 2020). Lastly, the role of ERK in regulating autophagy, the process of recycling internal cellular components into basic energetic building blocks as an alternative nutrient source, has been thoroughly described by multiple groups (Bryant et al., 2019; Kinsey et al., 2019; Yang et al., 2011). Pharmacologic inhibition of macropinocytosis and/or autophagy has been shown to impair the tumorigenic growth of KRAS-mutant PDAC in various preclinical mouse models (Commisso et al., 2013; Yang et al., 2011). However, there are currently no clinically useful inhibitors of macropinocytosis, and clinically available autophagy inhibitors (hydroxychloroquine, chloroquine) are limited by lack of potency and selectivity.

Given the extensive role of MYC in ERK-driven metabolic regulation, it is tempting to speculate that MYC-driven transcriptional programs may similarly regulate aspects of autophagy. As the dominant negative omoMYC enters clinical trials as a promising direct MYC inhibitor for the treatment of triple negative breast cancer and RAS-mutant lung and colorectal cancers, it will be interesting to see how patients respond. Targeting a transcription factor is notoriously difficult, yet neutralizing this versatile substrate of the ERK MAPK pathway may be critical to restraining the transforming power of RAS-ERK signaling.

Acknowledgments

We thank Anna Barkovskaya for artwork. Our research was supported by grants from the National Cancer Institute (NCI) to A.D.C. and/or C.J.D. from the NCI (R21CA 179193, R01CA42978, R01CA175747, R01CA223775, P50CA196510, U01CA199235, P01CA203657 and R35CA232113), and from the Pancreatic Cancer Action Network/AACR (15-90-25-DER), Department of Defense (W81XWH-15-1-0611), and the Lustgarten Foundation (388222) to C.J.D. J.N.D. and I.O.D. were supported by the Slomo and Cindy Silvian Foundation. J.N.D. was also supported by NCI T32CA071341 and F30CA243253. P.S.H. was supported by NCI T32CA071341. C.M.G. was supported by NCI T32CA009156 and F32CA221005.

Disclosure of potential conflicts of interest

C.J.D. is a consultant/advisory board member for Mirati Therapeutics and Deciphera Pharmaceuticals. C.J.D. has received research funding support from Mirati Therapeutics and Deciphera Pharmaceuticals. C.J.D. has consulted for Axon Advisors LLC, Eli Lilly, Jazz Therapeutics, Revolution Medicines, Ribometrix, SmartAnalyst, SVB Leerink, Third Bridge and Turning Point Therapeutics. A.D.C. has consulted for Eli Lilly and Mirati Therapeutics.

References

Adiseshaiah, P., Peddakama, S., Zhang, Q., Kalvakolanu, D. V., & Reddy, S. P. (2005). Mitogen regulated induction of FRA-1 proto-oncogene is controlled by the transcription factors binding to both serum and TPA response elements. *Oncogene*, *24*(26), 4193–4205. https://doi.org/10.1038/sj.onc.1208583.

Agarwal, E., Altman, B. J., Ho Seo, J., Bertolini, I., Ghosh, J. C., Kaur, A., et al. (2019). Myc regulation of a mitochondrial trafficking network mediates tumor cell invasion and metastasis. *Molecular and Cellular Biology*, *39*(14), e00109-19. https://doi.org/10.1128/MCB.00109-19.

Agarwal, E., Altman, B. J., Seo, J. H., Ghosh, J. C., Kossenkov, A. V., Tang, H. Y., et al. (2019). Myc-mediated transcriptional regulation of the mitochondrial chaperone TRAP1 controls primary and metastatic tumor growth. *The Journal of Biological Chemistry*, *294*(27), 10407–10414. https://doi.org/10.1074/jbc.AC119.008656.

Annibali, D., Whitfield, J. R., Favuzzi, E., Jauset, T., Serrano, E., Cuartas, I., et al. (2014). Myc inhibition is effective against glioma and reveals a role for Myc in proficient mitosis. *Nature Communications*, *5*, 4632. https://doi.org/10.1038/ncomms5632.

Arthur, J. S. (2008). MSK activation and physiological roles. *Frontiers in Bioscience*, *13*, 5866–5879. https://doi.org/10.2741/3122.

Ashford, A. L., Dunkley, T. P., Cockerill, M., Rowlinson, R. A., Baak, L. M., Gallo, R., et al. (2016). Identification of DYRK1B as a substrate of ERK1/2 and characterisation of the kinase activity of DYRK1B mutants from cancer and metabolic syndrome. *Cellular and Molecular Life Sciences*, *73*(4), 883–900. https://doi.org/10.1007/s00018-015-2032-x.

Aslam, B., Basit, M., Nisar, M. A., Khurshid, M., & Rasool, M. H. (2017). Proteomics: Technologies and their applications. *Journal of Chromatographic Science*, *55*(2), 182–196. https://doi.org/10.1093/chromsci/bmw167.

Awad, M. M., Liu, S., Rybkin, I. I., Arbour, K. C., Dilly, J., Zhu, V. W., et al. (2021). Acquired resistance to KRAS(G12C) inhibition in cancer. *The New England Journal of Medicine*, *384*(25), 2382–2393. https://doi.org/10.1056/NEJMoa2105281.

Basbous, J., Chalbos, D., Hipskind, R., Jariel-Encontre, I., & Piechaczyk, M. (2007). Ubiquitin-independent proteasomal degradation of Fra-1 is antagonized by Erk1/2 pathway-mediated phosphorylation of a unique C-terminal destabilizer. *Molecular and Cellular Biology*, *27*(11), 3936–3950. https://doi.org/10.1128/MCB.01776-06.

Beaulieu, M. E., Jauset, T., Masso-Valles, D., Martinez-Martin, S., Rahl, P., Maltais, L., et al. (2019). Intrinsic cell-penetrating activity propels Omomyc from proof of concept to viable anti-MYC therapy. *Science Translational Medicine*, *11*(484), eaar5012. https://doi.org/10.1126/scitranslmed.aar5012.

Belguise, K., Kersual, N., Galtier, F., & Chalbos, D. (2005). FRA-1 expression level regulates proliferation and invasiveness of breast cancer cells. *Oncogene*, *24*(8), 1434–1444. https://doi.org/10.1038/sj.onc.1208312.

Bensaddek, D., Nicolas, A., & Lamond, A. I. (2015). Evaluating the use of HILIC in large-scale, multi dimensional proteomics: Horses for courses? *International Journal of Mass Spectrometry*, *391*, 105–114. https://doi.org/10.1016/j.ijms.2015.07.029.

Berndt, N., Hamilton, A. D., & Sebti, S. M. (2011). Targeting protein prenylation for cancer therapy. *Nature Reviews. Cancer*, *11*(11), 775–791. https://doi.org/10.1038/nrc3151.

Blasco, R. B., Francoz, S., Santamaria, D., Canamero, M., Dubus, P., Charron, J., et al. (2011). c-Raf, but not B-Raf, is essential for development of K-Ras oncogene-driven non-small cell lung carcinoma. *Cancer Cell*, *19*(5), 652–663. https://doi.org/10.1016/j.ccr.2011.04.002.

Blumenschein, G. R., Jr., Smit, E. F., Planchard, D., Kim, D. W., Cadranel, J., De Pas, T., et al. (2015). A randomized phase II study of the MEK1/MEK2 inhibitor trametinib (GSK1120212) compared with docetaxel in KRAS-mutant advanced non-small-cell lung cancer (NSCLC)dagger. *Annals of Oncology*, *26*(5), 894–901. https://doi.org/10.1093/annonc/mdv072.

Bodoky, G., Timcheva, C., Spigel, D. R., La Stella, P. J., Ciuleanu, T. E., Pover, G., et al. (2012). A phase II open-label randomized study to assess the efficacy and safety of selumetinib (AZD6244 [ARRY-142886]) versus capecitabine in patients with advanced or metastatic pancreatic cancer who have failed first-line gemcitabine therapy. *Investigational New Drugs*, *30*(3), 1216–1223. https://doi.org/10.1007/s10637-011-9687-4.

Bryant, K. L., Mancias, J. D., Kimmelman, A. C., & Der, C. J. (2014). KRAS: Feeding pancreatic cancer proliferation. *Trends in Biochemical Sciences*, *39*(2), 91–100. https://doi.org/10.1016/j.tibs.2013.12.004.

Bryant, K. L., Stalnecker, C. A., Zeitouni, D., Klomp, J. E., Peng, S., Tikunov, A. P., et al. (2019). Combination of ERK and autophagy inhibition as a treatment approach for pancreatic cancer. *Nature Medicine*, *25*(4), 628–640. https://doi.org/10.1038/s41591-019-0368-8.

Burkhard, K. A., Chen, F., & Shapiro, P. (2011). Quantitative analysis of ERK2 interactions with substrate proteins: Roles for kinase docking domains and activity in determining binding affinity. *The Journal of Biological Chemistry*, *286*(4), 2477–2485. https://doi.org/10.1074/jbc.M110.177899.

Canon, J., Rex, K., Saiki, A. Y., Mohr, C., Cooke, K., Bagal, D., et al. (2019). The clinical KRAS(G12C) inhibitor AMG 510 drives anti-tumour immunity. *Nature*, *575*(7781), 217–223. https://doi.org/10.1038/s41586-019-1694-1.

Cargnello, M., & Roux, P. P. (2011). Activation and function of the MAPKs and their substrates, the MAPK-activated protein kinases. *Microbiology and Molecular Biology Reviews*, *75*(1), 50–83. https://doi.org/10.1128/MMBR.00031-10.

Carlson, S. M., Chouinard, C. R., Labadorf, A., Lam, C. J., Schmelzle, K., Fraenkel, E., et al. (2011). Large-scale discovery of ERK2 substrates identifies ERK-mediated transcriptional regulation by ETV3. *Science Signaling*, *4*(196), rs11. https://doi.org/10.1126/scisignal.2002010.

Castellano, E., & Downward, J. (2011). RAS interaction with PI3K: More than just another effector pathway. *Genes & Cancer*, *2*(3), 261–274. https://doi.org/10.1177/1947601911408079.

Chandarlapaty, S. (2012). Negative feedback and adaptive resistance to the targeted therapy of cancer. *Cancer Discovery*, *2*(4), 311–319. https://doi.org/10.1158/2159-8290.CD-12-0018.

Chen, S. H., Zhang, Y., Van Horn, R. D., Yin, T., Buchanan, S., Yadav, V., et al. (2016). Oncogenic BRAF deletions that function as homodimers and are sensitive to inhibition by RAF dimer inhibitor LY3009120. *Cancer Discovery*, *6*(3), 300–315. https://doi.org/10.1158/2159-8290.CD-15-0896.

Chi, S., Kitanaka, C., Noguchi, K., Mochizuki, T., Nagashima, Y., Shirouzu, M., et al. (1999). Oncogenic Ras triggers cell suicide through the activation of a caspase-independent cell death program in human cancer cells. *Oncogene*, *18*(13), 2281–2290. https://doi.org/10.1038/sj.onc.1202538.

Chinenov, Y., & Kerppola, T. K. (2001). Close encounters of many kinds: Fos-Jun interactions that mediate transcription regulatory specificity. *Oncogene*, *20*(19), 2438–2452. https://doi.org/10.1038/sj.onc.1204385.

Cisowski, J., Sayin, V. I., Liu, M., Karlsson, C., & Bergo, M. O. (2016). Oncogene-induced senescence underlies the mutual exclusive nature of oncogenic KRAS and BRAF. *Oncogene*, *35*(10), 1328–1333. https://doi.org/10.1038/onc.2015.186.

Cohen-Armon, M., Visochek, L., Rozensal, D., Kalal, A., Geistrikh, I., Klein, R., et al. (2007). DNA-independent PARP-1 activation by phosphorylated ERK2 increases Elk1 activity: A link to histone acetylation. *Molecular Cell*, *25*(2), 297–308. https://doi.org/10.1016/j.molcel.2006.12.012.

Collins, M. A., Bednar, F., Zhang, Y., Brisset, J. C., Galban, S., Galban, C. J., et al. (2012). Oncogenic Kras is required for both the initiation and maintenance of pancreatic cancer in mice. *The Journal of Clinical Investigation*, *122*(2), 639–653. https://doi.org/10.1172/JCI59227.

Collisson, E. A., Trejo, C. L., Silva, J. M., Gu, S., Korkola, J. E., Heiser, L. M., et al. (2012). A central role for RAF- ->MEK- ->ERK signaling in the genesis of pancreatic ductal adenocarcinoma. *Cancer Discovery*, *2*(8), 685–693. https://doi.org/10.1158/2159-8290.CD-11-0347.

Commisso, C., Davidson, S. M., Soydaner-Azeloglu, R. G., Parker, S. J., Kamphorst, J. J., Hackett, S., et al. (2013). Macropinocytosis of protein is an amino acid supply route in Ras-transformed cells. *Nature*, *497*(7451), 633–637. https://doi.org/10.1038/nature12138.

Courcelles, M., Fremin, C., Voisin, L., Lemieux, S., Meloche, S., & Thibault, P. (2013). Phosphoproteome dynamics reveal novel ERK1/2 MAP kinase substrates with broad spectrum of functions. *Molecular Systems Biology*, *9*, 669. https://doi.org/10.1038/msb.2013.25.

Cox, A. D., & Der, C. J. (2010). Ras history: The saga continues. *Small GTPases*, *1*(1), 2–27. https://doi.org/10.4161/sgtp.1.1.12178.

Cox, A. D., Der, C. J., & Philips, M. R. (2015). Targeting RAS membrane association: Back to the future for anti-RAS drug discovery? *Clinical Cancer Research*, *21*(8), 1819–1827. https://doi.org/10.1158/1078-0432.CCR-14-3214.

Cox, A. D., Fesik, S. W., Kimmelman, A. C., Luo, J., & Der, C. J. (2014). Drugging the undruggable RAS: Mission possible? *Nature Reviews. Drug Discovery*, *13*(11), 828–851. https://doi.org/10.1038/nrd4389.

Dang, C. V. (2012). MYC on the path to cancer. *Cell, 149*(1), 22–35. https://doi.org/10.1016/j.cell.2012.03.003.
Dang, C. V., Reddy, E. P., Shokat, K. M., & Soucek, L. (2017). Drugging the 'undruggable' cancer targets. *Nature Reviews. Cancer, 17*(8), 502–508. https://doi.org/10.1038/nrc.2017.36.
Dimitri, C. A., Dowdle, W., MacKeigan, J. P., Blenis, J., & Murphy, L. O. (2005). Spatially separate docking sites on ERK2 regulate distinct signaling events in vivo. *Current Biology, 15*(14), 1319–1324. https://doi.org/10.1016/j.cub.2005.06.037.
Doehn, U., Hauge, C., Frank, S. R., Jensen, C. J., Duda, K., Nielsen, J. V., et al. (2009). RSK is a principal effector of the RAS-ERK pathway for eliciting a coordinate promotile/invasive gene program and phenotype in epithelial cells. *Molecular Cell, 35*(4), 511–522. https://doi.org/10.1016/j.molcel.2009.08.002.
Duncan, J. S., Whittle, M. C., Nakamura, K., Abell, A. N., Midland, A. A., Zawistowski, J. S., et al. (2012). Dynamic reprogramming of the kinome in response to targeted MEK inhibition in triple-negative breast cancer. *Cell, 149*(2), 307–321. https://doi.org/10.1016/j.cell.2012.02.053.
Durrant, D. E., & Morrison, D. K. (2018). Targeting the Raf kinases in human cancer: The Raf dimer dilemma. *British Journal of Cancer, 118*(1), 3–8. https://doi.org/10.1038/bjc.2017.399.
Eblen, S. T. (2018). Extracellular-regulated kinases: Signaling from Ras to ERK substrates to control biological outcomes. *Advances in Cancer Research, 138*, 99–142. https://doi.org/10.1016/bs.acr.2018.02.004.
Eser, S., Reiff, N., Messer, M., Seidler, B., Gottschalk, K., Dobler, M., et al. (2013). Selective requirement of PI3K/PDK1 signaling for Kras oncogene-driven pancreatic cell plasticity and cancer. *Cancer Cell, 23*(3), 406–420. https://doi.org/10.1016/j.ccr.2013.01.023.
Fernandes, N., Bailey, D. E., Vanvranken, D. L., & Allbritton, N. L. (2007). Use of docking peptides to design modular substrates with high efficiency for mitogen-activated protein kinase extracellular signal-regulated kinase. *ACS Chemical Biology, 2*(10), 665–673. https://doi.org/10.1021/cb700158q.
Fernandes Neto, J. M., Nadal, E., Bosdriesz, E., Ooft, S. N., Farre, L., McLean, C., et al. (2020). Multiple low dose therapy as an effective strategy to treat EGFR inhibitor-resistant NSCLC tumours. *Nature Communications, 11*(1), 3157. https://doi.org/10.1038/s41467-020-16952-9.
Flaherty, K. T., Infante, J. R., Daud, A., Gonzalez, R., Kefford, R. F., Sosman, J., et al. (2012). Combined BRAF and MEK inhibition in melanoma with BRAF V600 mutations. *The New England Journal of Medicine, 367*(18), 1694–1703. https://doi.org/10.1056/NEJMoa1210093.
Foster, S. A., Whalen, D. M., Ozen, A., Wongchenko, M. J., Yin, J., Yen, I., et al. (2016). Activation mechanism of oncogenic deletion mutations in BRAF, EGFR, and HER2. *Cancer Cell, 29*(4), 477–493. https://doi.org/10.1016/j.ccell.2016.02.010.
Furuse, J., & Nagashima, F. (2017). Emerging protein kinase inhibitors for treating pancreatic cancer. *Expert Opinion on Emerging Drugs, 22*(1), 77–86. https://doi.org/10.1080/14728214.2017.1293648.
Gille, H., Sharrocks, A. D., & Shaw, P. E. (1992). Phosphorylation of transcription factor p62TCF by MAP kinase stimulates ternary complex formation at c-fos promoter. *Nature, 358*(6385), 414–417. https://doi.org/10.1038/358414a0.
Grandori, C., Cowley, S. M., James, L. P., & Eisenman, R. N. (2000). The Myc/Max/Mad network and the transcriptional control of cell behavior. *Annual Review of Cell and Developmental Biology, 16*, 653–699. https://doi.org/10.1146/annurev.cellbio.16.1.653.
Gurel, B., Iwata, T., Koh, C. M., Jenkins, R. B., Lan, F., Van Dang, C., et al. (2008). Nuclear MYC protein overexpression is an early alteration in human prostate carcinogenesis. *Modern Pathology, 21*(9), 1156–1167. https://doi.org/10.1038/modpathol.2008.111.

Haigis, K. M. (2017). KRAS alleles: The devil is in the detail. *Trends in Cancer, 3*(10), 686–697. https://doi.org/10.1016/j.trecan.2017.08.006.
Hallin, J., Engstrom, L. D., Hargis, L., Calinisan, A., Aranda, R., Briere, D. M., et al. (2020). The KRAS(G12C) inhibitor MRTX849 provides insight toward therapeutic susceptibility of KRAS-mutant cancers in mouse models and patients. *Cancer Discovery, 10*(1), 54–71. https://doi.org/10.1158/2159-8290.CD-19-1167.
Hanker, A. B., Kaklamani, V., & Arteaga, C. L. (2019). Challenges for the clinical development of PI3K inhibitors: Strategies to improve their impact in solid tumors. *Cancer Discovery, 9*(4), 482–491. https://doi.org/10.1158/2159-8290.CD-18-1175.
Hayes, T. K., Neel, N. F., Hu, C., Gautam, P., Chenard, M., Long, B., et al. (2016). Long-term ERK inhibition in KRAS-mutant pancreatic cancer is associated with MYC degradation and senescence-like growth suppression. *Cancer Cell, 29*(1), 75–89. https://doi.org/10.1016/j.ccell.2015.11.011.
Hingorani, S. R., Petricoin, E. F., Maitra, A., Rajapakse, V., King, C., Jacobetz, M. A., Ross, S., Conrads, T. P., Veenstra, T. D., Hitt, B. A., Kawaguchi, Y., Johann, D., Liotta, L. A., Crawford, H. C., Putt, M. E., Jacks, T., Wright, C. V., Hruban, R. H., Lowy, A. M., & Tuveson, D. A. (2003). Preinvasive and invasive ductal pancreatic cancer and its early detection in the mouse. Cancer Cell, 4(6), 437-450. https://doi.org/10.1016/s1535-6108(03)00309-x.
Hobbs, G. A., Baker, N. M., Miermont, A. M., Thurman, R. D., Pierobon, M., Tran, T. H., et al. (2020). Atypical KRAS(G12R) mutant is impaired in PI3K signaling and macropinocytosis in pancreatic cancer. *Cancer Discovery, 10*(1), 104–123. https://doi.org/10.1158/2159-8290.CD-19-1006.
Hong, D. S., Fakih, M. G., Strickler, J. H., Desai, J., Durm, G. A., Shapiro, G. I., et al. (2020). KRAS(G12C) inhibition with sotorasib in advanced solid tumors. *The New England Journal of Medicine, 383*(13), 1207–1217. https://doi.org/10.1056/NEJMoa1917239.
Hymowitz, S. G., & Malek, S. (2018). Targeting the MAPK pathway in RAS mutant cancers. *Cold Spring Harbor Perspectives in Medicine, 8*(11), a031492. https://doi.org/10.1101/cshperspect.a031492.
Infante, J. R., Somer, B. G., Park, J. O., Li, C. P., Scheulen, M. E., Kasubhai, S. M., et al. (2014). A randomised, double-blind, placebo-controlled trial of trametinib, an oral MEK inhibitor, in combination with gemcitabine for patients with untreated metastatic adenocarcinoma of the pancreas. *European Journal of Cancer, 50*(12), 2072–2081. https://doi.org/10.1016/j.ejca.2014.04.024.
Janne, P. A., Shaw, A. T., Pereira, J. R., Jeannin, G., Vansteenkiste, J., Barrios, C., Franke, F. A., Grinsted, L., Zazulina, V., Smith, P., Smith, I., & Crino, L. (2013). Selumetinib plus docetaxel for KRAS-mutant advanced non-small-cell lung cancer: A randomised, multicentre, placebo-controlled, phase 2 study. The Lancet Oncology, 14(1), 38-47. https://doi.org/10.1016/S1470-2045(12)70489-8.
Janne, P. A., van den Heuvel, M. M., Barlesi, F., Cobo, M., Mazieres, J., Crino, L., et al. (2017). Selumetinib plus docetaxel compared with docetaxel alone and progression-free survival in patients with KRAS-mutant advanced non-small cell lung cancer: The SELECT-1 randomized clinical trial. *JAMA, 317*(18), 1844–1853. https://doi.org/10.1001/jama.2017.3438.
Kanda, M., Matthaei, H., Wu, J., Hong, S. M., Yu, J., Borges, M., et al. (2012). Presence of somatic mutations in most early-stage pancreatic intraepithelial neoplasia. *Gastroenterology, 142*(4), 730–733 e739. https://doi.org/10.1053/j.gastro.2011.12.042.
Karreth, F. A., Frese, K. K., DeNicola, G. M., Baccarini, M., & Tuveson, D. A. (2011). C-Raf is required for the initiation of lung cancer by K-Ras(G12D). *Cancer Discovery, 1*(2), 128–136. https://doi.org/10.1158/2159-8290.CD-10-0044.

Kashatus, J. A., Nascimento, A., Myers, L. J., Sher, A., Byrne, F. L., Hoehn, K. L., et al. (2015). Erk2 phosphorylation of Drp1 promotes mitochondrial fission and MAPK-driven tumor growth. *Molecular Cell, 57*(3), 537–551. https://doi.org/10.1016/j.molcel.2015.01.002.

Kinsey, C. G., Camolotto, S. A., Boespflug, A. M., Guillen, K. P., Foth, M., Truong, A., et al. (2019). Protective autophagy elicited by RAF-->MEK-->ERK inhibition suggests a treatment strategy for RAS-driven cancers. *Nature Medicine, 25*(4), 620–627. https://doi.org/10.1038/s41591-019-0367-9.

Klemke, R. L., Cai, S., Giannini, A. L., Gallagher, P. J., de Lanerolle, P., & Cheresh, D. A. (1997). Regulation of cell motility by mitogen-activated protein kinase. *The Journal of Cell Biology, 137*(2), 481–492. https://doi.org/10.1083/jcb.137.2.481.

Klomp, J. E., Klomp, J. A., & Der, C. J. (2021). The ERK mitogen-activated protein kinase signaling network: The final frontier in RAS signal transduction. *Biochemical Society Transactions, 49*(1), 253–267. https://doi.org/10.1042/BST20200507.

Kosako, H., Yamaguchi, N., Aranami, C., Ushiyama, M., Kose, S., Imamoto, N., et al. (2009). Phosphoproteomics reveals new ERK MAP kinase targets and links ERK to nucleoporin-mediated nuclear transport. *Nature Structural & Molecular Biology, 16*(10), 1026–1035. https://doi.org/10.1038/nsmb.1656.

Lake, D., Correa, S. A., & Muller, J. (2016). Negative feedback regulation of the ERK1/2 MAPK pathway. *Cellular and Molecular Life Sciences, 73*(23), 4397–4413. https://doi.org/10.1007/s00018-016-2297-8.

Lamba, S., Russo, M., Sun, C., Lazzari, L., Cancelliere, C., Grernrum, W., et al. (2014). RAF suppression synergizes with MEK inhibition in KRAS mutant cancer cells. *Cell Reports, 8*(5), 1475–1483. https://doi.org/10.1016/j.celrep.2014.07.033.

Land, H., Parada, L. F., & Weinberg, R. A. (1983). Tumorigenic conversion of primary embryo fibroblasts requires at least two cooperating oncogenes. *Nature, 304*(5927), 596–602. https://doi.org/10.1038/304596a0.

Larkin, J., Ascierto, P. A., Dreno, B., Atkinson, V., Liszkay, G., Maio, M., et al. (2014). Combined vemurafenib and cobimetinib in BRAF-mutated melanoma. *The New England Journal of Medicine, 371*(20), 1867–1876. https://doi.org/10.1056/NEJMoa1408868.

Ledford, H. (2015). Cancer: The Ras renaissance. *Nature, 520*(7547), 278–280. https://doi.org/10.1038/520278a.

Lee, T., Hoofnagle, A. N., Kabuyama, Y., Stroud, J., Min, X., Goldsmith, E. J., Chen, L., Resing, K. A., & Ahn, N. G. (2004). Docking motif interactions in MAP kinases revealed by hydrogen exchange mass spectrometry. Molecular Cell, 14(1), 43-55. https://doi.org/10.1016/s1097-2765(04)00161-3.

Lewis, T. S., Hunt, J. B., Aveline, L. D., Jonscher, K. R., Louie, D. F., Yeh, J. M., Nahreini, T. S., Resing, K. A., & Ahn, N. G. (2000). Identification of novel MAP kinase pathway signaling targets by functional proteomics and mass spectrometry. Molecular Cell, 6(6), 1343-1354. https://doi.org/10.1016/s1097-2765(00)00132-5.

Lin, W. C., Rajbhandari, N., Liu, C., Sakamoto, K., Zhang, Q., Triplett, A. A., et al. (2013). Dormant cancer cells contribute to residual disease in a model of reversible pancreatic cancer. *Cancer Research, 73*(6), 1821–1830. https://doi.org/10.1158/0008-5472.CAN-12-2067.

Long, G. V., Stroyakovskiy, D., Gogas, H., Levchenko, E., de Braud, F., Larkin, J., et al. (2014). Combined BRAF and MEK inhibition versus BRAF inhibition alone in melanoma. *The New England Journal of Medicine, 371*(20), 1877–1888. https://doi.org/10.1056/NEJMoa1406037.

Lorenzo, C., & McCormick, F. (2020). SPRED proteins and their roles in signal transduction, development, and malignancy. *Genes & Development, 34*(21–22), 1410–1421. https://doi.org/10.1101/gad.341222.120.

Lou, K., Steri, V., Ge, A. Y., Hwang, Y. C., Yogodzinski, C. H., Shkedi, A. R., et al. (2019). KRAS(G12C) inhibition produces a driver-limited state revealing collateral dependencies. *Science Signaling*, *12*(583), eaaw9450. https://doi.org/10.1126/scisignal.aaw9450.

Ma, L., Chen, Z., Erdjument-Bromage, H., Tempst, P., & Pandolfi, P. P. (2005). Phosphorylation and functional inactivation of TSC2 by Erk implications for tuberous sclerosis and cancer pathogenesis. *Cell*, *121*(2), 179–193. https://doi.org/10.1016/j.cell.2005.02.031.

Marais, R., Wynne, J., & Treisman, R. (1993). The SRF accessory protein Elk-1 contains a growth factor-regulated transcriptional activation domain. Cell, 73(2), 381-393. https://doi.org/10.1016/0092-8674(93)90237-k.

Martz, C. A., Ottina, K. A., Singleton, K. R., Jasper, J. S., Wardell, S. E., Peraza-Penton, A., et al. (2014). Systematic identification of signaling pathways with potential to confer anticancer drug resistance. *Science Signaling*, 7(357), ra121. https://doi.org/10.1126/scisignal.aaa1877.

Merchant, M., Moffat, J., Schaefer, G., Chan, J., Wang, X., Orr, C., et al. (2017). Combined MEK and ERK inhibition overcomes therapy-mediated pathway reactivation in RAS mutant tumors. *PLoS One*, *12*(10), e0185862. https://doi.org/10.1371/journal.pone.0185862.

Morrison, D. K. (2012). MAP kinase pathways. *Cold Spring Harbor Perspectives in Biology*, *4*(11), a011254. https://doi.org/10.1101/cshperspect.a011254.

Moore, A. R., Rosenberg, S. C., McCormick, F., & Malek, S. (2020). RAS-targeted therapies: Is the undruggable drugged? *Nature Reviews. Drug Discovery*, *19*, 533–552. https://doi.org/10.1038/s41573-020-0068-6.

Murphy, L. O., MacKeigan, J. P., & Blenis, J. (2004). A network of immediate early gene products propagates subtle differences in mitogen-activated protein kinase signal amplitude and duration. *Molecular and Cellular Biology*, *24*(1), 144–153. https://doi.org/10.1128/MCB.24.1.144-153.2004.

Murphy, L. O., Smith, S., Chen, R. H., Fingar, D. C., & Blenis, J. (2002). Molecular interpretation of ERK signal duration by immediate early gene products. *Nature Cell Biology*, *4*(8), 556–564. https://doi.org/10.1038/ncb822.

Mylona, A., Theillet, F. X., Foster, C., Cheng, T. M., Miralles, F., Bates, P. A., et al. (2016). Opposing effects of Elk-1 multisite phosphorylation shape its response to ERK activation. *Science*, *354*(6309), 233–237. https://doi.org/10.1126/science.aad1872.

Nagdas, S., Kashatus, J. A., Nascimento, A., Hussain, S. S., Trainor, R. E., Pollock, S. R., et al. (2019). Drp1 promotes KRas-driven metabolic changes to drive pancreatic tumor growth. *Cell Reports*, *28*(7), 1845–1859 e1845. https://doi.org/10.1016/j.celrep.2019.07.031.

Nie, Z., Guo, C., Das, S. K., Chow, C. C., Batchelor, E., Simons, S. S. J., et al. (2020). Dissecting transcriptional amplification by MYC. *eLife*, *9*, e52483. https://doi.org/10.7554/eLife.52483.

Okazaki, K., & Sagata, N. (1995). The Mos/MAP kinase pathway stabilizes c-Fos by phosphorylation and augments its transforming activity in NIH 3T3 cells. *The EMBO Journal*, *14*(20), 5048–5059. https://www.ncbi.nlm.nih.gov/pubmed/7588633.

Ozkan-Dagliyan, I., Diehl, J. N., George, S. D., Schaefer, A., Papke, B., Klotz-Noack, K., et al. (2020). Low-dose vertical inhibition of the RAF-MEK-ERK cascade causes apoptotic death of KRAS mutant cancers. *Cell Reports*, *31*(11), 107764. https://doi.org/10.1016/j.celrep.2020.107764.

Palaskas, N., Larson, S. M., Schultz, N., Komisopoulou, E., Wong, J., Rohle, D., et al. (2011). 18F-fluorodeoxy-glucose positron emission tomography marks MYC-overexpressing human basal-like breast cancers. *Cancer Research*, *71*(15), 5164–5174. https://doi.org/10.1158/0008-5472.CAN-10-4633.

Pan, C., Olsen, J. V., Daub, H., & Mann, M. (2009). Global effects of kinase inhibitors on signaling networks revealed by quantitative phosphoproteomics. *Molecular & Cellular Proteomics*, *8*(12), 2796–2808. https://doi.org/10.1074/mcp.M900285-MCP200.
Papke, B., & Der, C. J. (2017). Drugging RAS: Know the enemy. *Science*, *355*(6330), 1158–1163. https://doi.org/10.1126/science.aam7622.
Prior, I. A., Hood, F. E., & Hartley, J. L. (2020). The frequency of Ras mutations in cancer. *Cancer Research*, *80*(14), 2969–2974. https://doi.org/10.1158/0008-5472.CAN-19-3682.
Roskoski, R., Jr. (2012). ERK1/2 MAP kinases: Structure, function, and regulation. *Pharmacological Research*, *66*(2), 105–143. https://doi.org/10.1016/j.phrs.2012.04.005.
Ryan, M. B., & Corcoran, R. B. (2018). Therapeutic strategies to target RAS-mutant cancers. *Nature Reviews. Clinical Oncology*, *15*(11), 709–720. https://doi.org/10.1038/s41571-018-0105-0.
Samatar, A. A., & Poulikakos, P. I. (2014). Targeting RAS-ERK signalling in cancer: Promises and challenges. *Nature Reviews. Drug Discovery*, *13*(12), 928–942. https://doi.org/10.1038/nrd4281.
Sanclemente, M., Francoz, S., Esteban-Burgos, L., Bousquet-Mur, E., Djurec, M., Lopez-Casas, P. P., et al. (2018). c-RAF ablation induces regression of advanced Kras/Trp53 mutant lung adenocarcinomas by a mechanism independent of MAPK signaling. *Cancer Cell*, *33*(2), 217–228 e214. https://doi.org/10.1016/j.ccell.2017.12.014.
Santana-Codina, N., Roeth, A. A., Zhang, Y., Yang, A., Mashadova, O., Asara, J. M., et al. (2018). Oncogenic KRAS supports pancreatic cancer through regulation of nucleotide synthesis. *Nature Communications*, *9*(1), 4945. https://doi.org/10.1038/s41467-018-07472-8.
Savino, M., Annibali, D., Carucci, N., Favuzzi, E., Cole, M. D., Evan, G. I., et al. (2011). The action mechanism of the Myc inhibitor termed Omomyc may give clues on how to target Myc for cancer therapy. *PLoS One*, *6*(7), e22284. https://doi.org/10.1371/journal.pone.0022284.
Schleger, C., Verbeke, C., Hildenbrand, R., Zentgraf, H., & Bleyl, U. (2002). c-MYC activation in primary and metastatic ductal adenocarcinoma of the pancreas: Incidence, mechanisms, and clinical significance. *Modern Pathology*, *15*(4), 462–469. https://doi.org/10.1038/modpathol.3880547.
Sears, R., Nuckolls, F., Haura, E., Taya, Y., Tamai, K., & Nevins, J. R. (2000). Multiple Ras-dependent phosphorylation pathways regulate Myc protein stability. *Genes & Development*, *14*(19), 2501–2514. https://doi.org/10.1101/gad.836800.
Serasinghe, M. N., Wieder, S. Y., Renault, T. T., Elkholi, R., Asciolla, J. J., Yao, J. L., et al. (2015). Mitochondrial division is requisite to RAS-induced transformation and targeted by oncogenic MAPK pathway inhibitors. *Molecular Cell*, *57*(3), 521–536. https://doi.org/10.1016/j.molcel.2015.01.003.
Sheridan, D. L., Kong, Y., Parker, S. A., Dalby, K. N., & Turk, B. E. (2008). Substrate discrimination among mitogen-activated protein kinases through distinct docking sequence motifs. *The Journal of Biological Chemistry*, *283*(28), 19511–19520. https://doi.org/10.1074/jbc.M801074200.
Siegel, R. L., Miller, K. D., Fuchs, H. E., & Jemal, A. (2021). Cancer statistics, 2021. *CA: A Cancer Journal for Clinicians*, *71*(1), 7–33. https://doi.org/10.3322/caac.21654.
Sodir, N. M., Kortlever, R. M., Barthet, V. J. A., Campos, T., Pellegrinet, L., Kupczak, S., et al. (2020). MYC instructs and maintains pancreatic adenocarcinoma phenotype. *Cancer Discovery*, *10*(4), 588–607. https://doi.org/10.1158/2159-8290.CD-19-0435.
Son, J., Lyssiotis, C. A., Ying, H., Wang, X., Hua, S., Ligorio, M., et al. (2013). Glutamine supports pancreatic cancer growth through a KRAS-regulated metabolic pathway. *Nature*, *496*(7443), 101–105. https://doi.org/10.1038/nature12040.

Soucek, L., Whitfield, J., Martins, C. P., Finch, A. J., Murphy, D. J., Sodir, N. M., et al. (2008). Modelling Myc inhibition as a cancer therapy. *Nature*, *455*(7213), 679–683. https://doi.org/10.1038/nature07260.

Soucek, L., Whitfield, J. R., Sodir, N. M., Masso-Valles, D., Serrano, E., Karnezis, A. N., et al. (2013). Inhibition of Myc family proteins eradicates KRas-driven lung cancer in mice. *Genes & Development*, *27*(5), 504–513. https://doi.org/10.1101/gad.205542.112.

Stefanovsky, V. Y., Pelletier, G., Hannan, R., Gagnon-Kugler, T., Rothblum, L. I., & Moss, T. (2001). An immediate response of ribosomal transcription to growth factor stimulation in mammals is mediated by ERK phosphorylation of UBF. Molecular Cell, 8(5), 1063-1073. https://doi.org/10.1016/s1097-2765(01)00384-7.

Sturgill, T. W., Ray, L. B., Erikson, E., & Maller, J. L. (1988). Insulin-stimulated MAP-2 kinase phosphorylates and activates ribosomal protein S6 kinase II. *Nature*, *334*(6184), 715–718. https://doi.org/10.1038/334715a0.

Suzuki, T., K-Tsuzuku, J., Ajima, R., Nakamura, T., Yoshida, Y., & Yamamoto, T. (2002). Phosphorylation of three regulatory serines of Tob by Erk1 and Erk2 is required for Ras-mediated cell proliferation and transformation. *Genes & Development*, *16*(11), 1356–1370. https://doi.org/10.1101/gad.962802.

Tanaka, N., Lin, J. J., Li, C., Ryan, M. B., Zhang, J., Kiedrowski, L. A., et al. (2021). Clinical acquired resistance to KRASG12C inhibition through a novel KRAS switch-II pocket mutation and polyclonal alterations converging on RAS-MAPK reactivation. *Cancer Discovery*, *11*(8), 1913–1922. https://doi.org/10.1158/2159-8290.CD-21-0365.

Tanoue, T., Adachi, M., Moriguchi, T., & Nishida, E. (2000). A conserved docking motif in MAP kinases common to substrates, activators and regulators. *Nature Cell Biology*, *2*(2), 110–116. https://doi.org/10.1038/35000065.

TCGA. (2017). Integrated genomic characterization of pancreatic ductal adenocarcinoma. *Cancer Cell*, *32*(2), 185–203 e113. https://doi.org/10.1016/j.ccell.2017.07.007.

Terrell, E. M., & Morrison, D. K. (2019). Ras-mediated activation of the Raf family kinases. *Cold Spring Harbor Perspectives in Medicine*, *9*(1), a033746. https://doi.org/10.1101/cshperspect.a033746.

The Pancreatic Cancer Working Group, & Clinical Trials and Translational Research Advisory Committee. (2013). *Pancreatic cancer scanning the horizon for focused interventions: A report to the director of the National Cancer Institute.*

Tsai, F. D., Lopes, M. S., Zhou, M., Court, H., Ponce, O., Fiordalisi, J. J., et al. (2015). K-Ras4A splice variant is widely expressed in cancer and uses a hybrid membrane-targeting motif. *Proceedings of the National Academy of Sciences of the United States of America*, *112*(3), 779–784. https://doi.org/10.1073/pnas.1412811112.

Unal, E. B., Uhlitz, F., & Bluthgen, N. (2017). A compendium of ERK targets. *FEBS Letters*, *591*(17), 2607–2615. https://doi.org/10.1002/1873-3468.12740.

Unni, A. M., Harbourne, B., Oh, M. H., Wild, S., Ferrarone, J. R., Lockwood, W. W., et al. (2018). Hyperactivation of ERK by multiple mechanisms is toxic to RTK-RAS mutation-driven lung adenocarcinoma cells. *eLife*, 7, e33718. https://doi.org/10.7554/eLife.33718.

Unni, A. M., Lockwood, W. W., Zejnullahu, K., Lee-Lin, S. Q., & Varmus, H. (2015). Evidence that synthetic lethality underlies the mutual exclusivity of oncogenic KRAS and EGFR mutations in lung adenocarcinoma. *eLife*, *4*, e06907. https://doi.org/10.7554/eLife.06907.

Uprety, D., & Adjei, A. A. (2020). KRAS: From undruggable to a druggable cancer target. *Cancer Treatment Reviews*, *89*, 102070. https://doi.org/10.1016/j.ctrv.2020.102070.

Vaseva, A. V., Blake, D. R., Gilbert, T. S. K., Ng, S., Hostetter, G., Azam, S. H., et al. (2018). KRAS suppression-induced degradation of MYC is antagonized by a MEK5-ERK5 compensatory mechanism. *Cancer Cell*, *34*(5), 807–822 e807. https://doi.org/10.1016/j.ccell.2018.10.001.

Villanueva, J., Vultur, A., Lee, J. T., Somasundaram, R., Fukunaga-Kalabis, M., Cipolla, A. K., et al. (2010). Acquired resistance to BRAF inhibitors mediated by a RAF kinase switch in melanoma can be overcome by cotargeting MEK and IGF-1R/PI3K. *Cancer Cell*, *18*(6), 683–695. https://doi.org/10.1016/j.ccr.2010.11.023.

Waskiewicz, A. J., Flynn, A., Proud, C. G., & Cooper, J. A. (1997). Mitogen-activated protein kinases activate the serine/threonine kinases Mnk1 and Mnk2. *The EMBO Journal*, *16*(8), 1909–1920. https://doi.org/10.1093/emboj/16.8.1909.

Waters, A. M., & Der, C. J. (2018). KRAS: The critical driver and therapeutic target for pancreatic cancer. *Cold Spring Harbor Perspectives in Medicine*, *8*(9), a031435. https://doi.org/10.1101/cshperspect.a031435.

Wood, K. C. (2015). Mapping the pathways of resistance to targeted therapies. *Cancer Research*, *75*(20), 4247–4251. https://doi.org/10.1158/0008-5472.CAN-15-1248.

Yang, J. Y., Zong, C. S., Xia, W., Yamaguchi, H., Ding, Q., Xie, X., et al. (2008). ERK promotes tumorigenesis by inhibiting FOXO3a via MDM2-mediated degradation. *Nature Cell Biology*, *10*(2), 138–148. https://doi.org/10.1038/ncb1676.

Yang, S., Wang, X., Contino, G., Liesa, M., Sahin, E., Ying, H., et al. (2011). Pancreatic cancers require autophagy for tumor growth. *Genes & Development*, *25*(7), 717–729. https://doi.org/10.1101/gad.2016111.

Yeh, E., Cunningham, M., Arnold, H., Chasse, D., Monteith, T., Ivaldi, G., et al. (2004). A signalling pathway controlling c-Myc degradation that impacts oncogenic transformation of human cells. *Nature Cell Biology*, *6*(4), 308–318. https://doi.org/10.1038/ncb1110.

Yen, I., Shanahan, F., Lee, J., Hong, Y. S., Shin, S. J., Moore, A. R., et al. (2021). ARAF mutations confer resistance to the RAF inhibitor belvarafenib in melanoma. *Nature*, *594*(7863), 418–423. https://doi.org/10.1038/s41586-021-03515-1.

Yen, I., Shanahan, F., Merchant, M., Orr, C., Hunsaker, T., Durk, M., et al. (2018). Pharmacological induction of RAS-GTP confers RAF inhibitor sensitivity in KRAS mutant tumors. *Cancer Cell*, *34*(4), 611–625 e617. https://doi.org/10.1016/j.ccell.2018.09.002.

Ying, H., Kimmelman, A. C., Lyssiotis, C. A., Hua, S., Chu, G. C., Fletcher-Sananikone, E., et al. (2012). Oncogenic Kras maintains pancreatic tumors through regulation of anabolic glucose metabolism. *Cell*, *149*(3), 656–670. https://doi.org/10.1016/j.cell.2012.01.058.

CHAPTER FIVE

Pan-RAS inhibitors: Hitting multiple RAS isozymes with one stone

Alexander B. Coley[a,b], Antonio Ward[a,b], Adam B. Keeton[c], Xi Chen[c], Yulia Maxuitenko[c], Aishwarya Prakash[b,d], Feng Li[c], Jeremy B. Foote[e], Donald J. Buchsbaum[f], and Gary A. Piazza[c,*]

[a]Department of Pharmacology, University of South Alabama, Mobile, AL, United States
[b]Mitchell Cancer Institute, Mobile, AL, United States
[c]Department of Drug Discovery and Development, Harrison School of Pharmacy, Auburn University, Auburn, AL, United States
[d]Department of Biochemistry & Molecular Biology, University of South Alabama, Mobile, AL, United States
[e]Department of Microbiology, University of Alabama at Birmingham, Birmingham, AL, United States
[f]Department of Radiation Oncology, University of Alabama at Birmingham, Birmingham, AL, United States
*Corresponding author: e-mail address: gap0034@auburn.edu

Contents

1. Introduction 132
2. PAN-RAS inhibitors 138
   2.1 Direct binding Pan-RAS inhibitors 141
   2.2 Indirect binding Pan-RAS inhibitors 149
   2.3 RAS pathway inhibitors 155
3. Discussion 156
4. Conclusion 159
Grant support 159
Disclosure of potential conflicts of interest 159
References 159

Abstract

Mutations in the three *RAS* oncogenes are present in approximately 30% of all human cancers that drive tumor growth and metastasis by aberrant activation of RAS-mediated signaling. Despite the well-established role of RAS in tumorigenesis, past efforts to develop small molecule inhibitors have failed for various reasons leading many to consider RAS as "undruggable." Advances over the past decade with KRAS(G12C) mutation-specific inhibitors have culminated in the first FDA-approved RAS drug, sotorasib. However, the patient population that stands to benefit from KRAS(G12C) inhibitors is inherently limited to those patients harboring KRAS(G12C) mutations. Additionally, both intrinsic and acquired mechanisms of resistance have been reported that indicate allele-specificity may afford disadvantages. For example, the compensatory

Advances in Cancer Research, Volume 153
ISSN 0065-230X
https://doi.org/10.1016/bs.acr.2021.07.009

activation of uninhibited wild-type (WT) NRAS and HRAS isozymes can rescue cancer cells harboring KRAS(G12C) mutations from allele-specific inhibition or the occurrence of other mutations in KRAS. It is therefore prudent to consider alternative drug discovery strategies that may overcome these potential limitations. One such approach is pan-RAS inhibition, whereby all RAS isozymes co-expressed in the tumor cell population are targeted by a single inhibitor to block constitutively activated RAS regardless of the underlying mutation. This chapter provides a review of past and ongoing strategies to develop pan-RAS inhibitors in detail and seeks to outline the trajectory of this promising strategy of RAS inhibition.

1. Introduction

Two viruses capable of causing tumor formation in rat models were identified by Harvey et al. and Kirsten et al. in the mid-1960s (Harvey, 1964; Kirsten & Mayer, 1967). The genetic components of these viruses would later be given the names *HRAS* (Harvey Rat Sarcoma) and *KRAS* (Kirsten Rat Sarcoma). It would require another two decades of research before the human proto-oncogene homologs of the viral oncogenes *HRAS* and *KRAS* were identified and validated as drivers of human oncogenesis (Der, Krontiris, & Cooper, 1982; Santos, Tronick, Aaronson, Pulciani, & Barbacid, 1982). Accompanied by the 1983 discovery of another *RAS* gene in neuroblastoma, *NRAS* (Neuroblastoma RAS), *RAS* genes were the first human oncogenes to be identified (Hall, Marshall, Spurr, & Weiss, 1983). Roughly four decades of research have since provided a wealth of knowledge about *RAS* genes and RAS protein biology. HRAS, NRAS, and the splice variants KRAS4A and KRAS4B constitute the four RAS isozymes and are collectively referred to as RAS (Lu et al., 2016; Nussinov, Tsai, Chakrabarti, & Jang, 2016). RAS proteins are small guanosine triphosphate hydrolyzing enzymes (GTPases) that share overlapping function as major hubs of signal transduction. In response to upstream signaling activation, guanosine triphosphate (GTP) is loaded onto RAS resulting in an active RAS conformation that will bind and activate a plethora of "pro-growth" signaling pathways essential for proliferation and survival of cancer cells (Campbell, Khosravi-Far, Rossman, Clark, & Der, 1998). Intrinsic or enzyme-mediated hydrolysis of the GTP γ-phosphate produces the guanosine diphosphate (GDP) loaded RAS conformation, which is inactive regarding its canonical function as a signal transducer. GDP release creates a transient nucleotide-free intermediate, which can bind the abundance of cellular GTP with picomolar affinity thus refreshing the

RAS nucleotide cycle (John et al., 1990). Nucleotide cycling is tied to the activation or inaction of downstream signaling pathways, where GTP-bound RAS facilitates the activation of RAS effector pathways, including the RAF-mitogen activated protein kinase (MAPK), phosphoinositide-3-kinase/protein kinase B (PI3K/AKT), and RALGDS signaling pathways (Gimple & Wang, 2019). It is understood that RAS activation is tightly regulated in healthy cells to maintain tissue homeostasis. RAS is therefore of immense oncologic importance with many labs studying various aspects of RAS biology or attempting to develop RAS inhibitors. There is perhaps no greater unmet medical need in oncology than for a broadly efficacious and safe RAS inhibitor.

The similar function shared by each RAS isozyme can be explained structurally by their nearly uniform sequence identity in the RAS G-domain (amino acids 1–166) (Hobbs, Der, & Rossman, 2016). This domain is further divided into the effector lobe (1–87) and the allosteric lobe (88–166). The effector lobe is responsible for the GTPase and signaling functions of RAS while the allosteric lobe is the site of allosteric interactions. Three regions within the effector lobe, the P-loop (10–17), Switch I (28–40), and Switch II (57–76) constitute the nucleotide-binding pocket and coordinate a magnesium (Mg^{2+}) cofactor necessary for GTPase function (Pantsar, 2020). Further, Switch I and Switch II are the primary sites for effector interaction and undergo significant conformational shifts depending on whether RAS is bound to GTP or GDP (Hall et al., 2002). At the C-terminus, amino acids 167–189 (HRAS, NRAS, KRAS4A) or 167–188 (KRAS4B) comprise the hypervariability region (HVR) (Lu, Jang, Gu, Zhang, & Nussinov, 2016). Indicated by its name, the HVR is the major site of sequence divergence between the four isozymes with <20% shared identity (Hobbs et al., 2016). The HVR determines how each isozyme is post-translationally modified via the addition of lipid moieties, a process that directs RAS to localize at the plasma membrane and is essential for RAS function (Cox, Der, & Philips, 2015). Each isozyme shares a terminal CAAX (cysteine, aliphatic residue, aliphatic residue, any residue) motif which is farnesylated at the cysteine residue and further modified to produce a methylated farnesylcysteine at the C-terminus. This "first signal" of membrane localization is not sufficient and must be accompanied by a "second signal" which is isozyme dependent. HRAS, NRAS, and KRAS4A are palmitoylated at multiple cysteine residues in the HVR to fulfill the requisite second signal (Hancock, Cadwallader, Paterson, & Marshall, 1991). KRAS4B, alternatively, contains a polybasic region in the HVR composed of 8 positively charged lysine residues that form electrostatic interactions with the plasma membrane, which bears a net negative charge (Ahearn, Haigis, Bar-Sagi, & Philips, 2012). Once properly localized to the plasma membrane, RAS functions as a master regulator of signal transduction to drive proliferation of both normal and neoplastic cells.

For a mutation to be oncogenic, it will arise in a gene that offers a significant advantage to cells in one of a small group of physiologic characteristics described as the hallmarks of cancer (Hanahan & Weinberg, 2011). These changes in cell behavior often accumulate at key axes of signaling pathways where they deregulate cellular homeostasis by flooding the cell with pro-growth signals (Stratton, Campbell, & Futreal, 2009). Therefore, it is understandable that *RAS* genes, the products of which are the key regulators of the extensive RAS signaling network, are the most frequently mutated oncogenes in cancer. A staggering ~30% of human cancer patients bear a mutation in one of three *RAS* genes (Kodaz, 2017). Approximately 98% of oncogenic RAS mutations occur at either G12 or G13 of Switch I or at Q61 of Switch II (Rajalingam, Schreck, Rapp, & Albert, 2007). These loci are critical for both intrinsic and GTPase activating protein (GAP)-mediated GTP hydrolysis, and consequently amino acid substitutions at these loci greatly inhibit this function. The acquisition of such mutations results in constitutively active RAS that sustains active RAS signaling networks, promoting tumor formation and progression by influencing all hallmarks of cancer (Fig. 1).

KRAS mutations command a dominant 85% of all *RAS* mutations, with just 11–15% of *RAS* mutations occurring in *NRAS* and 1% in *HRAS* (Kodaz, 2017). Studies of *RAS* mutations in human cancer patients have revealed recurring themes. First, RAS isozyme mutations are tissue specific. Over 90% of pancreatic cancer *RAS* mutations occur in *KRAS* with *HRAS* mutations being extremely rare. Conversely, *HRAS* mutations are far more frequent hepatocellular carcinomas, occurring in 29% of cases, while *KRAS* mutations account for only 0–5% (Kodaz, 2017). Second, the mutated *RAS* codon is isozyme and tissue specific. *KRAS* is most frequently mutated at codon 12, *NRAS* favors codon 61, and *HRAS* displays no preference. In lung cancers, *NRAS* mutations favor Q61, but in prostate cancer G12 is more frequently mutated (Prior, Lewis, & Mattos, 2012). Further, mutated *RAS* genes are typical of the most fatal cancers, being present in over 90% of patients with pancreatic cancer, 30% with colorectal cancer (CRC), and 40% with lung cancers (Simanshu, Nissley, & McCormick, 2017). The well-established role of *RAS* oncogenes in human cancer is accompanied by their status as greatly sought-after therapeutic targets.

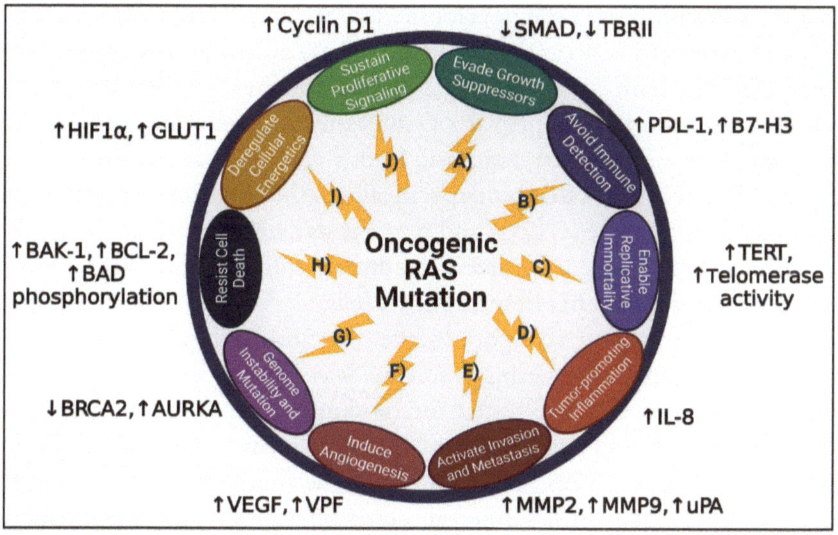

Fig. 1 The Hallmarks of Oncogenic RAS Mutations in Cancer. (A) Oncogenic RAS signaling enhances tumor evasion of growth suppressors by negatively regulating SMAD and reducing tumor growth factor β type II (TβRII) expression (Halder et al., 2011; Kretzschmar, Doody, Timokhina, & Massagué, 1999). (B) Elevated active RAS enhances tumor expression of PDL-1 and B7-H3 to suppress T-cell activity and avoid immune detection (Ward et al., 2020). (C) RAS mutations increase telomerase reverse transcriptase (TERT) expression and telomerase activity, contributing to replicative immortality by lengthening telomeres (Liu et al., 2017). (D) Elevated RAS signaling leads to increased interleukin 8 (IL-8) expression and secretion which recruits inflammatory cells to the tumor microenvironment (Ancrile, O'Hayer, & Counter, 2008). (E) RAS stimulates matrix metalloprotease 2 (MMP2) and MMP9 expression as well as urokinase-type plasminogen activator (u-PA) which together increase tumor invasion and metastasis (Kranenburg, Gebbink, & Voest, 2004; Smakman, Borel Rinkes, Voest, & Kranenburg, 2005; Testa, Medcalf, Cajot, Schleuning, & Sordat, 1989). (F) Oncogenic RAS signaling increases vascular endothelial growth factor (VEGF) and vascular permeability factor (VPF) expression to recruit vasculature to the tumor microenvironment (Kranenburg et al., 2004; Rak et al., 1995). (G) RAS signaling causes a decrease in breast cancer susceptibility gene 2 (BRCA2) expression and overexpression of aurora kinase A (AURKA) leading to increased genome instability and mutation incidence (Abulaiti, Fikaris, Tsygankova, & Meinkoth, 2006; Yang et al., 2013). (H) Active RAS signaling suppresses apoptosis in part by increasing inhibitory phosphorylation of the pro-apoptotic BAD protein, allowing RAS mutant tumors to resist cell death (Cox & Der, 2003). (I) Oncogenic RAS promotes HIF1α expression which increases GLUT1 production, enhancing glucose uptake and metabolism (Pylayeva-Gupta, Grabocka, & Bar-Sagi, 2011). (J) Major RAS signaling pathways including RAF-MAPK, PI3K, and RALGDS activate transcription factors that increase cyclin D1 production to constitutively drive cancer cell proliferation (Pylayeva-Gupta et al., 2011). Created with BioRender.com. *Adapted from Hanahan, D., Weinberg, R. A. (2011). Hallmarks of cancer: the next generation.* Cell, 144*(5), 646–674.*

Past shortcomings in the development of small molecule RAS inhibitors led many to regard RAS as the "Holy Grail" of cancer drug discovery (Cox, Fesik, Kimmelman, Luo, & Der, 2014). With the U.S. Food and Drug Administration (FDA)-approval of sotorasib, it could be said that the Holy Grail has been seized. Sotorasib leads a class of isozyme and mutation specific RAS inhibitors that bind specifically to the Switch II pocket (Switch II-P) of mutant KRAS(G12C) and prevent its activation. The Switch II-P was first identified in a veritable landmark study in the field of RAS drug discovery (Ostrem, Peters, Sos, Wells, & Shokat, 2013). With the purpose of identifying compounds that specifically bind KRAS(G12C), a high-throughput "tethering" screen was performed in which 480 disulfide compounds were tested for binding to the GDP-bound mutant KRAS(G12C). Co-crystallization of the high-affinity binding compounds with GDP-bound KRAS(G12C) revealed the mechanism of binding that involved a previously undescribed allosteric binding site near the nucleotide binding pocket. By occupying this site, termed switch II-P as it resides in the switch II region, allosteric KRAS(G12C) inhibitors displace amino acids critical for stabilizing the GTP-y phosphate and sterically clash with the γ-phosphate to prevent GTP binding. This dramatically shifts the affinity of KRAS(G12C) to favor GDP binding, reducing the population of active KRAS(G12C) and thereby suppressing oncogenic RAS signaling.

Several compounds designed and optimized based on this strategy have entered clinical trials. These compounds and corresponding trials include the recently FDA-approved sotorasib (AMG-510) (e.g., NCT03600883), as well as MRTX849 (e.g., NCT03785249), GDC-6036 (NCT04449874), and D-1553 (NCT04585035) (Chen, Smaill, Liu, Ding, & Lu, 2020). In September of 2020, the phase I clinical trial results for sotorasib treatment were published. Among 129 patients with KRAS(G12C) mutations enrolled in the trial, 59 were diagnosed with non-small cell lung cancer (NSCLC), 42 with colorectal cancer (CRC), and 28 with other cancers. Up to 88.1% of patients with NSCLC and 73.8% of patients with CRC had either stable disease or an objective response in response to sotorasib treatment (Hong et al., 2020). When looking just at objective response, patients with CRC (7.1%) were less responsive than patients with NSCLC (32.2%) for reasons unknown. Phase II sotorasib clinical trial results released in June of 2021 revealed an impressive median overall survival of 12.5 months in patients with previously treated NSCLC, and accelerated FDA-approval was granted for sotorasib in KRAS(G12C) mutated NSCLC (FDA, 2021; Skoulidis et al., 2021). These results represent a turning point in the struggle to bring RAS-targeting therapies to the clinic and have since provided clinical validation of KRAS as a drug target.

Several reports have suggested the potential for compensatory mechanisms that could lead to drug resistance among mutant specific inhibitors of RAS (Dunnett-Kane, Nicola, Blackhall, & Lindsay, 2021; Hata & Shaw, 2020). It has been established that not all KRAS mutant cancers are identical but rather that they vary in degree of dependence on their mutant RAS (Singh et al., 2009). From the start then, a patient cohort of KRAS(G12C) mutant tumors may include some cancers less dependent on the KRAS(G12C) mutation and therefore more resistant to therapy. In addition, Xue et al. identified a resistance mechanism by which cells overexpress constitutively active KRAS(G12C) (Xue et al., 2020). This newly expressed KRAS(G12C) is maintained in the GTP-bound conformation in part by EGFR-stimulated nucleotide exchange and aurora kinase A (AURKA)-mediated effector interactions. KRAS(G12C) inhibitors exclusively target GDP-bound RAS, enabling this adaptive response to produce KRAS(G12C) that is not susceptible to being drugged as it does not cycle through the GDP-bound conformation. An interplay between mutant KRAS and the remaining WT RAS isozymes mediates another form of resistance to KRAS(G12C) inhibition. Oncogenic mutant KRAS is dependent on dimerization for its function, and this need can be fulfilled by dimerizing with WT KRAS. WT KRAS thus synergizes with mutant KRAS to drive cancer progression and potentially can facilitate resistance to mutant KRAS inhibition (Ambrogio et al., 2018; Burgess et al., 2017). A study by Ryan et al. found that in a diverse panel of KRAS(G12C) mutant cell lines resistance was mediated by hyper-activation of WT HRAS and NRAS to subvert dependence on KRAS(G12C) signaling and rescue MAPK activation (Ryan et al., 2020). Awad et al. identified a resistance mechanism where tumors accumulate additional KRAS mutations to subvert the G12C specificity of allele-specific inhibitors (Awad et al., 2021). These publications highlight a few potential mechanisms of resistance to KRAS(G12C) inhibitors that merit further study. While some may be addressed in part by combination therapies in clinical trials, such as the combination of sotorasib and the MEK inhibitor trametinib, this pitfall of specific KRAS inhibition leaves open the possibility of greater sustained benefit for patients treated with inhibitors that target all RAS isozymes.

Another limitation faced by KRAS(G12C) inhibitors is the narrow target patient population. KRAS(G12C) mutations are present in only ~14% of NSCLC patients and 3% of CRC patients (Nassar, Adib, & Kwiatkowski, 2021). Furthermore, targeting the reactive cysteine is what enabled the development of these inhibitors in the first place, meaning that novel strategies will need to be developed if the goal is to discover mutation-specific inhibitors for each of the most common KRAS mutations. A pre-clinical data summary was released in 2020 regarding their KRAS(G12D) specific inhibitor MRTX1133, indicating that specific inhibition of KRAS mutants other than KRAS(G12C) is indeed feasible (Mirati Therapeutics Inc., 2020). However, this does not ensure that the more frequent KRAS mutations are readily druggable and the task of developing specific inhibitors for each may yet be fraught with difficulties. There is no doubt that KRAS(G12C) specific inhibitors represent the high-water mark for drugging RAS. Sotorasib's FDA-approval has renewed excitement in the field and has spurred a shift in attention onto the development of highly specific RAS inhibitors. Aware of the foreseeable pitfalls of this strategy, in this chapter we will focus on the past and ongoing efforts to discover broadly effective pan-RAS inhibitors and develop them into clinically pble RAS therapeutics.

2. PAN-RAS inhibitors

The overall strategy of pan-RAS inhibition has been divided into two subgroups of inhibitors in this review. The first group, direct binding pan-RAS inhibitors, bind to RAS proteins directly to suppress RAS signaling. These include inhibitors that target RAS binding sites common to all RAS isozymes, allowing for direct binding and suppression of the entire RAS signaling hub regardless of isozyme or mutated codon. Indirect binding pan-RAS inhibitors, on the other hand, bind a RAS-interacting protein and inhibit RAS protein-protein interactions (PPIs) thereby reducing RAS signaling. For example, inhibitors that target RAS-activating proteins with the purpose of suppressing RAS fit this second category. Finally, while not necessarily pan-RAS inhibitors, RAS pathway inhibitors include FDA-approved inhibitors that suppress RAS signaling by binding to RAS effectors downstream of RAS. Although these inhibitors fill important roles in other contexts of disease, their utility in RAS-driven cancers is limited. Perhaps their most promising application will be in combination with RAS inhibitors to improve pharmacological control over the extensive RAS pathway. Therefore, RAS pathway inhibitors are reviewed as a group as they will likely have an important role to play alongside the RAS inhibitors that gain FDA approval. The inhibitors selected for review include a diverse array of small-molecule compounds, monoclonal antibodies, and peptides to fill out a growing arsenal of pan-RAS inhibitors (Table 1).

Table 1 Compilation of the pan-RAS inhibitors reviewed in this chapter.

Binding site	Inhibitor	PDB ID	Mechanism of action	Selected references
NT Pocket	N2-substituted guanosine 5′-triphosphates		Prevent GTP binding	Noonan, Brown, Dudycz, and Wright (1991)
	Pyrasolo[3,4-*b*] quinoline ribofuranosides		Prevent GTP binding	Wolin et al. (1996)
	DABP-GTP	1CLU	Increase GTP hydrolysis	Ahmadian et al. (1999)
	CID1067700		Prevent GTP binding	Surviladze, Young, and Sklar (2012) and Hong et al. (2015)
Switch I	3144		Prevent effector interactions	Welsch et al. (2017)
Switch I/II	HBS3		Block SOS nucleotide exchange	Patgiri, Yadav, Arora, and Bar-Sagi (2011)
	SAH-SOS1$_{\alpha}$		Block SOS nucleotide exchange	Leshchiner et al. (2015)
	inRAS37		Prevent effector interactions	Shin et al. (2017) and Shin et al. (2020)
	Sulindac		Prevent effector interactions	Thompson et al. (1997) and Herrmann et al. (1998)
	Kobe0065	2LWI	Prevent effector interactions	Shima et al. (2013)
Switch I/II-P	DCAI	4DST	Block SOS Nucleotide exchange	Maurer et al. (2012)
	Ch-1/2/3	6GQW/ 6GQX/ 6GQY	Block SOS nucleotide exchange	Quevedo et al. (2018) and Cruz-Migoni et al. (2019)
	BI-2852	6GJ7	Block SOS nucleotide exchange	Kessler et al. (2019)

Continued

Table 1 Compilation of the pan-RAS inhibitors reviewed in this chapter.—cont'd

Binding site	Inhibitor	PDB ID	Mechanism of action	Selected references
FTase	Lonafarnib	1O5M	Prevent membrane localization	Sun, Zhou, Wang, Fu, and Khuri (2004) and Berndt, Hamilton, and Sebti (2011)
	Tipifarnib	1SA4	Prevent membrane localization	End et al. (2001) and Berndt et al. (2011)
PDEδ	Deltarasin	4JVF	Prevent membrane localization	Zimmermann et al. (2013)
	Deltazinone 1	5E80	Prevent membrane localization	Papke et al. (2016)
RBD	MCP compounds		Prevent Raf effector interactions	Kato-Stankiewicz et al. (2002) and González-Pérez et al. (2010)
	Rigosertib	5J18	Prevent effector interactions	Athuluri-Divakar et al. (2016) and Kowalczyk et al. (2021)
SOS1	BI-1701963	6SCM	Block SOS Nucleotide exchange	Gerlach et al. (2020) and Gort et al. (2020)
SHP2	II-B08	3O5X	Prevent SHP2 phosphatase	Bunda et al. (2015) and Zhang et al. (2010)
	RMC-4630		Prevent SHP2 phosphatase	Nichols et al. (2018) and Bunda et al. (2015)
	JAB-3068		Prevent SHP2 phosphatase	Bunda et al. (2015)
	TNO-155	7JVM	Prevent SHP2 phosphatase	Bunda et al. (2015)

2.1 Direct binding Pan-RAS inhibitors

2.1.1 Nucleotide pocket

One of the earliest strategies to directly inhibit RAS activation was the development of nucleotide analogue compounds that bind the RAS nucleotide binding site and prevent oncogenic RAS signaling (Cox et al., 2014). In 1991 and later in 1996, N2-substituted guanosine 5′-triphosphates and pyrazolo[3,4-*b*]quinoline ribofuranosides were synthesized for this purpose (Noonan et al., 1991; Wolin et al., 1996). Neither series of compounds, however, were able to sufficiently inhibit RAS activation. This can be attributed to the daunting task of direct competition with the picomolar affinity of guanosine nucleotides for the RAS nucleotide binding pocket (John et al., 1990). It quickly became clear that alternative approaches would have to be taken to either avoid competition with GDP/GTP or overcome it, and indeed many compounds detailed below rely on the identification of alternative target regions shared by RAS proteins to suppress RAS activation (Fig. 2).

In 1999, Ahmadian et al. proposed diaminobenzophenone-phosphoroamidate-GTP (DABP-GTP) as a GTP-analogue that was readily hydrolysable by mutant RAS proteins, the result of which would produce

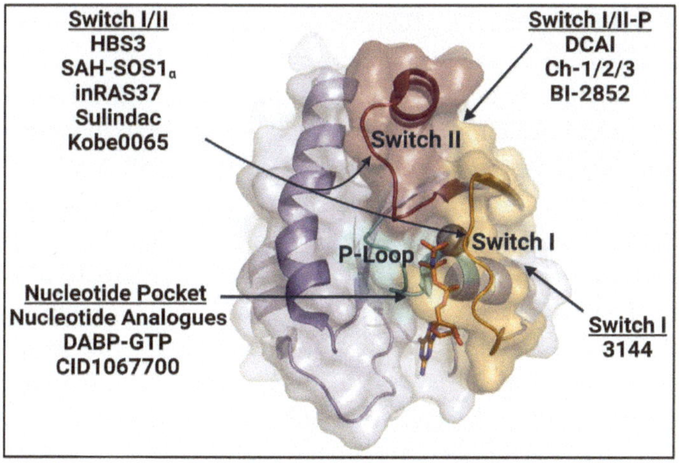

Fig. 2 Direct Binding Pan-RAS Inhibitors and Their RAS Binding Sites. The ligand binding sites for each direct pan-RAS inhibitor reviewed are indicated on a cartoon representation of the crystal structure of KRAS(Q61H) (PDB ID: 6GQW) (Cruz-Migoni et al., 2019). The P-loop (cyan), Switch I (orange), Switch II (crimson), and Mg^{2+} ion (gray sphere) are each indicated and the GppNHp is represented as a stick model. Created with PyMOL and with BioRender.com (Schrödinger & DeLano, 2020).

GDP-bound inactive RAS (Ahmadian et al., 1999). Oncogenic *RAS* mutations produce RAS proteins with a markedly reduced ability to hydrolyze GTP either intrinsically or through GAP stimulation. DABP-GTP contains a catalytic amine group that stabilizes the GTP hydrolysis transition state, restoring GTP hydrolysis capabilities to mutant RAS proteins. HRAS loaded with either DABP-GTP or GTP was analyzed by HPLC to measure the hydrolysis rate of each nucleotide. The turnover rate of DABP-GTP was found to be 10-fold faster than that of GTP. Mutant HRAS proteins also performed the hydrolysis reaction faster on DABP-GTP than GTP, validating the rationale for DABP-GTP's mechanism of action. This compound revealed that the impaired GTP hydrolysis of mutant RAS was reversible and could potentially be targeted as a strategy for selective mutant RAS inhibition (Ahmadian et al., 1999).

A high-throughput flow cytometry-based assay that was used to screen compounds for activity against GTPases identified the small compound CID1067700 as a RAB7 inhibitor (Surviladze et al., 2012). In BODIPY nucleotide-exchange assays, CID10667700 was found to inhibit fluorescent nucleotide loading onto members of the RAS and RHO protein families in addition to the RAB family and later studies revealed the compound to exert a pan-GTPase effect (Hong et al., 2015). These results were confirmed in cell monolayers by immunoblot where the compound suppressed RAB-mediated EGFR degradation, RHO signal transduction via integrin VLA-4, and RAS signaling via ERK phosphorylation (p-ERK). Interestingly, the suppression of BODIPY-GTP loading onto RAB by CID1067700 closely matched the effect of adding non-labeled GTP, suggesting that the compound competes for the RAS nucleotide-binding pocket. RAS and RHO proteins were less affected by CID1067700 treatment, leading the authors to conclude that subtle differences in the nucleotide-binding pockets of members of the RAS superfamily cause the observed differences in sensitivity to compound treatment. This conclusion is supported by molecular docking analysis that predicted favorable binding of the compound to the nucleotide-binding pocket of RAB7; however, a structural determination of the compound in complex with RAB is needed to confirm this mechanism (Agola et al., 2012). Several recent studies have used CID1067700 as a RAB7 inhibitor in animal models however not in the context of cancer (Lam et al., 2016; Qin et al., 2019). Regardless, these indicate that the compound is well tolerated within its therapeutic window and merits further study in cancer models. While additional structural determination of the mechanism of action is warranted to further optimize the compound and to confirm cellular permeability, CID1067700 is currently a useful molecular probe used to examine GTPase suppression.

2.1.2 Switch I

In an effort by Welsch et al. in 2017 to discover novel pan-RAS inhibitors, a library of compounds was generated to simultaneously target 2–3 adjacent sites on KRAS(G12D). Computational docking of this chemical library resulted in the identification of a lead compound, coded "3144" (Welsch et al., 2017). KRAS(G12D) binding was confirmed using NMR (nuclear magnetic resonance), isothermal titration calorimetry (ITC), and microscale thermophoresis (MST) with an approximate K_d of 20 μM (micromolar). Pan-RAS inhibition was confirmed in vitro where 3144 was found to selectively kill *KRAS* and *HRAS* mutant cells as well as DLD-1 *RAS* WT cells that had been sensitized using a pan-RAS siRNA. 3144 inhibited the proliferation of RAS-driven cell lines and caused significant tumor growth inhibition in patient-derived mouse xenografts. MAPK and PI3K signaling were inhibited following 3144 treatment both in vitro and in vivo. 3144 is currently limited by its low water solubility and potency as well as toxicity issues in vivo. The authors propose further chemical optimization to produce more drug-like 3144 derivatives to overcome the current limitations of the compound (Welsch et al., 2017).

2.1.3 Switch I/II

RAS activation is largely mediated by the guanine exchange factor (GEF) Son of Sevenless (SOS). SOS interacts with RAS proteins by way of its α-helix, which induces a conformational shift in the RAS switch regions that facilitates nucleotide release (Cherfils & Zeghouf, 2013). Patgiri et al. took advantage of this well-established PPI by engineering a SOS α-helix mimetic with enhanced stability using a hydrogen bond surrogate (HBS) scaffold (Patgiri et al., 2011). The lead mimetic peptide, HBS3, prevented GDP release from recombinant RAS in a cell-free assay. Cellular uptake is a characteristic hurdle encountered by would-be peptide drugs, and impressively HBS3 was efficiently taken up into HeLa cells cultured in monolayer. Active RAS pulldown revealed that HBS3 reduced EGF-stimulated RAS activation and suppressed EGF-stimulated ERK phosphorylation as measured by immunoblotting in HeLa cells. In contrast, no significant inhibition of EGFR phosphorylation by HBS3 was observed, indicating a mechanism of action that takes place at the SOS-RAS interaction. Conceptually, SOS α-helix mimetics would prevent SOS-mediated activation of all RAS isozymes, providing a pan-RAS mechanism that could serve as an effective therapeutic if the hurdles of small-peptide inhibitors can be overcome. Namely, peptide drugs suffer from poor oral bioavailability and stability due to the activity of bodily peptidases (Lau & Dunn, 2018). Encouragingly, HBS3 has shown promise in overcoming the characteristically poor cellular uptake of peptide drugs, yet further progress for this unique approach will require advancement in the field of peptide-based RAS inhibitors (Di, 2015).

One such advancement is the stapled peptide, which includes a covalent linker that provides stability to the alpha helix and enhances drug-like properties. In 2015, Leshchiner et al. developed SOS α-helix mimetics using the peptide stapling method (Leshchiner et al., 2015). A fluorescence polarization screen identified SAH-SOS1α as the lead inhibitor based on its low nanomolar EC_{50} values for binding WT KRAS as well as KRAS mutant variants (G12V, G12C, G12S, and Q61H). SAH-SOS1α inhibited the growth of a panel of 11 oncogenic KRAS mutant cell lines in a dose-dependent manner and reduced MAPK and AKT signaling in Panc 10.05 (KRAS G12D) and HeLa (WT KRAS) cell lines as detected by immunoblot. A *Drosophila melanogaster* model was employed to measure the effect of SAH-SOS1α on RAS signaling in vivo. This model uses an inducible system by which RU486 treatment drives organism-wide elevated expression of RAS85DV12, a mutant of the *Drosophila* homolog of human *KRAS*, *NRAS*, and *HRAS* genes. SAH-SOS1α treatment suppressed ERK and AKT phosphorylation when administered via injection or orally and produced no observable toxicity at the highest administered micromolar dose. The stapled peptide drug design afforded SAH-SOS1α sufficient cellular uptake to be used for treatment both in vitro and in vivo where it displayed a pan-RAS mechanism of action. The authors propose further optimization to improve cellular potency and to advance into *RAS* mutant murine models. In addition, further characterization of the pan-RAS mechanism in mutant *HRAS* and *NRAS* cell lines would afford a better understanding of SAH-SOS1α's scope and potential target patient population for future derivatives.

Another approach to inhibit RAS evolved from studies of a non-steroidal anti-inflammatory drug (NSAID), sulindac, that has well-documented antineoplastic properties, although the mechanism of action is poorly understood (Gurpinar, Grizzle, & Piazza, 2014; Labayle et al., 1991; Piazza et al., 1995,

1997; Shiff, Qiao, Tsai, & Rigas, 1995). Sulindac and other NSAIDs suppress inflammation by inhibiting cyclooxygenase (COX) and reducing prostaglandin levels, but the basis for their antineoplastic activity may involve additional or alternative mechanisms. For example, the dosages required to inhibit tumor growth exceed those effective for inhibiting inflammation and non-COX inhibitory derivatives of sulindac have been synthesized with improved anticancer activity (Gurpinar et al., 2014; Thompson et al., 2000). In 1997, Thompson et al. reported that sulindac and its non-COX inhibitory metabolite, sulindac sulfone, preferentially inhibited *RAS* mutant tumors in a carcinogen-induced rat model of breast cancer (Thompson et al., 1997). Subsequently, investigators reported that sulindac had modest RAS inhibitory activity to suppress MAPK signaling and RAS-dependent malignant transformation (Herrmann et al., 1998). Later studies generated sulindac derivatives with enhanced potency to inhibit cancer cell growth that were found to directly interact with RAS between the switch I and II regions as observed by HSQC (heteronuclear single quantum coherence) NMR chemical perturbation analysis (Waldmann et al., 2004). This connection between sulindac's anticancer effect and the RAS pathway has been further detailed in several studies, setting up sulindac as a candidate for derivatization to identify novel RAS inhibitors (Gala, Sun, & Yang, 2009; Herrmann et al., 1998; Karaguni et al., 2002; Müller et al., 2004; Taylor et al., 2000). Together, these publications provided the initial rationale for chemically modifying sulindac to enhance RAS binding affinity and improve antitumor efficacy, while blocking COX binding to reduce toxicities resulting from COX inhibition and the suppression of physiological prostaglandins.

A custom library of over 1500 compounds was synthesized that shared the indene scaffold of sulindac but had substitutions to the carboxylic acid moiety previously reported to be essential for COX binding and to other moieties to improve RAS binding affinity. The library was screened in full-dose response for differential growth inhibitory effects using mutant KRAS(G13D) HCT116 colon cancer cells vs HT29 colon cancer cells with WT RAS, but with a mutation in B Rapidly Accelerated Fibrosarcoma (BRAF) V600E downstream of RAS. This pair of cancer cell lines was used because of large differences in RAS-GTP levels, while both rely on MAPK signaling as evident by equal sensitivity to MAPK inhibitors. Following the identification of an initial hit and extensive chemical optimization with iterative screening, a lead compound, ADT 007, emerged with a growth inhibition IC_{50} value of $\sim$5 nM for mutant *KRAS* HCT116 cells, while the IC_{50} for WT *RAS* HT29 cells was $\sim$100-fold higher. Screening a panel of histologically diverse tumor cell lines with mutations in various RAS isozymes and codons for sensitivity to ADT-007 revealed that sensitive cell lines share similarly low nanomolar IC_{50} values and all bear mutations in *RAS* or upstream RAS-activating components (e.g., EGFR or NF1). Despite this broad pan-RAS activity, this class of RAS inhibitors does not affect the growth of cells from normal tissues, suggesting the potential for a high degree of cancer cell selectivity (Keeton et al., 2019; Mattox et al., 2019). Efforts to identify the mechanism of binding include HSQC NMR where treatment of WT KRAS produced chemical shift perturbations corresponding to amino acids that coordinate nucleotide binding. Nucleotide exchange assays using the fluorescently labeled GTP analogue *N*-methylanthraniloyl (MANT) GTP to measure RAS-GTP formation as an increase in fluorescence suggest that ADT-007 acts by inhibiting GTP binding. Cellular engagement of KRAS by ADT-007 was confirmed using the cellular thermal shift assay (CETSA) to measure protein-ligand interactions based on binding that protects proteins from thermal degradation. Other experiments revealed that ADT-007 suppresses RAS activation and MAPK/AKT signaling in RAS-dependent cell lines and mouse tumor models at concentrations or dosages that suppress growth in vitro and in vivo, respectively. Ongoing efforts to further define the mechanism of binding will aid further chemical optimization of ADT-007 toward a development candidate.

A computational screen carried out by Shima et al. in 2013 identified 97 potential active RAS binding compounds from which Kobe0065 emerged as the lone compound displaying potency to competitively inhibit HRAS binding to RAF (Shima et al., 2013). Consistent with the effects expected of a RAS inhibitor, Kobe0065 and the less potent analogue Kobe2602 suppressed MAPK and AKT activation in HRAS(G12V) driven NIH/3T3 cells. The compounds inhibited colony formation of cell lines that bear *RAS* mutations including both oncogenic *KRAS* and *NRAS*, and treatment of nude mice with SW480 xenografts caused tumor growth inhibition with no observable toxicity. A less potent but more water-soluble analogue, Kobe2601, was used to probe the Kobe family of compounds' mechanism of action. The NMR structure of Kobe2601 in complex with GTP-bound HRAS(T35S) revealed extensive contacts between the compound and the Switch I/II regions where RAS interacts with the RAS binding domains (RBDs) of downstream effectors. Therefore, the proposed mechanism of RAS inhibition is that Kobe compounds bind active RAS-GTP and prevent effector interactions such as RAF and PI3K activation (Shima et al., 2013). Further medicinal chemistry efforts to improve potency and in vivo efficacy are warranted; however, no studies continuing the development of Kobe compounds have been published.

As discussed in Chapter 8 "Targeting the "Undruggable" RAS with Biologics,"Shin et al. developed a second generation pan-RAS binding mAb referred to as "inRAS37" (Shin et al., 2017, 2020). inRAS37 is composed of three major functional groups to facilitate selective entry to tumor cells, escape the endosome, and specifically binds RAS-GTP to inhibits RAS signaling.

2.1.4 Switch I/II pocket

In 2012, Maurer et al. identified compounds that bound to a novel RAS binding site located between switch I and switch II called the switch I/II pocket (switch I/II-P). The group employed a fragment-based discovery method to screen 3300 compounds for affinity to KRAS(G12D) in an NMR-based binding assay (Maurer et al., 2012). The 240 initial hits were then further scrutinized using HSQC NMR to delineate only those compounds that cause chemical shift perturbations at loci corresponding to a single cohesive binding site on KRAS. The remaining 25 compounds were co-crystallized with KRAS(Q61H). 4,6-dichloro-2-methyl-3-aminoethyl-indole (DCAI) bound to the switch I/II-P and was selected as a lead compound for further analysis. KRAS, NRAS, and HRAS share identical amino acid sequences at the switch I/II-P indicating that DCAI would bind as a pan-RAS inhibitor. This was partially confirmed by NMR saturation transfer difference using HRAS to show DCAI binding, suggesting a pan-RAS mechanism of action. The compound binding site is adjacent to the PPI site where the SOS catalytic domain binds RAS-GDP to facilitate nucleotide exchange, indicating DCAI prevents this interaction to suppress RAS activation. This was supported using MANT-GTP to measure RAS-GTP formation. In the presence of catalytically active SOS, DCAI treatment inhibited RAS nucleotide release ($IC_{50} = 155 \pm 36\,\mu M$). This mechanism was found to be RAS-specific, as DCAI treatment was unable to prevent nucleotide release from other RAS superfamily proteins in the same experimental design. To examine whether DCAI RAS inhibition persists in a cellular context, a live-cell imaging model was designed using HEK-293T cells expressing a fluorescently labeled RAF RBD. DCAI treatment reduced recruitment of RAF to the plasma membrane thereby inhibiting RAS effector interactions. Treatment was also capable of inhibiting EGF-stimulated RAS activation in the MANT-GTP nucleotide exchange assay.

While DCAI itself lacks sufficient potency and cellular activity, its discovery resulted in the identification a novel druggable site on RAS that several groups would later use to generate more potent and drug-like compounds (Maurer et al., 2012).

A library of potential RAS PPI inhibitors was screened by Cruz-Migoni et al. in 2019 for binding against two control proteins as well as KRAS(G12V) in a surface plasmon resonance screening assay (Cruz-Migoni et al., 2019). Thirty compounds from this library successfully bound KRAS while not interacting with the control proteins, seven of which displayed RAS specificity. Two of these compounds, PPIN-1 and PPIN-2, were found to display sufficiently good interactions with KRAS(G12V) via waterLOGSY NMR to warrant further study. Crystal soaking experiments using KRAS(Q61H) revealed that PPIN-1 and PPIN-2 bind in the previously described switch I/II-P. PPIN-1 and PPIN-2, however, did not prevent RAS PPIs following switch I/II-P binding. To generate an inhibitor that disrupts RAS PPIs, an in silico chemistry approach was used to model a conjugation of the PPIN compounds with Abd-7, a RAS-binding intracellular antibody-derived compound previously discovered by this group (Quevedo et al., 2018). The resulting conjugated compounds, Ch-1, Ch-2, and Ch-3 were each found to bind to the switch I pocket of KRAS(Q61H) (Cruz-Migoni et al., 2019). A BRET-based assay found that C-RAF, PI3Kγ, and RALGDS interactions with RAS were each inhibited by these conjugated compounds in *KRAS*, *NRAS*, and *HRAS* mutant cell lines. Western blot analysis of DLD-1 CRC cells following treatment revealed a dose-dependent decrease of EGF-stimulated p-ERK and p-AKT from all three compounds. Of these, Ch-3 demonstrated the most potent signaling inhibition with a near total suppression of AKT phosphorylation at the 10 μM concentration. Ch-3 was also the most potent inhibitor of monolayer cell growth using DLD-1 cells with an IC_{50} value of 4.5 μM. The rationale for the combination of two less effective compounds into a single more beneficial compound was validated in this study, resulting in a novel group of pan-RAS PPI inhibitors (Cruz-Migoni et al., 2019). With a thoroughly characterized binding mechanism, it will be informative to determine the anticancer activity of these compounds in a panel of cancer cell lines with various *RAS* mutations or WT *RAS* to further assess compound potency and selectivity for cancer cells with activated RAS before moving into more clinically relevant cancer models.

A large-scale RAS switch I/II-P inhibitor discovery effort by Kessler et al. involving two fragment-based screens and structurally informed drug design culminated in the novel RAS inhibitor compound BI-2852

(Kessler et al., 2019). ITC was performed using BI-2852 and each RAS isozyme bound to either GDP or the stable GTP analog phosphomethylphosphonic acid guanylate ester (GCP). The results identified BI-2852 as a pan-RAS inhibitor, binding to all RAS isozymes with low or sub micromolar K_d values. The highest affinity interaction was with GCP-bound WT HRAS ($K_d = 570\,nM$) while GCP-bound WT KRAS ($K_d = 7.5\,\mu M$) and GDP-bound WT NRAS ($K_d = 8.3\,\mu M$) resulted in the lowest affinity of the tested proteins. BI-2852 was found to prevent KRAS PPIs with the SOS catalytic site, the SOS allosteric site, and downstream effectors (CRAF and PI3Kα) in a fluorescence resonance energy transfer (FRET)-based alpha screen assay. As RAS-SOS PPI inhibitor, BI-2852 was predictably unable to reduce RAS-GTP levels in growth factor stimulated NCI-H358 cells. When these cells were incubated in low serum media for 24 h the portion of GTP-bound RAS decreased and subsequent BI-2852 treatment prevented RAS-GTP re-formation, validating the mechanism of action. Treatment of NCI-H358 with BI-2852 inhibited p-ERK albeit with a rebound in ERK phosphorylation returning to pre-treatment levels approximately 20 h after initial treatment. Under low serum conditions in a 14-day soft agar assay, BI-2852 inhibited NCI-H358 proliferation with an $IC_{50} = 6.7\,\mu M$. No antiproliferative effect was observed when BI-2852 was used to treat NCI-H358 under adherent conditions. Using both growth assay designs, BI-2852 was shown to have no effect on proliferation or p-ERK of four WT *RAS*/mutant *B-RAF* cell lines demonstrating RAS pathway selectivity (Kessler et al., 2019). Along with other compounds that target the switch I/II-P, BI-2852 represents a promising novel strategy that merits optimization to become clinically viable. While *HRAS* mutant cancers are rare, the preferential HRAS affinity of BI-2852 would encourage further experimentation in head and neck squamous cell carcinomas (HNSCCs) or bladder carcinomas as they more frequently develop oncogenic *HRAS* mutations.

2.2 Indirect binding Pan-RAS inhibitors

2.2.1 Farnesyltransferase

Farnesyltransferase inhibitors (FTIs), one of the earliest attempts to drug RAS, focused on inhibiting a critical protein involved in RAS post-translational processing (Marín-Ramos, Ortega-Gutiérrez, & López-Rodríguez, 2019; Reid & Beese, 2004; Strickland et al., 1999). RAS proteins are translated as cytosolic proteins; however, RAS activation depends on its association with upstream activating enzymes linked to plasma membrane receptors (Ahearn et al., 2012). Thus, each RAS isozyme requires post-translational modifications (PTMs) to target them to the interior of the cell membrane where they can respond to growth stimuli and become activated. In general, these PTMs involve prenylation of the C-terminal CAAX motif shared by all RAS isozymes, removal of AAX by RAS-converting-enzyme 1 (RCE1), and methylation of the C-terminus by isoprenylcysteine carboxyl methyltransferase (ICMT) to counteract the negative charge of the prenylated cysteine and further increase affinity for cellular membranes (Berndt et al., 2011). Understanding the critical function of these RAS PTMs, one of the earliest attempts to indirectly inhibit all RAS isozymes involved inhibiting prenylation of the CAAX. Farnesyltransferase (FTase) catalyzes this reaction for all RAS isozymes, and therefore FTIs were developed to prevent RAS localization to the plasma membrane (Berndt et al., 2011). FTIs showed great promise in preclinical models both in vitro and in vivo where they inhibited RAS-driven proliferation and tumor growth. HRAS-driven tumors were especially sensitive as FTI treatment caused tumors to shrink in HRAS-transgenic mice. To date, four FTIs have been evaluated in clinical trials: tipifarnib, lonafarnib, BMS-214662 and L-778123 (Berndt et al., 2011). Despite preclinical success, the two FTIs that reached phase III, tipifarnib and lonafarnib, were unable to improve patient outcomes either as a monotherapy or in combination with carboplatin and paclitaxel or gemcitabine (End et al., 2001; Marín-Ramos, Ortega-Gutiérrez, & López-Rodríguez, 2019; Sun et al., 2004). It was discovered that upon FTase inhibition KRAS and NRAS isozymes were instead geranylgeranylated by geranylgeranyltransferase (GGTase) (Rowell, Kowalczyk, Lewis, & Garcia, 1997). This escape route provided a readily available adaptive resistance mechanism against FTIs and is the most likely explanation for their failure in clinical trials.

Alternative strategies have focused on GGTase, RCE1, or ICMT as targets to prevent RAS membrane localization (Marín-Ramos, Ortega-Gutiérrez, & López-Rodríguez, 2019). GGTase inhibitors often fail to sufficiently differentiate between FTase and GGTase, however a series of GGTase-specific inhibitors were identified from a library of allenoate-derived compounds in 2008 (Watanabe et al., 2008). RCE1 inhibition has been linked to lethal cardiomyopathy in mice as well as acceleration of some cancers; consequently, RCE1 inhibitors do not have a clear path to the clinic with our current understanding of RCE1 interactions (Bergo et al., 2004). Several ICMT inhibitors show promise in preclinical studies including a cysmethynil derivative with antitumor efficacy in immunodeficient mice bearing HepG2 xenografts (Lau et al., 2014; Marín-Ramos et al., 2019). RAS localization inhibitors, like direct binding RAS inhibitors, have benefited from improved understanding of RAS biology. Even FTIs, which failed initial clinical trials, may be useful for treatment of HRAS-driven cancer now that the GGTase escape route which HRAS is unable to access has been characterized. This path to the clinic may be short since lonafarnib has recently received FDA approval for another indication, Hutchinson-Gilford Progeria Syndrome (Dhillon, 2021). Together, RAS-localization inhibitors represent a pan-RAS inhibition strategy that is incorporating insights gained from its earlier defeats into better-informed strategies which are showing promise in pre-clinical models (Fig. 3).

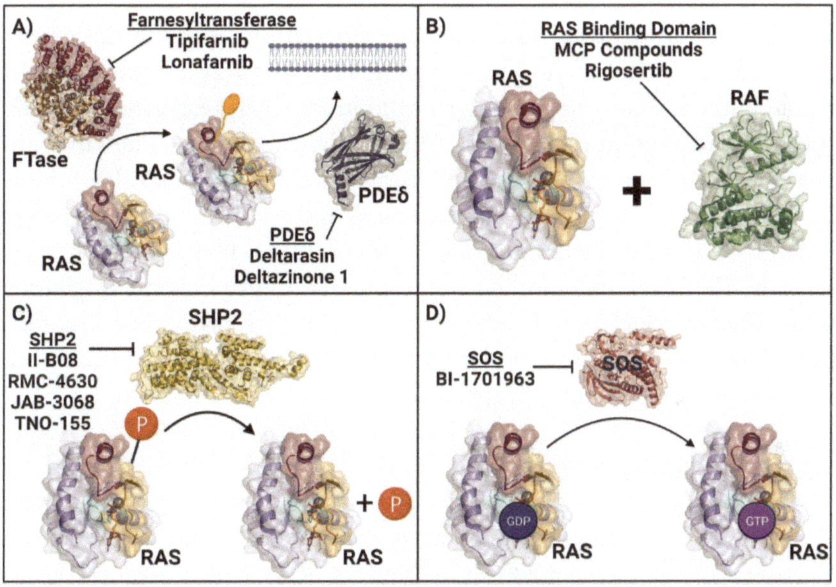

Fig. 3 Indirect Binding Pan-RAS Inhibitors and Their Protein Binding Sites. The ligand binding sites for each indirect pan-RAS inhibitor reviewed are indicated on cartoon representations of the crystal structures of target proteins. The P-loop (cyan), Switch I (orange), Switch II (crimson), and Mg^{2+} ion (gray) of KRAS(Q61H) (PDB ID: 6GQW) are indicated and the GppNHp is represented as a stick model (Cruz-Migoni et al., 2019). (A) Farnesyltransferase (dark red alpha subunit and gold beta subunit, PDB ID: 1SA4) prenylates RAS which is then guided to the plasma membrane by PDEδ (gray, PDB ID: 5X74) (Jiang et al., 2017; Reid & Beese, 2004). (B) RAS interacts with the RAS binding domain of RAF (green, PDB ID: 1UWH) (Wan et al., 2004). (C) SHP2 (yellow, PDB ID: 3B7O) removes an inhibitory phosphate from RAS Y32 (Barr et al., 2009). (D) SOS (pink, PDB ID: 6SCM) enhances RAS nucleotide exchange (Hofmann et al., 2021). Created with PyMOL and BioRender.com (Schrödinger & DeLano, 2020).

2.2.2 PDEδ

Another approach to prevent RAS localization is the inhibition of phosphodiesterase δ (PDEδ), a protein that binds prenylated RAS and facilitates delivery to the plasma membrane (Chandra et al., 2012). The PDEδ inhibitor Deltarasin was identified in 2013 using an alpha screen to identify compounds that bind to the PDEδ prenyl-binding pocket (Zimmermann et al., 2013). Structure-based chemical optimization of hit compounds to increase PDEδ affinity as well as solubility and membrane permeability resulted in the generation of Deltarasin as the lead PDEδ inhibitor. FRET measurements of tagged KRAS and PDEδ revealed near-total inhibition of the PDEδ-KRAS interaction within 1 min of Deltarasin treatment at a 5 μM concentration. Deltarasin treatment suppressed the proliferation of oncogenic KRAS-dependent Panc-Tu-I and Capan-1 cells and produced a dose-dependent decrease in ERK phosphorylation. In nude mice bearing subcutaneous Panc-Tu-I tumor xenografts, Deltarasin treatment dose-dependently caused tumor growth inhibition with near total suppression of growth at the highest dose of 10 mg/kg twice per day. The same group that produced Deltarasin generated a chemically distinct PDEδ inhibitor, Deltazinone 1, by using a similar screening and chemical optimization process as that used for Deltarasin (Papke et al., 2016). Deltazinone 1 targeted the same PDEδ prenyl-binding site as Deltrasin with enhanced selectivity. While Deltazinone 1 offers advantages in vitro, unfortunately it is rapidly metabolized in vivo and will require further chemical optimization to improve pharmacokinetic properties.

2.2.3 RAS binding domain

A subtractive yeast 2-hybrid screening assay designed to identify disruptors of the RAS/RAF interaction identified the MCP family of compounds (Kato-Stankiewicz et al., 2002). To determine at which level of the RAS pathway that MCP compounds exert their effect, a *Candida elegans* model was implemented where elevated RAS ortholog (let-60) signaling produces a multi-vulvar (Muv) phenotype (González-Pérez et al., 2010). MCP compounds were able to reduce Muv incidence in *C. elegans* expressing mutant let-60 and in *C. elegans* expressing constitutively active RAF, but not in *C. elegans* whose Muv phenotype was due to Ets-loss. These results pinpoint the MCP compounds' mechanism of action to the RAS/MAPK pathway and between the RAS and RAF interaction. MCP compounds revert *RAS* mutant but not *RAF* mutant/*RAS* WT human carcinoma cells to a flattened (epithelial) morphology and inhibit colony formation and invasion of *RAS* mutant human cancers in vitro (Kato-Stankiewicz et al., 2002). A dose-dependent decrease of MAPK signaling is observed upon MCP treatment of NIH/3T3 cells, and RAS/RAF colocalization to the plasma membrane is suppressed (González-Pérez et al., 2010). The potent MCP derivative MCP110 is well tolerated in murine models and maintains elevated plasma levels. Treatment of nude mice bearing lung adenocarcinoma caused significant tumor growth inhibition, and profoundly synergized with microtubule-targeting agent paclitaxel indicating promise for translation into human patients (Skobeleva, Menon, Weber, Golemis, & Khazak, 2007). It is still unclear whether MCP derivatives bind RAS, RAF, or the RAS/RAF PPI, and there have been no further publications regarding the compounds since 2010.

Rather than interfere with RAS signaling upstream of the RAS proteins, the compound Rigosertib (RGS) instead has been reported to bind to the RBD of downstream RAS effectors to prevent active RAS from binding and propagating RAS signaling (Athuluri-Divakar et al., 2016). Originally characterized as a polo-like kinase 1 (PLK1) inhibitor, RGS has since been attributed a myriad of mechanisms of action. Athuluri-Divakar et al. confirmed the RBDs of RAF, PI3K, and RALGDS as RGS targets using NMR and differential scanning fluorimetry (DSF). In vitro treatment with RGS reduces RAS-RBD binding as measured by RAS pulldown, but does not interfere with RAS activation confirming the mechanism of RAS signaling inhibition taking place at the RAS-RBD interaction. Further, HCT-116 and A549 xenografts as well as a KRAS(G12C) genetically engineered mouse models (GEMMs) exhibited greatly reduced tumor growth and MAPK/PI3K activation in response to RGS treatment. RGS is currently in phase 1/2 clinical trials in combination with the PD-1 inhibitor nivolumab with the target patient population being those with *KRAS* mutant cancers (Athuluri-Divakar et al., 2016). An alternative mechanism was described, however, in which RGS-mediated mitotic arrest leads to ROS-mediated JNK activation (Ritt et al., 2016). Hyperphosphorylation of SOS1, C/BRAF by activated JNK leads to reduced affinity for RAS and reduced ERK1/2 signaling in a manner analogous to ERK-mediated feedback inhibition. RGS's ever-evolving mechanism of action expanded in a 2021 publication focused on treating pediatric relapsed rhabdomyosarcomas and neuroblastomas which are characteristically RAS-dependent (Kowalczyk et al., 2021). When used to treat these cell lines in vitro, RGS displayed no anti-RAS signaling effect but instead directly bound to β-tubulin and caused microtubule instability. This effect could not be reproduced in vivo, a result which the authors attribute to early resistance or poor delivery in the murine model (Kowalczyk et al., 2021). Regardless, while the prospect of this pan-RAS inhibitor is promising it will be important to consider the broad mechanism of action of RGS when interpreting treatment effects or predicting cancer genotypes that would benefit from RGS treatment.

2.2.4 SOS

SOS inhibitors aim to deactivate the activator and prevent GEF-mediated RAS nucleotide exchange. Drug-bound SOS is unable to facilitate RAS activation for all RAS isozymes, suppressing pan-RAS signaling. The topic of inhibiting SOS-RAS interactions has been increasingly studied in recent years and as such is well reviewed (Lu, Jang, Zhang, & Nussinov, 2016). A recent SOS inhibitor that has progressed to clinical trials is the small proprietary compound BI 1701963 from Boehringer Ingelheim. BI 1701963 selectively binds the SOS1 catalytic domain and prevents it from associating with RAS-GDP (Gerlach et al., 2020). This leaves RAS bound to GDP, suppressing RAS activation and MAPK signaling. In *KRAS* mutant cell lines in vitro, BI 1701963 treatment reduced KRAS activation and suppressed MAPK signaling resulting in cytostasis, an effect was enhanced in vivo by combination with the MEK inhibitor trametinib. This combination dually controls both RAS activation through SOS1 inhibition and feedback regulated by MEK due to the resulting MAPK suppression. BI 1701963 and trametinib combination therapy as well as BI 1701963 monotherapy are currently being tested in clinical trials that are recruiting patients with advanced solid tumors bearing a *KRAS* mutation (NCT04627142, NCT04111458) (Gort et al., 2020).

2.2.5 SHP2

SHP2 is another RAS-interacting protein that has been developed into a tractable drug target for indirect pan-RAS inhibition (Yuan, Bu, Zhou, Yang, & Zhang, 2020). As a protein tyrosine phosphatase (PTP), SHP2 is essential for oncogenic RAS signaling. While the molecular mechanism of SHP2 tumor promotion is not fully understood, one emerging explanation is SHP2-mediated dephosphorylation of RAS Y32 (Bunda et al., 2015). Src-mediated phosphorylation of RAS Y32 is a RAS-inhibitory process that facilitates GTP hydrolysis and dissociation of RAS from RAF. SHP2 reverses this, resulting in increased active GTP-bound RAS. Targeting this protein is not restricted to benefit patients with mutations in the SHP2 gene

PTPN11, but also *RAS* mutant cancers broadly. Recent studies have shown that mutant *RAS* cancers can be dependent on functional SHP2 during carcinogenesis (Ruess et al., 2018). A study in 2010 by Zhang et al. found that U87 glioblastoma cells, which express elevated SHP2, treated with the SHP2 inhibitor II-B08 had decreased RAS-GTP formation and ERK phosphorylation (Bunda et al., 2015; Zhang et al., 2010). In vivo treatment of U87 xenograft mice with II-B08 resulted in significant tumor growth inhibition, paving the way for further development of SHP2 inhibitors. Allosteric SHP2 inhibitors have been successful in preclinical studies and are now undergoing clinical trials (Nichols et al., 2018). These include RMC-4630 (NCT03634982, NCT03989115), JAB-3068 (NCT03518554), and TNO-155 (NCT03114319, NCT04330664) either as monotherapies or in combination with RAS pathway inhibitors or KRAS(G12C) inhibitors.

2.3 RAS pathway inhibitors

RAF kinases are the first effectors downstream of RAS in the MAPK signaling cascade (Zhang & Liu, 2002). Active RAS recruits RAF to the plasma membrane where RAS binds to the RAF RBD and activates RAF (Hagemann & Rapp, 1999). Originally designed as therapies for *RAS* mutant tumors, RAF inhibitors were found to be limited due to their paradoxical response to inhibition. Drug-bound inactive RAF can bind and activate WT *RAF* in the presence of RAS-GTP resulting in an increase in RAF activation upon treatment with RAF inhibitors (Heidorn et al., 2010). FDA-approved RAF inhibitors such as vemurafenib are instead indicated for BRAF V600E/K melanoma (Karoulia, Gavathiotis, & Poulikakos, 2017). To date, RAF inhibitors have had limited success in cancers with oncogenic *RAS* mutations. Mitogen Activated Protein Kinase (MEK) inhibitors such as trametinib have had mixed results as either monotherapies or combination therapies for tumors with *RAS* mutations (Okumura & Jänne, 2014). MEK is the substrate of RAF and is immediately downstream of RAF in the MAPK signaling pathway (Zhang & Liu, 2002). Inhibiting MEK using selumetinib combined with docetaxel in the context of *KRAS* mutant NSCLC resulted in a modest increase in median overall survival that was not statistically significant from the placebo control (Jänne et al., 2013). This was followed up with a phase II trial that concluded there was no clinical benefit of MEK inhibitor and docetaxel combination treatment for advanced or metastatic NSCLC patients (Soria et al., 2017). This combination resulted in increased toxicity compared to docetaxel alone, indicating another obstacle for MEK inhibitors to overcome to become a viable therapy for *RAS* mutant cancers. ERK is the final protein in the MAPK pathway (Zhang & Liu, 2002). Once activated by MEK, a major function of ERK is the phosphorylation of transcription factors leading to an increase in cyclin D1 expression to drive cell-cycle progression (Buscà, Pouysségur, & Lenormand, 2016). While there are currently no FDA-approved ERK inhibitors, investigational compounds such as ulixertinib have shown promising preclinical and phase I clinical trial results (Germann et al., 2017; Sullivan et al., 2018). Unfortunately, treatment-related grade 3 toxic events were observed in 41% of enrolled patients (Sullivan et al., 2018).

A direct RAS-effector, Phosphoinositide 3 Kinase (PI3K) is activated when RAS binds to the PI3K RBD, also known as the p110 subunit (Zhang, Jang, & Nussinov, 2020). Once PI3K is activated by RAS it stimulates downstream proteins including Protein Kinase B (AKT) and mammalian target of rapamycin (mTOR), both of which regulate cell death, survival, and gene transcription (McCubrey et al., 2012). Recent efforts have been made to increase the efficacy of PI3K inhibitors by combining them with other RAS pathway inhibitors such as MAPK inhibitors. However, these efforts have been met with poor drug tolerability and toxicity issues in clinical trials. Alternative strategies combining AKT inhibitors (ipatasertib) and mTOR inhibitors (everolimus) with MAPK inhibitors to reduce RAS related feedback activation have produced similar results.

3. Discussion

The development of allele-specific RAS inhibitors has yielded the first ever FDA-approved RAS inhibitor. Precision medicines for patients with KRAS(G12C) mutant cancer are finally a reality, and inhibitors for other common *KRAS* mutations are in development to benefit a greater number of patients with *KRAS* mutations. While these are exciting advancements, the fact remains that allele-specific inhibitors are limited in scope. With a variety of inhibitors in pre-clinical and clinical development, pan-RAS inhibitors represent an alternative strategy that would provide a RAS-targeted therapy for all patients regardless of mutant isozyme or codon. A clinically viable pan-RAS inhibitor would wrest more total control over the RAS signaling network than allele-specific inhibitors. By suppressing all RAS isozymes, pan-RAS inhibitors could potentially avoid adaptive resistance mediated by compensatory activation of WT RAS proteins. The lack of dependency on a particular mutated codon would allow pan-RAS inhibitors to avoid intrinsic resistance arising from *RAS* mutation heterogeneity in the greater tumor microenvironment.

Just as with allele-specific inhibitors, pan-RAS inhibitors have foreseeable limitations. To offer a benefit over isozyme/mutation specific inhibitors regarding resistance and target patient population, pan-RAS inhibitors trade specificity for broader RAS pathway inhibition. This trade-off can result in toxicity issues if the lack of specificity produces unintended off-target inhibition. CID1067700 is an example where broader pan-GTPase inhibition resulted in a greater diversity of sensitive cancer mutational profiles but also introduced toxicity limitations (Agola et al., 2012; Hong et al., 2015). Rigosertib, an indirect pan-RAS inhibitor, has been ascribed alternative protein targets which have fortunately synergized to increase its ability to inhibit RAS-driven cancers. However, it would be going too far to suggest that pan-RAS inhibitors in general are plagued by insurmountable toxicity limitations. Many inhibitors reviewed here demonstrate RAS-specific mechanisms of action and are well-tolerated in animal models. While there are currently no FDA-approved pan-RAS inhibitors to support the claim that toxicity can be overcome, FDA-approved pan-inhibitors exist for several other drug targets and attest to the feasibility and advantages of the pan-inhibition strategy.

EGFR inhibitors including erlotonib and cetuximab suffer resistance from tumors that is attributed in part to redundant signaling by other HER-2 family receptors (Britten, 2004). Dacomitinib, a pan-HER-2 inhibitor, was found to suppress the growth of a greater number of HNSCC cell lines than erlotonib or cetuximab (Ather et al., 2013). Not surprisingly, the pan-HER-2 inhibitor exerted more complete suppression of EGFR, ERK, and AKT activity than did the EGFR-specific inhibitors. This mechanism by which dacomitinib controls multiple arms of the HER-2 growth signaling pathway was well tolerated in vivo, and dacomitinib would progress through clinical trials and gain FDA approval for first-line treatment of patients with metastatic NSCLC (FDA, 2018). Examples of other pan-inhibition targets in cancer include histone deacetylase (HDAC), mammalian target of rapamycin (mTOR), and PI3K at varying stages of preclinical and clinical development (Lemoine et al., 2012; Maira et al., 2012; Preuss, Hugle, Reimann, Schlecht, & Fulda, 2013). The precedent for effective and non-toxic pan-RAS inhibition becoming a viable treatment strategy has been set. Clearly then, pan-RAS inhibitors deserve further development to realize the potential increased efficacy and avoidance of drug resistance that they may offer over allele-specific RAS inhibitors.

Systemic delivery is one issue that presents a challenge in developing many of the novel small molecule pan-RAS inhibitors described above, particularly for the treatment of solid cancers. A potential solution is the development of pan-RAS inhibitor antibody-drug conjugates (ADC). Monoclonal antibodies can specifically target surface receptors on cancer cells to facilitate internalization, affording increased drug delivery to tumor cells. About 30 mAb have received approval from FDA for cancer indications (Chau, Steeg, & Figg, 2019). ADCs are comprised of a mAb to a tumor-associated antigen tethered to a cytotoxic drug or "warhead" via a chemical linker, and have been developed and successfully used for the treatment of Hodgkin's lymphoma, breast cancer, and colon cancer (Chau et al., 2019). One can envision the possibility of using a highly potent and selective pan-RAS inhibitor as the warhead in an ADC to enable highly selective targeting of RAS-driven cancer cells. This strategy could combat pan-RAS inhibitor toxicity by minimizing the inhibition of non-cancerous cells' normal physiological RAS-mediated processes.

Nanoparticles (NPs) are another promising delivery system for pan-RAS inhibitors to improve tumor targeting. NP formulations can be used to significantly improve drugs with low solubility, poor stability, or toxicity that prevents their clinical application. In the context of RAS-mutant cancer, several NP-based strategies can be leveraged to take advantage of RAS mutant cancer cells' increased dependence on macropinocytosis to scavenge nutrients from extracellular environment. These NPs include albumin conjugates and albumin NPs, exosomal nanocarriers, cell penetration peptides, biomimetic lipoprotein NPs, dendrimer conjugates, ricinB-quantum dot, and mesoporous silica NPs (Battah et al., 2007; Cullis et al., 2017; He, Li, & Huang, 2018; Huang et al., 2017; Iversen, Frerker, & Sandvig, 2012; Verdera, Gitz-Francois, Schiffelers, & Vader, 2017; Wang et al., 2018). An example this NP strategy in action was presented in a 2017 study by Kamerkar et al. where they showed that exosomes engineered with CD47 displayed enhanced delivery of their drug cargo into pancreatic cancers with oncogenic KRAS (Kamerkar et al., 2017). NP-mediated delivery may enhance therapeutic efficacy and minimize toxicity, thus greatly improving the efficacy of pan-RAS inhibitors.

With the hope that KRAS(G12C) therapies will improve the lives of patients, we also must prepare for the possibility that alternative approaches will be needed to bring RAS inhibitors to their greater potential. This chapter has detailed previous and ongoing pan-RAS inhibitor development with the purpose of informing novel pan-RAS targeted strategies and invigorating the scientific community's interest in pan-RAS inhibitors as an advantageous solution to the problem of drugging RAS.

4. Conclusion

Pan-RAS inhibitors take advantage of a mechanism that potentially allows for a more complete suppression of RAS signaling by inhibiting all RAS isozymes regardless of mutational status. Examples spanning small molecules, peptides, and monoclonal antibodies confirm that pan-RAS inhibition is not only lethal to RAS-driven cancers in vitro and in vivo, but that it can also retain cancer specificity. The most daunting obstacles facing specific RAS inhibitors are the adaptive resistance mechanisms that take advantage of the isozyme and mutation specificity as well as the limited percentage of patients that stand to benefit from treatment. Pan-RAS inhibitors can potentially overcome these difficulties, and thus warrant continued investigation as a mechanism with great potential to be broadly efficacious for RAS-driven malignancies.

Grant support

Funding provided by the National Cancer Institute, National Institutes of Health under Award Numbers R01 CA254197, R01 CA131378, and RO1 CA197147, and R01 CA254197 to G.A. Piazza. The content is solely the responsibility of the authors and does not necessarily represent the official views of the National Institutes of Health. A. Prakash is supported by a National Institutes of Environmental Health Sciences (NIEHS), Outstanding New Environmental Scientist (ONES) R01 ES030084 and NIEHS R35 ES031708 awarded to Dr. Joann Sweasy (U. of Arizona).

Disclosure of potential conflicts of interest

G.A. Piazza, A.B. Keeton, and X. Chen are co-founders of ADT Pharmaceuticals LLC and consultants.

References

Abulaiti, A., Fikaris, A. J., Tsygankova, O. M., & Meinkoth, J. L. (2006). Ras induces chromosome instability and abrogation of the DNA damage response. *Cancer Research*, *66*(21), 10505–10512.

Agola, J. O., Hong, L., Surviladze, Z., Ursu, O., Waller, A., Strouse, J. J., et al. (2012). A competitive nucleotide binding inhibitor: In vitro characterization of Rab7 GTPase inhibition. *ACS Chemical Biology*, 7(6), 1095–1108.

Ahearn, I. M., Haigis, K., Bar-Sagi, D., & Philips, M. R. (2012). Regulating the regulator: Post-translational modification of RAS. *Nature Reviews Molecular Cell Biology*, *13*(1), 39–51.

Ahmadian, M. R., Zor, T., Vogt, D., Kabsch, W., Selinger, Z., Wittinghofer, A., et al. (1999). Guanosine triphosphatase stimulation of oncogenic Ras mutants. *Proceedings of the National Academy of Sciences of the United States of America, 96*(12), 7065–7070.

Ambrogio, C., Köhler, J., Zhou, Z.-W., Wang, H., Paranal, R., Li, J., et al. (2018). KRAS dimerization impacts MEK inhibitor sensitivity and oncogenic activity of mutant KRAS. *Cell, 172*(4), 857–868.e15.

Ancrile, B. B., O'Hayer, K. M., & Counter, C. M. (2008). Oncogenic ras-induced expression of cytokines: A new target of anti-cancer therapeutics. *Molecular Interventions, 8*(1), 22–27. NIH Public Access.

Ather, F., Hamidi, H., Fejzo, M. S., Letrent, S., Finn, R. S., Kabbinavar, F., et al. (2013). Dacomitinib, an irreversible Pan-ErbB inhibitor significantly abrogates growth in head and neck Cancer models that exhibit low response to Cetuximab. *PLoS One, 8*(2), e56112.

Athuluri-Divakar, S. K., Vasquez-Del Carpio, R., Dutta, K., Baker, S. J., Cosenza, S. C., Basu, I., et al. (2016). A small molecule RAS-mimetic disrupts RAS Association with effector proteins to Block signaling. *Cell, 165*(3), 643–655.

Awad, M., Liu, S., Rybkin, I., Arbour, K., Dilly, J., et al. (2021). Acquired resistance to KRASG12C inhibition in Cancer. *The New England Journal of Medicine, 384*, 2382–2393.

Barr, A. J., Ugochukwu, E., Lee, W. H., King, O. N. F., Filippakopoulos, P., Alfano, I., et al. (2009). Large-scale structural analysis of the classical human protein tyrosine phosphatome. *Cell, 136*(2), 352–363.

Battah, S., Balaratnam, S., Casas, A., O'Neill, S., Edwards, C., et al. (2007). Macromolecular delivery of 5-aminolaevulinic acid for photodynamic therapy using dendrimer conjugates. *Molecular Cancer Therapeutics, 6*(3), 876–885.

Bergo, M. O., Lieu, H. D., Gavino, B. J., Ambroziak, P., Otto, J. C., Casey, et al. (2004). On the physiological importance of endoproteolysis of CAAX proteins. Heart-specific Rce1 knockout mice develop a lethal cardiomyopathy. *Journal of Biological Chemistry, 279*(6), 4729–4736.

Berndt, N., Hamilton, A. D., & Sebti, S. M. (2011). Targeting protein prenylation for cancer therapy. *Nature Reviews Cancer, 11*(11), 775–791.

Britten, C. D. (2004). Targeting ErbB receptor signaling: A pan-ErbB approach to cancer. *Molecular Cancer Therapeutics, 3*(10), 1335–1342.

Bunda, S., Burrell, K., Heir, P., Zeng, L., Alamsahebpour, A., Kano, Y., et al. (2015). Inhibition of SHP2-mediated dephosphorylation of Ras suppresses oncogenesis. *Nature Communications, 6*(1), 8859.

Burgess, M. R., Hwang, E., Mroue, R., Bielski, C. M., Wandler, A. M., Huang, B., et al. (2017). KRAS allelic imbalance enhances fitness and modulates MAP kinase dependence in cancer. *Cell, 168*(5), 817–829.e15.

Buscà, R., Pouysségur, J., & Lenormand, P. (2016). ERK1 and ERK2 map kinases: Specific roles or functional redundancy? *Frontiers in Cell and Development Biology, 4*(Issue JUN), 53. Frontiers Media S.A.

Campbell, S. L., Khosravi-Far, R., Rossman, K. L., Clark, G. J., & Der, C. J. (1998). Increasing complexity of Ras signaling. *Oncogene, 17*(11), 1395–1413.

Chandra, A., Grecco, H. E., Pisupati, V., Perera, D., Cassidy, L., Skoulidis, F., et al. (2012). The GDI-like solubilizing factor PDEδ sustains the spatial organization and signalling of Ras family proteins. *Nature Cell Biology, 14*(2), 148–158.

Chau, C., Steeg, P., & Figg, W. (2019). Antibody-drug conjugates for cancer. *Lancet, 394*(10200), 793–804.

Chen, H., Smaill, J. B., Liu, T., Ding, K., & Lu, X. (2020). Small-molecule inhibitors directly targeting KRAS as anticancer therapeutics. *Journal of Medicinal Chemistry, 63*(23), 14404–14424.

Cherfils, J., & Zeghouf, M. (2013). Regulation of small GTPases by GEFs, GAPs, and GDIs. *Physiological Reviews*, *93*(1), 269–309.
Cox, A. D., & Der, C. J. (2003). The dark side of Ras: Regulation of apoptosis. *Oncogene*, *22*(56 REV. ISS. 8), 8999–9006. Nature Publishing Group.
Cox, A. D., Der, C. J., & Philips, M. R. (2015). Targeting RAS membrane association: Back to the future for anti-RAS drug discovery? *Clinical Cancer Research: An Official Journal of the American Association for Cancer Research*, *21*(8), 1819.
Cox, A. D., Fesik, S. W., Kimmelman, A. C., Luo, J., & Der, C. J. (2014). Drugging the undruggable Ras: Mission possible? *Nature Reviews. Drug Discovery*, *13*(11), 828–851.
Cruz-Migoni, A., Canning, P., Quevedo, C., Bataille, C., Bery, N., Miller, A., et al. (2019). Structure-based development of new RAS-effector inhibitors from a combination of active and inactive RAS-binding compounds. *Proceedings of the National Academy of Sciences of the United States of America*, *116*, 2545–2550.
Cullis, J., Siolas, D., Avanzi, A., Barui, S., Maitra, A., et al. (2017). Macropinocytosis of nab-paclitaxel drives macrophage activation in pancreatic Cancer. *Cancer Immunology Research*, *5*(3), 182–190.
Der, C. J., Krontiris, T. G., & Cooper, G. M. (1982). Transforming genes of human bladder and lung carcinoma cell lines are homologous to the ras genes of Harvey and Kirsten sarcoma viruses. *Proceedings of the National Academy of Sciences of the United States of America*, *79*(11), 3637–3640.
Dhillon, S. (2021). Lonafarnib: First approval. *Drugs*, *81*(2), 283–289. https://doi.org/10.1007/s40265-020-01464-z.
Di, L. (2015). Strategic approaches to optimizing peptide ADME properties. *The AAPS Journal*, *17*(1), 134–143.
Dunnett-Kane, V., Nicola, P., Blackhall, F., & Lindsay, C. (2021). Mechanisms of resistance to KRASG12C inhibitors. *Cancers*, *13*(1).
End, D. W., Smets, G., Todd, A. V., Applegate, T. L., Fuery, C. J., Angibaud, P., et al. (2001). Characterization of the antitumor effects of the selective farnesyl protein transferase inhibitor R115777 in vivo and in vitro. *Cancer Research*, *61*(1), 131–137.
FDA. (2018). *FDA approves dacomitinib for metastatic non-small cell lung cancer*. https://www.fda.gov/drugs/drug-approvals-and-databases/fda-approves-dacomitinib-metastatic-non-small-cell-lung-cancer-0.
FDA. (2021). *FDA grants accelerated approval to sotorasib for KRAS G12C mutated NSCLC*. https://www.fda.gov/drugs/resources-information-approved-drugs/fda-grants-accelerated-approval-sotorasib-kras-g12c-mutated-nsclc.
Gala, M., Sun, R., & Yang, V. (2009). Inhibition of cell transformation by sulindac sulfide is confined to specific oncogenic pathways. *Cancer Letters*, *175*(1), 89–94.
Gerlach, D., Gmachl, M., Ramharter, J., Teh, J., Fu, S.-C., Trapani, F., et al. (2020). Abstract 1091: BI-3406 and BI 1701963: Potent and selective SOS1::KRAS inhibitors induce regressions in combination with MEK inhibitors or irinotecan. *Cancer Research*, *80*(16 Supplement), 1091.
Germann, U. A., Furey, B. F., Markland, W., Hoover, R. R., Aronov, A. M., Roix, J. J., et al. (2017). Targeting the MAPK signaling pathway in cancer: Promising preclinical activity with the novel selective ERK1/2 inhibitor BVD-523 (ulixertinib). *Molecular Cancer Therapeutics*, *16*(11), 2351–2363.
Gimple, R. C., & Wang, X. (2019). RAS: Striking at the core of the oncogenic circuitry. *Frontiers in Oncology*, *9*, 965.
González-Pérez, V., Reiner, D. J., Alan, J. K., Mitchell, C., Edwards, L. J., Khazak, V., et al. (2010). Genetic and functional characterization of putative Ras/Raf interaction inhibitors in C. elegans and mammalian cells. *Journal of Molecular Signaling*, *5*, 2.

Gort, E., Johnson, M. L., Hwang, J. J., Pant, S., Dünzinger, U., Riemann, K., et al. (2020). A phase I, open-label, dose-escalation trial of BI 1701963 as monotherapy and in combination with trametinib in patients with KRAS mutated advanced or metastatic solid tumors. *Journal of Clinical Oncology*, *38*(15_suppl), TPS3651.

Gurpinar, E., Grizzle, W. E., & Piazza, G. A. (2014). NSAIDs inhibit tumorigenesis, but how? *Clinical Cancer Research*, *20*(5), 1104–1113. American Association for Cancer Research Inc.

Hagemann, C., & Rapp, U. R. (1999). Isotype-specific functions of Raf kinases. *Experimental Cell Research*, *253*(1), 34–46.

Halder, S. K., Cho, Y. J., Datta, A., Anumanthan, G., Ham, A. J. L., Carbone, et al. (2011). Elucidating the mechanism of regulation of transforming growth factor β type II receptor expression in human lung cancer cell lines. *Neoplasia*, *13*(10), 912–922.

Hall, B. E., Bar-Sagi, D., & Nassar, N. (2002). The structural basis for the transition from Ras-GTP to Ras-GDP. *Proceedings of the National Academy of Sciences of the United States of America*, *99*(19), 12138–12142.

Hall, A., Marshall, C. J., Spurr, N. K., & Weiss, R. A. (1983). Identification of transforming gene in two human sarcoma cell lines as a new member of the ras gene family located on chromosome 1. *Nature*, *303*(5916), 396–400.

Hanahan, D., & Weinberg, R. A. (2011). Hallmarks of Cancer: The next generation. *Cell*, *144*(5), 646–674.

Hancock, J. F., Cadwallader, K., Paterson, H., & Marshall, C. J. (1991). A CAAX or a CAAL motif and a second signal are sufficient for plasma membrane targeting of ras proteins. *EMBO Journal*, *10*(13), 4033–4039.

Harvey, J. J. (1964). An unidentified virus which causes the rapid production of tumours in mice. *Nature*, *204*(4963), 1104–1105.

Hata, A. N., & Shaw, A. T. (2020). Resistance looms for KRAS G12C inhibitors. *Nature Medicine*, *26*(2), 169–170.

He, Y., Li, F., & Huang, Y. (2018). *Advances in protein chemistry and structural biology. Vol. 112* (pp. 183–220). Academic Press.

Heidorn, S. J., Milagre, C., Whittaker, S., Nourry, A., Niculescu-Duvas, I., Dhomen, N., et al. (2010). Kinase-dead BRAF and oncogenic RAS cooperate to drive tumor progression through CRAF. *Cell*, *140*(2), 209–221.

Herrmann, C., Block, C., Geisen, C., Haas, K., Weber, C., Winde, G., et al. (1998). Sulindac sulfide inhibits Ras signaling. *Oncogene*, *17*(14), 1769–1776.

Hobbs, G. A., Der, C. J., & Rossman, K. L. (2016). RAS isoforms and mutations in cancer at a glance. *Journal of Cell Science*, *129*(7), 1287–1292.

Hofmann, M. H., Gmachl, M., Ramharter, J., Savarese, F., Gerlach, D., Marszalek, J. R., et al. (2021). Bi-3406, a potent and selective sos1–kras interaction inhibitor, is effective in kras-driven cancers through combined mek inhibition. *Cancer Discovery*, *11*(1), 142–157.

Hong, D. S., Fakih, M. G., Strickler, J. H., Desai, J., Durm, G. A., Shapiro, et al. (2020). KRAS G12C inhibition with Sotorasib in advanced solid tumors. *New England Journal of Medicine*, *383*(13), 1207–1217.

Hong, L., Guo, Y., BasuRay, S., Agola, J. O., Romero, E., Simpson, D. S., et al. (2015). A Pan-GTPase inhibitor as a molecular probe. *PLoS One*, *10*(8), e0134317.

Huang, J., Jiang, G., Song, Q. X., Gu, X., Hu, M., et al. (2017). Lipoprotein-biomimetic nanostructure enables efficient targeting delivery of siRNA to Ras-activated glioblastoma cells via macropinocytosis. *Nature Communications*, *8*, 15144.

Iversen, T., Frerker, N., & Sandvig, K. (2012). Uptake of ricinB-quantum dot nanoparticles by a macropinocytosis-like mechanism. *Journal of Nanobiotechnology*, *10*, 33.

Jänne, P. A., Shaw, A. T., Pereira, J. R., Jeannin, G., Vansteenkiste, J., Barrios, C., et al. (2013). Selumetinib plus docetaxel for KRAS-mutant advanced non-small-cell lung cancer: A randomised, multicentre, placebo-controlled, phase 2 study. *The Lancet Oncology*, *14*(1), 38–47.

Jiang, Y., Zhuang, C., Chen, L., Lu, J., Dong, G., Miao, Z., et al. (2017). Structural biology-inspired discovery of novel KRAS-PDEδ inhibitors. *Journal of Medicinal Chemistry*, *60*(22), 9400–9406.

John, J., Sohmen, R., Feurstein, J., Linke, R., Wittinghofer, A., Goody, R. S., et al. (1990). Kinetics of interaction of nucleotides with nucleotide-free H-ras p21. *Proceedings of the National Academy of Sciences of the United States of America*, *29*, 6058–6065.

Kamerkar, S., LeBleu, V., Sugimoto, H., Yang, S., Ruivo, C., et al. (2017). Exosomes facilitate therapeutic targeting of oncogenic KRAS in pancreatic cancer. *Nature*, *546*(7659), 498–503.

Karaguni, I. M., Herter, P., Debryne, P., Chtarbova, S., Kasprzynski, A., et al. (2002). The new sulindac derivative IND 12 reverses Ras-induced cell transformation. *Cancer Research*, *62*(6), 1718–1723.

Karoulia, Z., Gavathiotis, E., & Poulikakos, P. I. (2017). New perspectives for targeting RAF kinase in human cancer. In *Vol. 17. Nature Reviews Cancer* (pp. 676–691). Nature Publishing Group. issue 11.

Kato-Stankiewicz, J., Hakimi, I., Zhi, G., Zhang, J., Serebriiskii, I., Guo, L., et al. (2002). Inhibitors of Ras/Raf-1 interaction identified by two-hybrid screening revert Ras-dependent transformation phenotypes in human cancer cells. *Proceedings of the National Academy of Sciences of the United States of America*, *99*(22), 14398–14403.

Keeton, A. B., Ward, A., Chen, X., Valiyaveettil, J., Zhu, B., Ramirez-Alcantara, V., et al. (2019). Abstract 2707: A novel RAS inhibitor, MCI-062, inhibits colon tumor growth in vivo and activates antitumor immunity. *Cancer Research*, *79*(13 Supplement), 2707.

Kessler, D., Gmachl, M., Mantoulidis, A., Martin, L. J., Zoephel, A., Mayer, M., et al. (2019). Drugging an undruggable pocket on KRAS. *Proceedings of the National Academy of Sciences of the United States of America*, *116*(32), 15823–15829.

Kirsten, W. H., & Mayer, L. A. (1967). Morphologic responses to a murine erythroblastosis virus. *JNCI: Journal of the National Cancer Institute*, *39*(2), 311–335.

Kodaz, H. (2017). Frequency of RAS mutations (KRAS, NRAS, HRAS) in human solid cancer. *Eurasian Journal of Medicine and Oncology*, *1*, 1–7.

Kowalczyk, J. T., Wan, X., Hernandez, E. R., Luo, R., Lyons, G. C., Wilson, K. M., et al. (2021). Rigosertib induces mitotic arrest and apoptosis in RAS-mutated rhabdomyosarcoma and neuroblastoma. *Molecular Cancer Therapeutics*, *20*(2), 307–319.

Kranenburg, O., Gebbink, M. F. B. G., & Voest, E. E. (2004). Stimulation of angiogenesis by Ras proteins. *Biochimica et Biophysica Acta, Reviews on Cancer*, *1654*(1), 23–37. Elsevier.

Kretzschmar, M., Doody, J., Timokhina, I., & Massagué, J. (1999). A mechanism of repression of TGFfβ/Smad signaling by oncogenic Ras. *Genes and Development*, *13*(7), 804–816.

Labayle, D., Fischer, D., Vielh, P., Drouhin, F., Pariente, A., Bories, C., et al. (1991). Sulindac causes regression of rectal polyps in familial adenomatous polyposis. *Gastroenterology*, *101*(3), 635–639.

Lam, T., Kulp, D. V., Wang, R., Lou, Z., Taylor, J., Rivera, C. E., et al. (2016). Small molecule inhibition of Rab7 impairs B cell class-switching and plasma cell survival to dampen the autoantibody response in murine lupus. *Journal of Immunology (Baltimore, Md. : 1950)*, *197*(10), 3792–3805.

Lau, J. L., & Dunn, M. K. (2018). Therapeutic peptides: Historical perspectives, current development trends, and future directions. *Bioorganic & Medicinal Chemistry*, *26*(10), 2700–2707.

Lau, H. Y., Ramanujulu, P. M., Guo, D., Yang, T., Wirawan, M., Casey, P. J., et al. (2014). An improved isoprenylcysteine carboxylmethyltransferase inhibitor induces cancer cell death and attenuates tumor growth in vivo. *Cancer Biology & Therapy*, *15*(9), 1280–1291.

Lemoine, M., Derenzini, E., Buglio, D., Medeiros, L. J., Davis, R. E., Zhang, J., et al. (2012). The pan-deacetylase inhibitor panobinostat induces cell death and synergizes with everolimus in Hodgkin lymphoma cell lines. *Blood*, *119*(17), 4017–4025.

Leshchiner, E. S., Parkhitko, A., Bird, G. H., Luccarelli, J., Bellairs, J. A., Escudero, S., et al. (2015). Direct inhibition of oncogenic KRAS by hydrocarbon-stapled SOS1 helices. *Proceedings of the National Academy of Sciences of the United States of America*, *112*(6), 1761–1766.

Liu, W., Yin, Y., Wang, J., Shi, B., Zhang, L., Qian, D., et al. (2017). Kras mutations increase telomerase activity and targeting telomerase is a promising therapeutic strategy for Kras-mutant NSCLC. *Oncotarget*, *8*(1), 179–190.

Lu, S., Jang, H., Gu, S., Zhang, J., & Nussinov, R. (2016). Drugging Ras GTPase: A comprehensive mechanistic and signaling structural view. *Chemical Society Reviews*, *45*(18), 4929.

Lu, S., Jang, H., Muratcioglu, S., Gursoy, A., Keskin, O., Nussinov, R., et al. (2016). Ras conformational ensembles, Allostery, and signaling. *Chemical Reviews*, *116*(11), 6607–6665.

Lu, S., Jang, H., Zhang, J., & Nussinov, R. (2016). Inhibitors of Ras-SOS interactions. *ChemMedChem*, *11*(8), 814–821.

Maira, S. M., Pecchi, S., Huang, A., Burger, M., Knapp, M., Sterker, D., et al. (2012). Identification and characterization of NVP-BKM120, an orally available pan-class I PI3-kinase inhibitor. *Molecular Cancer Therapeutics*, *11*(2), 317–328.

Marín-Ramos, N. I., Balabasquer, M., Ortega-Nogales, F. J., Torrecillas, I. R., Gil-Ordóñez, A., Marcos-Ramiro, B., et al. (2019). A potent Isoprenylcysteine carboxylmethyltransferase (ICMT) inhibitor improves survival in Ras-driven acute myeloid leukemia. *Journal of Medicinal Chemistry*, *62*(13), 6035–6046.

Marín-Ramos, N. I., Ortega-Gutiérrez, S., & López-Rodríguez, M. L. (2019). Blocking Ras inhibition as an antitumor strategy. *Seminars in Cancer Biology*, *54*, 91–100.

Mattox, T. E., Chen, X., Valiyaveettil, J., Maxuitenko, Y., Zhu, B., Ward, A., et al. (2019). Abstract 345: Novel RAS inhibitor, MCI-062, potently and selectively inhibits the growth of KRAS mutant pancreatic tumor cells by blocking GTP loading of RAS. *Cancer Research*, *79*(13 Supplement), 345.

Maurer, T., Garrenton, L. S., Oh, A., Pitts, K., Anderson, D. J., Skelton, N. J., et al. (2012). Small-molecule ligands bind to a distinct pocket in Ras and inhibit SOS-mediated nucleotide exchange activity. *Proceedings of the National Academy of Sciences of the United States of America*, *109*(14), 5299–5304.

McCubrey, J. A., Steelman, L. S., Chappell, W. H., Abrams, S. L., Franklin, R. A., Montalto, G., et al. (2012). Ras/Raf/MEK/ERK and PI3K/PTEN/Akt/mTOR cascade inhibitors: How mutations can result in therapy resistance and how to overcome resistance. *Oncotarget*, *3*(10), 1068–1111. Impact Journals LLC.

Mirati Therapeutics Inc. (2020). *Mirati Therapeutics Reports Investigational Adagrasib (MRTX849) Preliminary Data Demonstrating Tolerability and Durable Anti-Tumor Activity as well as Initial MRTX1133 Preclinical Data*. https://ir.mirati.com/press-releases/press-release-details/2020/Mirati-Therapeutics-Reports-Investigational-Adagrasib-MRTX849-Preliminary-Data-Demonstrating-Tolerability-and-Durable-Anti-Tumor-Activity-as-well-as-Initial-MRTX1133-Preclinical-Data/default.aspx.

Müller, O., Gourzoulidou, E., Carpintero, M., Karaguni, I. M., Langerak, A., et al. (2004). Identification of potent Ras signaling inhibitors by pathway-selective phenotype-based screening. *Angewandte Chemie (International Ed. in English)*, *43*(4), 450–454.

Nassar, A., Adib, E., & Kwiatkowski, D. (2021). Distribution of *KRAS*^{G12C} somatic mutations across race, sex, and Cancer type. *The New England Journal of Medicine*, *384*(2), 185–187.

Nichols, R. J., Haderk, F., Stahlhut, C., Schulze, C. J., Hemmati, G., Wildes, D., et al. (2018). RAS nucleotide cycling underlies the SHP2 phosphatase dependence of mutant BRAF-, NF1- and RAS-driven cancers. *Nature Cell Biology*, *20*(9), 1064–1073.

Noonan, T., Brown, N., Dudycz, L., & Wright, G. (1991). Interaction of GTP derivatives with cellular and oncogenic ras-p21 proteins. *Journal of Medicinal Chemistry*, *34*, 1302–1307.

Nussinov, R., Tsai, C.-J., Chakrabarti, M., & Jang, H. (2016). A new view of Ras isoforms in cancers. *Cancer Research*, *76*(1), 18–23.

Okumura, S., & Jänne, P. A. (2014). Molecular pathways: The basis for rational combination using MEK inhibitors in KRAS-mutant cancers. *Clinical Cancer Research*, *20*(16), 4193–4195.

Ostrem, J. M., Peters, U., Sos, M. L., Wells, J. A., & Shokat, K. M. (2013). K-Ras(G12C) inhibitors allosterically control GTP affinity and effector interactions. *Nature*, *503*(7477), 548–551.

Pantsar, T. (2020). The current understanding of KRAS protein structure and dynamics. *Computational and Structural Biotechnology Journal*, *18*, 189–198.

Papke, B., Murarka, S., Vogel, H. A., Martín-Gago, P., Kovacevic, M., Truxius, D. C., et al. (2016). Identification of pyrazolopyridazinones as PDEδ inhibitors. *Nature Communications*, 7(1), 1–9.

Patgiri, A., Yadav, K. K., Arora, P. S., & Bar-Sagi, D. (2011). An Orthosteric inhibitor of the Ras-Sos interaction. *Nature Chemical Biology*, 7(9), 585–587.

Piazza, G., Alberts, D., Hixson, L., Paranka, N., Li, H., et al. (1997). Sulindac sulfone inhibits Azoxymethane-induced Colon carcinogenesis in rats without reducing prostaglandin levels. *Cancer Research*, *57*(14), 2909–2915.

Piazza, G., Rahm, A., Krutzsch, M., Sperl, G., Paranka, N., et al. (1995). Antineoplastic drugs Sulindac sulfide and sulfone inhibit cell growth by inducing apoptosis. *Cancer Research*, *55*(14), 3110–3116.

Preuss, E., Hugle, M., Reimann, R., Schlecht, M., & Fulda, S. (2013). Pan-mammalian target of rapamycin (mTOR) inhibitor AZD8055 primes rhabdomyosarcoma cells for ABT-737- induced apoptosis by Down-regulating Mcl-1 protein. *Journal of Biological Chemistry*, *288*(49), 35287–35296.

Prior, I. A., Lewis, P. D., & Mattos, C. (2012). A comprehensive survey of Ras mutations in cancer. *Cancer Research*, *72*(10), 2457–2467.

Pylayeva-Gupta, Y., Grabocka, E., & Bar-Sagi, D. (2011). RAS oncogenes: Weaving a tumorigenic web. *Nature Reviews Cancer*, *11*(11), 761–774. NIH public access.

Qin, Y., He, Y., Zhu, Y., Li, M., Ni, Y., Liu, J., et al. (2019). CID1067700, a late endosome GTPase Rab7 receptor antagonist, attenuates brain atrophy, improves neurologic deficits and inhibits reactive astrogliosis in rat ischemic stroke. *Acta Pharmacologica Sinica*, *40*(6), 724–736.

Quevedo, C. E., Cruz-Migoni, A., Bery, N., Miller, A., Tanaka, T., Petch, D., et al. (2018). Small molecule inhibitors of RAS-effector protein interactions derived using an intracellular antibody fragment. *Nature Communications*, *9*(1), 3169.

Rajalingam, K., Schreck, R., Rapp, U. R., & Albert, Š. (2007). Ras oncogenes and their downstream targets. *Biochimica et Biophysica Acta (BBA) - Molecular Cell Research*, *1773*(8), 1177–1195.

Rak, J., Mitsuhashi, Y., Bayko, L., Filmus, J., Shirasawa, S., Sasazuki, T., et al. (1995). Mutant ras oncogenes upregulate VEGF/VPF expression: Implications for induction and inhibition of tumor angiogenesis. *Cancer Research*, *55*(20), 4575–4580.

Reid, T. S., & Beese, L. S. (2004). Crystal structures of the anticancer clinical candidates R115777 (Tipifarnib) and BMS-214662 complexed with protein farnesyltransferase suggest a mechanism of FTI selectivity. *Biochemistry*, *43*(22), 6877–6884.

Ritt, D., Abreu-Blanco, M., Bindu, L., Durrant, D., Zhou, M., et al. (2016). Inhibition of Ras/Raf/MEK/ERK pathway signaling by a stress-induced Phospho-regulatory circuit. *Molecular Cell, 64*(5), 875–887.

Rowell, C. A., Kowalczyk, J. J., Lewis, M. D., & Garcia, A. M. (1997). Direct demonstration of Geranylgeranylation and Farnesylation of Ki-Ras in vivo. *Journal of Biological Chemistry, 272*(22), 14093–14097.

Ruess, D. A., Heynen, G. J., Ciecielski, K. J., Ai, J., Berninger, A., Kabacaoglu, D., et al. (2018). Mutant KRAS-driven cancers depend on PTPN11/SHP2 phosphatase. *Nature Medicine, 24*(7), 954–960.

Ryan, M. B., de la Cruz, F., Phat, S., Myers, D. T., Wong, E., Shahzade, H. A., et al. (2020). Vertical pathway inhibition overcomes adaptive feedback resistance to KRASG12C inhibition. *Clinical Cancer Research: An Official Journal of the American Association for Cancer Research, 26*(7), 1633–1643.

Santos, E., Tronick, S. R., Aaronson, S. A., Pulciani, S., & Barbacid, M. (1982). T24 human bladder carcinoma oncogene is an activated form of the normal human homologue of BALB- and Harvey-MSV transforming genes. *Nature, 298*(5872), 343–347.

Schrödinger, L., & DeLano, W. (2020). *The PyMOL Molecular Graphics System, Version 2.0*. Retrieved from http://www.pymol.org/pymol.

Shiff, S. J., Qiao, L., Tsai, L. L., & Rigas, B. (1995). Sulindac sulfide, an aspirin-like compound, inhibits proliferation, causes cell cycle quiescence, and induces apoptosis in HT-29 colon adenocarcinoma cells. *Journal of Clinical Investigation, 96*(1), 491–503.

Shima, F., Yoshikawa, Y., Ye, M., Araki, M., Matsumoto, S., Liao, J., et al. (2013). In silico discovery of small-molecule Ras inhibitors that display antitumor activity by blocking the Ras-effector interaction. *Proceedings of the National Academy of Sciences of the United States of America, 110*(20), 8182–8187.

Shin, S.-M., Choi, D.-K., Jung, K., Bae, J., Kim, J., Park, S., et al. (2017). Antibody targeting intracellular oncogenic Ras mutants exerts anti-tumour effects after systemic administration. *Nature Communications, 8*(1), 15090.

Shin, S.-M., Kim, J.-S., Park, S.-W., Jun, S.-Y., Kweon, H.-J., Choi, D.-K., et al. (2020). Direct targeting of oncogenic RAS mutants with a tumor-specific cytosol-penetrating antibody inhibits RAS mutant–driven tumor growth. *Science Advances, 6*(3), eaay2174.

Simanshu, D. K., Nissley, D. V., & McCormick, F. (2017). RAS proteins and their regulators in human disease. *Cell, 170*(1), 17–33.

Singh, A., Greninger, P., Rhodes, D., Koopman, L., Violette, S., Bardeesy, N., et al. (2009). A gene expression signature associated with "K-Ras addiction" reveals regulators of EMT and tumor cell survival. *Cancer Cell, 15*(6), 489–500.

Skobeleva, N., Menon, S., Weber, L., Golemis, E. A., & Khazak, V. (2007). In vitro and in vivo synergy of MCP compounds with MAPK pathway- and microtubule-targeting inhibitors. *Molecular Cancer Therapeutics, 6*(3), 898–906.

Skoulidis, F., Li, B., Dy, G., Price, T., Falchook, G., et al. (2021). Sotorasib for lung cancers with *KRAS* p.G12C mutation. *The New England Journal of Medicine, 384*(25), 2371–2381.

Smakman, N., Borel Rinkes, I. H. M., Voest, E. E., & Kranenburg, O. (2005). Control of colorectal metastasis formation by K-Ras. *Biochimica et Biophysica Acta, Reviews on Cancer, 1756*(2), 103–114. Elsevier.

Soria, J. C., Fülöp, A., Maciel, C., Fischer, J. R., Girotto, G., Lago, S., et al. (2017). SELECT-2: A phase II, double-blind, randomized, placebo-controlled study to assess the efficacy of selumetinib plus docetaxel as a second-line treatment of patients with advanced or metastatic non-small-cell lung cancer. *Annals of Oncology, 28*(12), 3028–3036.

Stratton, M. R., Campbell, P. J., & Futreal, P. A. (2009). The cancer genome. *Nature, 458*(7239), 719–724.

Strickland, C. L., Weber, P. C., Windsor, W. T., Wu, Z., Le, H. V., Albanese, M. M., et al. (1999). Tricyclic farnesyl protein transferase inhibitors: Crystallographic and calorimetric studies of structure-activity relationships. *Journal of Medicinal Chemistry, 42*(12), 2125–2135.

Sullivan, R. J., Infante, J. R., Janku, F., Lee Wong, D. J., Sosman, J. A., Keedy, V., et al. (2018). First-in-class ERK1/2 inhibitor ulixertinib (BVD-523) in patients with MAPK mutant advanced solid tumors: Results of a phase I dose-escalation and expansion study. *Cancer Discovery, 8*(2), 184–195.

Sun, S. Y., Zhou, Z., Wang, R., Fu, H., & Khuri, F. R. (2004). The farnesyltransferase inhibitor Lonafarnib induces growth arrest or apoptosis of human lung cancer cells without downregulation of Akt. *Cancer Biology and Therapy, 3*(11), 1092–1098.

Surviladze, Z., Young, S. M., & Sklar, L. A. (2012). High throughput flow cytometry bead-based multiplex assay for identification of rho GTPase inhibitors. *Methods in Molecular Biology (Clifton, N.J.), 827*, 253–270.

Taylor, M. T., Lawson, K. R., Ignatenko, N. A., Marek, S. E., Stringer, D. E., Skovan, B. A., et al. (2000). Sulindac sulfone inhibits K-ras-dependent Cyclooxygenase-2 expression in human Colon Cancer cells. *Cancer Research, 60*(23), 6607–6610.

Testa, J. E., Medcalf, R. L., Cajot, J.-F., Schleuning, W.-D., & Sordat, B. (1989). Urokinase-type plasminogen activator biosynthesis is induced by the EJ-ha-ras oncogene in CL26 mouse colon carcinoma cells. *International Journal of Cancer, 43*(5), 816–822.

Thompson, H. J., Jiang, C., Lu, J., Mehta, R. G., Piazza, G. A., Paranka, N. S., et al. (1997). Sulfone metabolite of Sulindac inhibits mammary carcinogenesis. *Cancer Research, 57*(2), 267–271.

Thompson, W., Piazza, G., Li, H., Liu, L., Fetter, J., et al. (2000). Exisulind induction of apoptosis involves guanosine 3′,5′-cyclic monophosphate phosphodiesterase inhibition, protein kinase G activation, and attenuated beta-catenin. *Cancer Research, 60*(13), 3338–3342.

Verdera, H., Gitz-Francois, J., Schiffelers, R., & Vader, P. (2017). Cellular uptake of extracellular vesicles is mediated by clathrin-independent endocytosis and micropinocytosis. *Journal of Controlled Release, 266*, 100–108.

Waldmann, H., Karaguni, I.-M., Carpintero, M., Gourzoulidou, E., Herrmann, C., Brockmann, C., et al. (2004). Sulindac-derived Ras pathway inhibitors target the Ras-Raf interaction and downstream effectors in the Ras pathway. *Angewandte Chemie (International Ed. in English), 43*(4), 454–458.

Wan, P. T. C., Garnett, M. J., Roe, S. M., Lee, S., Niculescu-Duvaz, D., Good, V. M., et al. (2004). Mechanism of activation of the RAF-ERK signaling pathway by oncogenic mutations of B-RAF. *Cell, 116*(6), 855–867.

Wang, X., Sheng, W., Wang, Y., Li, L., Li, Y., et al. (2018). A macropinocytosis-intensifying albumin domain-based scFv antibody and its conjugate directed against K-Ras mutant pancreatic Cancer. *Molecular Pharmaceutics, 15*(6), 2403–2412.

Ward, A. B., Keeton, A. B., Chen, X., Mattox, T. E., Coley, A. B., Maxuitenko, Y. Y., et al. (2020). Enhancing anticancer activity of checkpoint immunotherapy by targeting RAS. *MedComm, 1*(2), 121–128.

Watanabe, M., Fiji, H. D. G., Guo, L., Chan, L., Kinderman, S. S., Slamon, D. J., et al. (2008). Inhibitors of protein geranylgeranyltransferase I and Rab geranylgeranyltransferase identified from a library of allenoate-derived compounds. *Journal of Biological Chemistry, 283*(15), 9571–9579.

Welsch, M. E., Kaplan, A., Chambers, J. M., Stokes, M. E., Bos, P. H., Zask, A., et al. (2017). Multivalent small molecule pan-RAS inhibitors. *Cell, 168*(5), 878–889.e29.

Wolin, R., Wang, D., Kelly, J., Afonso, A., James, L., Kirschmeier, P., et al. (1996). Synthesis and evaluation of pyrazolo[3,4-b]quinoline ribofuranosides and their derivatives as inhibitors of oncogenic Ras. *Bioorganic & Medicinal Chemistry Letters, 6*(2), 195–200.

Xue, J. Y., Zhao, Y., Aronowitz, J., Mai, T. T., Vides, A., Qeriqi, B., et al. (2020). Rapid non-uniform adaptation to conformation-specific KRAS(G12C) inhibition. *Nature, 577*(7790), 421–425.

Yang, G., Mercado-Uribe, I., Multani, A. S., Sen, S., Shih, I. M., Wong, K. K., et al. (2013). RAS promotes tumorigenesis through genomic instability induced by imbalanced expression of Aurora-A and BRCA2 in midbody during cytokinesis. *International Journal of Cancer, 133*(2), 275–285.

Yuan, X., Bu, H., Zhou, J., Yang, C. Y., & Zhang, H. (2020). Recent advances of SHP2 inhibitors in Cancer therapy: Current development and clinical application. *Journal of Medicinal Chemistry, 63*(20), 11368–11396. American Chemical Society.

Zhang, X., He, Y., Liu, S., Yu, Z., Jiang, Z. X., Yang, Z., et al. (2010). Salicylic acid based small molecule inhibitor for the oncogenic src homology-2 domain containing protein tyrosine phosphatase-2 (SHP2). *Journal of Medicinal Chemistry, 53*(6), 2482–2493.

Zhang, M., Jang, H., & Nussinov, R. (2020). PI3K inhibitors: Review and new strategies. *Chemical Science, 11*(23), 5855–5865.

Zhang, W., & Liu, H. T. (2002). MAPK signal pathways in the regulation of cell proliferation in mammalian cells. *Cell Research, 12*(1), 9–18.

Zimmermann, G., Papke, B., Ismail, S., Vartak, N., Chandra, A., Hoffmann, M., et al. (2013). Small molecule inhibition of the KRAS-PDEδ interaction impairs oncogenic KRAS signalling. *Nature, 497*(7451), 638–642.

CHAPTER SIX

Targeting RAS oncogenesis with SOS1 inhibitors

Roman Christian Hillig[a,]* and Benjamin Bader[b]
[a]Structural Biology, Lead Discovery, Nuvisan ICB GmbH, Berlin, Germany
[b]Screening, Lead Discovery, Nuvisan ICB GmbH, Berlin, Germany
*Corresponding author: e-mail address: roman.hillig@nuvisan.com

Contents

1. Introduction: RAS proteins and their Guanine Nucleotide Exchange Factors (GEFs) 170
2. SOS1 as a molecular target for cancer therapy 174
3. Discovery of the hydrophobic pocket on SOS1 175
4. More binders for the hydrophobic pocket 183
5. The long way from activators to inhibitors 185
6. Molecular basis for SOS1 activation versus inhibition 189
7. Selectivity against SOS2 and other RasGEFs 190
8. Activity of SOS1 inhibitors in cellular assays 191
9. SOS1 inhibition affects RAS mutations with residual intrinsic GTPase activity 193
10. SOS1 inhibitors in RASopathies 193
11. SOS1 inhibitors in combination therapy 194
12. Clinical studies 196
13. Conclusions 196
References 197

Abstract

RAS proteins play major roles in many human cancers, but programs to develop direct RAS inhibitors so far have only been successful for the oncogenic KRAS mutant G12C. As an alternative approach, inhibitors for the RAS guanine nucleotide exchange factor SOS1 have been investigated by several academic groups and companies, and major progress has been achieved in recent years in the optimization of small molecule activators and inhibitors of SOS1. Here, we review the discovery and development of small molecule modulators of SOS1 and their molecular binding modes and modes of action. As targeting the RAS pathway is expected to result in the development of resistance mechanisms, SOS1 inhibitors will most likely be best applied in vertical combination approaches where two nodes of the RAS signaling pathway are hit simultaneously. We summarize the current understanding of which combination partners may be most beneficial for patients with RAS driven tumors.

Advances in Cancer Research, Volume 153
ISSN 0065-230X
https://doi.org/10.1016/bs.acr.2021.07.001

1. Introduction: RAS proteins and their Guanine Nucleotide Exchange Factors (GEFs)

The *RAS* genes (*KRAS*, *NRAS* and *HRAS*) form the most frequently mutated gene family in cancers and mutations in *KRAS* are known drivers of three of the most lethal cancers, lung cancer, colorectal cancer and pancreatic cancer (Moore, Rosenberg, McCormick, & Malek, 2020; Stalnecker & Der, 2020). The RAS proteins act as molecular switches that cycle between the GTP-bound On-state and the GDP-bound Off-state (Simanshu, Nissley, & McCormick, 2017; Wittinghofer & Vetter, 2011). Two classes of regulatory proteins help RAS to cycle between its two binding states. The GTPase-activating proteins (GAP) complement the catalytic site of RAS by inserting an arginine residue and thereby accelerate the intrinsically very weak GTP hydrolysis reaction and help RAS to return to its Off-state (Scheffzek et al., 1997). The guanine nucleotide exchange factors (GEF) help RAS to release the otherwise tightly bound GDP and thereby return RAS to its activated GTP bound On-state (Bos, Rehmann, & Wittinghofer, 2007). The RAS switch can be aberrantly activated by oncogenic mutations of RAS itself, but also by deregulated upstream signaling of growth factor receptors or the protein tyrosine phosphatase SHP2 (Nichols et al., 2018), or by loss of GAP activity or gain of function mutations in a GEF (Gray, von Delft, & Brennan, 2020). All these events result in increased levels of GTP-bound RAS and constitutive downstream signaling (Cox, Fesik, Kimmelman, Luo, & Der, 2014). Due to the relatively small size of the RAS proteins, the lack of any deeper binding pocket (except for the nucleotide binding site) and the picomolar affinity for both GDP and GTP, combined with micromolar concentrations of GTP in the cell, the RAS proteins were considered "undruggable" for a long period of time. This has changed only recently with the discovery of the switch II pocket in KRAS (Ostrem, Peters, Sos, Wells, & Shokat, 2013) and the optimization of covalent inhibitors for KRAS-G12C which, at least for this specific oncogenic RAS mutant, can bind into this pocket and form a covalent bond to Cys12 (Canon et al., 2019; Hallin et al., 2020; Janes et al., 2018; Kettle et al., 2020; Mortier et al., 2020). The molecules sotorasib/AMG-510 (Canon et al., 2019) and adagrasib/MRTX849 (Hallin et al., 2020) are currently the most advanced molecules and have shown very promising results in clinical trials (Amgen, 2021; Riely et al., 2021).

However, this major progress so far only opens the door to an effective treatment for patients with this particular oncogenic G12C mutation, which is most frequently observed in non-small-cell lung carcinoma (NSCLC) but otherwise occurs relatively rarely (Moore et al., 2020). There is still a massive medical need to inhibit signaling caused by other oncogenic driver mutations in RAS, in particular G12D and G12V (Moore et al., 2020). For this reason, alternative approaches have been investigated which involve targeting other sites on RAS, the membrane association of RAS and targeting RAS using biologicals (reviewed in (Khan, Rhett, & O'Bryan, 2020; Mullard, 2019; Nyiri, Koppany, & Vertessy, 2020)), as well as targeting the GAPs and GEFs (Gray et al., 2020). In this review we will summarize the progress that has been made towards inhibitors of Son of Sevenless 1 (SOS1) as the major GEF for KRAS.

The protein SOS1 (UniProtKB Q07889) was discovered in 1993 to act as a GEF for RAS proteins (Chardin et al., 1993). Mammalian cells contain two related but distinct SOS proteins called SOS1 and SOS2. Both are large proteins of about 1330 residues which share the same multi-domain architecture (Sondermann et al., 2004) (Fig. 1A). The N-terminal Histone-like fold domain (ca. 200 residues) contains two tandem histone folds and is followed by a Dbl homology (DH) domain (ca. 200 residues) and a pleckstrin homology (pH) domain (ca. 150 residues) which together are involved in a tight autoinhibitory regulation of the catalytic activity of SOS1 (Gureasko et al., 2010). Next are two domains, the RAS Exchanger Motif (REM) domain (ca. 200 residues) and the catalytic Cdc25 domain (named after its homolog cdc25, the RAS activator in yeast, ca. 300 residues) which together form the catalytic unit of SOS1. The remaining approx. 250 C-terminal residues form docking sites for adaptor proteins such as Grb2 (Buday & Downward, 1993; Egan et al., 1993; Gale, Kaplan, Lowenstein, Schlessinger, & Bar-Sagi, 1993; Li et al., 1993) and SHP2 (Nichols et al., 2018) and link the localization and activation state of SOS1 to the signaling of growth factor receptors such as the epidermal growth factor receptor (EGFR) (Moore et al., 2020) or the fibroblast growth factor receptor (FGFR) (Ong et al., 2001).

Erroneous activation of RAS via SOS1 would have dramatic consequences for the cell. The activity of the RAS GEFs therefore needs to be tightly regulated and controlled. How this control is executed on the molecular level is now well understood for SOS1, where crystal structures covering a series of truncation variants have helped to identify the contributions of all domains of this multi-domain protein. The molecular basis of the catalytic activity, the nucleotide-exchange of RAS by SOS1, was revealed by the crystal structure of HRAS in complex with the catalytic unit of SOS1 solved by the teams of John Kuriyan and Dafna Bar-Sagi (Boriack-Sjodin, Margarit,

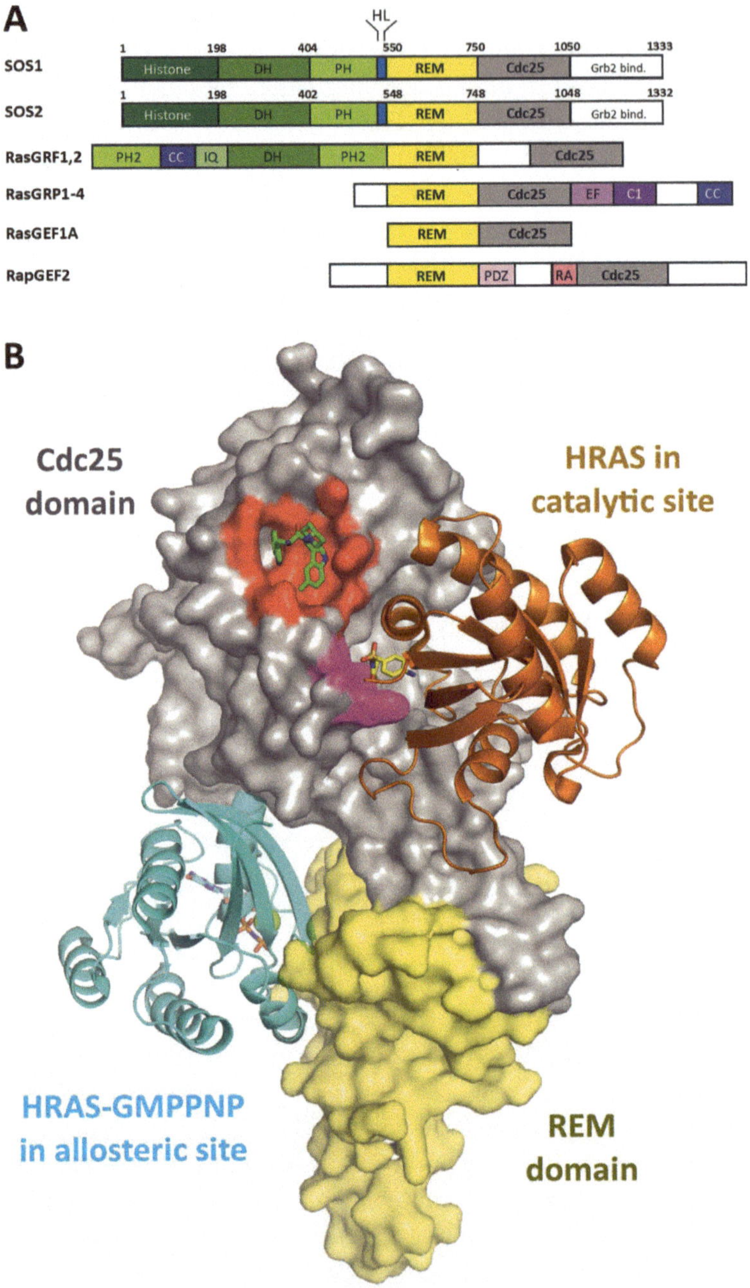

Fig. 1 Domain structures of Ras GEFs and surface pockets in the catalytic unit of SOS1. (A) Domain structure of the different RAS GEF protein families (Histone = Histone-like fold; DH = Dbl homology domain; pH = Pleckstrin homology domain; HL = helical linker;

Bar-Sagi, & Kuriyan, 1998). The structure showed that SOS1 binds RAS via a large and mainly hydrophobic interaction surface which enables helix αH (residues 930–943) in the Cdc25 domain of SOS1 to insert into the RAS nucleotide binding site. In doing so, it preys open the switch I region and displaces the catalytical Mg^{2+} ion of RAS. The resulting massive conformational change of RAS opens up the nucleotide binding pocket and triggers the release of any bound nucleotide. Bar-Sagi and Kuriyan subsequently discovered that the REM domain contains a second allosteric RAS binding site (Margarit et al., 2003) (Fig. 1B). This site preferentially binds GTP-bound RAS which, in turn, stabilizes SOS1 in its active conformation and thereby further accelerates nucleotide exchange of the RAS-GDP molecule bound at the catalytic site. The crystal structure of a SOS1 construct spanning both the DH-PH tandem domain and the catalytic unit revealed that the DH-PH domain blocks the allosteric RAS binding site resulting in an auto-inhibited ground state of SOS1 (Sondermann et al., 2004). Full activation of SOS1 involves the release of the DH-PH domain, followed by binding of first RAS-GDP to the catalytic site and then RAS-GTP to the allosteric site. Finally, the crystal structure of a SOS1 construct covering additionally the histone-like fold domain showed that this domain adds another layer of autoinhibition by occluding the allosteric RAS-GTP binding site and stabilizing the DH-PH tandem domain in its inhibitory conformation (Gureasko et al., 2010).

The role of SOS2 is less well understood. We determined the crystal structure of the catalytic unit of SOS2 in its apo form (Hillig et al., 2019) which confirmed the same overall domain architecture as observed before for SOS1. Recent work by Sheffels et al. suggests that the different RAS isoforms have a different requirement for the expression and activity of

---

REM = RAS Exchanger Motif domain; Cdc25 = Cdc25 homology domain (catalytic GEF domain); CC = coiled-coil domain; IQ, Calmodulin binding domain; EF = Ca^{2+}-binding EF hand, C1 = diacylglycerol-binding C1 domain; RA = Ras associating domain). (B) Crystal structure of the catalytic unit of SOS1 (PDB accession code 4NYI, Cdc25 domain in gray and REM domain in yellow surface representation) in complex with nucleotide-free HRAS (orange) bound to the catalytic RAS binding site and GMPPNP-bound $HRAS^{Y64A}$ (cyan) bound to the allosteric RAS binding site of SOS1. The hydrophobic pocket (surface shown in red) in the Cdc25 domain of SOS1, immediately adjacent to RAS bound to the catalytic site is illustrated by the co-crystallized small molecule (compound 3 (Burns et al., 2014), carbon atoms in green). The interface binding site between SOS1 and HRAS is shown in magenta with a small molecule binder (compound 6, from superimposition with 4URZ (Winter et al., 2015) depicted as stick model with yellow carbon atoms).

SOS2 and that for KRAS, activation by SOS2 may be more important for signaling via the Phosphoinositide 3-kinase (PI3K)-AKT pathway than the RAS/mitogen-activated protein kinase (MAPK) pathway (Sheffels et al., 2018; Sheffels, Sealover, Theard, & Kortum, 2021).

In addition to the RasGEFs SOS1 and SOS2, there are two further RAS-GEF families, the Ras Guanyl Releasing Proteins (RasGRP1–4) and the RAS-specific Guanine nucleotide-Releasing Factors RasGRF1 and RasGRF2, and the proteins RasGEF1A and RapGEF2 (Rojas, Oliva, & Santos, 2011; Vigil, Cherfils, Rossman, & Der, 2010) (Fig. 1A). They all share the catalytic unit formed by the REM and Cdc25 domains, but differ in the type and arrangement of the adjacent regulatory domains and thus in the type of signals that release the specific autoinhibitory mechanism and trigger activation of these GEFs (Fig. 1A). The RasGRP proteins play important roles in immune and bleeding disorders and certain cancers (Jun, Rubio, & Roose, 2013; Vercoulen et al., 2017). Crystal structures have been reported for RasGRP1 (4L9M, (Iwig et al., 2013)), RasGRP2 (6AXF, (Vercoulen et al., 2017)) and RasGRP4 (6AXG, (Vercoulen et al., 2017)) which confirmed that the Cdc25 and REM domains of these proteins indeed adopt the same overall fold as observed before for SOS1 and SOS2. No structures are yet available for the other human RAS GEFs.

2. SOS1 as a molecular target for cancer therapy

The idea of targeting SOS1 as an indirect way to inhibit RAS signaling can be traced back to the discovery of the molecular mode of action of the fungal toxin brefeldin A (BFA). BFA targets a related pair of a GTPase with its GEF. It binds to the complex of the small GTPase ARF1 and its guanine nucleotide exchange factor Sec7 and the crystal structures of the ARF1-Sec7-BFA complex (Mossessova, Corpina, & Goldberg, 2003; Renault, Guibert, & Cherfils, 2003) revealed that BFA was bound within the interface between the GTPase and its GEF and stabilized the complex in a non-productive conformation. BFA thus acted as molecular glue, stabilized ARF in a dead-end complex and thereby blocked the activation of ARF (Vigil et al., 2010).

The RAS-SOS1 interaction and SOS1 as a molecular target moved further into the focus of interest when in 2012 the team of Bar-Sagi could show that the SOS1-mediated cross activation of wild-type RAS by oncogenic mutant RAS was essential for tumorigenesis (Jeng, Taylor, & Bar-Sagi, 2012): The shRNA-mediated silencing of SOS1 reduced tumor growth in cellular and in vivo experiments, suggesting that the cross-activation of WT KRAS (and also HRAS and NRAS) via binding of oncogenic mutant KRAS-GTP to the allosteric site of SOS1 may be a major step in mutant KRAS driven tumorigenesis. These findings indicated inhibition of SOS1 as a potential new targeting strategy for RAS-driven tumors.

3. Discovery of the hydrophobic pocket on SOS1

The first small molecule modulators for SOS1, however, were not identified in a SOS1 directed approach, but by serendipity. The Fesik team noticed during optimization of previously discovered RAS binding fragments (Sun et al., 2012) that the observed structure activity relationship could not be explained by RAS binding anymore (Burns et al., 2014). Co-crystallization with the HRAS-SOS1 complex revealed that these first binders, represented by the aminopiperidine indole compound 4 (API, Table 1) and by the closely related compound 3 (Table 1, Figs. 1B and 2A), did not bind to RAS anymore. Instead, these compounds bind to a surface pocket in the Cdc25 domain of SOS1 which is located immediately adjacent to the catalytic RAS binding site (Fig. 1B). Adopting a horseshoe-like conformation, the compounds inserted their two indole moieties into a mainly hydrophobic pocket lined by the side chains of Phe890, Lys898 and Leu901, and formed one central hydrogen bond between one indole nitrogen and the back bone carbonyl group of Met878 (Fig. 2A). This new binding site on SOS1 was later termed the hydrophobic pocket (highlighted in red in Fig. 1B). The compounds are not in direct contact with any RAS residue as the nearest RAS residue, Arg73^{RAS}, is in about 8 Å distance. However, the guanidinium headgroup of Arg73^{RAS} forms a hydrogen bond to the backbone oxygen of Asn879 and a π-π stacking interaction with the side chain of Tyr884. This tyrosine residue is, in turn, in direct contact with the linker between the two indole moieties of this ligand series (Fig. 2A). Unexpectedly, these first SOS1 binders increased the cellular RAS-GTP levels, so acted biochemically as activators of the SOS1 mediated nucleotide exchange of RAS. In the crystal structures, HRAS can indeed still bind in the presence of these molecules and the observed interaction of Tyr884 with these compounds may be the cause for the increased RAS activation as it may result in a stabilization of the Tyr884-Arg73^{RAS} π-π stacking interaction and thereby of the RAS-SOS1 interaction (Fig. 2A). Remarkably, in the cellular context

Table 1 Selected small molecule modulators of the catalytic domain of SOS1.

Compound	Structure	Mode of action	Binding pocket	$IC_{50}/EC_{50}/K_D$	PDB code	References
Compound 3	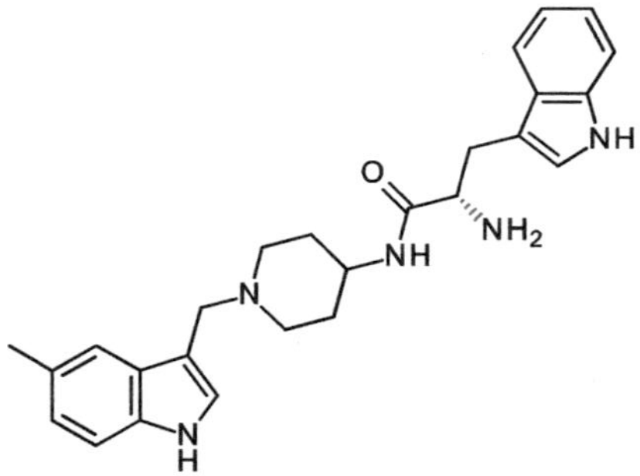	Activator	Hydrophobic pocket (Phe-in)	EC_{50} = 25 μM	4NYI	Burns et al. (2014)
Compound 4, API	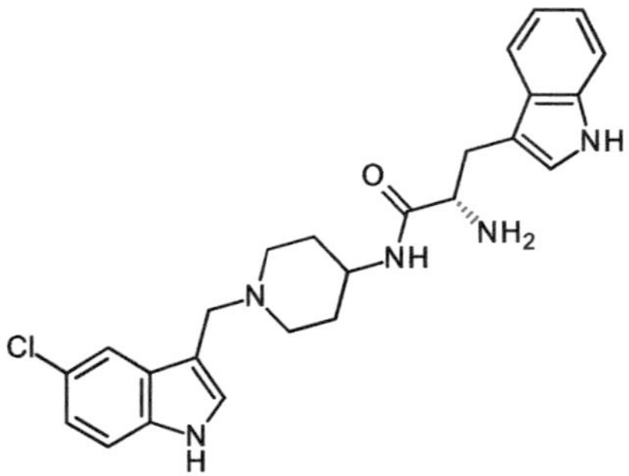	Activator	Hydrophobic (Phe-in)	EC_{50} = 14 μM	–	Burns et al. (2014)

Compound 17	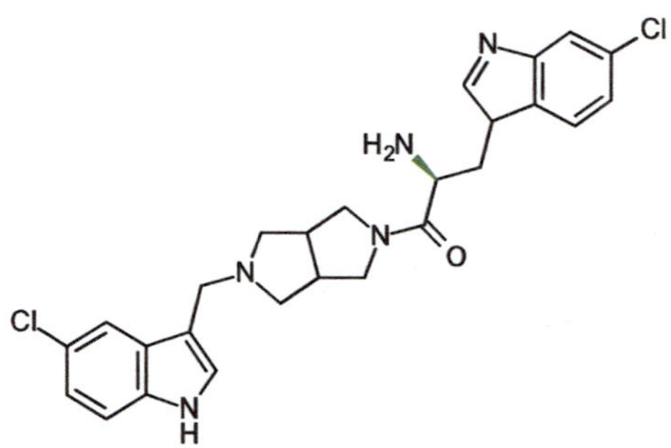	Activator	Hydrophobic pocket(Phe-in)	EC_{50} = 0.8 μM	6BVM (compound 15d, close derivative of 17)	Abbott, Hodges et al. (2018)
Compound 6	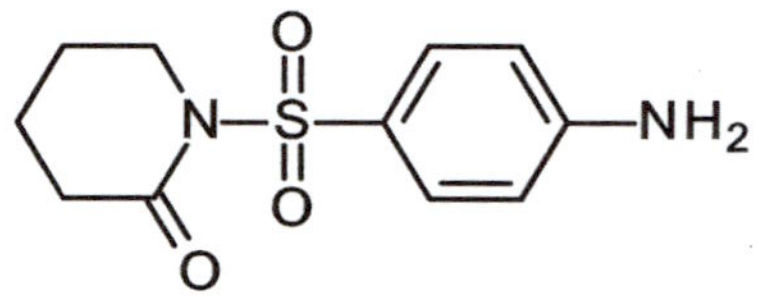	No biol. effect	Interfacial site	K_D = 3 mM	4URZ	Winter et al. (2015)
Compound 4a	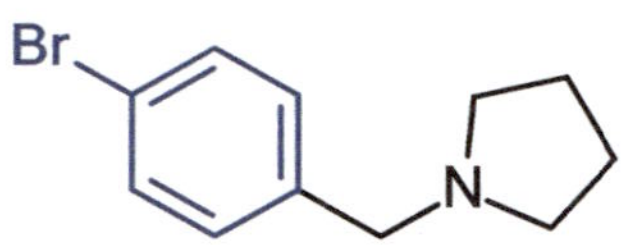	No biol. effect	Hydrophobic pocket (Phe-out)	K_D = 1.3 mM	4URX	Winter et al. (2015)
Compound 4b	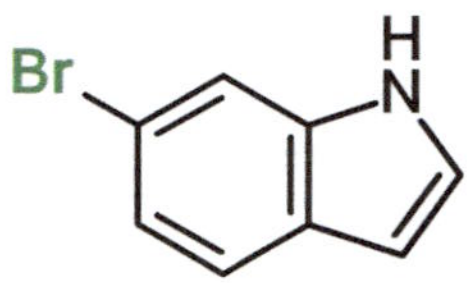	No biol. effect	Hydrophobic pocket (Phe-out)	K_D = 6.0 mM	4URX	Winter et al. (2015)

Continued

Table 1 Selected small molecule modulators of the catalytic domain of SOS1.—cont'd

Compound	Structure	Mode of action	Binding pocket	$IC_{50}/EC_{50}/K_D$	PDB code	References
Fragment F3	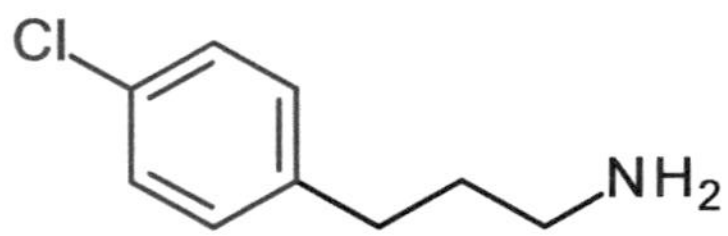	Stabilizer	Hydrophobic pocket (Phe-out)	n.d.	6EPO	Hillig et al. (2019)
Fragment F1	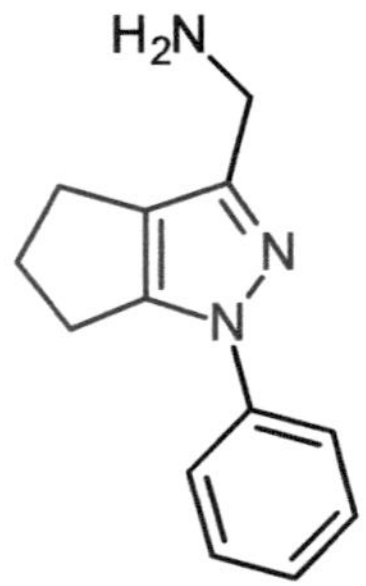	Stabilizer	Hydrophobic pocket (Phe-out)	n.d.	6EPM	Hillig et al. (2019)
Compound 65	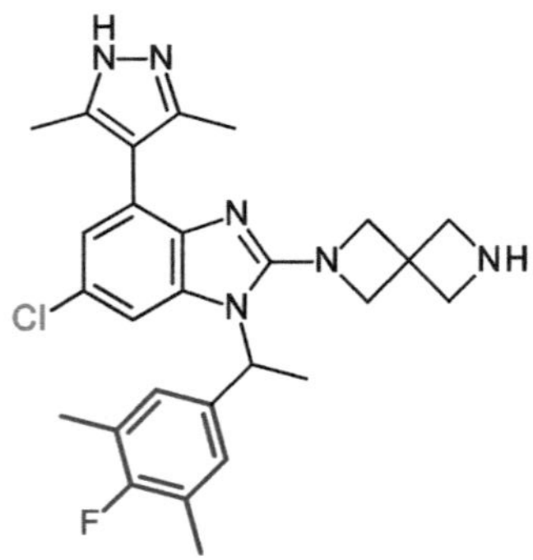	Activator	Hydrophobic pocket (Phe-out)	$K_D =$ 9 nM	6D56	Hodges et al. (2018)

Compound 6b	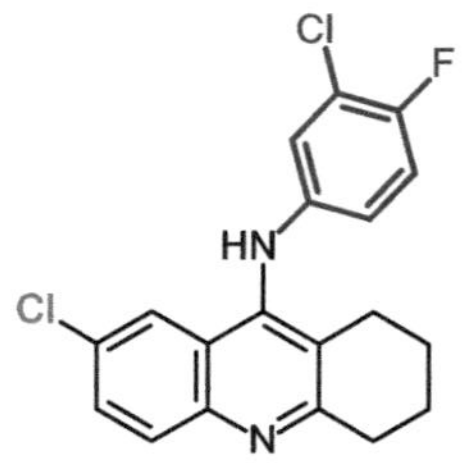	Activator	Hydrophobic pocket (Phe-out)	EC_{50} = 1.5 μM	5WFQ	Burns et al. (2018)
Compound 34	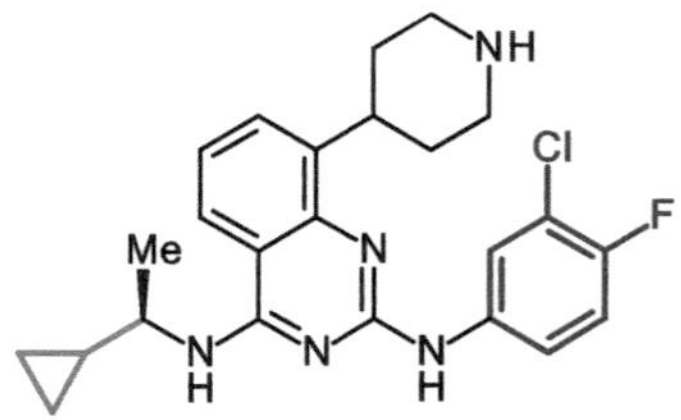	Activator	Hydrophobic pocket (Phe-out)	EC_{50} = 1.8 μM	6CUR	Abbott, Patel et al. (2018)
Compound 2d	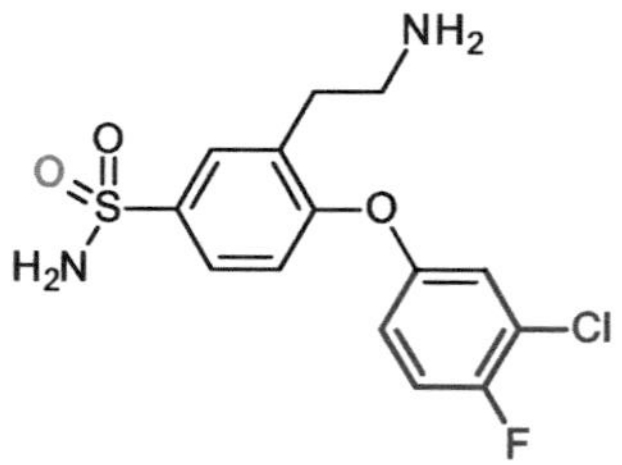	Activator	Hydrophobic pocket (Phe-out)	K_D = 4.5 μM	6V9J	Sarkar et al. (2020)

Continued

Table 1 Selected small molecule modulators of the catalytic domain of SOS1.—cont'd

Compound	Structure	Mode of action	Binding pocket	$IC_{50}/EC_{50}/K_D$	PDB code	References
BAY-293	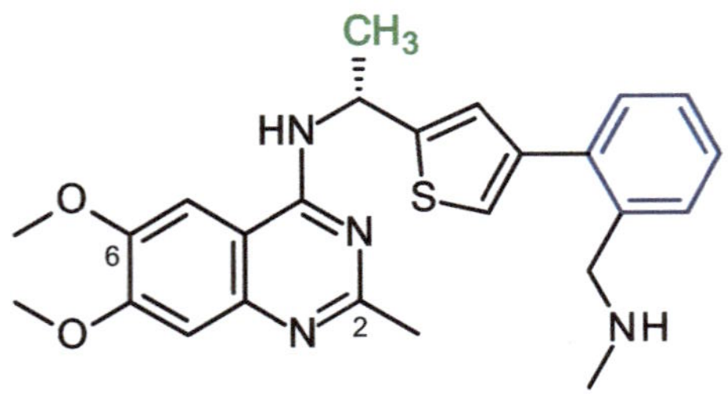	Inhibitor	Hydrophobic pocket (Phe-out) & His-Tyr sandwich	IC_{50} = 21 nM	5OVI	Hillig et al. (2019)
BI-3406	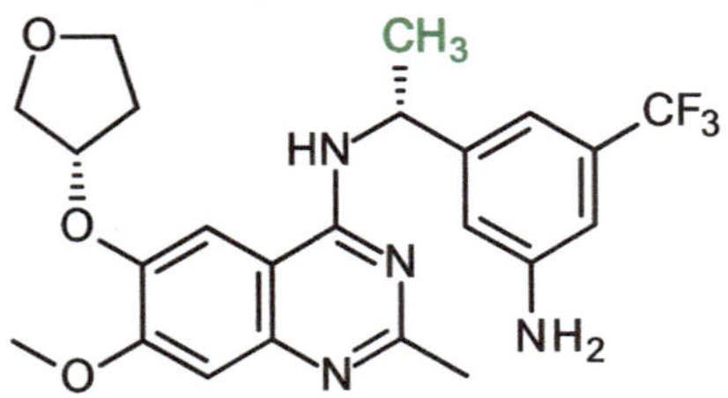	Inhibitor	Hydrophobic pocket (Phe-in) & His-Tyr sandwich	IC_{50} = 5 nM	6SCM	Hofmann et al. (2020)

Compound 2	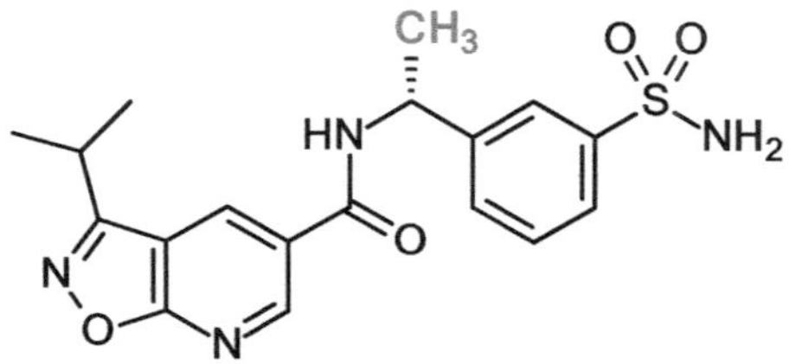	Inhibitor	Hydrophobic pocket (Phe-in) & His-Tyr sandwich	$K_D =$ 937 nM	7AVI	Ramharter et al. (2021)
RMC-0331	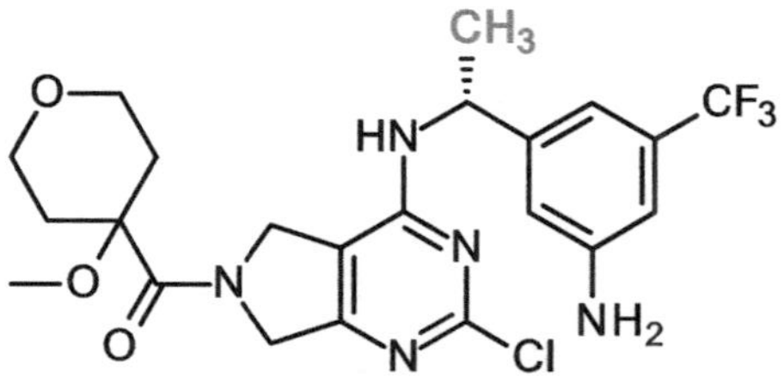	Inhibitor	Hydrophobic pocket (Phe-in) & His-Tyr sandwich	$IC_{50} =$ 71 nM	n.a.	Buckl et al. (2021)

Highlighted in *blue* are the moieties which insert into the Phe-out pocket; highlighted in *green* are the groups which occupy the Asn879 niche.

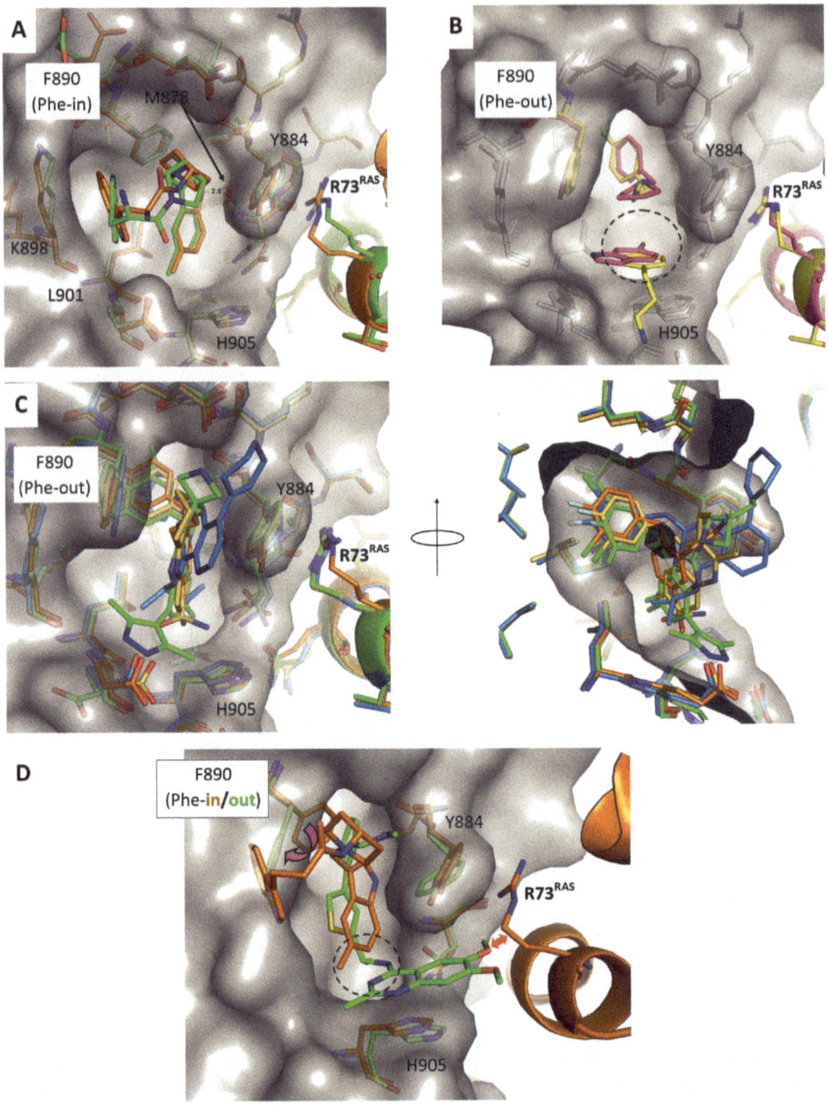

Fig. 2 Different binding modes within the hydrophobic pocket of SOS1. (A) Example of the first series of SOS1 binders (activator compound 3, PDB accession code 4NYI.pdb (Burns et al., 2014), orange carbon atoms) and one of the most advanced variants of this series (compound 15d, 6BVM.pdb (Abbott, Hodges, et al., 2018), green carbon atoms). Both molecules bind to the Phe-in conformation and activate RAS nucleotide exchange. Note the π-π stacking interaction between Tyr884^{SOS1} and Arg73^{RAS} (B) The first binder that opened the deeper pocket of the Phe-out conformation (fragment 4a, co-bound with second fragment 4b, 4URX.pdb (Winter et al., 2015), magenta carbon atoms) and superimposed the closely related fragment F3 (6EPO.pdb (Hillig et al., 2019), yellow carbon atoms, bound twice in binding site) that also opens the

API disrupted MAPK and PI3K signaling (Burns et al., 2014), though with only micromolar potencies. The Fesik team later reported the optimization of representatives of this indole series to compounds with sub-micromolar potencies (compound 17, Table 1) (Abbott, Hodges, et al., 2018) which adopted a very similar binding mode, still acted as a SOS1 activator and also retained the $Tyr884^{SOS1}$-$Arg73^{RAS}$ stacking interaction (see the crystal structure with closely related compound 15d in Fig. 2A).

4. More binders for the hydrophobic pocket

Next, a fragment-based screening approach for binders of the HRAS-SOS1 complex reported by AstraZeneca identified further binders of SOS1 (Winter et al., 2015). Conceptionally, Winter et al. explicitly aimed for stabilizers of the RAS-SOS1 interaction as an approach to prevent RAS activation by SOS1. This line of attack was considered a more attractive approach to interfere with RAS signaling than targeting the relatively featureless KRAS protein directly or to inhibit other protein-protein interactions, such as the KRAS-effector complexes. Using the published HRAS: SOS1 crystallization system (Boriack-Sjodin et al., 1998), Winter et al. carried out a crystallographic fragment screen and indeed identified binders both in the interface between SOS1 and HRAS and in the hydrophobic pocket previously described by the Fesik team (Burns et al., 2014). The conceptionally most exciting pocket, in the interface between SOS1 and HRAS (highlighted as magenta surface in Fig. 1B), was targeted by two related sulfonamide-containing fragment hits (represented by compound 6, Table 1). Typical for initial fragment hits, the top binder of these two compounds featured a K_D value for the interaction with SOS1 of only

deeper Phe-out pocket. (C) The final compounds from different optimization campaigns: SOS1 with compound 65 (6D56.pdb (Hodges et al., 2018), green carbon atoms), superimposed with SOS1 crystal structures with compound 6b (5WFQ.pdb (Burns et al., 2018)), compound 34 (6CUR.pdb (Abbott, Patel, et al., 2018), blue carbon atoms) and compound 2d (6V9J.pdb (Sarkar et al., 2020), orange carbon atoms). All these optimized compounds bind to the Phe-out conformation and act as activators. The right panel provides a better view into the deeper subpocket opened by the Phe-out conformational change. (D) The first SOS1 inhibitor, BAY-293, bound to SOS1 (6OVI.pdb (Hillig et al., 2019), green carbon atoms) superimposed with one of the first activators, compound 3 bound to the SOS1-HRAS complex (4NYI.pdb (Burns et al., 2014), orange carbon atoms, HRAS as orange ribbon). Expected steric clash with $Arg73^{RAS}$ (from 4NYI.pdb) indicated by red arrow. Small indentation next to Asn879 (the Asn niche) highlighted by a black dotted circle in (B) and (D).

3 mM. It was found in the interface between HRAS and SOS1 where it inserted into the "primary site" surface pocket on HRAS. This RAS binding site had first been discovered independently by Genentech (Maurer et al., 2012) and the Fesik team (Sun et al., 2012). On the SOS1 site, the compound contacted a relatively flat surface patch (magenta in Fig. 2B). Optimization of these interfacial binders to higher potency, however, failed. But five further fragment hits turned out to target the hydrophobic binding pocket in the Cdc25 domain. One of them, fragment 4a (Table 1 and Fig. 2B) was also the first SOS1 binder that revealed the possibility of a deeper subpocket to open, triggered by a rotation of the side chain of Phe890 (later termed the Phe-out conformation). Its bromo-phenyl moiety inserts into this mainly hydrophobic, deeper pocket within the hydrophobic pocket. The co-complex structure of fragment 4a with HRAS-SOS1 was determined from a data set collected from a cocktail soak where several fragments were soaked simultaneously into HRAS-SOS1 crystals. The resulting crystal structure indeed showed a second fragment, bromoindole 4b (Table 1) co-bound with fragment 4a (Fig. 2B). Fragment 4b inserts with its bromium atom into a small surface indentation between Asn879, His905 and Leu901 (dotted circle in Fig. 2B and D). As this small pocket next to Asn879 was later found to be an important interaction site for most optimized SOS1 binders (see below), we will term it the Asn niche in the following. Fragment 4b revealed another important interaction point with SOS1 as its indole scaffold is sandwiched between the side chain of His905 and the pyrrolidine ring of fragment 4a. Such a π-π stacking interaction to the side chain of His905 later turned out to be crucial for the development of SOS1 inhibitors (see below). However, despite several rounds of chemistry optimization none of these fragment hits could be further improved, and the work was stopped, in part also because Fesik's activators suggested the hydrophobic pocket may be a target for SOS1 activation rather than for SOS1 inhibition (Winter et al., 2015).

In a second fragment screening approach, our team at Bayer screened a fragment library for stabilizers of the KRAS-G12C:SOS1 interaction (Hillig et al., 2019). All fragment hits identified in this screen turned out to bind into the hydrophobic pocket on SOS1. Most targeted the Phe-in conformation, but a smaller subset also opened the deeper subpocket of the Phe-out conformation. In a series of biophysical and biochemical assays we established that the best two fragment hits, fragment F1 (Table 1) and fragment F3 (Table 1 and Fig. 2B) both were able to stabilize the interaction between SOS1 and KRAS-G12C, in line with the concept of finding dead-end stabilizers of the RAS-SOS interaction. Fragment F3 turned out to be a close variant of the first Phe-out binder fragment 4a from Winter et al., and it inserts with a chlorophenyl moiety into the deeper subpocket in a very similar fashion as was observed previously for the bromo-phenyl moiety of compound 4a (Winter et al., 2015) (Fig. 2B). Fragment F3 binds twice within the hydrophobic pocket and, as observed before with the bromoindole 4b co-bound with compound 4a, the second copy of F3 forms a π-π stacking interaction with His905 and inserts a halogen atom into the Asn niche. However, despite intensive efforts, we were not able to optimize these fragment hits to better than millimolar potencies (Hillig et al., 2019).

The Fesik team kept screening for binders in this hydrophobic pocket which resulted in the discovery and optimization of four further chemical classes, the benzimidazoles represented by compound 65 (Table 1, (Hodges et al., 2018)), a tetrahydroacridine series represented by compound 6b (Table 1, (Burns et al., 2018)), a quinazoline series (compound 34, Table 1 (Abbott, Patel, et al., 2018)) and a sulfonamide-based series (compound 2d, Table 1 (Sarkar et al., 2020)). All these compounds acted as activators of the SOS1-catalyzed nucleotide exchange of KRAS. Remarkably, though representing very different chemical scaffolds, they all share three features in their binding modes: All open the deeper Phe-out pocket and insert a halogenated phenyl ring into it (Fig. 2C, right panel; substituents also highlighted in blue in Table 1); all fill the space between Phe890 and Tyr884 and most likely thereby stabilize the π-π stacking interaction of Arg73^{RAS} with Tyr884; and all place a suitable small moiety into the Asn879 niche (highlighted in green in the compound structures in Table 1).

5. The long way from activators to inhibitors

All SOS1-binding molecules identified in the studies reported above acted as agonists of the SOS1 catalyzed nucleotide exchange and thus resulted in an increased activation of RAS (Abbott, Hodges, et al., 2018; Abbott, Patel, et al., 2018; Burns et al., 2014, 2018; Hodges et al., 2018; Sarkar et al., 2020) or stabilization of the RAS-SOS1 interaction (Hillig et al., 2019). The effect of these activators on the MAPK pathway versus the PI3K/AKT pathway downstream of RAS have been further characterized (Akan et al., 2019). The agonistic effect on the RAS-SOS1 interaction was also confirmed in biochemical KRAS-SOS interaction assays where compound 4 and compound 2, respectively, of Fesik's initial activator series (Burns et al., 2014) acted as stabilizers of the RAS-SOS1 interaction

(Hillig et al., 2019; Hofmann et al., 2020). However, when potent enough to show an effect in cellular assays, these compounds triggered a biphasic response where activated RAS-GTP levels and levels of phosphorylated ERK (pERK) initially went up, but at higher concentrations and later time points pERK levels were found to be downregulated (Burns et al., 2014). Howes et al. explain this initially counter-intuitive finding by a negative feedback loop where SOS1 activation first results in increased RAS-GTP levels and an increase of pERK levels downstream of RAS (Howes et al., 2018). In this model, the initially activated pERK phosphorylates SOS1 at Ser1178 in its C-terminal domain which disrupts the association between SOS1 and GRB2 and thus interferes with the membrane localization of SOS1, resulting in a downregulation of RAS-GTP levels and consequently pERK levels at later stages (Howes et al., 2018).

The story of the hydrophobic pocket on SOS1 took a turn when in 2019 and 2020, our team at Bayer (Hillig et al., 2019) and Hofmann et al. at Boehringer Ingelheim (Hofmann et al., 2020) independently reported aminoquinazolines as bona fide inhibitors and disruptors of the KRAS-SOS1 interaction. Remarkably, these new inhibitors still targeted the same hydrophobic pocket which previously resulted in activation of RAS, but now disrupted the KRAS-SOS1 interaction. Our quinazoline inhibitor series, represented by BAY-293 (Table 1, Figs. 2D and 3A) was identified in a high throughput screen for inhibitors of the SOS1 catalyzed nucleotide exchange reaction on GDP-bound KRAS-G12C. A series of additional biochemical and biophysical assays were applied to distinguish the mode of action of the hits from this screen which, in principle, could bind to either KRAS-G12C, or to SOS1, or even act as a molecular glue to form a dead-end complex in analogy to BFA. Here, thermal shift assays and isothermal titration calorimetry experiments finally identified SOS1 (and not KRAS) as the molecular target, and a biochemical interaction assay as well as surface plasmon resonance experiments revealed that the mode of action was the disruption of the KRAS-SOS1 complex. Subsequently, crystal structures with SOS1 showed that the central quinazoline scaffold inserts between the side chains of His905 and Tyr884 and forms a π-π stacking interaction which we will term the His-Tyr sandwich in the following. The adjacent amino group forms a crucial central hydrogen bond to the side chain oxygen of Asn879 which is locked in the required rotamer conformation by a second hydrogen bond to His905 (Fig. 3A). The (*R*)-configuration at the central stereocenter turned out to be crucial for activity as it allowed the methyl group at this stereo center to insert into the Asn niche. During

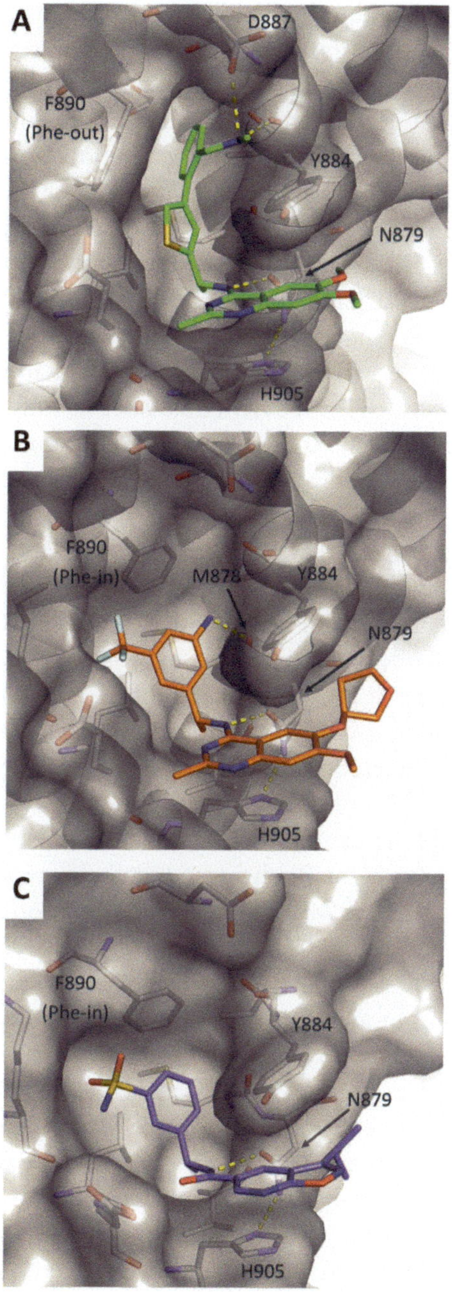

Fig. 3 Binding modes of three SOS1 inhibitors. (A) and (B), the two first SOS1 probe molecules, (A) BAY-293 (PDB accession code 6OVI.pdb) and (B) BI-3406 (6SCM.pdb), in complex with the catalytic unit of SOS1. (C) Compound 2 (7AVI.pdb (Ramharter et al., 2021)) as a representative of a further chemical series of SOS1 inhibitors, adopting a very similar binding mode. Shown is the inhibitor binding site on SOS1 in surface representation with residues lining the binding site depicted in stick representation.

optimization, we picked up the Phe-out concept learnt from the fragment hits from AstraZeneca and our own fragment screen and grew the initial HTS hit series into the deeper pocket opened by the Phe-out conformation. The outer methyl-aminomethyl side chain of the phenyl ring in the deeper pocket allowed to pick up two more hydrogen bonds, to Asp887 and to the backbone carbonyl group of Tyr884.

The aminoquinazoline scaffold is part of clinically approved EGFR inhibitors such as erlotinib (Stamos, Sliwkowski, & Eigenbrot, 2002), gefitinib (Yun et al., 2007) and afatinib (Solca et al., 2012) and forms a crucial hydrogen bond to the hinge region of this protein kinase. Indeed, the initial HTS hit series that led to BAY-293 still showed inhibitory activity on EGFR. This kinase off target activity could later be fully abolished by introduction of the methyl moiety in position 2 of the quinazoline ring system (Table 1) which is tolerated by SOS1 but sterically clashes with the kinase hinge region (Hillig et al., 2019). The resulting optimized compound BAY-293 (Figs. 2D and 3A) was the first selective and potent inhibitor of the SOS1-KRAS interaction.

The superimposition of BAY-293 with the first activator (Fig. 2D) shows that BAY-293 extends away from the hydrophobic pocket and into the RAS binding epitope at the catalytic site of SOS1. The methoxy moiety in position 6 of the quinazoline scaffold (Table 1) would clash with the Tyr884-Arg73^{RAS} interaction and thus drives the inhibitory effect, which was confirmed by loss of inhibitory activity for variants without this methoxy moiety (Hillig et al., 2019).

Later, Hofmann et al. at Boehringer Ingelheim reported the discovery of BI-3406 (Table 1, Fig. 3B) of the same aminoquinazoline class via high-throughput screening using a RAS-G12D SOS1 proximity assay (Hofmann et al., 2020). Fig. 3 shows that BAY-293 and BI-3406 feature very similar binding modes, with the common aminoquinazoline scaffold sandwiched between His905 and Tyr884, while the amine side substituents extend into the hydrophobic pocket, for BI3406 without opening the Phe-out subpocket while for BAY-293 with the opening of this pocket. Both molecules share the central hydrogen bond to Asn879 and insert a methyl group into the Asn niche. By introduction of the amino group in the phenyl side chain of BI-3406 this compound picks up the same hydrogen bond interaction with the backbone carbonyl oxygen of Met878 which had been identified as an interaction point already in the first Fesik activator series (Fig. 2A). The methoxy group of position 6 of BAY-293 (Table 1) which triggers the inhibitory effect via direct clash with RAS is replaced with a tetrahydrofuryl substituent in BI-3406. This substituent additionally improved solubility and metabolic stability and showed an improved interaction with Tyr884 (Hofmann et al., 2020).

Recently, Boehringer Ingelheim reported a second series of SOS1 inhibitors represented by compound 2 from (Ramharter et al., 2021) in Table 1. The co-crystal structure with SOS1 (Fig. 3C) shows that it adopts a binding mode similar to BAY-293 and BI-3406. It also fills the hydrophobic pocket (Phe-In conformation), forms the central hydrogen bond to Asn879, establishes the sandwich interaction between Tyr884 and His905 and extends a suitable substituent (here an isopropyl moiety) into the RAS binding interface to interrupt the interaction of Tyr884 with Arg^{73RAS} (Ramharter et al., 2021). Very recently, Revolution Medicine revealed the structure of their SOS1 inhibitor RMC-0331 (Buckl et al., 2021). The co-complex structure with SOS1 is not yet available from the Protein Data Bank, but a comparison of the chemical structure of RMC-0331 with those of BAY-293 and BI-3406 (Table 1) suggests that this molecule as well follows the design principles derived from the first two probe molecules. It features suitable moieties to fill the hydrophobic pocket and the Asn niche and to donate a hydrogen bond to Asn879, it can establish the His-Tyr sandwich interaction and it possesses a substituent that extends into the RAS binding epitope.

6. Molecular basis for SOS1 activation versus inhibition

The first crystal structures of the quinazoline series together with the associated SAR (Hillig et al., 2019) led us to suggest that the molecular basis for activation versus inhibition of the SOS1-RAS complex formation could be two-fold: Firstly, stabilizers such as API/compound 4 may activate the RAS-nucleotide exchange by sterically stabilizing the side chain of Tyr884 in an orientation that enables its interaction with RAS residue $Arg73^{RAS}$ (Fig. 2D). In contrast, inhibitors of the quinazoline series pull this Tyr884 side chain away from its Ras-interacting conformation and into the His-Tyr sandwich with the quinazoline scaffold (Fig. 2D) (Hillig et al., 2019). The importance of the Tyr884-$Arg73^{RAS}$ interaction for the strength of the overall SOS1-RAS interaction was confirmed by a Tyr884Ala point mutation which showed 30-fold reduced binding affinity to KRAS (Ramharter et al., 2021). Secondly, activators bind in the hydrophobic pocket, relatively far away from the adjacent RAS binding epitope on SOS1 and therefore sterically do not clash with RAS in the catalytic site on SOS1. Inhibitors, in contrast, extend from the hydrophobic pocket via the His-Tyr sandwich into the RAS binding site. As discussed above, one of the two methoxy side chains in BAY-293 reaches into the KRAS binding interface and would sterically clash with KRAS (Fig. 2D). Replacement of the methoxy moiety by hydrogen completely abolished the disrupting effect on the RAS-SOS complex (Hillig et al., 2019). These findings were recently confirmed when the Boehringer Ingelheim team reported that in this very position of the quinazoline scaffold of BI-4306, exchange of a chlorine atom with a smaller fluorine or a hydrogen atom transforms an inhibitor into an activator (Ramharter et al., 2021).

7. Selectivity against SOS2 and other RasGEFs

The knockout of SOS1 in mice is embryonically lethal (Qian et al., 2000; Wang et al., 1997) whereas the conditional knockout is viable (Baltanas et al., 2013) and the knockout of SOS2 results in viable and fertile animals (Esteban et al., 2000). Animals with a SOS2 knockout and an additional conditional SOS1 knockout develop severe phenotypes with body size reduction, leucocyte reduction and liver failure and lethality 15 days post induction (Baltanas et al., 2013). These findings support that panSOS1/2 inhibitors may be toxic, whereas selective inhibition of SOS1 may be a viable therapeutic modality.

Remarkably, despite the high sequence identity of about 80% between SOS1 and SOS2 in the catalytic domain, the aminoquinazoline inhibitors BAY-293 and BI-3406 do not inhibit SOS2 (Hillig et al., 2019; Hofmann et al., 2020). The molecular basis for this selectivity was revealed by the crystal structure of SOS2 (Hillig et al., 2019) which showed that the inhibitor binding pocket on SOS1 is in principle also present in SOS2 and that most residues lining it are indeed conserved, but with one crucial exception: The SOS1 residue His905 which contributes to the His-Tyr sandwich coordination of the quinazoline scaffold of BAY-293 and BI-3406 (Fig. 3) is replaced by a valine residue in SOS2 and can therefore not act as a π-π stacking interaction partner anymore. The relevance of this exchange was confirmed by mutational studies where the H905A mutation strongly reduced the inhibitory effect of a precursor of BAY-293 (Hillig et al., 2019) and the H905V mutation abrogated the ability of BI-3406 to inhibit phosphorylation of ERK and cell proliferation (Hofmann et al., 2020).

The catalytic domain formed by the REM and CDC25 domains is also present in the more distantly related RAS GEFs of the Ras Guanyl Releasing Protein family (RasGRP1–4) and the Ras-specific guanine nucleotide-releasing factors RasGRF1 and RasGRF2. But the catalytic domains of these proteins show a much lower sequence identity of around 50% to SOS1, and inspection of the available crystal structures of RasGRP1 (PDB accession code 4L9M, (Iwig et al., 2013)), RasGRP2 (6AXF, (Vercoulen et al., 2017)) and RasGRP4 (6AXG, (Vercoulen et al., 2017)) revealed that here, not only SOS1 residue His905 but also the other two crucial residues, Tyr884 and Asn879, are not conserved. With both constitutive parts of the His-Tyr sandwich interaction missing, it appears very unlikely that the binding mode observed in SOS1 will be possible in the other RAS GEFs. This further supports that BAY-293 and BI-3406 are indeed highly selective SOS1 inhibitors.

8. Activity of SOS1 inhibitors in cellular assays

The in vitro probe BAY-293 was the first small molecule SOS1 inhibitor showing specific and efficient inhibition of the RAS-pathway in vitro. Amongst the most proximal readouts for SOS1 inhibitors are assays quantifying the concentration of GTP-bound RAS in cells. BAY-293 inhibited active RAS-GTP levels in HeLa (RAS-WT, IC50 = 410 nM) and Calu-1 cells (RAS-G12C, 200 nM). Similarly, BI-3406 was reported to inhibit RAS GTP levels with an IC50 of 83 to 231 nM in NCI-358 cell (RAS-G12C) and in A549 cells (RAS-G12S) (Hofmann et al., 2020). Hofmann et al. tested for the influence of SOS2 (with a NCI-H358 cell line where SOS2 was genetically inactivated) and found that the effect on RAS-GTP levels was significantly more pronounced in NCI-H358 cells harboring the SOS2 knockout compared with the parental cell line.

To investigate the effect on downstream signaling into the MAPK pathway (Fig. 4) the SOS1 inhibitors were tested for the ability to inhibit the phosphorylation of ERK. BAY-293 potently blocked phospho-ERK (pERK) levels in K-562 cells, a tumor cell line for which sensitivity to SOS1 knock-out has been reported (Wang et al., 2015). Interestingly, the effects of BAY-293 on pERK levels in tumor cell lines with RAS mutations were different compared to wildtype RAS cells. BAY-293 showed dose-dependent inhibition in KRAS-G12C mutant Calu-1 cells, albeit with partial efficacy resulting in approximately 50% residual pERK levels after treatment for 24 h. The observation was in line with a stronger antiproliferative potency of BAY-293 when tested in wildtype compared to mutated RAS cells. Similarly, Hofmann et al. reported that BI-3406 could bring down pERK levels with submicromolar IC50 values in cell lines with

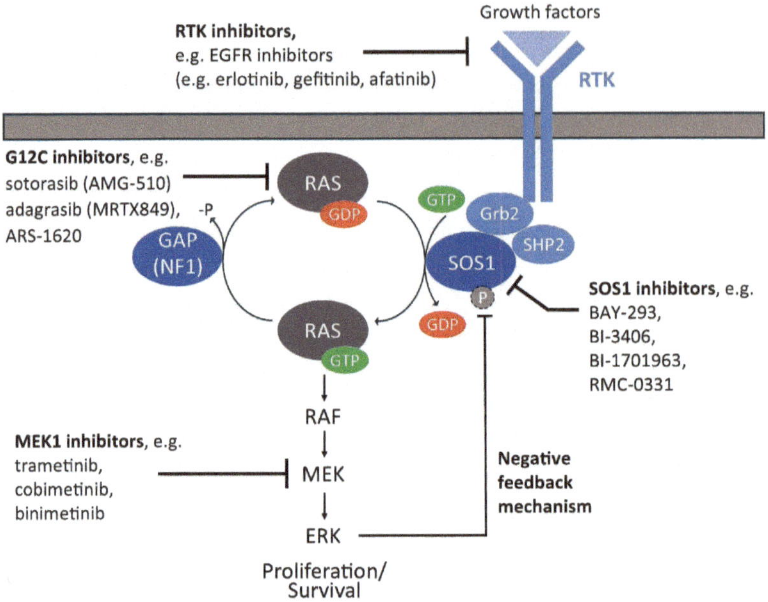

Fig. 4 The KRAS MAPK signaling pathway. Shown are the interference points of SOS1 inhibitors and of potential combination partners for vertical combination therapy approaches with SOS1 inhibitors. RTK = Receptor Tyrosine Kinase, P = phosphate.

WT RAS and in certain KRAS mutant cell lines (G12S, G13D, G12C), but also only to between 40 and 60%; however, this effect was enhanced in SOS2 null-cells (Hofmann et al., 2020). These results suggest that at least in cell lines with RAS mutations, some remaining exchange activity appears to be mediated by SOS2.

Further characterization of the effect of SOS1 inhibition on tumor cell proliferation confirmed sensitivity of hematopoietic tumor cell lines. Tested in a panel of 60 cancer cell lines in two-dimensional (2-D) monolayer proliferation assays, the racemate of BAY-293 (compound 22) displayed a relatively broad inhibition spectrum, with hematopoietic cells (K-562, KG-1, MOLM-13, and THP-1) being most vulnerable to SOS1 inhibition (Hillig et al., 2019). Hofmann et al. switched to three-dimensional (3-D) growth assays and could demonstrate that BI-3406 inhibited the 3-D growth of four KRAS-mutant cancer cell lines while the 3-D growth of two KRAS wild-type cell lines was not appreciably affected (Hofmann et al., 2020). Notably, the inhibition curves in these assays displayed only partial responses of 30–50%, suggesting that even in 3-D assays SOS1 inhibition alone cannot fully suppress RAS signaling. Nevertheless, in a drug sensitivity profiling approach Hofmann et al. could show with such 3-D assays that the antiproliferative potency of SOS1 inhibitor correlated with KRAS mutational status in a panel of 40 cell lines (Hofmann et al., 2020). Sensitive cell lines harbored a broad range of KRAS-mutant alleles including KRAS-G12C, G12V, G12S, G12A, and G13D mutations.

9. SOS1 inhibition affects RAS mutations with residual intrinsic GTPase activity

Mutations in the codons 12, 13 and 61 of oncogenic RAS mutants disrupt the GTP hydrolysis of RAS proteins. The GTPase activity of these mutated proteins cannot be activated by GAPs anymore, rendering them in a constitutively active On-state. However, recent research has revealed that not all RAS mutants are equal. In particular, certain mutants keep some remaining intrinsic hydrolysis capability so that in these, RAS occasionally returns to the GDP-bound Off-state and oncogenic growth then must depend on SOS1 to bring RAS back to its On-state (reviewed in (Moore et al., 2020)). This is particularly true for G12C and G12D, which therefore, in principle, should be more sensitive to inhibition of SOS1. At the other end of the scale are the mutations in codon 61 with very low intrinsic GTPase activity, and increased levels of SOS-independent nucleotide exchange. In line with these considerations, Hofmann et al. explored the effect of BI-3406 on different KRAS mutations in a panel of isogenic cell lines and discovered that pERK levels were reduced by treatment with BI-3406 for wild-type RAS and for the mutations G12C, G12V, G12D, but only weak pERK inhibition was observed for Q61H and no modulation of pERK levels was observed for G12R (Hofmann et al., 2020).

10. SOS1 inhibitors in RASopathies

RASopathies are a group of human diseases caused by germline mutations in genes that code for components or regulators of the RAS-MAPK pathway (Rauen, 2013). These disorders include, amongst others, neurofibromatosis type 1 (NF1) and the Noonan syndrome (NS). NF1 is caused by mutations in the gene coding for the protein Neurofibromin (NF1) which acts as a tumor suppressor and is a GAP for RAS. The mutations result in a loss of function of NF1 and thus in an overall increase of GTP-bound RAS. The Noonan syndrome can be caused by a gain of function mutation in SOS1 located in the helical linker between the PH and the REM domain

(Fig. 1A). The mutation (R552G) disrupts the internal docking of the histone fold-like domain onto the rest of SOS1 and thereby destabilizes the autoinhibitory effect of the N-terminal regulatory domains on the catalytic unit of SOS1 (Gureasko et al., 2008; Roberts et al., 2007; Tartaglia et al., 2007). Conceptionally, such patients with reduced deactivation of the RAS On-state (NF1) or with reduced autoinhibition of SOS1 and therefor increased exchange activity of SOS1 (NS) could benefit from a SOS1 inhibitor blocking the reactivation of RAS-GDP to RAS-GTP. Indeed, Hofmann et al. observed that seven of 14 cell lines with NF1 aberrations were sensitive to BI-3406 treatment (Hofmann et al., 2020).

Activating mutations in the DH domain of SOS2 have also been reported to be associated with certain variants of the Noonan syndrome (Cordeddu et al., 2015; Yamamoto et al., 2015), suggesting that SOS2 inhibitors as well could play a valuable role in understanding RASopathies and ultimately in helping patients. However, both available SOS1 probe molecules, BAY-293 and BI-4306, are selective for SOS1 and do not inhibit SOS2.

11. SOS1 inhibitors in combination therapy

While the initial in vitro findings in cell lines with NF1 aberrations look promising for the application of SOS1 inhibitors in monotherapy, results reported for the treatment of direct RAS-G12C inhibitors strongly suggest that inhibitors targeting the RAS pathway inevitably will lead to resistance (Dunnett-Kane, Nicola, Blackhall, & Lindsay, 2021; Molina-Arcas et al., 2019; van Maldegem & Downward, 2020). Consequently, for the direct KRAS-G12C inhibitors sotorasib/AMG-510 and adagrasib/MRTX849 various combination strategies have been investigated including co-inhibition of EGFR, PI3K, SHP2, AKT, mTOR, or MEK which improved responses in lung and colon tumor xenograft models (Canon et al., 2019; Hallin et al., 2020; van Maldegem & Downward, 2020). Results reported by Ryan et al. highlight the potential importance of vertical inhibition strategies, i.e., simultaneously hitting two nodes in the KRAS signaling pathway (Ryan et al., 2020). Similarly, combination approaches for SOS1 inhibitors have been investigated (Fig. 4) which will be discussed in the following.

Covalent RAS-G12C inhibitors such as ARS-853 (Lito, Solomon, Li, Hansen, & Rosen, 2016; Patricelli et al., 2016), ARS-1620 (Janes et al., 2018), sotorasib/AMG-510 (Canon et al., 2019) and adagrasib/MRTX849 (Hallin et al., 2020) all target the same switch II pocket, first identified by Kevan Shokat's team (Ostrem et al., 2013). However, while the G12C mutation was generally assumed to keep KRAS constitutively active in the GTP-bound On-state, these inhibitors sterically can only bind to the GDP-bound Off-state (Fig. 4). This seemingly paradoxical situation was resolved when Patricelli et al. could show that KRAS-G12C has some remaining GAP-independent intrinsic GTP hydrolysis capacity and occasionally returns to the Off-state (Patricelli et al., 2016). They hypothesized that a SOS1 inhibitor should act synergistically with covalent KRAS-G12C inhibitors as it would slow down the re-activation of RAS-GDP by SOS1 and thereby should widen the therapeutic window for G12C inhibitors. We could later confirm this when the discovery of BAY-293 provided the first SOS1 inhibitor and in vitro combination experiments with ARS-853 indeed showed strong synergism for the combination treatment with BAY-293 in NCI-H358 cells (Hillig et al., 2019). This concept was further confirmed by in vitro combination studies with BI-3406 and AMG-510 in NCI-H358 cells where the combination resulted in in a stronger and more prolonged suppression of pERK than observed in the monotherapies (Hofmann et al., 2020).

It is noteworthy that SOS1 inhibitors can mechanistically be considered as very similar to SHP2 inhibitors. The protein tyrosine phosphatase SHP2 plays a role in recruiting SOS1 to the membrane and is therefore thought to act via activation of SOS1-mediated RAS-GTP loading (Fig. 4). In two recent studies Fedele et al. discovered that SHP2 inhibitors increased the life time of the GDP bound off-state, thereby increasing the effect of RAS-G12C inhibitors (Fedele et al., 2018, 2021). SHP2 and RAS-G12C inhibitors alone had minimal effects on murine pancreatic ductal adenocarcinoma (PDAC) and NSCLC models whereas the combination resulted in extended survival with no evidence of toxicity. SHP2 inhibitors are being developed by several companies (Chen et al., 2016; Nichols et al., 2018; Ou et al., 2020) and first published data from a phase I clinical study with RMC-4630 show promising results for patients with a RAS-G12C mutation (Ou et al., 2020).

A second potential combination approach for SOS1 inhibitors are ***MEK1 inhibitors***. MEK1 inhibitors turned out to unexpectedly trigger increased RAS activation, which was later traced back to a negative feedback loop (Fig. 4) where activation of ERK by MEK results in an increased phosphorylation of SOS1 which downregulates its nucleotide exchange activity (Lake, Correa, & Muller, 2016). Inhibition of MEK reliefs this negative feedback loop, resulting in an increased activation of RAS by SOS1 and increased MEK1 phosphorylation. Based on this mechanistic hypothesis, Hoffmann et al. tested the combination of BI-3406 with the MEK inhibitor trametinib, which resulted in an improved reduction of pERK compared to the two monotherapies in MIA PaCa-2 cells and in a near-complete reduction of pERK phosphorylation compared to the partial effects observed in monotherapy (Hofmann et al., 2020).

Testing of BI 3406 in a panel of 40 cancer cell lines revealed that a fraction of NSCLC cell lines driven by EGFR mutations responded to BI-3406 treatment, suggesting that ***inhibitors of oncogenic receptor tyrosine kinases (RTK)*** upstream of RAS (Fig. 4) can confer sensitivity to SOS1 inhibition (Hofmann et al., 2020). This concept was further supported by an in vitro combination study by Theard et al. where combination of EGFR inhibitors and the SOS1 inhibitor BAY-293 showed strong synergistic effects (Theard et al., 2020), and was recently confirmed in a xenograft in vivo experiment where the combination of the SOS1 inhibitor BI-3406 with the EGFR inhibitor afatinib resulted in stronger antitumor efficacy in KRAS-mutant tumors than treatment with the two inhibitors alone (Ramharter et al., 2021).

12. Clinical studies

The first SOS1 inhibitor reported to have entered clinical trials is Boehringer Ingelheim's BI-1701963 (Gort et al., 2020). To our knowledge, the structure of BI-1701963 has not yet been disclosed but it is most likely chemically related to the probe compound BI-3406. It is currently being tested in two clinical studies: The first trial tests BI-1701963 in monotherapy and combined with the MEK inhibitor trametinib in patients with solid tumors with KRAS mutations (NCT04111458). In the second trial, it is given in combination with the topoisomerase 1 inhibitor irinotecan to patients with metastatic colorectal cancer with a KRAS mutation (NCT04627142). Based on the synergistic effects observed preclinically between SOS1 inhibitors and KRAS-G12C inhibitors Mirati announced a clinical collaboration with Boehringer Ingelheim to evaluate their KRAS-G12C inhibitor adagrasib/MRTX849 in combination with BI-1701963 (Mirati, 2020).

13. Conclusions

In the recent years large progress has been made towards the discovery and optimization of small molecules that can bind to and modulate the nucleotide exchange activity of the RAS GEF SOS1. Driven by screening efforts and structural biology insights, a hydrophobic pocket adjacent to the RAS-binding epitope on SOS1 was discovered and binders to this pocket were identified and optimized, giving rise to activators as well as inhibitors with low nanomolar binding affinities. In addition to filling the hydrophobic pocket, all SOS1 inhibitors insert an aromatic scaffold between the side chains of His905 and Tyr884 to form the His-Tyr sandwich, donate a hydrogen bond to Asn879 and insert a small moiety into the Asn niche. The molecular basis for the mode of action of these inhibitors, the disruption of the RAS-SOS1 complex, is now clearly understood and consists of a steric clash with RAS, caused by a substituent of the inhibitors that reaches into the RAS binding epitope at the catalytic RAS binding site on SOS1.

Two selective and highly potent SOS1 inhibitors, the in vitro probe BAY-293 and the structurally closely related in vivo probe BI-3406 have enabled testing of the effects of SOS1 inhibition in vitro and in vivo. An effect in monotherapy may be expected for patients suffering from RASopathies where the cause for the disease lays directly in the over-activation of SOS1 or the inactivation of the GAP NF1. The current focus in the development of SOS1 inhibitors is, however, very strongly on identifying the most promising combination approaches to translate SOS1 inhibition into sustained inhibition of the KRAS MAPK pathway. Combination strategies with allele-specific direct KRAS-G12C inhibitors, MEK inhibitors and EGFR inhibitors are mechanistically well understood and already show promising results in vitro and in vivo. The first clinical trials with a SOS1 inhibitor have been started and give rise to hope to provide benefits for cancer patients.

Note: A further review on SOS1 inhibitors by Kessler, Gerlach, Kraut, & McConnell, 2021 has been published while this manuscript was being completed. An overview of the patent literature regarding small molecule SOS1 inhibitors (Thompson, Buckl, Dossetter, Griffen, & Gill, 2021) has been published while this manuscript was being reviewed.

References

Abbott, J. R., Hodges, T. R., Daniels, R. N., Patel, P. A., Kennedy, J. P., Howes, J. E., et al. (2018). Discovery of aminopiperidine indoles that activate the guanine nucleotide exchange factor SOS1 and modulate RAS signaling. *Journal of Medicinal Chemistry*, *61*(14), 6002–6017. https://doi.org/10.1021/acs.jmedchem.8b00360.

Abbott, J. R., Patel, P. A., Howes, J. E., Akan, D. T., Kennedy, J. P., Burns, M. C., et al. (2018). Discovery of quinazolines that activate SOS1-mediated nucleotide exchange on RAS. *ACS Medicinal Chemistry Letters*, *9*(9), 941–946. https://doi.org/10.1021/acsmedchemlett.8b00296.

Akan, D. T., Howes, J. E., Sai, J., Arnold, A. L., Beesetty, Y., Phan, J., et al. (2019). Small molecule SOS1 agonists modulate MAPK and PI3K signaling via independent cellular responses. *ACS Chemical Biology*, *14*(3), 325–331. https://doi.org/10.1021/acschembio.8b00869.

Amgen. (2021). *FDA grants sotorasib priority review designation for the treatment of patients with KRAS G12C-mutated locally advanced or metastatic non-small cell lung cancer [Press release]*. Retrieved from https://www.amgen.com/newsroom/press-releases/2021/02/fda-grants-sotorasib-priority-review-designation-for-the-treatment-of-patients-with-kras-g12c-mutated-locally-advanced-or-metastatic-non-small-cell-lung-cancer.

Baltanas, F. C., Perez-Andres, M., Ginel-Picardo, A., Diaz, D., Jimeno, D., Liceras-Boillos, P., et al. (2013). Functional redundancy of Sos1 and Sos2 for lymphopoiesis and organismal homeostasis and survival. *Molecular and Cellular Biology*, *33*(22), 4562–4578. https://doi.org/10.1128/MCB.01026-13.

Boriack-Sjodin, P. A., Margarit, S. M., Bar-Sagi, D., & Kuriyan, J. (1998). The structural basis of the activation of Ras by Sos. *Nature*, *394*(6691), 337–343. https://doi.org/10.1038/28548.

Bos, J. L., Rehmann, H., & Wittinghofer, A. (2007). GEFs and GAPs: Critical elements in the control of small G proteins. *Cell*, *129*(5), 865–877. https://doi.org/10.1016/j.cell.2007.05.018.

Buckl, A., Quintana, E., Grace, J. L., Shifrin, N., Zhong, M., Garrenton, L. S., et al. (2021). Discovery of a potent, selective and orally bioavailable SOS1, inhibitor, RMC-0331 (RM-023), an in vivo tool compound that Blocks RAS activation via disruption of the RAS-SOS1 interaction. In *Paper presented at the AACR Annual Meeting 2021, Virtual Meeting*.

Buday, L., & Downward, J. (1993). Epidermal growth factor regulates p21ras through the formation of a complex of receptor, Grb2 adapter protein, and Sos nucleotide exchange factor. *Cell*, *73*(3), 611–620. https://doi.org/10.1016/0092-8674(93)90146-h.

Burns, M. C., Howes, J. E., Sun, Q., Little, A. J., Camper, D. V., Abbott, J. R., et al. (2018). High-throughput screening identifies small molecules that bind to the RAS:SOS:RAS complex and perturb RAS signaling. *Analytical Biochemistry*, *548*, 44–52. https://doi.org/10.1016/j.ab.2018.01.025.

Burns, M. C., Sun, Q., Daniels, R. N., Camper, D., Kennedy, J. P., Phan, J., et al. (2014). Approach for targeting Ras with small molecules that activate SOS-mediated nucleotide exchange. *Proceedings of the National Academy of Sciences of the United States of America*, *111*(9), 3401–3406. https://doi.org/10.1073/pnas.1315798111.

Canon, J., Rex, K., Saiki, A. Y., Mohr, C., Cooke, K., Bagal, D., et al. (2019). The clinical KRAS(G12C) inhibitor AMG 510 drives anti-tumour immunity. *Nature*, *575*(7781), 217–223. https://doi.org/10.1038/s41586-019-1694-1.

Chardin, P., Camonis, J. H., Gale, N. W., van Aelst, L., Schlessinger, J., Wigler, M. H., et al. (1993). Human Sos1: A guanine nucleotide exchange factor for Ras that binds to GRB2. *Science*, *260*(5112), 1338–1343. https://doi.org/10.1126/science.8493579.

Chen, Y. N., LaMarche, M. J., Chan, H. M., Fekkes, P., Garcia-Fortanet, J., Acker, M. G., et al. (2016). Allosteric inhibition of SHP2 phosphatase inhibits cancers driven by receptor tyrosine kinases. *Nature*, *535*(7610), 148–152. https://doi.org/10.1038/nature18621.

Cordeddu, V., Yin, J. C., Gunnarsson, C., Virtanen, C., Drunat, S., Lepri, F., et al. (2015). Activating mutations affecting the Dbl homology domain of SOS2 cause noonan syndrome. *Human Mutation*, *36*(11), 1080–1087. https://doi.org/10.1002/humu.22834.

Cox, A. D., Fesik, S. W., Kimmelman, A. C., Luo, J., & Der, C. J. (2014). Drugging the undruggable RAS: Mission possible? *Nature Reviews. Drug Discovery*, *13*(11), 828–851. https://doi.org/10.1038/nrd4389.

Dunnett-Kane, V., Nicola, P., Blackhall, F., & Lindsay, C. (2021). Mechanisms of resistance to KRAS(G12C) inhibitors. *Cancers (Basel)*, *13*(1), 151. https://doi.org/10.3390/cancers13010151.

Egan, S. E., Giddings, B. W., Brooks, M. W., Buday, L., Sizeland, A. M., & Weinberg, R. A. (1993). Association of Sos Ras exchange protein with Grb2 is implicated in tyrosine kinase signal transduction and transformation. *Nature*, *363*(6424), 45–51. https://doi.org/10.1038/363045a0.

Esteban, L. M., Fernandez-Medarde, A., Lopez, E., Yienger, K., Guerrero, C., Ward, J. M., et al. (2000). Ras-guanine nucleotide exchange factor sos2 is dispensable for mouse growth and development. *Molecular and Cellular Biology*, *20*(17), 6410–6413. https://doi.org/10.1128/mcb.20.17.6410-6413.2000.

Fedele, C., Li, S., Teng, K. W., Foster, C. J. R., Peng, D., Ran, H., et al. (2021). SHP2 inhibition diminishes KRASG12C cycling and promotes tumor microenvironment remodeling. *The Journal of Experimental Medicine*, *218*(1), e20201414. https://doi.org/10.1084/jem.20201414.

Fedele, C., Ran, H., Diskin, B., Wei, W., Jen, J., Geer, M. J., et al. (2018). SHP2 inhibition prevents adaptive resistance to MEK inhibitors in multiple cancer models. *Cancer Discovery*, *8*(10), 1237–1249. https://doi.org/10.1158/2159-8290.CD-18-0444.

Gale, N. W., Kaplan, S., Lowenstein, E. J., Schlessinger, J., & Bar-Sagi, D. (1993). Grb2 mediates the EGF-dependent activation of guanine nucleotide exchange on Ras. *Nature*, *363*(6424), 88–92. https://doi.org/10.1038/363088a0.

Gort, E., Johnson, M. L., Hwang, J. J., Pant, S., Dünzinger, U., Riemann, K., et al. (2020). A phase I, open-label, dose-escalation trial of BI 1701963 as monotherapy and in combination with trametinib in patients with KRAS mutated advanced or metastatic solid tumors. *Journal of Clinical Oncology*, *38*(15_suppl), TPS3651. https://doi.org/10.1200/JCO.2020.38.15_suppl.TPS3651.

Gray, J. L., von Delft, F., & Brennan, P. E. (2020). Targeting the small GTPase superfamily through their regulatory proteins. *Angewandte Chemie (International Ed. in English)*, *59*(16), 6342–6366. https://doi.org/10.1002/anie.201900585.

Gureasko, J., Galush, W. J., Boykevisch, S., Sondermann, H., Bar-Sagi, D., Groves, J. T., et al. (2008). Membrane-dependent signal integration by the Ras activator Son of sevenless. *Nature Structural & Molecular Biology*, *15*(5), 452–461. https://doi.org/10.1038/nsmb.1418.

Gureasko, J., Kuchment, O., Makino, D. L., Sondermann, H., Bar-Sagi, D., & Kuriyan, J. (2010). Role of the histone domain in the autoinhibition and activation of the Ras activator Son of Sevenless. *Proceedings of the National Academy of Sciences of the United States of America*, *107*(8), 3430–3435. https://doi.org/10.1073/pnas.0913915107.

Hallin, J., Engstrom, L. D., Hargis, L., Calinisan, A., Aranda, R., Briere, D. M., et al. (2020). The KRAS(G12C) inhibitor MRTX849 provides insight toward therapeutic susceptibility of KRAS-mutant cancers in mouse models and patients. *Cancer Discovery*, *10*(1), 54–71. https://doi.org/10.1158/2159-8290.CD-19-1167.

Hillig, R. C., Sautier, B., Schroeder, J., Moosmayer, D., Hilpmann, A., Stegmann, C. M., et al. (2019). Discovery of potent SOS1 inhibitors that block RAS activation via disruption of the RAS-SOS1 interaction. *Proceedings of the National Academy of Sciences of the United States of America*, *116*(7), 2551–2560. https://doi.org/10.1073/pnas.1812963116.

Hodges, T. R., Abbott, J. R., Little, A. J., Sarkar, D., Salovich, J. M., Howes, J. E., et al. (2018). Discovery and structure-based optimization of benzimidazole-derived activators of SOS1-mediated nucleotide exchange on RAS. *Journal of Medicinal Chemistry*, *61*(19), 8875–8894. https://doi.org/10.1021/acs.jmedchem.8b01108.

Hofmann, M. H., Gmachl, M., Ramharter, J., Savarese, F., Gerlach, D., Marszalek, J. R., et al. (2020). BI-3406, a potent and selective SOS1-KRAS interaction inhibitor, is effective in KRAS-driven cancers through combined MEK inhibition. *Cancer Discovery*, *11*(1), 142–157. https://doi.org/10.1158/2159-8290.CD-20-0142.

Howes, J. E., Akan, D. T., Burns, M. C., Rossanese, O. W., Waterson, A. G., & Fesik, S. W. (2018). Small molecule-mediated activation of RAS elicits biphasic modulation of phospho-ERK levels that are regulated through negative feedback on SOS1. *Molecular Cancer Therapeutics*, *17*(5), 1051–1060. https://doi.org/10.1158/1535-7163.MCT-17-0666.

Iwig, J. S., Vercoulen, Y., Das, R., Barros, T., Limnander, A., Che, Y., et al. (2013). Structural analysis of autoinhibition in the Ras-specific exchange factor RasGRP1. *eLife*, *2*, e00813. https://doi.org/10.7554/eLife.00813.

Janes, M. R., Zhang, J., Li, L. S., Hansen, R., Peters, U., Guo, X., et al. (2018). Targeting KRAS mutant cancers with a covalent G12C-specific inhibitor. *Cell*, *172*(3), 578–589 e517. https://doi.org/10.1016/j.cell.2018.01.006.

Jeng, H. H., Taylor, L. J., & Bar-Sagi, D. (2012). Sos-mediated cross-activation of wild-type Ras by oncogenic Ras is essential for tumorigenesis. *Nature Communications*, *3*, 1168. https://doi.org/10.1038/ncomms2173.

Jun, J. E., Rubio, I., & Roose, J. P. (2013). Regulation of ras exchange factors and cellular localization of ras activation by lipid messengers in T cells. *Frontiers in Immunology*, *4*, 239. https://doi.org/10.3389/fimmu.2013.00239.

Kessler, D., Gerlach, D., Kraut, N., & McConnell, D. B. (2021). Targeting son of sevenless 1: The pacemaker of KRAS. *Current Opinion in Chemical Biology*, *62*, 109–118. https://doi.org/10.1016/j.cbpa.2021.02.014.

Kettle, J. G., Bagal, S. K., Bickerton, S., Bodnarchuk, M. S., Breed, J., Carbajo, R. J., et al. (2020). Structure-based design and pharmacokinetic optimization of covalent allosteric inhibitors of the mutant GTPase KRAS(G12C). *Journal of Medicinal Chemistry*, *63*(9), 4468–4483. https://doi.org/10.1021/acs.jmedchem.9b01720.

Khan, I., Rhett, J. M., & O'Bryan, J. P. (2020). Therapeutic targeting of RAS: New hope for drugging the "undruggable". *Biochimica et Biophysica Acta, Molecular Cell Research*, *1867*(2), 118570. https://doi.org/10.1016/j.bbamcr.2019.118570.

Lake, D., Correa, S. A., & Muller, J. (2016). Negative feedback regulation of the ERK1/2 MAPK pathway. *Cellular and Molecular Life Sciences*, *73*(23), 4397–4413. https://doi.org/10.1007/s00018-016-2297-8.

Li, N., Batzer, A., Daly, R., Yajnik, V., Skolnik, E., Chardin, P., et al. (1993). Guanine-nucleotide-releasing factor hSos1 binds to Grb2 and links receptor tyrosine kinases to Ras signalling. *Nature*, *363*(6424), 85–88. https://doi.org/10.1038/363085a0.

Lito, P., Solomon, M., Li, L. S., Hansen, R., & Rosen, N. (2016). Allele-specific inhibitors inactivate mutant KRAS G12C by a trapping mechanism. *Science*, *351*(6273), 604–608. https://doi.org/10.1126/science.aad6204.

Margarit, S. M., Sondermann, H., Hall, B. E., Nagar, B., Hoelz, A., Pirruccello, M., et al. (2003). Structural evidence for feedback activation by Ras.GTP of the Ras-specific nucleotide exchange factor SOS. *Cell*, *112*(5), 685–695. https://doi.org/10.1016/s0092-8674(03)00149-1.

Maurer, T., Garrenton, L. S., Oh, A., Pitts, K., Anderson, D. J., Skelton, N. J., et al. (2012). Small-molecule ligands bind to a distinct pocket in Ras and inhibit SOS-mediated nucleotide exchange activity. *Proceedings of the National Academy of Sciences of the United States of America*, *109*(14), 5299–5304. https://doi.org/10.1073/pnas.1116510109.

Mirati. (2020). *Boehringer Ingelheim and Mirati Therapeutics announce clinical collaboration to study BI 1701963, A SOS1::pan-KRAS inhibitor in combination with MRTX849, A KRAS G12C selective inhibitor [Press release]*. Retrieved from https://ir.mirati.com/press-releases/press-release-details/2020/Boehringer-Ingelheim-And-Mirati-Therapeutics-Announce-Clinical-Collaboration-To-Study-BI-1701963-A-SOS1pan-KRAS-Inhibitor-In-Combination-With-MRTX849-A-KRAS-G12C-Selective-Inhibitor/default.aspx.

Molina-Arcas, M., Moore, C., Rana, S., van Maldegem, F., Mugarza, E., Romero-Clavijo, P., et al. (2019). Development of combination therapies to maximize the impact of KRAS-G12C inhibitors in lung cancer. *Science Translational Medicine*, *11*(510), eaaw7999. https://doi.org/10.1126/scitranslmed.aaw7999.

Moore, A. R., Rosenberg, S. C., McCormick, F., & Malek, S. (2020). RAS-targeted therapies: Is the undruggable drugged? *Nature Reviews. Drug Discovery*, *19*(8), 533–552. https://doi.org/10.1038/s41573-020-0068-6.

Mortier, J., Friberg, A., Badock, V., Moosmayer, D., Schroeder, J., Steigemann, P., et al. (2020). Computationally empowered workflow identifies novel covalent allosteric binders for KRAS(G12C). *ChemMedChem*, *15*(10), 827–832. https://doi.org/10.1002/cmdc.201900727.

Mossessova, E., Corpina, R. A., & Goldberg, J. (2003). Crystal structure of ARF1*Sec7 complexed with Brefeldin A and its implications for the guanine nucleotide exchange mechanism. *Molecular Cell*, *12*(6), 1403–1411. https://doi.org/10.1016/s1097-2765(03)00475-1.

Mullard, A. (2019). Cracking KRAS. *Nature Reviews. Drug Discovery*, *18*(12), 887–891. https://doi.org/10.1038/d41573-019-00195-5.

Nichols, R. J., Haderk, F., Stahlhut, C., Schulze, C. J., Hemmati, G., Wildes, D., et al. (2018). RAS nucleotide cycling underlies the SHP2 phosphatase dependence of mutant BRAF-, NF1- and RAS-driven cancers. *Nature Cell Biology*, *20*(9), 1064–1073. https://doi.org/10.1038/s41556-018-0169-1.

Nyiri, K., Koppany, G., & Vertessy, B. G. (2020). Structure-based inhibitor design of mutant RAS proteins-a paradigm shift. *Cancer Metastasis Reviews*, *39*(4), 1091–1105. https://doi.org/10.1007/s10555-020-09914-6.

Ong, S. H., Hadari, Y. R., Gotoh, N., Guy, G. R., Schlessinger, J., & Lax, I. (2001). Stimulation of phosphatidylinositol 3-kinase by fibroblast growth factor receptors is mediated by coordinated recruitment of multiple docking proteins. *Proceedings of the National Academy of Sciences of the United States of America*, *98*(11), 6074–6079. https://doi.org/10.1073/pnas.111114298.

Ostrem, J. M., Peters, U., Sos, M. L., Wells, J. A., & Shokat, K. M. (2013). K-Ras(G12C) inhibitors allosterically control GTP affinity and effector interactions. *Nature*, *503*(7477), 548–551. https://doi.org/10.1038/nature12796.

Ou, S. I., Koczywas, M., Ulahannan, S., Janne, P., Burris, H., McCoach, C., et al. (2020). A12 the SHP2 inhibitor RMC-4630 in patients with KRAS-mutant non-small cell lung cancer: Preliminary evaluation of a first-in-man phase 1 clinical trial. *Journal of Thoracic Oncology*, *15*, S15–S16. https://doi.org/10.1016/j.jtho.2019.12.041.

Patricelli, M. P., Janes, M. R., Li, L. S., Hansen, R., Peters, U., Kessler, L. V., et al. (2016). Selective inhibition of oncogenic KRAS output with small molecules targeting the inactive state. *Cancer Discovery*, *6*(3), 316–329. https://doi.org/10.1158/2159-8290.CD-15-1105.

Qian, X., Esteban, L., Vass, W. C., Upadhyaya, C., Papageorge, A. G., Yienger, K., et al. (2000). The Sos1 and Sos2 Ras-specific exchange factors: Differences in placental expression and signaling properties. *The EMBO Journal*, *19*(4), 642–654. https://doi.org/10.1093/emboj/19.4.642.

Ramharter, J., Kessler, D., Ettmayer, P., Hofmann, M. H., Gerstberger, T., Gmachl, M., et al. (2021). One atom makes all the difference: Getting a foot in the door between SOS1 and KRAS. *Journal of Medicinal Chemistry*, *64*, 6569–6580. https://doi.org/10.1021/acs.jmedchem.0c01949.

Rauen, K. A. (2013). The RASopathies. *Annual Review of Genomics and Human Genetics*, *14*, 355–369. https://doi.org/10.1146/annurev-genom-091212-153523.

Renault, L., Guibert, B., & Cherfils, J. (2003). Structural snapshots of the mechanism and inhibition of a guanine nucleotide exchange factor. *Nature*, *426*(6966), 525–530. https://doi.org/10.1038/nature02197.

Riely, G. J., Ou, S. H. I., Rybkin, I., Spira, A., Papadopoulos, K., Sabari, J. K., et al. (2021). 99O_PR KRYSTAL-1: Activity and preliminary pharmacodynamic (PD) analysis of adagrasib (MRTX849) in patients (Pts) with advanced non–small cell lung cancer (NSCLC) harboring KRASG12C mutation. *Journal of Thoracic Oncology*, *16*(4, Suppl), S751–S752. https://doi.org/10.1016/S1556-0864(21)01941-9.

Roberts, A. E., Araki, T., Swanson, K. D., Montgomery, K. T., Schiripo, T. A., Joshi, V. A., et al. (2007). Germline gain-of-function mutations in SOS1 cause Noonan syndrome. *Nature Genetics*, *39*(1), 70–74. https://doi.org/10.1038/ng1926.

Rojas, J. M., Oliva, J. L., & Santos, E. (2011). Mammalian son of sevenless Guanine nucleotide exchange factors: Old concepts and new perspectives. *Genes & Cancer*, *2*(3), 298–305. https://doi.org/10.1177/1947601911408078.

Ryan, M. B., Fece de la Cruz, F., Phat, S., Myers, D. T., Wong, E., Shahzade, H. A., et al. (2020). Vertical pathway inhibition overcomes adaptive feedback resistance to KRAS(G12C) inhibition. *Clinical Cancer Research*, *26*(7), 1633–1643. https://doi.org/10.1158/1078-0432.CCR-19-3523.

Sarkar, D., Olejniczak, E. T., Phan, J., Coker, J. A., Sai, J., Arnold, A., et al. (2020). Discovery of sulfonamide-derived agonists of SOS1-mediated nucleotide exchange on RAS using fragment-based methods. *Journal of Medicinal Chemistry*, *63*(15), 8325–8337. https://doi.org/10.1021/acs.jmedchem.0c00511.

Scheffzek, K., Ahmadian, M. R., Kabsch, W., Wiesmuller, L., Lautwein, A., Schmitz, F., et al. (1997). The Ras-RasGAP complex: Structural basis for GTPase activation and its loss in oncogenic Ras mutants. *Science*, *277*(5324), 333–338. https://doi.org/10.1126/science.277.5324.333.

Sheffels, E., Sealover, N. E., Theard, P. L., & Kortum, R. L. (2021). Anchorage-independent growth conditions reveal a differential SOS2 dependence for transformation and survival in RAS-mutant cancer cells. *Small GTPases*, *12*(1), 67–78. https://doi.org/10.1080/21541248.2019.1611168.

Sheffels, E., Sealover, N. E., Wang, C., Kim, D. H., Vazirani, I. A., Lee, E., et al. (2018). Oncogenic RAS isoforms show a hierarchical requirement for the guanine nucleotide exchange factor SOS2 to mediate cell transformation. *Science Signaling*, *11*(546), eaar8371. https://doi.org/10.1126/scisignal.aar8371.

Simanshu, D. K., Nissley, D. V., & McCormick, F. (2017). RAS proteins and their regulators in human disease. *Cell*, *170*(1), 17–33. https://doi.org/10.1016/j.cell.2017.06.009.

Solca, F., Dahl, G., Zoephel, A., Bader, G., Sanderson, M., Klein, C., et al. (2012). Target binding properties and cellular activity of afatinib (BIBW 2992), an irreversible ErbB family blocker. *The Journal of Pharmacology and Experimental Therapeutics*, *343*(2), 342–350. https://doi.org/10.1124/jpet.112.197756.

Sondermann, H., Soisson, S. M., Boykevisch, S., Yang, S. S., Bar-Sagi, D., & Kuriyan, J. (2004). Structural analysis of autoinhibition in the Ras activator Son of sevenless. *Cell*, *119*(3), 393–405. https://doi.org/10.1016/j.cell.2004.10.005.

Stalnecker, C. A., & Der, C. J. (2020). RAS, wanted dead or alive: Advances in targeting RAS mutant cancers. *Science Signaling*, *13*(624), eaay6013. https://doi.org/10.1126/scisignal.aay6013.

Stamos, J., Sliwkowski, M. X., & Eigenbrot, C. (2002). Structure of the epidermal growth factor receptor kinase domain alone and in complex with a 4-anilinoquinazoline inhibitor. *The Journal of Biological Chemistry*, *277*(48), 46265–46272. https://doi.org/10.1074/jbc.M207135200.

Sun, Q., Burke, J. P., Phan, J., Burns, M. C., Olejniczak, E. T., Waterson, A. G., et al. (2012). Discovery of small molecules that bind to K-Ras and inhibit Sos-mediated activation. *Angewandte Chemie (International Ed. in English)*, *51*(25), 6140–6143. https://doi.org/10.1002/anie.201201358.

Tartaglia, M., Pennacchio, L. A., Zhao, C., Yadav, K. K., Fodale, V., Sarkozy, A., et al. (2007). Gain-of-function SOS1 mutations cause a distinctive form of Noonan syndrome. *Nature Genetics*, *39*(1), 75–79. https://doi.org/10.1038/ng1939.

Theard, P. L., Sheffels, E., Sealover, N. E., Linke, A. J., Pratico, D. J., & Kortum, R. L. (2020). Marked synergy by vertical inhibition of EGFR signaling in NSCLC spheroids shows SOS1 is a therapeutic target in EGFR-mutated cancer. *eLife*, *9*, e58204. https://doi.org/10.7554/eLife.58204.

Thompson, S. K., Buckl, A., Dossetter, A. G., Griffen, Ed, & Gill, A. (2021). Small molecule Son of Sevenless 1 (SOS1) inhibitors: A review of the patent literature. *Expert Opinion on Therapeutic Patents*. https://doi.org/10.1080/13543776.2021.1952984.
van Maldegem, F., & Downward, J. (2020). Mutant KRAS at the heart of tumor immune evasion. *Immunity, 52*(1), 14–16. https://doi.org/10.1016/j.immuni.2019.12.013.
Vercoulen, Y., Kondo, Y., Iwig, J. S., Janssen, A. B., White, K. A., Amini, M., et al. (2017). A Histidine pH sensor regulates activation of the Ras-specific guanine nucleotide exchange factor RasGRP1. *eLife, 6*, e29002. https://doi.org/10.7554/eLife.29002.
Vigil, D., Cherfils, J., Rossman, K. L., & Der, C. J. (2010). Ras superfamily GEFs and GAPs: Validated and tractable targets for cancer therapy? *Nature Reviews. Cancer, 10*(12), 842–857. https://doi.org/10.1038/nrc2960.
Wang, T., Birsoy, K., Hughes, N. W., Krupczak, K. M., Post, Y., Wei, J. J., et al. (2015). Identification and characterization of essential genes in the human genome. *Science, 350*(6264), 1096–1101. https://doi.org/10.1126/science.aac7041.
Wang, D. Z., Hammond, V. E., Abud, H. E., Bertoncello, I., McAvoy, J. W., & Bowtell, D. D. (1997). Mutation in Sos1 dominantly enhances a weak allele of the EGFR, demonstrating a requirement for Sos1 in EGFR signaling and development. *Genes & Development, 11*(3), 309–320. https://doi.org/10.1101/gad.11.3.309.
Winter, J. J., Anderson, M., Blades, K., Brassington, C., Breeze, A. L., Chresta, C., et al. (2015). Small molecule binding sites on the Ras:SOS complex can be exploited for inhibition of Ras activation. *Journal of Medicinal Chemistry, 58*(5), 2265–2274. https://doi.org/10.1021/jm501660t.
Wittinghofer, A., & Vetter, I. R. (2011). Structure-function relationships of the G domain, a canonical switch motif. *Annual Review of Biochemistry, 80*, 943–971. https://doi.org/10.1146/annurev-biochem-062708-134043.
Yamamoto, G. L., Aguena, M., Gos, M., Hung, C., Pilch, J., Fahiminiya, S., et al. (2015). Rare variants in SOS2 and LZTR1 are associated with Noonan syndrome. *Journal of Medical Genetics, 52*(6), 413–421. https://doi.org/10.1136/jmedgenet-2015-103018.
Yun, C. H., Boggon, T. J., Li, Y., Woo, M. S., Greulich, H., Meyerson, M., et al. (2007). Structures of lung cancer-derived EGFR mutants and inhibitor complexes: Mechanism of activation and insights into differential inhibitor sensitivity. *Cancer Cell, 11*(3), 217–227. https://doi.org/10.1016/j.ccr.2006.12.017.

CHAPTER SEVEN

Inhibition of SHP2 as an approach to block RAS-driven cancers

Yu-Ting Chou and Trever G. Bivona*
Department of Medicine, Division of Hematology and Oncology, and The Helen Diller Comprehensive Cancer Center, University of California, San Francisco, CA, United States
*Corresponding author: e-mail address: trever.bivona@ucsf.edu

Contents

1. Introduction 206
2. Biology of RTK/RAS signaling 207
3. General understanding of RAS-driven cancer and treatment 209
   3.1 *RAS* mutations and diversity in distinct tissues 209
   3.2 MAPK signaling as a key effector in RAS-driven cancer 210
4. The biological and oncogenic role of SHP2 211
   4.1 Classification and regulation of SHP2 211
   4.2 The substrates of SHP2 215
   4.3 SHP2 in cancer 216
5. SHP2 inhibitors as potential therapies in RAS-driven cancers 220
   5.1 Traditional SHP2-PTP inhibitors 220
   5.2 Allosteric SHP2 inhibitors 221
6. Conclusions and perspectives 226
Acknowledgments 227
Potential competing interests 228
References 228

Abstract

The non-receptor protein tyrosine phosphatase SHP2 (encoded by *PTPN11*) is a critical component of RAS/MAPK signaling by acting upstream of RAS to promote oncogenic signaling and tumor growth. Over three decades, SHP2 was considered "undruggable" because enzymatic active-site inhibitors generally showed off-target inhibition of other proteins and low membrane permeability. More recently, allosteric SHP2 inhibitors with striking inhibitory potency have been developed. These small molecules effectively block the signal transduction between receptor tyrosine kinases (RTKs) and RAS/MAPK signaling and show efficacy in preclinical cancer models. Moreover, clinical evaluation of these allosteric SHP2 inhibitors is ongoing. RAS proteins which harbor

Advances in Cancer Research, Volume 153
ISSN 0065-230X
https://doi.org/10.1016/bs.acr.2021.07.002

transforming properties by gain-of-function mutations are present in various cancer types. While inhibitors of KRASG12C show early clinical promise, resistance remains a challenge and other forms of oncogenic RAS remain to be selectively inhibited. Here, we summarize the role of SHP2 in RAS-driven cancers and the therapeutic potential of allosteric SHP2 inhibitors as a strategy to block RAS-driven cancers.

1. Introduction

Protein-tyrosine kinases (PTKs) and protein-tyrosine phosphatases (PTPs) control diverse cellular processes including differentiation, proliferation, survival, migration and apoptosis. Dysregulated tyrosyl-phosphorylation contributes to various human disease and cancer types. Aberrant hyper-activation of PTKs via gain-of-function mutation, overexpression and/or gene amplification promotes tumorigenesis. Although PTPs are important negative regulators of PTK signaling in many cases owing to their opposite action of PTKs, some PTPs can function as oncogenes, as exemplified by the Src homology-2 domain-containing phosphatase SHP2 (encoded by *PTPN11*) which is an authentic proto-oncogene and also a critical downstream target of other oncogenes (Chan, Kalaitzidis, & Neel, 2008).

Expression of SHP2 is ubiquitous in all human tissues. It functions as both a scaffolding adaptor protein and PTP. SHP2 is composed of two N-terminal Src homology-2 (SH2) domains (termed N-SH2 and C-SH2), a catalytic PTP domain as well as a C-terminal 66 amino acid segment with two tyrosyl-phosphorylation residues (Tyr542 and Tyr 580) and a proline rich domain (Neel, Gu, & Pao, 2003). In most RTK signaling pathways, SHP2 is essential for full activation of RAS/ERK signaling which is a major cascade in cancer biology. In 2001, germline *PTPN11* mutations were observed in around half of the cases of Noonan syndrome (NS), an autosomal dominant developmental disorder characterized by facial dysmorphism, short stature and cardiac defects (especially pulmonic stenosis and hypertrophic cardiomyopathy) (Tartaglia & Gelb, 2005; Tartaglia et al., 2001). Noonan syndrome was found to be attributed to hyper-activated *KRAS* and the RAS guanine nucleotide exchange factor (GEF) *SOS1* mutations in germline (Roberts et al., 2007; Schubbert et al., 2006; Tartaglia et al., 2007). These studies strongly suggest that Noonan syndrome is caused by gain-of-function mutations in SHP2 and RAS signaling proteins. In contrast to Noonan syndrome, somatic *PTPN11* mutations account for approximately 35% of sporadic juvenile myelomonocytic leukemia (JMML) which is a disorder with excessive proliferation of myelomonocytic cells. At lower incidence, somatic *PTPN11* mutations also occur in other myeloid neoplasms, such as acute myelogenous leukemia (AML), chronic myelomonocytic leukemia (CMML), and myelodysplastic syndrome (MDS). Remarkably, the JMML patients who without *PTPN11* mutations usually have homozygous deficiency (loss-of-function) of *Neurofibromatosis type 1* (*NF1*, encode a RAS-GAP) or activating RAS (*NRAS* or *KRAS*) mutations (Bentires-Alj et al., 2004; Loh, Reynolds, et al., 2004; Loh, Vattikuti, et al., 2004; Tartaglia et al., 2005). Given the role of wild type (WT) SHP2 upstream of RAS, these findings also suggest that SHP2, RAS and NF1 comprise a core signaling module in JMML pathogenesis. Although *PTPN11* mutations rarely occur in solid tumors, WT SHP2 has been considered as a critical component of various oncogenic pathways including RTK/RAS signaling.

SHP2 was considered as an "undruggable" target. However, in 2016, the first allosteric SHP2 inhibitor SHP099 was identified by a group from Novartis and provided a new opportunity to inhibit malignant growth through inhibition of RTK-MAPK signaling (Chen et al., 2016; Garcia Fortanet et al., 2016). In contrast to the previous strategy of targeting catalytic activity of the PTP domain, structure-based allosteric SHP2 inhibitors were identified that act as "molecular glue" owing to their ability to stabilize the basal auto-inhibited conformation of SHP2 (Ran, Tsutsumi, Araki, & Neel, 2016). Currently, several small molecules are being developed including TNO155, RMC-4630, BBP-398, JAB-3068, JAB-3312 and RLY-1971 which are under investigation in a wide range of cancer patients. Here, we will discuss the current knowledge of SHP2 biology in RAS signaling, the prospect of allosteric SHP2 inhibitors, and the potential for combination therapy in RAS-driven cancers.

2. Biology of RTK/RAS signaling

RTK/RAS signaling is one of the most well studied and conserved signaling cascades in normal cells and human pathologies. RTKs are indispensable in a wide range of cellular processes including differentiation, proliferation, motility, migration and metabolism. The process of signal transduction through RTKs plays a vital role for controlling embryonic development, synaptic plasticity, angiogenesis, growth of an organism and oncogenesis (Du & Lovly, 2018; McKay & Morrison, 2007). RTKs are transmembrane proteins which comprised of a N-terminal extracellular ligand binding domain, a transmembrane helix, and a cytosolic portion that contains a juxtamembrane regulatory region along with an intracellular tyrosine kinase domain (TKD) followed by carboxy (C-) terminal tail (Hubbard, 1999). RTKs are activated by binding of receptor-specific ligands. Except the insulin receptor (IR) family, most RTKs exist as inactive monomers and are activated by ligand-induced dimerization or oligomerization. The dimerization of RTKs leads to activation and autophosphorylation of TKD which provides phosphotyrosine (pTyr) residues as specific binding sites for cytosolic proteins containing Src homology-2 (SH2) and phosphotyrosine-binding (PTB) domains (Pawson, 2004; Schlessinger & Lemmon, 2003). Certain recruited effector proteins which have intrinsic enzymatic activity are phosphorylated and translocated to plasma membrane, such as phospholipase Cγ (PLCγ), PI-3′ kinase (PI3K) and Src homology-2 domain-containing phosphatase (SHP2). In addition to the protein with enzymatic activity, SH2/PTB domains are also present in adaptor proteins, such as Growth factor receptor-bound protein 2 (GRB2). GRB2 is recruited to binds to pTyr of RTKs through SH2 domain and induces the guanine nucleotide exchange factor (GEF), Son of Sevenless (SOS), to translocate from cytosol to the plasma membrane to stimulate exchange of GTP for GDP on RAS to activate RAS signaling (Lemmon & Schlessinger, 2010; Pawson, 2004; Schlessinger & Lemmon, 2003).

RAS proteins contribute to signaling transduction events that control normal cellular homeostasis and multiple biological process. These proteins belong to the larger small guanosine triphosphatases (GTPases) family and act as binary switches during active guanosine triphosphate (GTP)-bound and inactive guanosine diphosphate (GDP)-bound states. RAS proteins are tightly regulated by guanine nucleotide exchange factors (GEFs) and GTPase-activating proteins (GAPs). RAS activation by GEFs, such as SOS, catalyze the release of GDP and help convert RAS to the GTP-bound state. Active RAS then binds to effectors, such as RAF kinase, p110 subunit of PI3K and RAL GEFs. Conversely, RAS signaling is terminated by GAPs which inactivate RAS by increasing hydrolysis of GTP (Bos, Rehmann, & Wittinghofer, 2007; Cherfils & Zeghouf, 2013). To fully activate the RTK/RAS/mitogen-activated protein kinase (MAPK) cascade, SHP2 (encoded by *PTPN11*) is required due to dual roles: a scaffolding adaptor protein and a protein-tyrosine phosphatase (PTP) (Barford & Neel, 1998). Activated RTKs or scaffold proteins provide pTyr as specific binding site for SHP2 SH2 domain. This interaction initiates a SHP2 conformational change and increases PTP activity. SHP2 undergoes tyrosyl phosphorylation on its two C-terminal tyrosines and subsequently recruits GRB2/SOS and also deactivates negative regulators of RAS/MAPK signaling (Dance, Montagner, Salles, Yart, & Raynal, 2008).

3. General understanding of RAS-driven cancer and treatment

3.1 *RAS* mutations and diversity in distinct tissues

The three *RAS* genes (*KRAS*, *NRAS* and *HRAS*) were collectively mutated and recognized to be a human oncogene as early as the 1980s (Der, Krontiris, & Cooper, 1982; Parada, Tabin, Shih, & Weinberg, 1982; Tabin et al., 1982). These genes encode 4 highly homologous proteins: HRAS, NRAS, KRAS4A and KRAS4B. KRAS4A and KRAS4B are alternative RNA splicing variants which are encoded by the *KRAS* gene containing a different exon 4. Oncogenic mutations in *RAS* genes occur in numerous cancer types and with RAS allelic specificity across different tissue types. The highest frequency of *KRAS* mutations, especially the isoform KRAS4B, are present in 86% pancreatic ductal adenocarcinoma (PDAC), 42% colorectal adenocarcinoma (CRC) and 32% lung adenocarcinoma (LUAD; a subtype of non-small-cell lung cancer (NSCLC)). *NRAS* mutation are common in melanoma and hematopoietic malignancies. Compared with mutations in *KRAS* or *NRAS*, *HRAS* mutations are observed less frequently but are considered a driver event in a subset of bladder cancers and head and neck squamous cell carcinoma (Moore, Rosenberg, McCormick, & Malek, 2020; Pylayeva-Gupta, Grabocka, & Bar-Sagi, 2011). The oncogenic mutations in *RAS* are most commonly present in three missense-mutation hotspots including G12, G13, or Q61. These mutations disrupt the GTP hydrolysis and consequently cause the incessant activation of RAS-dependent pathways (Buhrman, Wink, & Mattos, 2007; Burd et al., 2014; Hunter et al., 2015; Smith, Neel, & Ikura, 2013). G12 is the most frequent mutation (83%) in *KRAS*, followed by G13 mutations (14%), and Q61 mutations (2%). Intriguingly, in contrast to KRAS, Q61 is the predominantly mutation in NRAS, followed by G12 and G13. A similar frequency of G12, G13 and Q61 mutations as in KRAS are found in HRAS. Furthermore, the mutation within each RAS isoform can show significant differences between cancer types. For examples, KRAS G12D mutations are observed predominantly in PDAC and CRC, followed by G12V mutations. However, the major substitution is G12C in NSCLC.

These patterns suggest the possibility that different RAS mutations contribute to different functional consequences and signaling properties that result in diverse oncogenic functions in the particular tissue of origin (Cox, Fesik, Kimmelman, Luo, & Der, 2014; Hobbs, Der, & Rossman, 2016).

Ideally, it should be possible to design a small molecule to block the GTP-binding site on RAS and inhibit activity. However, for almost four decades, tremendous efforts to target active RAS faced several challenges: (1) GTP has picomolar affinity for RAS and the dissociation constant (Kd) is around 10^{-11} mol/L; and (2) There was a perceived lack of druggable binding pockets in RAS for small-molecule binding (Cox et al., 2014; Liu, Wang, & Li, 2019; Uras, Moll, & Casanova, 2020). More recently, the exciting development of KRAS^{G12C} inhibitors represent a compelling therapeutic opportunit. These drugs such as AMG 510 (ClinicalTrials.gov Identifier: NCT03600883) and MRTX849 (ClinicalTrials.gov Identifier: NCT04613596) bind in the switch II pocket and lock KRAS^{G12C} in an inactive GDP-bound state and are in clinical trials with promising initial clinical impact. However, other forms of oncogenic RAS are still undruggable (Canon et al., 2019; Hallin et al., 2020; Seton-Rogers, 2020). Moreover, the efficacy selective covalent inhibitors of KRAS^{G12C} is challenged by adaptive and genetic reactivation of the MAPK cascade (Ryan et al., 2020; Xue et al., 2020). Given this concern, both AMG 510 and MRTX849 are being investigated in various combination therapy strategies, including co-inhibition of MEK, SHP2, PI3K or AKT which improve responses in lung and colon tumor xenograft models (Canon et al., 2019; Hallin et al., 2020; McCormick, 2020). Further studies to identify other potential targets that can complement G2C inhibitors are warranted.

3.2 MAPK signaling as a key effector in RAS-driven cancer

Because of the historical challenges in pharmacological targeting of oncogenic RAS, many efforts have centered on inhibiting downstream effector pathways, in particular MAPK signaling. By genetic experiments, the required components of the MAPK pathway have been dissected particularly in KRAS-mutant tumors. MEK1 and MEK2 individual isoforms are dispensable for the initiation of lung adenocarcinomas driven by KRAS^{G12V}. *Mek2* is also redundant for embryonic development and adult homeostasis in mouse models (Belanger et al., 2003). However, Mek1/2 double knockout mice die after birth within 24h which suggests Mek is required for mouse growth (Scholl et al., 2007). In KRAS-driven NSCLC, complete inactivation of both MEK1/2 efficiently blocked tumor formation (Blasco et al., 2011). These genetic studies revealed the importance of MEK activity in biological homeostasis and cancer development (Drosten & Barbacid, 2020; Uras et al., 2020). Accordingly, allosteric MEK inhibitors have been considered as an attractive therapeutic target for RAS-driven cancers. Trametinib (Mekinist) in combination with selective $BRAF^{V600E}$ inhibitor, dabrafenib, is approved for BRAF-mutant metastatic melanoma (Yaeger & Corcoran, 2019). Another MEK inhibitor binimetinib shows similar results with encorafenib (Rose, 2019). However, using MEK inhibitors as a single agent or combined with chemotherapy was ineffective in RAS-mutant cancer patients (Blumenschein et al., 2015; Engelman et al., 2008; Haura et al., 2010; Janne et al., 2017).

Parallel efforts have focused on ERK. There is no effect on initiation of LUAD by individual ablation of *Erk1* or *Erk2*, whereas elimination both completely prevents pulmonary tumor formation. Furthermore, systemic elimination of *ERK1* and *ERK2* in adult mice causes rapid death as a consequence of severe defects as in *MEK1/2*-dificient mice (Blasco et al., 2011). Unfortunately, inhibition of ERK has not shown robust clinical efficacy as of yet in RAS-mutant cancer patients. Reactivation of RAS pathway signaling may underlie the lack of therapeutic efficacy, as with MEK inhibition (Cox et al., 2014). Recently, a significant potential advantage of SHP2 inhibition has been identified. It was reported that inhibition of SHP2 induces a senescence response, prevents feedback activation of RTKs and overcomes the adaptive resistance of MEK inhibitors in RAS-driven cancers (Ahmed et al., 2019; Mainardi et al., 2018), as discussed below in more detail.

4. The biological and oncogenic role of SHP2

4.1 Classification and regulation of SHP2

The main catalytic activity of SHP2 is as a PTP. The human PTP superfamily contains 125 genes which were classified by the following criteria: (1) harboring the PTP domain defined by structural evaluation; or (2) presence of a CX_5R (C represents the catalytic Cys, X represents any amino acid, and R represents an Arg) "conserved catalytic motif" within the non-PTP domain; or (3) displaying validated Tyr phosphatase activity; or (4) containing a highly similar sequence to members with demonstrated Tyr phosphatase activity. According to the identity of the nucleophilic catalytic residue, this superfamily was further classified as Cys-based (116 genes), His-based (5 genes) and Asp-based (4 genes) Tyr phosphatases. Most of the PTP members are Cys-based Tyr phosphatases which fall into 3 classes (classes I-III). Class I includes the most PTPs comprising the 37 classical PTPs and 64 dual-specificity phosphatase (DUSPs) and 10 with other features. Classical PTPs can be subdivided into transmembrane receptor-like PTPs (RPTPs) and intracellular non-receptor PTPs (NRPTPs) (Fig. 1, left panel) (Alonso, Nunes-Xavier, Bayon, & Pulido, 2016). DUSPs (Fig. 1, right panel) are also called VH1-like phosphatases since the first phosphatase which was identified in this group was vaccinia poxvirus named VH1 (Guan, Broyles, & Dixon, 1991). DUSPs show a great diversity of substrates including phospho-tyrosine-, phospho-serine- and phospho-threonine-containing targets. One of the well-known lipid phosphatases in this group is the tumor suppressor PTEN which can dephosphorylate the critical second messenger phosphatidylinositol-3, 4, 5-phosphate (PIP3) and reduce signal transduction (Pulido, Stoker, & Hendriks, 2013).

The SH2 domain-containing PTPs (SHPs) belong to class I NRPTPs. There are two SHPs, SHP1 and SHP2, in vertebrates as well as one SHP ortholog, Corkscrew (Csw) and Ptp-2, in *Drosophila melanogaster* and *Caenorhabditis elegans*, respectively. Since SHPs have both SH2 and PTP domains which can bind to pTyr and integrate various growth and differentiation signaling pathways, SHPs have been studied in human health and disease (Feng, 1999; Neel et al., 2003; Van Vactor, O'Reilly, & Neel, 1998).

SHP2 promotes RAS/MAPK pathways activation through its PTP activity (Fig. 2A). In the basal state, SHP2 is maintained as an auto-inhibited protein by back-side loop insertion of N-SH2 through Asp61 and Tyr62 into the catalytic PTP domain. This closed conformation can be opened and activated by SHP2 binding to a protein containing 2 pTyr sites. Upon stimulation of receptors by growth factors or cytokines, the pTyr residues on RTKs, cytokine receptors and scaffold adaptors are provided as binding sites. The C-SH2 is the initial contact domain and engaged by a pTyr residue. The resultant increase in the pTyr peptide for the N-SH2 domain overcomes mutual allosteric inhibition and enables access to the opened active site. In addition, phosphorylation of the C-terminal tail at Tyr 542 and Tyr 580 is an optional process that confers non SHP2 an adaptor function for recruiting GRB2 (Barford & Neel, 1998; Kerr, Haderk, & Bivona, 2021; Mohi & Neel, 2007). Dysregulation of this mechanism by *PTPN11* mutation to activate SHP2 causes human diseases including cancer.

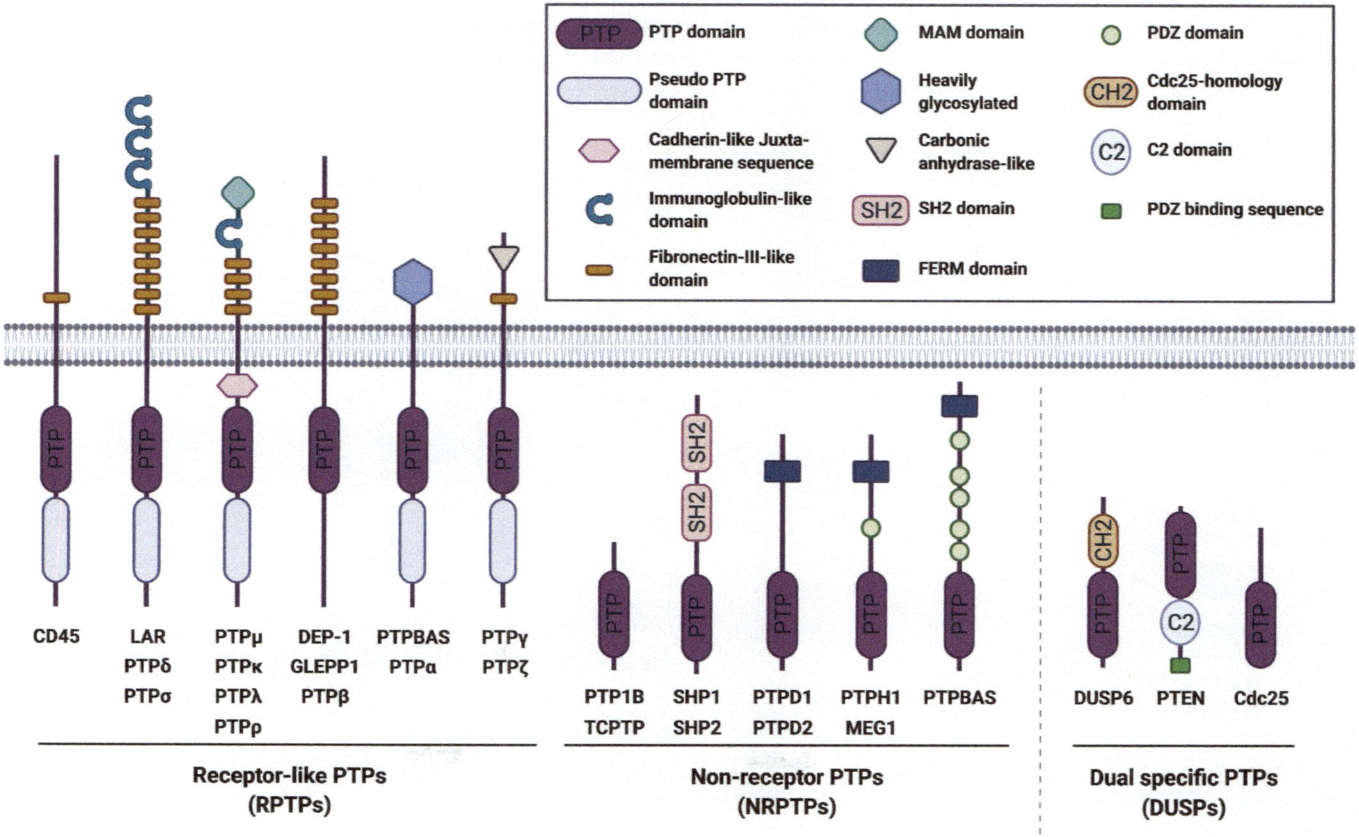

Fig. 1 Schematic representations of Class I PTPs. The class I PTP superfamily includes classical PTPs which can be further subdivided into RPTPs and NRPTPs as well as DUSPs. Some RPTPs contain both PTP and pseudo PTP domains that have catalytic and residual catalytic activity, respectively. NRPTPs are intrinsic PTPs which include SHP2. DUSPs are different from classical PTPs in that they can dephosphorylate phospho-tyrosine and phospho-serine and phospho-threonine. *Abbreviations*: *MAM*, Mephrin/A5/μ domain; *SH2*, Src-homology 2; *FERM*, 4.1 protein/ezrin/radixin/moesin domain; *PDZ*, PSD95/Dig1/Zo-1 domain.

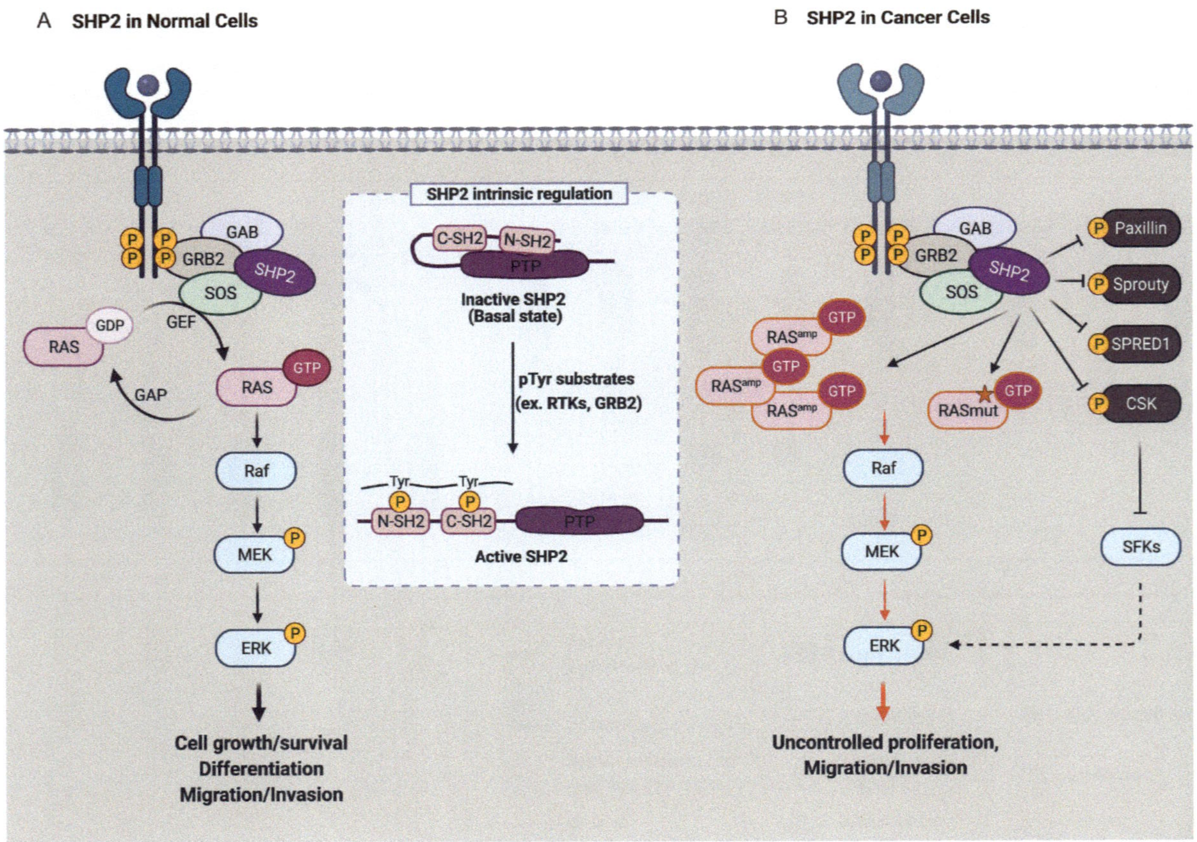

Fig. 2 SHP2 regulation in normal and cancer cells. SHP2 is a transducer of signals from RTKs to promote RAS activation. (A) In normal cells, SHP2 is under control and shifts between auto-inhibited basal state and active conformations. The closed (inactive) SHP2 can be opened and activated by binding to two pTyr substrates. (B) In cancer cells, SHP2 plays the role of a required component of oncogenic driver signaling to activate RAS/MAPK signaling. By dephosphorylating potential targets including Paxillin, Sprouty, SPRED1 and CSK each of which have inhibitory functions of RAS/MAPK pathway, SHP2 could also be a mechanism to induce uncontrolled oncogenic pathway activation.

4.2 The substrates of SHP2

Although it is known that SHP2 is required for activating RAS/MAPK signaling through PTP activity, the specific targets remain under active investigation. SHP2 has a dual role to activate the RAS/MAPK pathway: (1) direct interaction with activated RTKs, or (2) indirect interaction with the receptor scaffolding complexes. SHP2 can de-phosphorylate certain RTKs to decrease RAS-GAP recruitment to the plasma membrane, thereby increasing RAS activity (Agazie & Hayman, 2003). Beyond RTKs, adaptor proteins such as GRB2-associated binding (GAB), are also targets of SHP2 (Montagner et al., 2005). SHP2 was found to dephosphorylate docking sites on the scaffolding proteins Paxillin and phosphoprotein associated with glycosphingolipid microdomains 1 (PAG1). These phosphorylation sites are recruit and activate c-src tyrosine kinase (CSK) which suppresses RTK-activated SRC family kinases (SFKs) and ERK signaling (Ren et al., 2004; Zhang et al., 2004). Tyrosine-phosphorylated RTK inhibitory proteins Sprouty and Sprouty-related Ena/vasodilator-stimulated phosphoprotein homology 1-domain-containing protein (SPRED1) are interesting targets of SHP2, as they function as negative modulators of the RAS/ERK pathway. Membrane translocation and phosphorylation are required for the inhibitor activity of Sprouty. By binding to GRB2, Sprouty interrupts GRB2-SOS complex formation and blunts activation of ERK (Hanafusa, Torii, Yasunaga, & Nishida, 2002). Dysregulation of Sprouty has been shown in several cancer types (Masoumi-Moghaddam, Amini, & Morris, 2014). SPRED1 is a specific dephosphorylation target of SHP2 at Tyr-420. This evidence shows that SPRED1 is a promising substrate of SHP2 (Quintanar-Audelo, Yusoff, Sinniah, Chandramouli, & Guy, 2011). RAS per se is also identified as a SHP2 substrate. The non-receptor tyrosine kinase SRC has preferential binding to tyrosyl-phosphorylated, GTP-loaded RAS at Tyr32 within the switch I region. This interaction increases RAS-GAP binding and subsequent GTP hydrolysis instead of RAF recruitment, thereby inactivating RAS signaling (Bunda et al., 2014). SHP2 was further found to dephosphorylate RAS at Tyr32 to promote RAS activation (Bunda et al., 2015). These studies provided direct evidence that SHP2 is critical for RAS activation. In addition to the RAS/MAPK pathway, SHP2 has also been shown to regulate PI3K/AKT, JAK/STAT, JNK, and NF-kB signaling (Frankson et al., 2017; Huang et al., 2014).

Moreover, SHP2 also plays a critical role in various immune cell types present in the tumor microenvironment, such as those expressing programed cell death 1 pathway (PD-1). PD-1 is an immune checkpoint protein. SHP2 is activated and could cause immune evasion in cancer by interacting via its SH2 domains with immune receptor tyrosine–based inhibitory motifs (ITIMs) and immune receptor tyrosine–based switch motifs (ITSMs) present in PD1 (Marasco et al., 2020). Tumor-associated macrophages (TAMs) generally play a pro-tumor role to initiate tumorigenesis, angiogenesis, invasion and immunosuppression. The proliferation and M2-type polarization of macrophages are induced by colony stimulating factor-1 (CSF-1) through GRB2-mediated signaling which is known to interact with SHP2. In a mouse model, SHP2-deficient bone marrow macrophages (BMMs) showed reduced proliferation in response to CSF-1. Therefore, it seems SHP2 can modulate macrophages and could be a therapeutic target to address immune evasion of cancers (Noy & Pollard, 2014; Song et al., 2021). The substrates and role(s) of SHP2 in tumor microenvironment cells is an active area of investigation and could identify new therapeutic opportunities for allosteric SHP2 inhibitors.

4.3 SHP2 in cancer

Mutational activation of SHP2 increases its catalytic PTP activity in both hereditary and sporadic leukemia. Somatic *PTPN11* mutations are rare in solid tumors, although SHP2 promotes malignancies with other known oncogenic drivers, such as those with RAS hyperactivation via various mechanisms (Bentires-Alj et al., 2004; Mainardi et al., 2018; Martinelli et al., 2006) (Fig. 2B). Activation of chimeric fusion-PTKs (e.g., BCR/ABL), dysregulation of growth factors, RTKs, or scaffolding adapters cause tumorigenesis in part by driving inappropriate SHP2 activation (Agazie & Hayman, 2003; Chen et al., 2016; Gu et al., 2018; Hao et al., 2019; Sattler et al., 2002; Sha et al., 2013). The RTK/RAS/MAPK pathway is frequently activated in cancers owing to a variety of genetic mutations or amplifications in pathway components. In this section, the roles of SHP2 in connection with dysregulated RTK/RAS signaling in different cancer types are summarized.

4.3.1 Non-small cell lung cancer (NSCLC)

A high frequency of oncogenic activation of the RAS/MAPK cascade is present in NSCLC. Examples includes oncogenic alterations in *KRAS* or upstream or downstream components including RTKs, *BRAF* and neurofibromin 1 (*NF1*) mutation (Cancer Genome Atlas Research Network, 2014). Because SHP2 is a critical component to activate RTK/RAS/MAPK signaling, allosteric SHP2 inhibitors were tested in RTK-driven cancers in preclinical experiments. SHP2 inhibition was effective in RTK-driven preclinical models (e.g., EGFR-mutant ALK fusion positive), including those resistant to current clinical RTK inhibitors (Dardaei et al., 2018). Inactivation of SHP2 effectively reversed MEK inhibitor resistance in lung cancer model as well (Mainardi et al., 2018; Ruess et al., 2018). Other than oncogenic RAS, SHP2 promotes RAS/MAPK hyper-activation and cancer growth through other RAS-GTP dependent oncogenic effectors as well, such as class 3 BRAF mutants and mutant NF1. Inhibition of SHP2 disrupts the SHP2/GRB2/SOS1 complex and suppresses GTP loading of RAS in this context, as well as in oncogenic KRAS variants such as $KRAS^{G12C}$ that remain sensitive to upstream activating signals. (Nichols et al., 2018). These collective studies indicated that SHP2 is a vital regulator of oncogenic RTK/RAS/MAPK-driven NSCLC and it may be a promising therapeutic approach, particularly in combination therapy settings.

4.3.2 Breast cancer

In 2008, Zhou and Agazie revealed that SHP2 is upregulated in breast cancer cells. Genetic inhibition of SHP2 abolished EGF-induced ERK1/2 and AKT activation as well as up-regulated E-cadherin, which is an epithelial marker. They proposed that SHP2 is required for promoting epithelial to mesenchymal transition (EMT), whereas inhibition of SHP2 results in mesenchymal to epithelial transition (MET) to help reinstate a more normal breast epithelial phenotype (Zhou & Agazie, 2008). In breast cancer, human epidermal growth factor (EGF) receptor 2 (HER2) is a well-known marker. Interestingly, the RAS-GAP docking site at Tyr1023 on HER2 is found to be regulated by SHP2. This suggests that SHP2 can activate HER2-induced signaling through dephosphorylation of this negative regulatory auto-phosphorylation site (Zhou & Agazie, 2009). In a subset of human primary breast tumors which are associated with invasive ability and poor prognosis, "SHP2 signature" genes were demonstrated to be simultaneously activated. SHP2 depletion prevented invasion, breast tumor growth and attenuated metastasis (Aceto et al., 2012). Furthermore, the scaffolding adaptor protein growth factor receptor-bound protein 2 (GAB2) can contribute to breast cancer development. A pTyr residue of GAB2 promotes SHP2 recruitment, which helps to activate downstream RAS/MAPK signaling. In this context, p90 ribosomal S6 kinase (RSK) plays a role as negative modulator by phosphorylating GAB2 at specific Ser/Thr residues to inhibit SHP2 and GAB2 interaction, ultimately suppressing RAS/MAPK signaling (Zhang et al., 2013).

4.3.3 Pancreatic ductal adenocarcinoma (PDAC)

Activation of *KRAS* by missense mutation at codon G12 is broadly observed in PDAC. This mutation impairs GTPase activity of KRAS and interrupts binding between KRAS and RAS GAPs, resulting in hyperactivation of KRAS to drive PDAC (Eser, Schnieke, Schneider, & Saur, 2014). In 2016, a genome-wide association study identified germline variation of a long intergenic noncoding RNA (lincRNA), LINC00673, is associated with PDAC risk. LINC00673 is a potential tumor suppressor because of it's ability to reinforce the interaction of SHP2 with an E3 ubiquitin ligase to promote *PTPN11* degradation through ubiquitination. In PDAC, lower LINC00673 levels were found compared with adjacent normal tissue, along with higher ERK oncogenic signaling (Zheng et al., 2016). Using immunohistochemistry another study also reported higher expression levels of SHP2 in over half of PDAC tissues (44/79). Moreover, PDAC patients with higher cytoplasmic expression levels of SHP2 experienced shorter overall survival as compared with the cases with lower SHP2 expression (Zheng et al., 2016). SHP2 was also reported as a required component in KRAS-driven PDAC murine model. Genetic deletion of *Ptpn11* significantly inhibited pancreatic carcinogenesis and sensitized tumors to MEK inhibition (Ruess et al., 2018). These studies suggest that the expression level of SHP2 might be a prognostic marker and potential target in PDAC.

4.3.4 Colorectal cancer (CRC) and gastric cancer

Sporadic CRC frequently involves alterations in adenomatous polyposis coli (APC), KRAS, and p53, ultimately driving dysplastic adenoma formation and colorectal cancer (Smith et al., 2002). Gain-of-function mutations in *PTPN11* at the E76 residue which is located within PTP domain also have been reported in certain sporadic CRC cases by DNA sequencing (Bentires-Alj et al., 2004). In mouse models, although expression of an activated SHP2 E76K in intestinal epithelial cells (IEC) was not sufficient to induce tumorigenesis, it promoted tumor growth markedly on the $Apc^{Min/+}$ (Min, multiple intestinal neoplasia) background. Compared to paired adjacent normal tissue, greater transcript levels of SHP2 were observed in colorectal adenomas and stage 1 CRC (Gagne-Sansfacon et al., 2016). In addition to CRC, the expression levels of SHP2 were also found to be higher in gastric cancer and correlated with differentiation, clinical classification and metastasis (Dong et al., 2012; Zhang, Zhang, & Niu, 2015). Cytotoxin-associated antigen A (CagA)-positive *Helicobacter pylori* infection is associated with gastric carcinoma. By the bacterial type IV secretion system, CagA is secreted into gastric epithelial cells to form a complex with SHP2 and facilitate SHP2/ERK signal transduction as well as induce a growth factor-like response in gastric epithelial cells. Conversely, disruption of the CagA/SHP2 complex abolished the CagA-mediated hummingbird phenotype which is characterized by elongation and spreading of AGS cells (Hatakeyama, 2004; Higashi et al., 2002; Lee et al., 2010). Moreover, SHP2 was involved in an adaptive response to MEK inhibition in *KRAS*-amplified gastric cancer. Both genetical and pharmacological inhibition of SHP2 regained sensitivity to MEK inhibition in *KRAS*-amplified models (Wong et al., 2018). These results suggest that SHP2 could be a potential target in CRC and gastric cancer.

4.3.5 Glioblastoma (GBM)

GBM is the most malignant and lethal primary brain tumor in adults. Comprehensive genomic characterization showed genetic alteration in three crucial signaling pathways including RTK/RAS cascades. For instance, recurrent oncogenic alterations in PDGFR, which is directly modulated by SHP2 at Y771, were identified. Moreover, pharmacological inhibition of SHP2 suppressed PDGFRα-stimulated soft agar growth of *INK4A/ARF*-deficient glioma cells and cellular transformation. These data indicate that SHP2 plays a role as a mediator of oncogenic PDGFRα signaling in glioma formation (Ekman, Kallin, Engstrom, Heldin, & Ronnstrand, 2002; Liu et al., 2011; Tannous & Badr, 2019; Verhaak et al., 2010).

4.3.6 Hepatocellular carcinoma (HCC)

The molecular mechanisms underlying the development of HCC are complicated and heterogeneous. Intriguingly, SHP2 was reported to play a tumor-inhibitory role in liver cancer. Hepatocyte-specific ablation of SHP2 triggered hepatocellular adenoma (HCA) in aged mice and enhanced diethylenenitrite (DEN)-induced HCC development (Bard-Chapeau et al., 2011). Also, genetic deletion both of SHP2 and PTEN in hepatocytes accelerated non-alcoholic steatohepatitis (NASH) and hepatic tumorigenesis (Luo et al., 2016). However, SHP2 levels were also increased in 65.9% (394/598) of human HCCs. Overexpression of SHP2 was positively correlated with malignant clinical pathological characteristics and associated with poor prognosis of HCC patients. (Han et al., 2015). By single-strand conformation polymorphism (SSCP) analysis, the SHP2 mutation at T507K was identified in part of case of liver cancer (Miyamoto et al., 2008). Furthermore, SHP2 was required in mouse hepatic tumor models which were induced by both human MET (hMET) and PIK3CA or oncogenic β-catenin. SHP2 promoted RTK signaling and SHP2 deletion resulted in downregulation of proliferative signals (Liu et al., 2018). Although FDA-approved, the efficacy of the multi-kinase inhibitor sorafenib in HCC is limited by adaptive resistance. Recent studies showed SHP2 was markedly increased in sorafenib-resistant cell lines and patient-derived xenografts. The SHP2 inhibitor SHP099 in combination with sorafenib suppressed adaptive resistance by prohibiting RTK-induced RAS/MAPK and AKT signaling reactivation (Leung et al., 2020). Since the dual opposite roles of SHP2 in liver cancer and the molecular mechanisms are still obscure, whether SHP2 is a compelling potential therapeutic target requires further investigation.

5. SHP2 inhibitors as potential therapies in RAS-driven cancers

5.1 Traditional SHP2-PTP inhibitors

SHP2 is considered a compelling target in several cancers due to its critical roles in promoting RAS signaling. It has also been shown that SHP2 can contribute to both intrinsic and acquired resistance to current targeted therapies, and regulate T-cell activation by being a mediator of immune checkpoint receptor signaling involving PD-1 and B- and T-lymphocyte attenuator (BTLA) (Gavrieli, Watanabe, Loftin, Murphy, & Murphy, 2003; Hui et al., 2017; Okazaki, Chikuma, Iwai, Fagarasan, & Honjo, 2013; Prahallad et al., 2015). In the light of the highly successful general effort of targeting the ATP-binding pocket in kinases, this active-site oriented standard approach was used initially to develop SHP2 inhibitors. These traditional inhibitors bind to the PTP domain and prevent access of the substrates containing pTyr to the catalytic sites, thereby inhibiting the phosphatase activity of SHP2. However, the conserved active sites within phosphatase families such as SHP1, SHP2 and PTP1B precluded the design the specific inhibitors (Kohn, 2020). Furthermore, the pTyr mimetic is contained in most of PTP inhibitors and known to possess poor bioavailability (Yuan, Bu, Zhou, Yang, & Zhang, 2020). Furthermore, off-target effects of these small molecules against other targets have been reported (Tsutsumi, Ran, & Neel, 2018).

By screening the diversity set chemical library from the US National Cancer Institute (NCI), a sulfonic acid-based compound, NSC-87877, was identified as a potent SHP2 inhibitor (IC_{50} = 0.32 μM) (Chen et al., 2006). It inhibited the activity of PTP and RAS/RAF/ERK1/2 signaling in tumor cells. Unfortunately, the inhibitory effect on EGF-stimulated ERK activation in non-tumorous HEK293T cells could not be reproduced. NSC-87877 also showed a similar potency against SHP1 which is a negative regulator of cytokine signaling. These results argue that NSC-87877 might not be an ineffective small molecule to selectively interfere with SHP2 functions (Tsutsumi et al., 2018). In addition to NSC-87877, other active site SHP2 inhibitors identified include IIB-08, 11a-1, and GS-493. Each also showed off-target issues. In MEFs, suppression of PDGFRβ phosphorylation by these inhibitors was found to be independent of SHP2 (Tsutsumi et al., 2018). Moreover, many of these small molecule SHP2 active site inhibitors contain negatively charged carboxyl groups that conferred poor cell membrane permeability (Song et al., 2021). Because of these various challenges clinical development of SHP2 active site inhibitors stalled.

5.2 Allosteric SHP2 inhibitors

As an alternative to the development of active-site SHP2 inhibitors, allosteric small molecules inhibitors have been developed over the last several years. These compounds provide selective inhibition of SHP2 by stabilizing the basal auto-inhibited conformation of SHP2 (Fig. 3).

5.2.1 [SHP099]

In 2016, a group from Novartis introduced a new concept to inhibit SHP2 through allosteric small compounds (Garcia Fortanet et al., 2016). SHP099 is the first allosteric SHP2 inhibitor which was identified as a highly potent (IC_{50} = 71 nM), selective and orally bioavailable small molecule. This compound contains a pyrazine ring and binds to the central tunnel formed by the confluence of three domains of the auto-inhibited conformation of SHP2. Through hydrogen bonds, SHP099 provides main interactions with the linker between N-SH2 and C-SH2 domains at Arg111 and Phe113, as well as the PTP domain at Glu250. Unlike the sulfonic acid-based compound NSC-87877, SHP099 showed no substantial interaction with SHP1. The significantly larger central tunnel in SHP1 changes the orientation of C-SH2 domain and repositions the interaction with SHP099 through the linker between two SH2 domains at Arg109. This binding difference probably underlies the selectivity of SHP099 for SHP2 over SHP1. The efficacy of SHP099 further showed attenuating the proliferation of RTK-driven human cancers in both in vitro and in vivo by suppressing

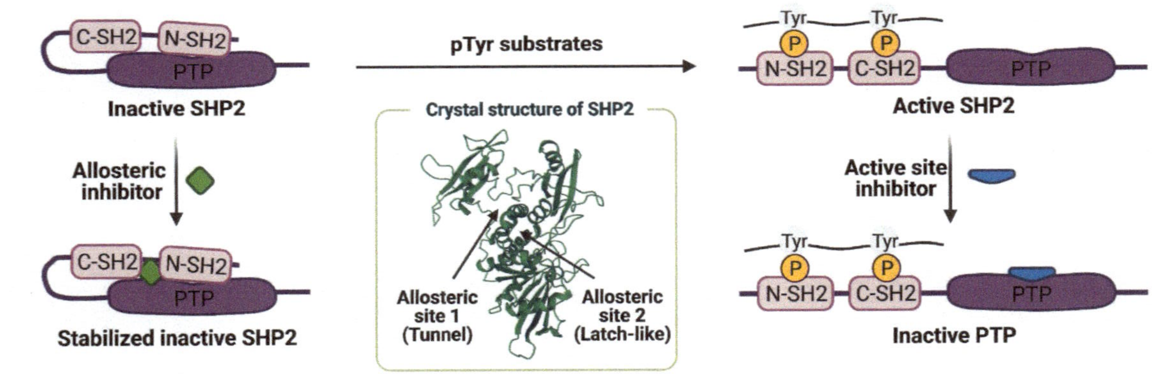

B

	Allosteric site 1			Allosteric site 2
SHP2 allosteric compound	SHP099	RMC-4550	IACS-13909	SHP244
Structure	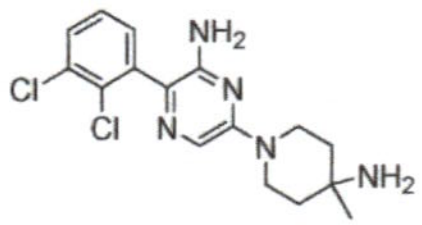	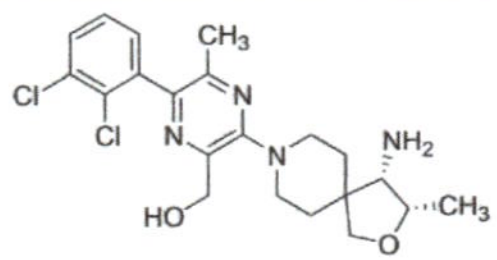	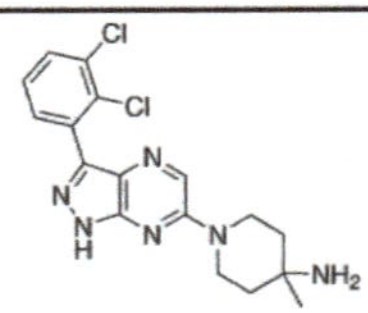	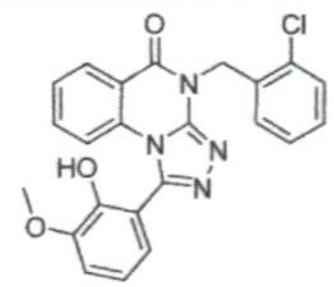
Clinical derivative	TNO155	RMC-4630	BBP-398	
NCT identifier	NCT03114319 NCT04000529 NCT04330664 NCT04294160	NCT03634982 NCT04185883 NCT03989115 NCT04418661	NCT04528836	

Fig. 3 Allosteric inhibitors of SHP2. (A) By binding to allosteric sites, the small molecules (green, left) lock SHP2 in the stabilized inactive conformation and inhibit PTP activity. Two allosteric sites are shown in the crystal structure of SHP2. The SHP2 inhibitors which are in clinical trials (TNO155, RMC4630 and BBP-398) are designed to bind allosteric site 1 (tunnel site). Traditional SHP2 inhibitors (blue, right) are designed to bind to PTP active site and inactivate phosphatase activity. The crystal structure is analyzed by Protein Data Bank (PDB) according to the CDS sequence from NCBI (NM_002834.5). (B) Shown is the table of SHP2 allosteric compounds.

RAS/MAPK signaling (Chen et al., 2016). Moreover, SHP099 was shown to prevent MEK inhibitor-induced adaptive resistance and anaplastic lymphoma kinase (ALK) inhibitor drug resistance. These data suggest that SHP099 holds promise to be an approach as a combination therapy together with MEK or ALK inhibitors in *KRAS*- or *ALK*-driven cancers (Dardaei et al., 2018; Lu et al., 2019; Ruess et al., 2018; Torres-Ayuso & Brognard, 2018).

5.2.2 [RMC-4550]

On the basis of SHP099, RMC-4550 is the representative SHP2 inhibitor from another series of compounds described by Revolution Medicines. As a small molecule SHP2 allosteric inhibitor, RMC-4550 showed a high inhibitory potency ($IC_{50} = 0.58\,nM$) and selectivity for SHP2 compared to 14 other phosphatases including SHP1 and PTP1B ($IC_{50} > 10\,\mu M$) (Nichols et al., 2018; Yuan et al., 2020). Inhibition of SHP2 by RMC-4550 suppressed tumor growth and RAS/MAPK signaling in cellular and xenograft cancer models bearing $KRAS^{G12C}$, the class 3 BRAF mutation or NF1 loss-of-function driver alterations. A similar efficacy was observed in patient-derived xenograft (PDX) models (Nichols et al., 2018). Interestingly, RMC-4550 showed no inhibitory effect on tumor growth in CT26 tumor–bearing RAG2–deficient mice as well as immunocompetent mice depleted of CD4+ and CD8+ T cells. These findings suggest that certain immune effector cells may be important for RMC-4550–mediated antitumor efficacy. Moreover, RMC-4550 contributed to improve CD8+ T cell infiltration in tumors, increase pro-inflammatory M1-like macrophages, and inhibited CSF-1 signaling in monocytes to enhance antitumor immunity in PD-1 sensitive tumors (Quintana et al., 2020; Yuan et al., 2020).

5.2.3 [IACS-13909]

In 2020, another allosteric SHP2 inhibitor IACS-13909 was identified by the University of Texas MD Anderson Cancer Center and Navire Pharma, Inc. IACS-13909 potently suppressed SHP2 ($IC_{50} = 15.7\,nM$) and showed high selectivity. This small molecule showed antitumor activity and suppressed the MAPK pathway in RTK-dependent cancers, both in vitro and in vivo. The long-term effectiveness of osimertinib is limited by acquired resistance which usually caused by EGFR mutations or activation of the MAPK pathway through alternative RTK signaling. As a single agent or in combination with osimertinib, preclinical results showed inhibition of SHP2 via IACS-13909 was efficient to suppress tumor cell proliferation and tumor regression in EGFR-mutant osimertinib-resistant NSCLC models (Sun et al., 2020).

5.2.4 [SHP244 and its derivatization]

Beyond the tunnel site, a second small molecule allosteric site at the interface of the N-SH2 and PTP domains was explored. In 2018, LaMarche et al. discovered compound SHP244 which was identified as a weak inhibitor against SHP2 ($SHP2^{1-525}$ $IC_{50} = 60\,\mu M$) by binding to the latch-like pocket. Structure-based design was used to derive optimized small molecules from SHP244 such as SHP844 ($SHP2^{1-525}$ $IC_{50} = 18.9\,\mu M$) and SHP504 ($SHP2^{1-525}$ $IC_{50} = 21\,\mu M$). These compounds provide an increase in SHP2 thermal stabilization and inhibitory activity. Dual specificity phosphatase 6 (DUSP6), a downstream MAPK pathway signaling biomarker was downregulated in KYSE-520 cancer cells upon SHP2 inhibitor treatment. Furthermore, this secondary allosteric binding site permits a dual inhibitory strategy by treating with two SHP2 inhibitors simultaneously. Inhibition of both of the allosteric regions showed enhanced pharmacological pathway inhibition (Fodor et al., 2018).

5.2.5 [TNO155]

TNO155 is an allosteric SHP2 inhibitor which was developed by Novartis from the initial compound SHP099 (biochemical $IC_{50} = 0.071\,\mu M$, p-ERK $IC_{50} = 0.250\,\mu M$, anti-proliferation $IC_{50} = 2.5\,\mu M$) and was the first compound to enter clinical studies. TNO155 has retention of biochemical potency ($IC_{50} = 0.011\,\mu M$) and improved cellular potency (p-ERK $IC_{50} = 0.011\,\mu M$, anti-proliferation $IC_{50} = 0.100\,\mu M$) compared to SHP099. Notably, unlike SHP099, TNO155 shows no measurable hERG activity ($IC_{50} > 30\,\mu M$) and was completely selective over phosphatases and kinases based on comprehensive screening assays (LaMarche et al., 2020). TNO155 has entered clinical trials for safety and tolerability studies. TNO155 alone or in combination with EGF816 (Nazartinib) is under phase I trial and in advanced EGFR-mutant NSCLC and $KRAS^{G12}$ mutant NSCLC, Esophageal Squamous Cell Cancer (SCC), Head/Neck SCC (HNSCC) and Melanoma (ClinicalTrials.gov identifier: NCT03114319). In parallel, a phase I/II trial of MRTX849 in combination with TNO155 is being tested in advanced solid tumors with the $KRAS^{G12C}$ mutation (ClinicalTrials.gov Identifier: NCT04330664). In $KRAS^{G12C}$ cancer cells, $KRAS^{G12C}$ inhibitors frequently induce feedback activation of wild-type

KRAS or other RAS isoforms. TNO155 was found to effectively block this feedback regulation and enhance the efficacy of $KRAS^{G12C}$ inhibitors in lung and colorectal cancer cells. Moreover, cyclin D is a downstream target gene of the MAPK pathway and it enables CDK4/6 to promote cell cycle progression. One of the CDK4/6 inhibitors, Ribociclib, is considered as a potential agent for poly-therapy with TNO155. The combination benefit from TNO155 and Ribociclib was shown in HCC366 ($KRAS^{WT}$), A-427 ($KRAS^{G12D/+}$), and NCI-H747 ($KRAS^{G13D/+}$) cancer cells in vitro and KRAS-mutated lung and colorectal cancer PDX models (Liu et al., 2021). This combination strategy is currently in phase Ib clinical testing (ClinicalTrials.gov Identifier: NCT04000529).

5.2.6 [RMC-4630]

RMC-4630 (clinical derivative of RMC-4550) was identified as a potent and selective inhibitor of SHP2 by Revolution Medicines/Sanofi. By daily administration of RMC-4630, dose-dependent tumor growth inhibition was observed in PDX models ($KRAS^{G12C}$ and class 3 BRAF-mutated NSCLC) (Nichols et al., 2018). In an ongoing phase I dose-escalation clinical trial, RMC-4630 monotherapy shows anti-tumor activity in participants with advanced relapsed or refractory solid tumors (Koczywas et al., 2021). Further interim studies on a subset of patients in the phase 1 trial reveal clinical immune-oncology effect of RMC-4630. Decrease of M2 macrophage and PDL-1 were observed in tumor from a patient who has previous response to anti-PD1 regimen (PR) and then progressed. In addition to tumor, reduction of monocytes and monocytic myeloid-derived suppressor cells (mMDSCs) were found in blood during RMC-4630 administration. These results showed the potential ability of RMC-4630 to modulate innate and adaptive anti-tumor immunity in patients (Chen et al., 2021). This compound is being investigated against class 3 BRAF mutant, $NF1^{LOF}$ as well as mutated RAS-driven NSCLC including $KRAS^{G12X}$ and KRAS amplification (ClinicalTrials.gov Identifier: NCT03634982). The analysis of pERK in peripheral blood cells also showed RMC-4630 inhibits RAS signaling in paired pre-treatment and post-treatment tumor biopsies.

AMG 510 (Sotorasib) is the first covalent $KRAS^{G12C}$ inhibitor to enter clinical trials (Amgen) for $KRAS^{G12C}$-driven cancers. While monotherapy activity was observed in NSCLC patients, innate and acquired resistance and heterogeneous responses have been observed in other cancer types and in NSCLC as well due to feedback activation of RTK/RAS and aurora kinase signaling (Xue et al., 2020). To address this, AMG 510 and RMC-4630 are being tested in a phase I clinical trial in advanced solid tumors with $KRAS^{G12C}$ mutation (ClinicalTrials.gov Identifier: NCT04185883). Another approach to treat oncogenic RAS-driven cancers is using a MEK inhibitor to interrupt RAS/MAPK pathway signaling output. Unfortunately, single-agent MEK inhibitor efficacy has been disappointing since it often induces rapid resistance through upstream activation of RTKs (Caunt, Sale, Smith, & Cook, 2015). A low-dose combination of RMC-4630 (10 mg/kg) with the MEK inhibitor Cobimetinib (2.5 mg/kg) showed enhanced efficacy in $KRAS^{G12C}$ xenograft models and prevented tumor progression (Nichols et al., 2018). In the sixth AACR-IASLC international joint conference in January 2020, initial findings from a phase Ib/II clinical trial testing RMC-4630 combined with Cobimetinib (NCT03989115) showed anti-tumor activity in KRAS-mutant cancer. In this study, intermittent RMC-4630 plus daily or intermittent Cobimetinib showed an acceptable tolerability in patients with tumors harboring RAS pathway alterations ($KRAS^{G12X}$, $KRAS^{amp}$, $NF1^{LOF}$ and Class 3 BRAF mutation) (Ou et al., 2020). The collective preclinical data and these early clinical findings support the continued combination testing of current targeted therapies together with allosteric SHP2 inhibitors in RAS-driven cancers.

6. Conclusions and perspectives

Targeting oncogenic RAS or downstream signaling proteins such as MEK has received much effort in RAS-driven cancers. Encouraged by emerging studies, allosteric inhibition of SHP2 provides a promising area for drug development to treat RAS-driven cancers. Six allosteric SHP2 inhibitors (TNO155 (Novartis), RMC-4630 (Revolution Medicines/Sanofi), BBP-398 (Navire Pharma Inc.), JAB-3068 (Jacobio), JAB-3312 (Jacobio) and RLY-1971 (Relay Therapeutics)) are currently being developed and have entered in clinical trials for various solid tumors. Notably, both TNO155 and RMC-4630 are in phase I trials as either a single agent or in combination with KRAS inhibitors or MEK inhibitor treatment for advanced $KRAS^{G12C}$ or $KRAS^{amp}$ tumors (Fig. 4).

To move forward, the full biological functions and mechanisms of SHP2 still need to be addressed as does SHP2's role in immune system modulation. The on-target toxicity of SHP2 inhibition needs to be addressed carefully and thoughtfully via different dosing and scheduling regimens in the clinic. Moreover, the mechanisms underlying resistance to allosteric

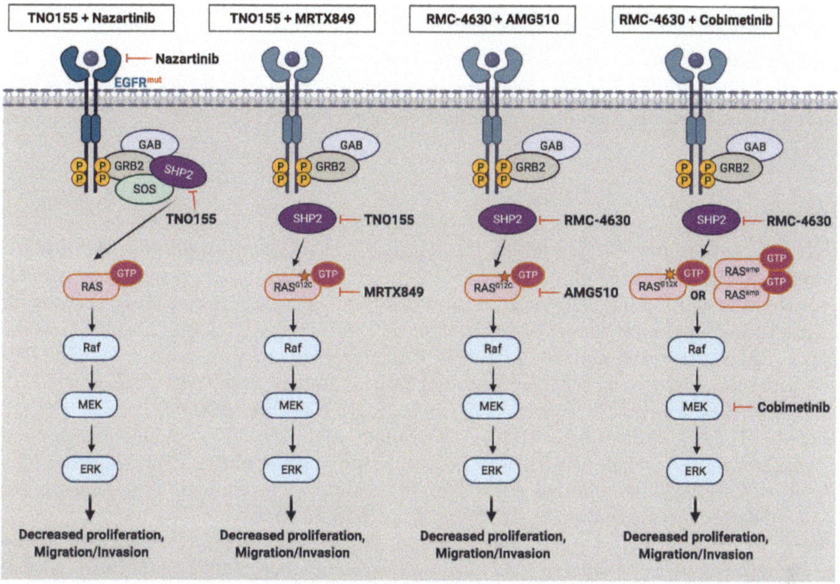

Fig. 4 Combination therapy of allosteric SHP2 inhibitors in RAS-driven cancers. Shown is a schematic model of allosteric SHP2 inhibitors combined with other clinical targeted therapies. These applications are currently under evaluation in various clinical trials. TNO155+Nazartinib: NCT03114319, TNO155+MRTX849: NCT04330664, RMC-4630 +AMG510: NCT04185883, RMC-4630+Cobimetinib: NCT03989115

SHP2 inhibitors should attract much attention. Although combinatorial inhibition of MEK and SHP2 provides an opportunity for RAS^{G12X} tumors, expression of RAS^{G13D} and RAS^{Q61X} and certain RTKs can activate RAS through SHP2-independent signaling. Adaptive resistance may also involve RTKs such as fibroblast growth factor receptor (FGFR) which could drive feedback activation of RAS through a SHP2-independent mechanism (Ahmed et al., 2019; Lu et al., 2020). Understanding how tumors evade SHP2 inhibition will be critical for the continued successful clinical development of allosteric SHP2 inhibitor for the treatment of RAS-driven cancers.

Acknowledgments

Figure illustrations were created with BioRender.com. Crystal structure depicted in the figure was analyzed through Protein Data Bank (PDB) which is supported by the corresponding structure publication "Toward a common library and tools for web molecular graphics" (Sehnal, Rose, Koca, Burley, & Velankar, 2018). Funding is acknowledged from the National Institutes of Health (R01CA231300, U54CA224081, R01CA204302, R01CA211052 and R01CA169338 to T.G.B,) and the Pew and Stewart Foundations (to T.G.B).

Potential competing interests

T.G.B. is an advisor to Array Biopharma/Pfizer, Revolution Medicines, Relay, Rain, Novartis, Astrazeneca, Takeda, Springworks, Jazz Pharmaceuticals, and receives research funding from Novartis, Revolution Medicines, Strategia, and Kinnate.

References

Aceto, N., Sausgruber, N., Brinkhaus, H., Gaidatzis, D., Martiny-Baron, G., Mazzarol, G., et al. (2012). Tyrosine phosphatase SHP2 promotes breast cancer progression and maintains tumor-initiating cells via activation of key transcription factors and a positive feedback signaling loop. *Nature Medicine*, *18*(4), 529–537. https://doi.org/10.1038/nm.2645.

Agazie, Y. M., & Hayman, M. J. (2003). Molecular mechanism for a role of SHP2 in epidermal growth factor receptor signaling. *Molecular and Cellular Biology*, *23*(21), 7875–7886. https://doi.org/10.1128/mcb.23.21.7875-7886.2003.

Ahmed, T. A., Adamopoulos, C., Karoulia, Z., Wu, X., Sachidanandam, R., Aaronson, S. A., et al. (2019). SHP2 drives adaptive resistance to ERK signaling inhibition in molecularly defined subsets of ERK-dependent tumors. *Cell Reports*, *26*(1), 65–78.e65. https://doi.org/10.1016/j.celrep.2018.12.013.

Alonso, A., Nunes-Xavier, C. E., Bayon, Y., & Pulido, R. (2016). The extended family of protein tyrosine phosphatases. *Methods in Molecular Biology*, *1447*, 1–23. https://doi.org/10.1007/978-1-4939-3746-2_1.

Bard-Chapeau, E. A., Li, S., Ding, J., Zhang, S. S., Zhu, H. H., Princen, F., et al. (2011). Ptpn11/Shp2 acts as a tumor suppressor in hepatocellular carcinogenesis. *Cancer Cell*, *19*(5), 629–639. https://doi.org/10.1016/j.ccr.2011.03.023.

Barford, D., & Neel, B. G. (1998). Revealing mechanisms for SH2 domain mediated regulation of the protein tyrosine phosphatase SHP-2. *Structure*, *6*(3), 249–254. https://doi.org/10.1016/s0969-2126(98)00027-6.

Belanger, L. F., Roy, S., Tremblay, M., Brott, B., Steff, A. M., Mourad, W., et al. (2003). Mek2 is dispensable for mouse growth and development. *Molecular and Cellular Biology*, *23*(14), 4778–4787. https://doi.org/10.1128/mcb.23.14.4778-4787.2003.

Bentires-Alj, M., Paez, J. G., David, F. S., Keilhack, H., Halmos, B., Naoki, K., et al. (2004). Activating mutations of the Noonan syndrome-associated SHP2/PTPN11 gene in human solid tumors and adult acute myelogenous leukemia. *Cancer Research*, *64*(24), 8816–8820. https://doi.org/10.1158/0008-5472.CAN-04-1923.

Blasco, R. B., Francoz, S., Santamaria, D., Canamero, M., Dubus, P., Charron, J., et al. (2011). C-Raf, but not B-Raf, is essential for development of K-Ras oncogene-driven non-small cell lung carcinoma. *Cancer Cell*, *19*(5), 652–663. https://doi.org/10.1016/j.ccr.2011.04.002.

Blumenschein, G. R., Jr., Smit, E. F., Planchard, D., Kim, D. W., Cadranel, J., De Pas, T., et al. (2015). A randomized phase II study of the MEK1/MEK2 inhibitor trametinib (GSK1120212) compared with docetaxel in KRAS-mutant advanced non-small-cell lung cancer (NSCLC)dagger. *Annals of Oncology*, *26*(5), 894–901. https://doi.org/10.1093/annonc/mdv072.

Bos, J. L., Rehmann, H., & Wittinghofer, A. (2007). GEFs and GAPs: Critical elements in the control of small G proteins. *Cell*, *129*(5), 865–877. https://doi.org/10.1016/j.cell.2007.05.018.

Buhrman, G., Wink, G., & Mattos, C. (2007). Transformation efficiency of RasQ61 mutants linked to structural features of the switch regions in the presence of Raf. *Structure*, *15*(12), 1618–1629. https://doi.org/10.1016/j.str.2007.10.011.

Bunda, S., Burrell, K., Heir, P., Zeng, L., Alamsahebpour, A., Kano, Y., et al. (2015). Inhibition of SHP2-mediated dephosphorylation of Ras suppresses oncogenesis. *Nature Communications*, *6*, 8859. https://doi.org/10.1038/ncomms9859.

Bunda, S., Heir, P., Srikumar, T., Cook, J. D., Burrell, K., Kano, Y., et al. (2014). Src promotes GTPase activity of Ras via tyrosine 32 phosphorylation. *Proceedings of the National Academy of Sciences of the United States of America*, *111*(36), E3785–E3794. https://doi.org/10.1073/pnas.1406559111.

Burd, C. E., Liu, W., Huynh, M. V., Waqas, M. A., Gillahan, J. E., Clark, K. S., et al. (2014). Mutation-specific RAS oncogenicity explains NRAS codon 61 selection in melanoma. *Cancer Discovery*, *4*(12), 1418–1429. https://doi.org/10.1158/2159-8290.CD-14-0729.

Cancer Genome Atlas Research Network. (2014). Comprehensive molecular profiling of lung adenocarcinoma. *Nature*, *511*(7511), 543–550. https://doi.org/10.1038/nature13385.

Canon, J., Rex, K., Saiki, A. Y., Mohr, C., Cooke, K., Bagal, D., et al. (2019). The clinical KRAS(G12C) inhibitor AMG 510 drives anti-tumour immunity. *Nature*, *575*(7781), 217–223. https://doi.org/10.1038/s41586-019-1694-1.

Caunt, C. J., Sale, M. J., Smith, P. D., & Cook, S. J. (2015). MEK1 and MEK2 inhibitors and cancer therapy: The long and winding road. *Nature Reviews. Cancer*, *15*(10), 577–592. https://doi.org/10.1038/nrc4000.

Chan, G., Kalaitzidis, D., & Neel, B. G. (2008). The tyrosine phosphatase Shp2 (PTPN11) in cancer. *Cancer Metastasis Reviews*, *27*(2), 179–192. https://doi.org/10.1007/s10555-008-9126-y.

Chen, A., Haura, E., Pacheco, J., Koczywas, M., Gordon, M., Ulahannan, S., et al. (2021). Modulation of innate and adaptive immunity in blood and tumor of patients receiving the SHP2 inhibitor RMC-4630 [abstract]. In *American association for cancer research annual meeting; 2021 April 10–15; virtual meeting I: AACR; 2021. Abstract nr LB050.*

Chen, Y. N., LaMarche, M. J., Chan, H. M., Fekkes, P., Garcia-Fortanet, J., Acker, M. G., et al. (2016). Allosteric inhibition of SHP2 phosphatase inhibits cancers driven by receptor tyrosine kinases. *Nature*, *535*(7610), 148–152. https://doi.org/10.1038/nature18621.

Chen, L., Sung, S. S., Yip, M. L., Lawrence, H. R., Ren, Y., Guida, W. C., et al. (2006). Discovery of a novel shp2 protein tyrosine phosphatase inhibitor. *Molecular Pharmacology*, *70*(2), 562–570. https://doi.org/10.1124/mol.106.025536.

Cherfils, J., & Zeghouf, M. (2013). Regulation of small GTPases by GEFs, GAPs, and GDIs. *Physiological Reviews*, *93*(1), 269–309. https://doi.org/10.1152/physrev.00003.2012.

Cox, A. D., Fesik, S. W., Kimmelman, A. C., Luo, J., & Der, C. J. (2014). Drugging the undruggable RAS: Mission possible? *Nature Reviews. Drug Discovery*, *13*(11), 828–851. https://doi.org/10.1038/nrd4389.

Dance, M., Montagner, A., Salles, J. P., Yart, A., & Raynal, P. (2008). The molecular functions of Shp2 in the Ras/mitogen-activated protein kinase (ERK1/2) pathway. *Cellular Signalling*, *20*(3), 453–459. https://doi.org/10.1016/j.cellsig.2007.10.002.

Dardaei, L., Wang, H. Q., Singh, M., Fordjour, P., Shaw, K. X., Yoda, S., et al. (2018). SHP2 inhibition restores sensitivity in ALK-rearranged non-small-cell lung cancer resistant to ALK inhibitors. *Nature Medicine*, *24*(4), 512–517. https://doi.org/10.1038/nm.4497.

Der, C. J., Krontiris, T. G., & Cooper, G. M. (1982). Transforming genes of human bladder and lung carcinoma cell lines are homologous to the ras genes of Harvey and Kirsten sarcoma viruses. *Proceedings of the National Academy of Sciences of the United States of America*, *79*(11), 3637–3640. https://doi.org/10.1073/pnas.79.11.3637.

Dong, S., Li, F. Q., Zhang, Q., Lv, K. Z., Yang, H. L., Gao, Y., et al. (2012). Expression and clinical significance of SHP2 in gastric cancer. *The Journal of International Medical Research*, *40*(6), 2083–2089. https://doi.org/10.1177/030006051204000605.

Drosten, M., & Barbacid, M. (2020). Targeting the MAPK pathway in KRAS-driven tumors. *Cancer Cell, 37*(4), 543–550. https://doi.org/10.1016/j.ccell.2020.03.013.

Du, Z., & Lovly, C. M. (2018). Mechanisms of receptor tyrosine kinase activation in cancer. *Molecular Cancer, 17*(1), 58. https://doi.org/10.1186/s12943-018-0782-4.

Ekman, S., Kallin, A., Engstrom, U., Heldin, C. H., & Ronnstrand, L. (2002). SHP-2 is involved in heterodimer specific loss of phosphorylation of Tyr771 in the PDGF beta-receptor. *Oncogene, 21*(12), 1870–1875. https://doi.org/10.1038/sj.onc.1205210.

Engelman, J. A., Chen, L., Tan, X., Crosby, K., Guimaraes, A. R., Upadhyay, R., et al. (2008). Effective use of PI3K and MEK inhibitors to treat mutant Kras G12D and PIK3CA H1047R murine lung cancers. *Nature Medicine, 14*(12), 1351–1356. https://doi.org/10.1038/nm.1890.

Eser, S., Schnieke, A., Schneider, G., & Saur, D. (2014). Oncogenic KRAS signalling in pancreatic cancer. *British Journal of Cancer, 111*(5), 817–822. https://doi.org/10.1038/bjc.2014.215.

Feng, G. S. (1999). Shp-2 tyrosine phosphatase: Signaling one cell or many. *Experimental Cell Research, 253*(1), 47–54. https://doi.org/10.1006/excr.1999.4668.

Fodor, M., Price, E., Wang, P., Lu, H., Argintaru, A., Chen, Z., et al. (2018). Dual allosteric inhibition of SHP2 phosphatase. *ACS Chemical Biology, 13*(3), 647–656. https://doi.org/10.1021/acschembio.7b00980.

Frankson, R., Yu, Z. H., Bai, Y., Li, Q., Zhang, R. Y., & Zhang, Z. Y. (2017). Therapeutic targeting of oncogenic tyrosine phosphatases. *Cancer Research*, 77(21), 5701–5705. https://doi.org/10.1158/0008-5472.CAN-17-1510.

Gagne-Sansfacon, J., Coulombe, G., Langlois, M. J., Langlois, A., Paquet, M., Carrier, J., et al. (2016). SHP-2 phosphatase contributes to KRAS-driven intestinal oncogenesis but prevents colitis-associated cancer development. *Oncotarget*, 7(40), 65676–65695. https://doi.org/10.18632/oncotarget.11601.

Garcia Fortanet, J., Chen, C. H., Chen, Y. N., Chen, Z., Deng, Z., Firestone, B., et al. (2016). Allosteric inhibition of SHP2: Identification of a potent, selective, and orally efficacious phosphatase inhibitor. *Journal of Medicinal Chemistry, 59*(17), 7773–7782. https://doi.org/10.1021/acs.jmedchem.6b00680.

Gavrieli, M., Watanabe, N., Loftin, S. K., Murphy, T. L., & Murphy, K. M. (2003). Characterization of phosphotyrosine binding motifs in the cytoplasmic domain of B and T lymphocyte attenuator required for association with protein tyrosine phosphatases SHP-1 and SHP-2. *Biochemical and Biophysical Research Communications, 312*(4), 1236–1243. https://doi.org/10.1016/j.bbrc.2003.11.070.

Gu, S., Sayad, A., Chan, G., Yang, W., Lu, Z., Virtanen, C., et al. (2018). SHP2 is required for BCR-ABL1-induced hematologic neoplasia. *Leukemia, 32*(1), 203–213. https://doi.org/10.1038/leu.2017.250.

Guan, K. L., Broyles, S. S., & Dixon, J. E. (1991). A Tyr/Ser protein phosphatase encoded by vaccinia virus. *Nature, 350*(6316), 359–362. https://doi.org/10.1038/350359a0.

Hallin, J., Engstrom, L. D., Hargis, L., Calinisan, A., Aranda, R., Briere, D. M., et al. (2020). The KRAS(G12C) inhibitor MRTX849 provides insight toward therapeutic susceptibility of KRAS-mutant cancers in mouse models and patients. *Cancer Discovery, 10*(1), 54–71. https://doi.org/10.1158/2159-8290.CD-19-1167.

Han, T., Xiang, D. M., Sun, W., Liu, N., Sun, H. L., Wen, W., et al. (2015). PTPN11/Shp2 overexpression enhances liver cancer progression and predicts poor prognosis of patients. *Journal of Hepatology, 63*(3), 651–660. https://doi.org/10.1016/j.jhep.2015.03.036.

Hanafusa, H., Torii, S., Yasunaga, T., & Nishida, E. (2002). Sprouty1 and Sprouty2 provide a control mechanism for the Ras/MAPK signalling pathway. *Nature Cell Biology, 4*(11), 850–858. https://doi.org/10.1038/ncb867.

Hao, H. X., Wang, H., Liu, C., Kovats, S., Velazquez, R., Lu, H., et al. (2019). Tumor intrinsic efficacy by SHP2 and RTK inhibitors in KRAS-mutant cancers. *Molecular Cancer Therapeutics*, *18*(12), 2368–2380. https://doi.org/10.1158/1535-7163.MCT-19-0170.

Hatakeyama, M. (2004). Oncogenic mechanisms of the helicobacter pylori CagA protein. *Nature Reviews. Cancer*, *4*(9), 688–694. https://doi.org/10.1038/nrc1433.

Haura, E. B., Ricart, A. D., Larson, T. G., Stella, P. J., Bazhenova, L., Miller, V. A., et al. (2010). A phase II study of PD-0325901, an oral MEK inhibitor, in previously treated patients with advanced non-small cell lung cancer. *Clinical Cancer Research*, *16*(8), 2450–2457. https://doi.org/10.1158/1078-0432.CCR-09-1920.

Higashi, H., Tsutsumi, R., Muto, S., Sugiyama, T., Azuma, T., Asaka, M., et al. (2002). SHP-2 tyrosine phosphatase as an intracellular target of helicobacter pylori CagA protein. *Science*, *295*(5555), 683–686. https://doi.org/10.1126/science.1067147.

Hobbs, G. A., Der, C. J., & Rossman, K. L. (2016). RAS isoforms and mutations in cancer at a glance. *Journal of Cell Science*, *129*(7), 1287–1292. https://doi.org/10.1242/jcs.182873.

Huang, W. Q., Lin, Q., Zhuang, X., Cai, L. L., Ruan, R. S., Lu, Z. X., et al. (2014). Structure, function, and pathogenesis of SHP2 in developmental disorders and tumorigenesis. *Current Cancer Drug Targets*, *14*(6), 567–588. https://doi.org/10.2174/1568009614666140717105001.

Hubbard, S. R. (1999). Structural analysis of receptor tyrosine kinases. *Progress in Biophysics and Molecular Biology*, *71*(3–4), 343–358. https://doi.org/10.1016/s0079-6107(98)00047-9.

Hui, E., Cheung, J., Zhu, J., Su, X., Taylor, M. J., Wallweber, H. A., et al. (2017). T cell costimulatory receptor CD28 is a primary target for PD-1-mediated inhibition. *Science*, *355*(6332), 1428–1433. https://doi.org/10.1126/science.aaf1292.

Hunter, J. C., Manandhar, A., Carrasco, M. A., Gurbani, D., Gondi, S., & Westover, K. D. (2015). Biochemical and structural analysis of common cancer-associated KRAS mutations. *Molecular Cancer Research*, *13*(9), 1325–1335. https://doi.org/10.1158/1541-7786.MCR-15-0203.

Janne, P. A., van den Heuvel, M. M., Barlesi, F., Cobo, M., Mazieres, J., Crino, L., et al. (2017). Selumetinib plus docetaxel compared with docetaxel alone and progression-free survival in patients with KRAS-mutant advanced non-small cell lung cancer: The SELECT-1 randomized clinical trial. *JAMA*, *317*(18), 1844–1853. https://doi.org/10.1001/jama.2017.3438.

Kerr, D. L., Haderk, F., & Bivona, T. G. (2021). Allosteric SHP2 inhibitors in cancer: Targeting the intersection of RAS, resistance, and the immune microenvironment. *Current Opinion in Chemical Biology*, *62*, 1–12. https://doi.org/10.1016/j.cbpa.2020.11.007.

Koczywas, M., Haure, E., Janne, P., Pacheco, J., Ulahannan, S., Wang, J. S., et al. (2021). Anti-tumor activity and tolerability of the SHP2 inhibitor RMC-4630 as a single agent in patients with RAS-addicted solid cancers. In *American association for cancer research annual meeting; 2021 April 10–15; virtual meeting I: AACR; 2021. Presentation nr LB001.*

Kohn, M. (2020). Turn and face the strange: A new view on phosphatases. *ACS Central Science*, *6*(4), 467–477. https://doi.org/10.1021/acscentsci.9b00909.

LaMarche, M. J., Acker, M., Argintaru, A., Bauer, D., Boisclair, J., Chan, H., et al. (2020). Identification of TNO155, an allosteric SHP2 inhibitor for the treatment of cancer. *Journal of Medicinal Chemistry*, *63*(22), 13578–13594. https://doi.org/10.1021/acs.jmedchem.0c01170.

Lee, I. O., Kim, J. H., Choi, Y. J., Pillinger, M. H., Kim, S. Y., Blaser, M. J., et al. (2010). Helicobacter pylori CagA phosphorylation status determines the gp130-activated

SHP2/ERK and JAK/STAT signal transduction pathways in gastric epithelial cells. *The Journal of Biological Chemistry*, *285*(21), 16042–16050. https://doi.org/10.1074/jbc.M110.111054.

Lemmon, M. A., & Schlessinger, J. (2010). Cell signaling by receptor tyrosine kinases. *Cell*, *141*(7), 1117–1134. https://doi.org/10.1016/j.cell.2010.06.011.

Leung, C. O. N., Tong, M., Chung, K. P. S., Zhou, L., Che, N., Tang, K. H., et al. (2020). Overriding adaptive resistance to sorafenib through combination therapy with Src homology 2 domain-containing phosphatase 2 blockade in hepatocellular carcinoma. *Hepatology*, *72*(1), 155–168. https://doi.org/10.1002/hep.30989.

Liu, K. W., Feng, H., Bachoo, R., Kazlauskas, A., Smith, E. M., Symes, K., et al. (2011). SHP-2/PTPN11 mediates gliomagenesis driven by PDGFRA and INK4A/ARF aberrations in mice and humans. *The Journal of Clinical Investigation*, *121*(3), 905–917. https://doi.org/10.1172/JCI43690.

Liu, J. J., Li, Y., Chen, W. S., Liang, Y., Wang, G., Zong, M., et al. (2018). Shp2 deletion in hepatocytes suppresses hepatocarcinogenesis driven by oncogenic beta-catenin, PIK3CA and MET. *Journal of Hepatology*, *69*(1), 79–88. https://doi.org/10.1016/j.jhep.2018.02.014.

Liu, C., Lu, H., Wang, H., Loo, A., Zhang, X., Yang, G., et al. (2021). Combinations with allosteric SHP2 inhibitor TNO155 to block receptor tyrosine kinase signaling. *Clinical Cancer Research*, *27*(1), 342–354. https://doi.org/10.1158/1078-0432.CCR-20-2718.

Liu, P., Wang, Y., & Li, X. (2019). Targeting the untargetable KRAS in cancer therapy. *Acta Pharmaceutica Sinica B*, *9*(5), 871–879. https://doi.org/10.1016/j.apsb.2019.03.002.

Loh, M. L., Reynolds, M. G., Vattikuti, S., Gerbing, R. B., Alonzo, T. A., Carlson, E., et al. (2004). PTPN11 mutations in pediatric patients with acute myeloid leukemia: Results from the Children's Cancer Group. *Leukemia*, *18*(11), 1831–1834. https://doi.org/10.1038/sj.leu.2403492.

Loh, M. L., Vattikuti, S., Schubbert, S., Reynolds, M. G., Carlson, E., Lieuw, K. H., et al. (2004). Mutations in PTPN11 implicate the SHP-2 phosphatase in leukemogenesis. *Blood*, *103*(6), 2325–2331. https://doi.org/10.1182/blood-2003-09-3287.

Lu, H., Liu, C., Huynh, H., Le, T. B. U., LaMarche, M. J., Mohseni, M., et al. (2020). Resistance to allosteric SHP2 inhibition in FGFR-driven cancers through rapid feedback activation of FGFR. *Oncotarget*, *11*(3), 265–281. https://doi.org/10.18632/oncotarget.27435.

Lu, H., Liu, C., Velazquez, R., Wang, H., Dunkl, L. M., Kazic-Legueux, M., et al. (2019). SHP2 inhibition overcomes RTK-mediated pathway reactivation in KRAS-mutant tumors treated with MEK inhibitors. *Molecular Cancer Therapeutics*, *18*(7), 1323–1334. https://doi.org/10.1158/1535-7163.MCT-18-0852.

Luo, X., Liao, R., Hanley, K. L., Zhu, H. H., Malo, K. N., Hernandez, C., et al. (2016). Dual Shp2 and Pten deficiencies promote non-alcoholic steatohepatitis and genesis of liver tumor-initiating cells. *Cell Reports*, *17*(11), 2979–2993. https://doi.org/10.1016/j.celrep.2016.11.048.

Mainardi, S., Mulero-Sanchez, A., Prahallad, A., Germano, G., Bosma, A., Krimpenfort, P., et al. (2018). SHP2 is required for growth of KRAS-mutant non-small-cell lung cancer in vivo. *Nature Medicine*, *24*(7), 961–967. https://doi.org/10.1038/s41591-018-0023-9.

Marasco, M., Berteotti, A., Weyershaeuser, J., Thorausch, N., Sikorska, J., Krausze, J., et al. (2020). Molecular mechanism of SHP2 activation by PD-1 stimulation. *Science Advances*, *6*(5), eaay4458. https://doi.org/10.1126/sciadv.aay4458.

Martinelli, S., Carta, C., Flex, E., Binni, F., Cordisco, E. L., Moretti, S., et al. (2006). Activating PTPN11 mutations play a minor role in pediatric and adult solid tumors. *Cancer Genetics and Cytogenetics*, *166*(2), 124–129. https://doi.org/10.1016/j.cancergencyto.2005.10.003.

Masoumi-Moghaddam, S., Amini, A., & Morris, D. L. (2014). The developing story of Sprouty and cancer. *Cancer Metastasis Reviews*, *33*(2–3), 695–720. https://doi.org/10.1007/s10555-014-9497-1.

McCormick, F. (2020). Sticking it to KRAS: Covalent inhibitors enter the clinic. *Cancer Cell*, *37*(1), 3–4. https://doi.org/10.1016/j.ccell.2019.12.009.

McKay, M. M., & Morrison, D. K. (2007). Integrating signals from RTKs to ERK/MAPK. *Oncogene*, *26*(22), 3113–3121. https://doi.org/10.1038/sj.onc.1210394.

Miyamoto, D., Miyamoto, M., Takahashi, A., Yomogita, Y., Higashi, H., Kondo, S., et al. (2008). Isolation of a distinct class of gain-of-function SHP-2 mutants with oncogenic RAS-like transforming activity from solid tumors. *Oncogene*, *27*(25), 3508–3515. https://doi.org/10.1038/sj.onc.1211019.

Mohi, M. G., & Neel, B. G. (2007). The role of Shp2 (PTPN11) in cancer. *Current Opinion in Genetics & Development*, *17*(1), 23–30. https://doi.org/10.1016/j.gde.2006.12.011.

Montagner, A., Yart, A., Dance, M., Perret, B., Salles, J. P., & Raynal, P. (2005). A novel role for Gab1 and SHP2 in epidermal growth factor-induced Ras activation. *The Journal of Biological Chemistry*, *280*(7), 5350–5360. https://doi.org/10.1074/jbc.M410012200.

Moore, A. R., Rosenberg, S. C., McCormick, F., & Malek, S. (2020). RAS-targeted therapies: Is the undruggable drugged? *Nature Reviews. Drug Discovery*, *19*(8), 533–552. https://doi.org/10.1038/s41573-020-0068-6.

Neel, B. G., Gu, H., & Pao, L. (2003). The 'Shp'ing news: SH2 domain-containing tyrosine phosphatases in cell signaling. *Trends in Biochemical Sciences*, *28*(6), 284–293. https://doi.org/10.1016/S0968-0004(03)00091-4.

Nichols, R. J., Haderk, F., Stahlhut, C., Schulze, C. J., Hemmati, G., Wildes, D., et al. (2018). RAS nucleotide cycling underlies the SHP2 phosphatase dependence of mutant BRAF-, NF1- and RAS-driven cancers. *Nature Cell Biology*, *20*(9), 1064–1073. https://doi.org/10.1038/s41556-018-0169-1.

Noy, R., & Pollard, J. W. (2014). Tumor-associated macrophages: From mechanisms to therapy. *Immunity*, *41*(1), 49–61. https://doi.org/10.1016/j.immuni.2014.06.010.

Okazaki, T., Chikuma, S., Iwai, Y., Fagarasan, S., & Honjo, T. (2013). A rheostat for immune responses: The unique properties of PD-1 and their advantages for clinical application. *Nature Immunology*, *14*(12), 1212–1218. https://doi.org/10.1038/ni.2762.

Ou, S. I., Koczywas, M., Ulahannan, S., Janne, P., Pacheco, J., Burris, H., et al. (2020). The SHP2 inhibitor RMC-4630 in patients with KRAS-mutant non-small cell lung cancer: Preliminary evaluation of a first-in-man phase 1 clinical trial. *Journal of Thoracic Oncology*, *15*(2), S15–S16. Retrieved from <Go to ISI>://WOS:000509464500041.

Parada, L. F., Tabin, C. J., Shih, C., & Weinberg, R. A. (1982). Human EJ bladder carcinoma oncogene is homologue of Harvey sarcoma virus ras gene. *Nature*, *297*(5866), 474–478. https://doi.org/10.1038/297474a0.

Pawson, T. (2004). Specificity in signal transduction: From phosphotyrosine-SH2 domain interactions to complex cellular systems. *Cell*, *116*(2), 191–203. https://doi.org/10.1016/s0092-8674(03)01077-8.

Prahallad, A., Heynen, G. J., Germano, G., Willems, S. M., Evers, B., Vecchione, L., et al. (2015). PTPN11 is a central node in intrinsic and acquired resistance to targeted cancer drugs. *Cell Reports*, *12*(12), 1978–1985. https://doi.org/10.1016/j.celrep.2015.08.037.

Pulido, R., Stoker, A. W., & Hendriks, W. J. (2013). PTPs emerge as PIPs: Protein tyrosine phosphatases with lipid-phosphatase activities in human disease. *Human Molecular Genetics*, *22*(R1), R66–R76. https://doi.org/10.1093/hmg/ddt347.

Pylayeva-Gupta, Y., Grabocka, E., & Bar-Sagi, D. (2011). RAS oncogenes: Weaving a tumorigenic web. *Nature Reviews. Cancer*, *11*(11), 761–774. https://doi.org/10.1038/nrc3106.

Quintana, E., Schulze, C. J., Myers, D. R., Choy, T. J., Mordec, K., Wildes, D., et al. (2020). Allosteric inhibition of SHP2 stimulates antitumor immunity by transforming the immunosuppressive environment. *Cancer Research, 80*(13), 2889–2902. https://doi.org/10.1158/0008-5472.CAN-19-3038.

Quintanar-Audelo, M., Yusoff, P., Sinniah, S., Chandramouli, S., & Guy, G. R. (2011). Sprouty-related Ena/vasodilator-stimulated phosphoprotein homology 1-domain-containing protein (SPRED1), a tyrosine-protein phosphatase non-receptor type 11 (SHP2) substrate in the Ras/extracellular signal-regulated kinase (ERK) pathway. *The Journal of Biological Chemistry, 286*(26), 23102–23112. https://doi.org/10.1074/jbc.M110.212662.

Ran, H., Tsutsumi, R., Araki, T., & Neel, B. G. (2016). Sticking it to cancer with molecular glue for SHP2. *Cancer Cell, 30*(2), 194–196. https://doi.org/10.1016/j.ccell.2016.07.010.

Ren, Y., Meng, S., Mei, L., Zhao, Z. J., Jove, R., & Wu, J. (2004). Roles of Gab1 and SHP2 in paxillin tyrosine dephosphorylation and Src activation in response to epidermal growth factor. *The Journal of Biological Chemistry, 279*(9), 8497–8505. https://doi.org/10.1074/jbc.M312575200.

Roberts, A. E., Araki, T., Swanson, K. D., Montgomery, K. T., Schiripo, T. A., Joshi, V. A., et al. (2007). Germline gain-of-function mutations in SOS1 cause Noonan syndrome. *Nature Genetics, 39*(1), 70–74. https://doi.org/10.1038/ng1926.

Rose, A. A. N. (2019). Encorafenib and binimetinib for the treatment of BRAF V600E/K-mutated melanoma. *Drugs Today (Barc), 55*(4), 247–264. https://doi.org/10.1358/dot.2019.55.4.2958476.

Ruess, D. A., Heynen, G. J., Ciecielski, K. J., Ai, J., Berninger, A., Kabacaoglu, D., et al. (2018). Mutant KRAS-driven cancers depend on PTPN11/SHP2 phosphatase. *Nature Medicine, 24*(7), 954–960. https://doi.org/10.1038/s41591-018-0024-8.

Ryan, M. B., Fece de la Cruz, F., Phat, S., Myers, D. T., Wong, E., Shahzade, H. A., et al. (2020). Vertical pathway inhibition overcomes adaptive feedback resistance to KRAS (G12C) inhibition. *Clinical Cancer Research, 26*(7), 1633–1643. https://doi.org/10.1158/1078-0432.CCR-19-3523.

Sattler, M., Mohi, M. G., Pride, Y. B., Quinnan, L. R., Malouf, N. A., Podar, K., et al. (2002). Critical role for Gab2 in transformation by BCR/ABL. *Cancer Cell, 1*(5), 479–492. https://doi.org/10.1016/s1535-6108(02)00074-0.

Schlessinger, J., & Lemmon, M. A. (2003). SH2 and PTB domains in tyrosine kinase signaling. *Science's STKE, 2003*(191), RE12. https://doi.org/10.1126/stke.2003.191.re12.

Scholl, F. A., Dumesic, P. A., Barragan, D. I., Harada, K., Bissonauth, V., Charron, J., et al. (2007). Mek1/2 MAPK kinases are essential for mammalian development, homeostasis, and Raf-induced hyperplasia. *Developmental Cell, 12*(4), 615–629. https://doi.org/10.1016/j.devcel.2007.03.009.

Schubbert, S., Zenker, M., Rowe, S. L., Boll, S., Klein, C., Bollag, G., et al. (2006). Germline KRAS mutations cause Noonan syndrome. *Nature Genetics, 38*(3), 331–336. https://doi.org/10.1038/ng1748.

Sehnal, D., Rose, A., Koca, J., Burley, S., & Velankar, S. (2018). *Workshop on Molecular Graphics and Visual Analysis of Molecular Data*. The Eurographics Association.

Seton-Rogers, S. (2020). KRAS-G12C in the crosshairs. *Nature Reviews. Cancer, 20*(1), 3. https://doi.org/10.1038/s41568-019-0228-3.

Sha, F., Gencer, E. B., Georgeon, S., Koide, A., Yasui, N., Koide, S., et al. (2013). Dissection of the BCR-ABL signaling network using highly specific monobody inhibitors to the SHP2 SH2 domains. *Proceedings of the National Academy of Sciences of the United States of America, 110*(37), 14924–14929. https://doi.org/10.1073/pnas.1303640110.

Smith, G., Carey, F. A., Beattie, J., Wilkie, M. J., Lightfoot, T. J., Coxhead, J., et al. (2002). Mutations in APC, Kirsten-ras, and p53—Alternative genetic pathways to colorectal cancer. *Proceedings of the National Academy of Sciences of the United States of America, 99*(14), 9433–9438. https://doi.org/10.1073/pnas.122612899.

Smith, M. J., Neel, B. G., & Ikura, M. (2013). NMR-based functional profiling of RASopathies and oncogenic RAS mutations. *Proceedings of the National Academy of Sciences of the United States of America, 110*(12), 4574–4579. https://doi.org/10.1073/pnas.1218173110.

Song, Z., Wang, M., Ge, Y., Chen, X. P., Xu, Z., Sun, Y., et al. (2021). Tyrosine phosphatase SHP2 inhibitors in tumor-targeted therapies. *Acta Pharmaceutica Sinica B, 11*(1), 13–29. https://doi.org/10.1016/j.apsb.2020.07.010.

Sun, Y., Meyers, B. A., Czako, B., Leonard, P., Mseeh, F., Harris, A. L., et al. (2020). Allosteric SHP2 inhibitor, IACS-13909, overcomes EGFR-dependent and EGFR-independent resistance mechanisms toward osimertinib. *Cancer Research, 80*(21), 4840–4853. https://doi.org/10.1158/0008-5472.CAN-20-1634.

Tabin, C. J., Bradley, S. M., Bargmann, C. I., Weinberg, R. A., Papageorge, A. G., Scolnick, E. M., et al. (1982). Mechanism of activation of a human oncogene. *Nature, 300*(5888), 143–149. https://doi.org/10.1038/300143a0.

Tannous, B. A., & Badr, C. E. (2019). An allosteric inhibitor of SHP2 effectively targets PDGFRalpha-driven glioblastoma. *Neuro-Oncology, 21*(11), 1348–1349. https://doi.org/10.1093/neuonc/noz176.

Tartaglia, M., & Gelb, B. D. (2005). Noonan syndrome and related disorders: Genetics and pathogenesis. *Annual Review of Genomics and Human Genetics, 6*, 45–68. https://doi.org/10.1146/annurev.genom.6.080604.162305.

Tartaglia, M., Martinelli, S., Iavarone, I., Cazzaniga, G., Spinelli, M., Giarin, E., et al. (2005). Somatic PTPN11 mutations in childhood acute myeloid leukaemia. *British Journal of Haematology, 129*(3), 333–339. https://doi.org/10.1111/j.1365-2141.2005.05457.x.

Tartaglia, M., Mehler, E. L., Goldberg, R., Zampino, G., Brunner, H. G., Kremer, H., et al. (2001). Mutations in PTPN11, encoding the protein tyrosine phosphatase SHP-2, cause Noonan syndrome. *Nature Genetics, 29*(4), 465–468. https://doi.org/10.1038/ng772.

Tartaglia, M., Pennacchio, L. A., Zhao, C., Yadav, K. K., Fodale, V., Sarkozy, A., et al. (2007). Gain-of-function SOS1 mutations cause a distinctive form of Noonan syndrome. *Nature Genetics, 39*(1), 75–79. https://doi.org/10.1038/ng1939.

Torres-Ayuso, P., & Brognard, J. (2018). Shipping out MEK inhibitor resistance with SHP2 inhibitors. *Cancer Discovery, 8*(10), 1210–1212. https://doi.org/10.1158/2159-8290.CD-18-0915.

Tsutsumi, R., Ran, H., & Neel, B. G. (2018). Off-target inhibition by active site-targeting SHP2 inhibitors. *FEBS Open Bio, 8*(9), 1405–1411. https://doi.org/10.1002/2211-5463.12493.

Uras, I. Z., Moll, H. P., & Casanova, E. (2020). Targeting KRAS mutant non-small-cell lung cancer: Past, present and future. *International Journal of Molecular Sciences, 21*(12), 4325. https://doi.org/10.3390/ijms21124325.

Van Vactor, D., O'Reilly, A. M., & Neel, B. G. (1998). Genetic analysis of protein tyrosine phosphatases. *Current Opinion in Genetics & Development, 8*(1), 112–126. https://doi.org/10.1016/s0959-437x(98)80070-1.

Verhaak, R. G., Hoadley, K. A., Purdom, E., Wang, V., Qi, Y., Wilkerson, M. D., et al. (2010). Integrated genomic analysis identifies clinically relevant subtypes of glioblastoma characterized by abnormalities in PDGFRA, IDH1, EGFR, and NF1. *Cancer Cell, 17*(1), 98–110. https://doi.org/10.1016/j.ccr.2009.12.020.

Wong, G. S., Zhou, J., Liu, J. B., Wu, Z., Xu, X., Li, T., et al. (2018). Targeting wild-type KRAS-amplified gastroesophageal cancer through combined MEK and SHP2 inhibition. *Nature Medicine*, *24*(7), 968–977. https://doi.org/10.1038/s41591-018-0022-x.

Xue, J. Y., Zhao, Y., Aronowitz, J., Mai, T. T., Vides, A., Qeriqi, B., et al. (2020). Rapid non-uniform adaptation to conformation-specific KRAS(G12C) inhibition. *Nature*, *577*(7790), 421–425. https://doi.org/10.1038/s41586-019-1884-x.

Yaeger, R., & Corcoran, R. B. (2019). Targeting alterations in the RAF-MEK pathway. *Cancer Discovery*, *9*(3), 329–341. https://doi.org/10.1158/2159-8290.CD-18-1321.

Yuan, X., Bu, H., Zhou, J., Yang, C. Y., & Zhang, H. (2020). Recent advances of SHP2 inhibitors in cancer therapy: Current development and clinical application. *Journal of Medicinal Chemistry*, *63*(20), 11368–11396. https://doi.org/10.1021/acs.jmedchem.0c00249.

Zhang, X., Lavoie, G., Fort, L., Huttlin, E. L., Tcherkezian, J., Galan, J. A., et al. (2013). Gab2 phosphorylation by RSK inhibits Shp2 recruitment and cell motility. *Molecular and Cellular Biology*, *33*(8), 1657–1670. https://doi.org/10.1128/MCB.01353-12.

Zhang, S. Q., Yang, W., Kontaridis, M. I., Bivona, T. G., Wen, G., Araki, T., et al. (2004). Shp2 regulates SRC family kinase activity and Ras/Erk activation by controlling Csk recruitment. *Molecular Cell*, *13*(3), 341–355. https://doi.org/10.1016/s1097-2765(04)00050-4.

Zhang, J., Zhang, F., & Niu, R. (2015). Functions of Shp2 in cancer. *Journal of Cellular and Molecular Medicine*, *19*(9), 2075–2083. https://doi.org/10.1111/jcmm.12618.

Zheng, J., Huang, S., Huang, Y., Song, L., Yin, Y., Kong, W., et al. (2016). Expression and prognosis value of SHP2 in patients with pancreatic ductal adenocarcinoma. *Tumour Biology*, *37*(6), 7853–7859. https://doi.org/10.1007/s13277-015-4675-5.

Zheng, J., Huang, X., Tan, W., Yu, D., Du, Z., Chang, J., et al. (2016). Pancreatic cancer risk variant in LINC00673 creates a miR-1231 binding site and interferes with PTPN11 degradation. *Nature Genetics*, *48*(7), 747–757. https://doi.org/10.1038/ng.3568.

Zhou, X. D., & Agazie, Y. M. (2008). Inhibition of SHP2 leads to mesenchymal to epithelial transition in breast cancer cells. *Cell Death and Differentiation*, *15*(6), 988–996. https://doi.org/10.1038/cdd.2008.54.

Zhou, X., & Agazie, Y. M. (2009). Molecular mechanism for SHP2 in promoting HER2-induced signaling and transformation. *The Journal of Biological Chemistry*, *284*(18), 12226–12234. https://doi.org/10.1074/jbc.M900020200.

CHAPTER EIGHT

Targeting the "undruggable" RAS with biologics

Michael Whaby[a], Imran Khan[a,b], and John P. O'Bryan[a,b,*]

[a]Department of Cell and Molecular Pharmacology and Experimental Therapeutics, Hollings Cancer Center, Medical University of South Carolina, Charleston, SC, United States
[b]Ralph H. Johnson VA Medical Center, Charleston, SC, United States
*Corresponding author: e-mail address: obryanjo@musc.edu

Contents

1. Introduction 237
2. Anti-RAS biologics 240
   2.1 Immunoglobulin-based biologics 240
   2.2 Non-immunoglobulin-based biologics 246
   2.3 Synthetic peptides 252
3. Targeting RAS using peptide-PROTACs: A new frontier for RAS inhibition 255
4. Conclusion and future outlook 259
Acknowledgments 259
Disclosures 260
References 260

Abstract

RAS proteins represent critical drivers of tumor development and thus are the focus of intense efforts to pharmacologically inhibit these proteins in human cancer. Although recent success has been attained in developing clinically efficacious inhibitors to $KRAS^{G12C}$, there remains a critical need for developing approaches to inhibit additional mutant RAS proteins. A number of anti-RAS biologics have been developed which reveal novel and potentially therapeutically targetable vulnerabilities in oncogenic RAS. This review will discuss the growing field of anti-RAS biologics and potential development of these reagents into new anti-RAS therapies.

1. Introduction

The *RAS* family of proto-oncogenes (*HRAS, KRAS, and NRAS*) represent the most frequently mutated oncogenes in cancer, with *KRAS* being mutated in nearly 100% of pancreatic cancers (Moore, Rosenberg, McCormick, & Malek, 2020; Ryan & Corcoran, 2018; Waters & Der, 2018). The discovery of *RAS* as the human homolog of viral-Rat sarcoma

Advances in Cancer Research, Volume 153
ISSN 0065-230X
https://doi.org/10.1016/bs.acr.2021.07.006

(v-Ras) in the 1980s led to an explosion of efforts to understand its biology and role in tumor development and progression (Karnoub & Weinberg, 2008). Three decades of dedicated efforts to therapeutically target RAS passed without clinical success until the NCI RAS Initiative was launched in 2013 (Ryan & Corcoran, 2018). Now the fruits of those efforts are coming to bear. Several RAS-specific pharmacological inhibitors have entered the clinic with sotorasib (LUMAKRAS) gaining FDA accelerated approval for treatment of $KRAS^{G12C}$ mutant non-small cell lung cancers in May 2021 (Hong et al., 2020). Unfortunately, these anti-RAS drugs inhibit only a single mutant RAS protein which is present in a small fraction of RAS-driven cancers (Zuberi, Khan, & O'Bryan, 2020). Thus, continued efforts to drug the "undruggable" RAS remain paramount.

RAS proteins are small molecular weight GTPases that function as regulated molecular switches (Fig. 1). RAS cycles between an inactive GDP-loaded state and the active GTP-loaded state. Although RAS possesses intrinsic GTPase activity, it is a rather poor enzyme and cooperates with GTPase activating/accelerating proteins (GAPs) to enhance the cleavage of GTP to GDP. Activation of RAS occurs through the action of guanine nucleotide exchange factors (GEFs) which bind to GDP-loaded RAS and destabilize the nucleotide binding pocket resulting in release of GDP. Given the high cellular concentration of GTP vs GDP, coupled with the high affinity of RAS for nucleotides, RAS reloads with GTP (Traut, 1994).

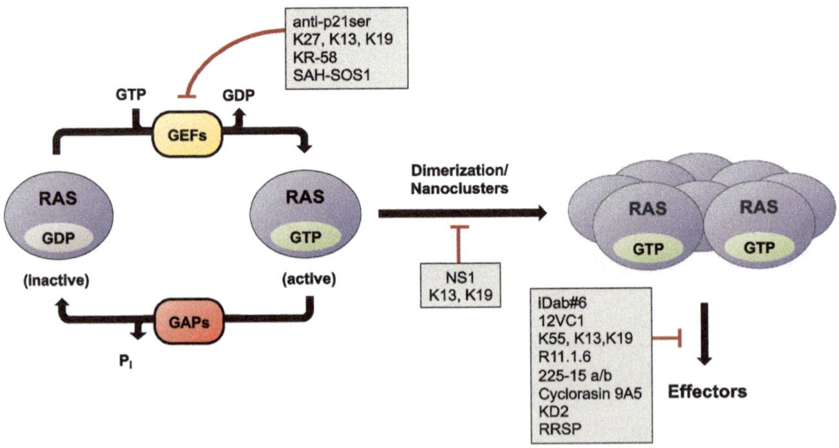

Fig. 1 Targeting RAS with biologics. The cycling of RAS between the inactive GDP-bound state and active GTP-bound state is regulated by the action of GAPs and GEFs, respectively. Once activated, RAS assembles into nanoclusters resulting in recruitment and activation of effectors. Various anti-RAS biologics target different aspects of this cycle.

GTP-loaded RAS couples to targets containing RAS binding domains (RBDs) or RAS association domains (RA). Although these domains share limited sequence homology, they all adopt a similar ubiquitin-like topology. This interaction activates pathways such as the RAF–MEK–ERK mitogen-activated protein kinase (MAPK) and PI3K–AKT–mTOR pathways, resulting in enhanced cell proliferation and survival (Karnoub & Weinberg, 2008). In normal cells, RAS is activated upon stimulation of membrane bound receptors such as receptor tyrosine kinases (RTKs). However, mutations in RAS, predominantly at codons 12, 13, and 61, stabilize the GTP-bound state resulting in chronic activation of its downstream target pathways leading to uncontrolled cell proliferation and ultimately cancer.

Despite great efforts over the past three decades, pharmacological inhibition of RAS has proven difficult. Although pharmacological inhibition of kinases has been achieved through design of ATP analogs, pharmacological targeting of the nucleotide pocket of RAS has not seen success due to the picomolar affinity of RAS for GTP/GDP and the high cellular concentration of guanine nucleotides. Furthermore, structural analysis of RAS revealed a lack of deep hydrophobic pockets for binding of small molecules suggesting that RAS may be "undruggable". The discovery of compounds that covalently bind and inhibit RAS^{G12C} by Kevan Shokat and colleagues has dispelled this premise and ushered in a newfound interest in pharmacologically targeting RAS (Ostrem, Peters, Sos, Wells, & Shokat, 2013). The success of these inhibitors is due to the unique chemical reactivity of the Cys12 mutation that allowed for development of chemically reactive warheads that irreversibly bind and lock $KRAS^{G12C}$ in an inactive GDP-bound state. Based on these results, both Amgen and Mirati have developed $KRAS^{G12C}$-specific inhibitors, sotarasib (Fig. 2A) and adagrasib, respectively, with both demonstrating promising clinical results (Canon et al., 2019; Fell et al., 2020; Hallin et al., 2020; Hong et al., 2020; Skoulidis et al., 2021). Unfortunately, these drugs are limited to treating only $KRAS^{G12C}$ mutant cancers. Thus, development of inhibitors targeting additional oncogenic mutants that lack similar chemical vulnerabilities remains a continuing challenge and unmet need.

As an alternative to small molecule pharmacologics, a number of groups have utilized biologics as an approach to identify vulnerabilities in RAS that can potentially be exploited to inhibit oncogenic RAS. Such biologics often inhibit their target by providing steric interference with critical protein–protein interactions (PPIs) (Hurd, Mott, & Owen, 2020;

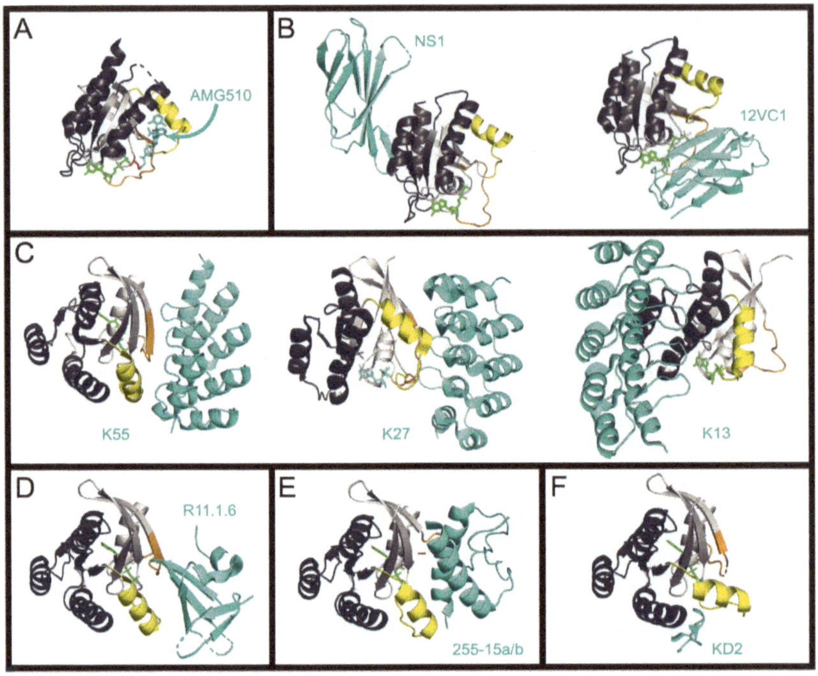

Fig. 2 Binding interfaces of various RAS inhibitors. The crystal structures of RAS in complex with various inhibitors. For all RAS structures the effector lobe is highlighted *light gray* and the allosteric lobe, *dark gray*, Switch 1 is highlighted *orange* and Switch 2 highlighted *yellow*. All inhibitors are shown in *cyan*. Nucleotide is colored *green*. Metal ions and water molecules have been removed from structures. (A) Structure of KRAS^{G12C} bound to Sotorasib (AMG-510) (PDB 6OIM). The Cys12 is highlighted in *red*. (B) RAS Monobodies. HRAS:NS1, left (PDB 5E95), and HRAS(G12C):12VC1 (7L0G). (C) RAS DARPins. KRAS^{G12V}:K55, left (PDB 5MLA); KRAS^{G12V}:K27, center (PDB 5O2S); KRAS^{G12V}:K13, right (PDB 6H46). (D) Sso7d protein R11.1.6. KRAS^{G12D}:R11.1.6 (PDB RUFQ). (E) Miniprotein 225-15a/b. RAS^{G12V}:225-15a/b (PDB 5WLB). (F) KD2 cyclic peptide. KRAS^{G12D}:KD2 (PDB 6WGN).

Pei, Chen, & Liao, 2018). In this review, we will describe the growing number of anti-RAS biologics that have been isolated, their mechanisms of action, and their potential for development into anti-RAS therapeutics.

2. Anti-RAS biologics

2.1 Immunoglobulin-based biologics

2.1.1 Antibodies

Antibodies have been harnessed in the therapeutic arena since 1986, when muromonab-CD3 became the first monoclonal antibody (mAb) to gain

FDA approval for treatment of acute transplant rejection (Lu et al., 2020). Since then, there have been approximately 80 mAbs to receive FDA-approval, with an increasing number being evaluated in clinical trials (Lu et al., 2020). Therapeutic mAbs have several potential modes of action, including disruption of protein–protein or protein–ligand interactions essential to the function of the antigen (Weiner, 2007), allosteric inhibition, alteration of the subcellular localization of target antigens (Koschubs et al., 2012; Marschall & Dubel, 2016), and initiation of cell signaling cascades via interacting with receptor antigens (Weiner, 2007). These characteristics highlight the rationale, and versatility, of utilizing mAbs therapeutically.

The first mAbs generated against RAS were discovered in the lab of Edward Scolnick (Furth, Davis, Fleurdelys, & Scolnick, 1982). These mAbs were produced in rats bearing HRAS-induced tumors and further isolated from eight rat lymphocyte–myeloma hybrid cell lines (Furth et al., 1982). Three of these mAbs also recognized KRAS as well other mammalian RAS homologs (Furth et al., 1982). This work set the stage for the development of therapeutic mAbs against RAS.

Antibody-driven inhibition of KRAS-induced cellular transformation was first observed by Mulcahy et al. by microinjecting a KRAS-specific mAb, Y13-259 (Table 1), into KRAS-transformed NIH 3T3 cells (Mulcahy et al., 1985). Y13-259 decreased cell division whereas a control mAb had no effect (Mulcahy et al., 1985). Y13-259 inhibited RAS function without altering nucleotide binding suggesting that it recognized a region distinct from the nucleotide binding pocket of RAS (Lacal & Aaronson, 1986). Seminal work from the lab of Frank McCormick led to the generation of anti-p21ser (Table 1), an inhibitory antibody specific for amino acid 12 of the oncogenic mutant $KRAS^{G12S}$ (Clark et al., 1985; Feramisco et al., 1985). Surprisingly, anti-p21ser competed with GDP or GTP for binding to $KRAS^{G12S}$ (Clark et al., 1985). These studies provided critical evidence as to the location of the nucleotide binding pocket of RAS and supported the notion that GTP-binding was required for the oncogenic activity of RAS (Clark et al., 1985; Feramisco et al., 1985). In addition, these results suggested that targeting the nucleotide-free state of RAS might provide a feasible approach to inhibit oncogenic RAS.

The high target specificity and ease of production make mAbs an enticing therapeutic modality; however, certain properties limit their clinical application. First, mAbs are cell impermeable and as such are almost exclusively limited to extracellular targets (Lin, Chen, Hu, Chen, & Zhang, 2020; Lu et al., 2020; Marschall & Dubel, 2016; Stocks, 2004). Second, mAbs rely

Table 1 Anti-RAS biologics.

Biologic	Name	Mechanism of action	Binding specificity	Affinity (K_D)	References
Antibodies	Y13-259	RAS sequestration, delocalization from membrane	a.a. 70–81 of HRAS	N/A	Mulcahy, Smith, and Stacey (1985)
	anti-p21ser	Inhibition of nucleotide binding	Ser12 of KRAS (G12S)	N/A	Clark, Wong, Arnheim, Nitecki, and McCormick (1985)
Intrabodies	scFv Y13-259	RAS sequestration, delocalization from membrane	a.a. 70–81 of HRAS	N/A	Cardinale, Lener, Messina, Cattaneo, and Biocca (1998)
	iDab#6	Blocks effector binding	Switch regions of GTP-loaded HRAS	6.2 nM	Tanaka, Lobato, and Rabbitts (2003)
	KGHV500	Cytokine-induced killer (CIK) cell-mediated, tumor-specific delivery	pan-RAS	N/A	Wang et al. (2018)
	inRas37	Tumor-specific, cell-penetrating, full-length antibody	GTP-loaded RAS	N/A	Shin et al. (2020)
Monobodies	NS1	Allosteric disruption of dimerization/nanoclustering; inhibits GTP loading on WT	α4-α5 region of H/KRAS	13–16 nM (HRAS); 60–67 nM (KRAS)	Spencer-Smith et al. (2017)
	12VC1	Blocks effector binding	Switch regions and mutant residues of KRAS (G12V/C)	100 nM	Teng et al. (2021)
	RasIn2	Blocks effector binding	Switch I of GTP-loaded RAS	120 nM	Cetin et al. (2017)
DARPins	K27	Blocks nucleotide exchange	GDP-loaded RAS	4 nM	Guillard et al. (2017)
	K55	Blocks effector binding	GTP-loaded RAS	167 nM	Guillard et al. (2017)

	K13	Blocks effector binding, nucleotide exchange, and dimerization/nanoclustering	α3-α4 (His95) region of KRAS	30 nM	Bery et al. (2019)
	K19	Blocks effector binding, nucleotide exchange, and dimerization/nanoclustering	α3-α4 (His95) region of KRAS	10 nM	Bery et al. (2019)
Sso7d	R11.1.6	Blocks effector binding	Switch regions; specific RAS (G12) mutants	2–6 nM	Kauke et al. (2017)
Miniproteins	225-15 a/b	Blocks effector binding	Binds as a dimer to RAS	60 ± 20 pM for KRAS-GppNHp	McGee et al. (2018)
Toxin Peptidases	RRSP	Cleaves Switch I region of RAS; blocks effector binding	Cleaves between Y32 and D33 of Switch I	N/A	Antic, Biancucci, Zhu, Gius, and Satchell (2015)
Cyclic Peptides	Compound 12	Blocks effector binding	Not specified	0.83 μM for KRAS	Wu, Upadhyaya, Villalona-Calero, Briesewitz, and Pei (2013)
	Cyclorasin 9A5	Blocks effector binding	Switch I of GTP-loaded RAS	0.44 μM for GTP-bound KRAS; 2.5 μM for GDP-bound KRAS	Upadhyaya et al. (2015)
	KS-58	Blocks nucleotide exchange/binding; competes with SOS1 GEF binding	KRAS (G12D)-selective	N/A	Sakamoto, Masutani, and Hirokawa (2020)
	KD2	Blocks effector binding	Switch II groove of GTP-loaded KRAS (G12D)	N/A	Zhang et al. (2020)
Stapled peptides	SAH-SOS1	Blocks nucleotide exchange/binding	Switch regions	100–175 nM for KRAS	Leshchiner et al. (2015)

on disulfide bonds for stability and thus are generally unstable in the reducing environment of the cytosol (Cardinale et al., 1998; Lin et al., 2020; Lu et al., 2020; Marschall & Dubel, 2016; Stocks, 2004). Therefore, simply employing cell-penetrating peptide (CPP) sequences or other methods of cellular entry cannot ensure their efficacy in targeting intracellular antigens, such as RAS. However, methods to generate cell-penetrating IgG antibodies have been described (Choi et al., 2014) and used to target RAS (Shin et al., 2017). Shin et al. isolated a heavy chain variable fragment specific to activated KRAS and then used this to replace the V_H region of a cell penetrating antibody, TMab4, to generate RT11, a cell-penetrating and PPI-interfering mAb against the GTP-bound state of all three RAS isoforms. RT11 bound active RAS isoforms with K_d values ranging from 4 to 17 nM (Shin et al., 2017). RT11 was further modified to confer tumor specificity by the addition of RGD10—a cyclic peptide that recognizes tumor-associated antigens—to the N-terminus of the light chain (Shin et al., 2017). This modified RT11 (RT11-i) preferentially localized to tumors *in vivo* after tail vein injection into athymic nude mice bearing SW480 xenografts (Shin et al., 2017). RT11-i was modified further (to inRas37; Table 1) to improve the biophysical and biochemical properties of the anti-RAS V_H, increase cell penetration, and enhance endosomal escape (Shin et al., 2020). inRas37 exhibited 1.6–3.5-fold enhanced antitumor activity compared to RT11-i and demonstrated favorable pharmacokinetic and pharmacodynamic profiles (Shin et al., 2020). These studies provide a potential new therapeutic approach to inhibit oncogenic RAS *in vivo* using a pan-RAS inhibitor mAb. However, some caution remains as this antibody targeted both WT and mutant forms of GTP-loaded RAS, suggesting the potential for systemic toxicity, although no systemic toxicity was noted (Shin et al., 2020).

2.1.2 Intrabodies

Intrabodies are essentially any antibody (e.g., inRas37) or antibody fragment capable of acting intracellularly (Stocks, 2004) such as single-chain variable fragments (scFvs) consisting of both the heavy (V_H) and light (V_L) chain variable fragments. Huston et al. were the first to describe the generation of scFvs as modified versions of mAbs (Huston et al., 1988). By using a peptide linker to connect the V_H and V_L chains, scFvs eliminated the need for disulfide bonds while maintaining specificity and allowing for genetic encoding (Huston et al., 1988). Using this methodology, Biocca et al. developed the first anti-RAS scFvs using the mAb Y13-259 (Biocca, Pierandrei-Amaldi,

Campioni, & Cattaneo, 1994; Biocca, Pierandrei-Amaldi, & Cattaneo, 1993). scFv Y13-259 (Table 1) inhibited insulin-stimulated meiosis of *Xenopus laevis,* colocalized with RAS at the cytosolic face of the plasma membrane (Biocca et al., 1993, 1994), and demonstrated anti-RAS activity in both mammalian cells (Cardinale et al., 1998; Montano & Jimenez, 1995; Werge, Baldari, & Telford, 1994) and tumor xenografts (Cochet et al., 1998). Subsequent studies suggested that scFv Y13-259 functioned by sequestering RAS into intracellular cytoplasmic aggregates (Cardinale et al., 1998).

Work from the laboratory of Terrence Rabbitts led to the discovery of intracellular single variable domains termed iDabs that maintain solubility, stability, and specificity in cells (Tanaka et al., 2003). Using antibody capture technology (Tse et al., 2002; Visintin et al., 2002), anti-RAS iDabs were identified that inhibited RAS-dependent transformation of NIH/3T3 cells (Tanaka & Rabbitts, 2003; Tanaka et al., 2003). Subsequent studies identified the iDab#6 V_H intrabody which bound and inhibited $HRAS^{G12V}$ (Table 1) (Tanaka, Williams, & Rabbitts, 2007). X-ray crystal structure analysis revealed that iDab#6 interacts with the Switch I (SW1) and Switch 2 (SW2) regions of $HRAS^{G12V}$ (Tanaka et al., 2007). This was further corroborated using surface plasmon resonance (SPR), where iDab#6 bound GST-HRAS(GTPγS) in a concentration-dependent manner, but not with GDP-loaded HRAS (Tanaka et al., 2007). Collectively, iDab#6 inhibited RAS function by disrupting effector binding to the SW1 and SW2 regions (Tanaka & Rabbitts, 2010; Tanaka et al., 2007).

More recently, Pan et al. developed a dual-promotor-regulated recombinant oncolytic adenovirus, KGHV300, expressing a pan-RAS scFv (Yang et al., 2016; Pan et al., 2017). In this construct, expression of the E1a and E1b viral replication genes were regulated by the telomerase promotor and the hypoxia response element, respectively, both of which are upregulated in cancer cells, with the pan-RAS scFv gene controlled by the CMV promotor. Although KGHV300 only replicated in cancer cells, it infected all cell types and therefore required intratumoral administration. To increase cancer cell-specificity of KGHV300, its cilia gene was modified to bind cytokine-induced killer (CIK) cells to enhance tumor cell homing ability (Wang et al., 2018). Thus, the newly modified recombinant adenovirus (now referred to as KGHV500; Table 1) was hypothesized to essentially "hitch a ride" from the CIK cells to the tumor microenvironment (Wang et al., 2018). Injection of CIK cells carrying KGHV500 into mice bearing SGC7901 (RAS^{WT}, human gastric cancer) xenografts reduced tumor growth compared to injection of KGHV500 without CIK cells (Wang et al., 2018). Furthermore, both the pan-RAS scFv and the adenovirus hexon protein were restricted to tumor and spleen in the KGHC500 + CIK injected samples whereas KGHV500 injection alone resulted in wide expression of both proteins, demonstrating tissue-specificity of the treatment (Wang et al., 2018). Thus, this approach may lead to new therapeutic approaches to treat RAS-mutant cancers.

2.2 Non-immunoglobulin-based biologics

2.2.1 Monobodies

Monobodies (Mbs) are high affinity engineered proteins derived from the fibronectin type III (FN3) domain (Sha, Salzman, Gupta, & Koide, 2017). Mbs bind critical sites of their target and often inhibit the function of that target. The isolation of Mbs involves selection by phage display and yeast display formats from large combinatorial libraries, and this art was pioneered and refined by Shohei Koide and colleagues (Biancucci et al., 2018; Koide, Bailey, Huang, & Koide, 1998; Koide, Wojcik, Gilbreth, Hoey, & Koide, 2012). The smaller size and lack of disulfide bonds provides an advantage for the use of Mbs as genetically encoded reagents and facilitates their production in *E. coli*.

Our laboratory, in collaboration with the Koide laboratory, has applied this technology to identify anti-RAS biologics (Spencer-Smith et al., 2017; Teng et al., 2021). Our first success came with the isolation of the NS1 Mb which bound the α4-α5 interface on the allosteric lobe of RAS, a region implicated in RAS self-association and nanoclustering (Table 1; Fig. 2B) (Khan, Spencer-Smith, & O'Bryan, 2019). Indeed, ectopic expression of NS1 inhibited RAS dimerization and nanoclustering and inhibited activation of ERK and AKT by both growth factors and oncogenic RAS. NS1 bound HRAS and KRAS with low nanomolar affinity but did not interact with NRAS. This selectivity was due to a single amino acid difference in α4 helix of NRAS (K135) vs H- and KRAS (R135) resulting in disruption of multiple Mb-RAS contacts. In addition to inhibiting RAS signaling, NS1 inhibited the oncogenic activity of H/KRAS mutants but not NRAS or downstream oncogenic kinases such as BRAF(V600E) and MEK(DD) (Spencer-Smith et al., 2017).

Recently, we found that RAS inhibition with NS1 enhanced the anti-tumor immune response (Khan et al., 2021). Using PDAC tumor cells derived from the KPC mouse model (Kras$^{G12D/+}$; TP53$^{R172H/+}$) (Hingorani et al., 2005), we established a subclone (KPC^{NS1}) that expressed

NS1 in a chemically regulated manner. Injection of KPC^{NS1} cells into the pancreas resulted in rapid tumor development which was dramatically reduced upon induction of NS1 expression. Furthermore, we observed an increase in tumor infiltrating CD4+ T-cells upon NS1 induction. Surprisingly, there was no change in the number of CD8+ cytotoxic T-cells in the NS1-expressing cells versus controls. Numerous pre-clinical and clinical studies have demonstrated effective anti-tumor immunity of CD4+ cells without engaging CD8+ T-cells (Galaine et al., 2019; Kreiter et al., 2015; Linnemann et al., 2015; Ott et al., 2017; Sahin et al., 2017; Tay, Richardson, & Toh, 2021; Zhang et al., 2018). Interestingly, in a DOX-inducible $KRAS^{G12D}$ colorectal cancer mouse model, extinguishing KRAS signaling upon DOX withdrawal led to increased CD4+ T cell infiltration (Liao et al., 2019). It is possible that different KRAS mutations mount different immunological responses in terms of T-cell signatures. This may account for the difference in T-cell response to NS1 vs $KRAS^{G12C}$ inhibitors (Canon et al., 2019). Indeed, recent studies illustrate that different KRAS mutations provoke distinct biochemical, cellular, and pathological responses (Hobbs et al., 2020; Poulin et al., 2019).

There are more than 130 different reported missense mutations in the *RAS* oncogene in various cancers (Hobbs, Der, & Rossman, 2016). Thus, a pan-RAS inhibitory agent like NS1, rather than a mutant selective inhibitor, may represent a more efficacious approach to treat diverse RAS mutant cancer types. The broad-spectrum nature of such an inhibitor may be therapeutically important in light of reports describing the presence of multiple RAS mutations in a single patient and heterogenous mutations between primary and metastatic tumor (Kordiak et al., 2019; Lamy et al., 2011; Oltedal et al., 2011; Richman et al., 2011). As with other RAS biologics, specificity to differentiate oncogenic mutants from WT RAS, which is essential for cells, raises the question of possible toxicity with NS1. Although recent results suggest that NS1 lacks non-specific "off-target" toxicity, it remains possible that residual NRAS expression may be sufficient to avoid such toxicity (Khan et al., 2021).

Although the main mode of action of NS1 appears to be inhibition of RAS self-association/nanoclustering, additional studies reveal added complexity in the inhibitory action of NS1 (Spencer-Smith et al., 2019). For example, NS1 reduced interaction of oncogenic KRAS, but not HRAS, with RAF and reduced KRAS localization to the plasma membrane. The isoform specific effects of NS1 on RAS–RAF association were mediated by the distinct hypervariable regions of RAS isoforms. Further, NS1 reduced GTP loading on WT but not oncogenic RAS in cells suggesting that NS1 binding may alter the orientation of the G-domain with the plasma membrane, hence, perturbing RAS interaction with SOS (Spencer-Smith et al., 2019).

Following on the success with NS1, Teng et al. recently reported the isolation of a mutation selective Mb to RAS (Teng et al., 2021). This Mb, termed 12VC1, selectively bound and inhibited $KRAS^{G12C}$ and $KRAS^{G12V}$ (Table 1; Fig. 2B). Structural analysis revealed that 12VC1 directly recognized the mutant residues along with the Switch regions to inhibit KRAS–effector interaction. Regulated expression of 12VC1 also reduced the growth of PATU8902 PDAC cells ($KRAS^{G12V}$ mutant) tumor cells in athymic nude mice. However, they further enhanced the activity of this Mb through fusion to the von Hippel Lindau E3 ligase (VHL) to generate a mutant selective degrading biologic (see Section 3 for further discussion). Indeed, this modified Mb resulted in proteosome-mediated degradation of $KRAS^{G12C}$ and $KRAS^{G12V}$ but not $KRAS^{G12D}$ or WT RAS. Further, this RAS degrader resulted in sustained pERK inhibition compared to the isolate 12VC1 Mb, further establishing the efficacy of this RAS degrader.

Another genre of RAS-specific Mbs, RasIns, were identified by Cetin et al. (Cetin et al., 2017) using a Mb mRNA display selection against GTPγS-bound RAS, where the BC loop of the Mb was based on mutagenized sequences of the complementary determining region of iDab#6 (Tanaka et al., 2007). RasIn1 was GTP-state-selective for HRAS and KRAS and bound to the Switch I region of RAS (Cetin et al., 2017). Affinity maturation of RasIn1 produced RasIn2 (Table 1). While RasIn1 bound GTP-loaded $HRAS^{G12V}$ with a K_D of 2.1 μM, RasIn2 bound with a K_D of 120 nM (Cetin et al., 2017). RasIn1 and RasIn2 also co-localized with RAS in COS-7 cells; however, their ability to inhibit RAS-mediated signaling was not assessed (Cetin et al., 2017). Together, these studies highlight the power of Mbs in both understanding RAS and inhibiting its oncogenic activity *in vivo*.

2.2.2 DARPins

Designed ankyrin repeat proteins (DARPins) represent an additional engineered affinity reagent that has been used to inhibit RAS. DARPins are single domain proteins (14 kDa) that are derived from ankyrin repeat binding proteins, which are naturally occurring in eukaryotic cells (Stumpp, Binz, & Amstutz, 2008). Like Mbs, they lack cysteine residues and hence are suitable as genetically encodable intracellular reagents

(Kummer et al., 2012; Parizek et al., 2012). The first RAS-specific DARPin, K55, is a pan-RAS binder that preferentially binds to the active GTP-loaded state of RAS and inhibits RAS effector interactions (Table 1; Fig. 2C) (Guillard et al., 2017). K27 DARPin is also a pan-RAS binder but in contrast to K55 prefers the GDP-loaded state and inhibits RAS mainly by preventing nucleotide exchange (Table 1; Fig. 2C) (Guillard et al., 2017). Building on these results, Bery et al. isolated two KRAS specific DARPins, K13 (Table 1; Fig. 2C) and K19 (Table 1), that bind the α3-α4 allosteric interface encompassing the region around His95, an amino acid unique to KRAS (Bery et al., 2019). K13 and K19 inhibited KRAS by impairing effector interactions, preventing nucleotide exchange, and inhibiting KRAS dimerization at the plasma membrane (Bery et al., 2019). Though K13 and K19 did not discriminate between mutant and WT KRAS, they did not bind to NRAS and HRAS owing to the selective interaction with His95. Thus, these DARPins provide a potential approach to selectively inhibit KRAS while sparing the function of both N- and HRAS thereby reducing the potential for toxicity due to inhibition of all RAS isoforms.

2.2.3 Affibodies

Affibodies are biologically active engineered protein reagents of approximately 6 kDa based on a scaffold of the three-α-helix bundle *Z*-domain (Friedman & Stahl, 2009; Nilsson & Tolmachev, 2007; Nygren, 2008). Affibodies typically lack cysteine residues and thus are immune to the redox environment of the cell (Nygren, 2008). HRAS-specific affibodies have been isolated that bound with high nanomolar affinity and inhibited HRAS–RAF interaction *in vitro* (Grimm et al., 2010). These affibodies inhibited cell proliferation and production of inflammatory mediators in synovial cells by impairing RAS activation of the MAPK signaling cascade (Shibasaki et al., 2014). Unlike Mbs and DARPins, the efficacy of affibodies in RAS addicted oncogenic lines or tumor models has yet to be demonstrated. Thus, further work is needed to demonstrate their utility as RAS targeting biologics *in vivo*.

2.2.4 Sso7d scaffold proteins

Antibodies and intrabodies, as highlighted above, are excellent candidates as therapeutic agents. However, the aforementioned hurdles thwart the development of effective antibody-based therapeutics against many sought-after intracellular targets such as RAS. Ideal binding peptides would exhibit smaller molecular masses, cell-penetrability, and lack disulfide bonds and glycosylation sites while maintaining stability, solubility, and specificity for their intended target (Traxlmayr et al., 2016). An example of a peptide binding scaffold displaying many of these qualities is Sso7d from the hyperthermophilic archaea *Sulfolobus solfataricus* (Kauke et al., 2017; Traxlmayr et al., 2016). In contrast to Mbs where the binding sites are typically located within flexible loops, the target binding site of Sso7d lies on the surface of a ridged β-sheet (Traxlmayr et al., 2016).

Kauke et al. isolated a RAS-specific Sso7d, R11.1 using GppNHp-loaded $KRAS^{G12D}$ as bait (Traxlmayr et al., 2016). Affinity maturation of R11.1 yielded R11.1.6 (Table 1; Fig. 2D) which was eight-fold more selective for $KRAS^{G12D}$ than WT KRAS with single-digit nanomolar affinity. However, this selectivity—but not affinity—was lost in the GDP-loaded state suggesting that R11.1.6 did not directly contact the mutant residue, but rather recognized a specific conformation of $KRAS^{G12D}$ in the GppNHp-loaded state (Traxlmayr et al., 2016). Accordingly, R11.1.6 bound other GppNHp-loaded KRAS oncogenic mutants (G12V and G12C), as well as HRAS and NRAS, with comparable affinity to GppNHp-loaded $KRAS^{G12D}$. Structural analysis indicated that R11.1.6 bound distal to the nucleotide pocket, further supporting the hypothesis R11.1.6 binding is due to the conformation of mutant KRAS in the GppNHp-loaded state. The binding interface of R11.1.6/GppNHp–$KRAS^{G12D}$ consisted of Switch II in a conformation that exposes hydrophobic residues that interact with aromatic residues of R11.1.6. This interaction at Switch II alters the orientation of Tyr32 and Glu61 at the active site of $KRAS^{G12D}$, rendering it catalytically incompetent. This was supported by the observation that R11.1.6 reduced the intrinsic hydrolysis of $KRAS^{G12D}$. Although the GTP-loaded state of $KRAS^{G12D}$ would presumably be stabilized by its interaction with R11.1.6, this is compensated for by the ability of R11.1.6 to disrupt RAS effector interactions. Additionally, in the R11.1.6/GppNHp–$KRAS^{G12D}$ complex, the conformation of Switch I of $KRAS^{G12D}$ places Glu37 and Glu63 in a position that favors interaction with Lys32 and Lys40 of R11.1.6. Switch I of wild-type KRAS, in contrast, appears to orient Glu63 away from Lys32, thus making the R11.1.6/wild-type KRAS complex less favorable.

2.2.5 Miniproteins

McGee et al. employed a randomized yeast surface display screening approach, utilizing a select number of small, conformationally stable "miniproteins" based on the avian pancreatic polypeptide (aPP) miniprotein scaffold to identify peptides that bound specifically to KRAS (McGee et al., 2018). Three hits were obtained and subjected to directed evolution to optimize the affinity for KRAS resulting in two improvements: addition of PRR to the N-terminus and incorporation of a Y7C mutation. The former was due to primer slipping during PCR; the latter was discarded to avoid unwanted disulfide interactions. Following a third round of directed evolution, miniprotein 225-3 was chosen for further optimization. 225-3 bound GppNHp-loaded H, K, and NRAS^{G12V} with mid-nanomolar affinity, competed with RAF-RBD binding, and impaired nucleotide release.

Due to solubility issues, 225-3 was replaced with 225-1. HSQC NMR spectroscopy revealed that 225-1 interacted with the effector binding domain of RAS. Alanine scanning mutagenesis revealed that nearly every effector domain residue contributed to 225-1 interaction suggesting a complex binding mechanism. Furthermore, subsequent NMR analyses suggested that the miniprotein possibly bound KRAS as a dimer. Re-incorporation of Y7C to generate 255-11 resulted in a dimeric miniprotein that bound KRAS more strongly in the absence vs presence of reducing agent. However, DTT had no effect on 255-11–KRAS interaction when the C7 residue was replaced with selenocysteine. Crystal structure analysis revealed that 225-11 formed a disulfide-stabilized dimer, of which one protomer bound KRAS in an extended, open conformation that distorted the Switch I loop and β-strand 2 of the KRAS effector domain. Further optimization of 225-11 by A30R mutation led to a twofold selectivity of GppNHp-loaded over GDP-loaded KRAS.

The authors hypothesized that breaking the symmetry between the 255-11 dimer might improve RAS binding and therefore, allowed each protomer to evolve independently (McGee et al., 2018). The resulting miniprotein 225-15 a/b heterodimer (Table 1; Fig. 2E) bound GppNHp-loaded KRAS >10-fold better than 225-11 (K_d of ~60 nM) and interacted with all 3 RAS isoforms (H, K, and NRAS) as well as various oncogenic KRAS mutants (G12V, G13D, and Q61H) with subnanomolar affinity. However, the ability of these RAS-targeting miniproteins to interfere with RAS function in cells was not examined. Thus, further work will be needed to determine the usefulness as anti-RAS biologics.

2.2.6 Toxin peptidases

The multifunctional-autoprocessing repeats-in-toxin (MARTX) is a bacterial toxin capable of penetrating eukaryotic cell membranes and inducing cytotoxicity (Antic et al., 2015; Kwak, Jeong, & Satchell, 2011). Kwak et al. showed that the most virulent strains of MARTX produced by *Vibrio vulnificus* contained an effector domain, DUF5$_{Vv}$, consisting of two subdomains: the membrane anchoring C1 subdomain and the cytotoxic

C2 domain (Antic et al., 2015). To identify potential targets of $DUF5_{Vv}$, a genome-wide, non-essential gene deletion library screen was performed on yeast strains expressing the cytotoxic C2 subdomain. Many of the hits from the yeast strains that survived were connected to the MAPK pathway. Indeed, $DUF5_{Vv}$ decreased ERK1/2 phosphorylation in HeLa cells. It was therefore postulated that a likely target of $DUF5_{Vv}$ was RAS, given the membrane localization of $DUF5_{Vv}$ via its C1 subdomain. It was further shown that $DUF5_{Vv}$ cleaves RAS (and RAP1) between Y32 and D33 in the Switch I region and was therefore dubbed a RAS/RAP1-specific protease (RRSP) (Table 1). Surprisingly, RAS nucleotide binding was not affected by RRSP cleavage; however, effector and GEF interactions were disrupted (Biancucci et al., 2018). Lastly, Vidimar et al. engineered a cell-penetrating RRSP by linking it to a catalytically inactive diphtheria toxin (RRSP-DT_B), which binds the heparin-binding epidermal growth factor-like growth factor (HB-EGF) cell surface receptor and promotes cell uptake via endocytosis (Vidimar et al., 2020). When injected into athymic nude mice harboring human HB-EGF positive, triple negative breast cancer xenografts (WT RAS), RRSP-DT_B effectively decreased tumor growth (Vidimar et al., 2020). Further studies will be necessary to confer mutant or isoform specificity as RRSP cleaves all RAS isoforms as well as RAP1.

2.3 Synthetic peptides

2.3.1 Cyclic peptides

Cyclic peptides, as opposed to traditional biologics, offer greater therapeutic potential as they are more amenable to modifications that increase bioavailability (Philippe, Craik, & Henriques, 2021). By combining the biophysical properties of small molecules and the high affinity of biologics, peptide cyclization may yield greater chemical stability and protease resistance, potentially increasing systemic half-life, as well as increased affinity for target ligand (Philippe et al., 2021). Importantly, cyclic peptides still retain the ability to bind and disrupt PPI interfaces (Santini and Zacharias, 2020). Properties like these increase the likelihood of developing desirable therapeutics such as Cyclosporin A, an orally available cyclic peptide used for treatment of immunoregulatory disorders (Nielsen et al., 2017; Philippe et al., 2021).

Wu et al. discovered the first cyclic peptide targeting RAS, Compound 12 (Table 1) (Wu et al., 2013). Although this molecule blocked RAS interaction with RBDs of various effectors, Compound 12 was unable to decrease MEK or ERK phosphorylation, likely due to poor membrane permeability (Wu et al., 2013). Upadhyaya et al. utilized a general methodology for synthesizing and screening bicyclic peptides to isolated $KRAS^{G12V}$ binding compounds, termed Cyclorasins (Upadhyaya, Qian, Habir, & Pei, 2014). Although two such compounds disrupted the RAS–RAF interaction and competed with Compound 12 for RAS binding suggesting similar target sites on RAS, they failed to confer any effects in cells, again due to poor membrane permeability (Upadhyaya et al., 2014). Further refinements to this approach led to the isolation of Cyclorasin 9A5 (Table 1) with improved cell permeability and affinity for KRAS (Upadhyaya et al., 2015). Indeed, fluorescein isothiocyanate (FITC)-labeled Cyclorasin 9A5 displayed diffuse fluorescence throughout the cytoplasm along with weaker, punctate fluorescence indicative of endosomal entrapment (Upadhyaya et al., 2015). FITC-labeled Cyclorasin 9A5 bound GTP-, GppNHp-, and GDP-loaded $KRAS^{G12V}$ with K_d values of 0.44, 0.64, and 2.5 μM, respectively, suggesting that Cyclorasin 9A5 preferentially binds KRAS-GTP (Upadhyaya et al., 2015). ^{1}H—^{15}N heteronuclear single quantum correlation (HSQC) NMR spectroscopy revealed that Cyclorasin 9A5 interacted with the Switch I loop of KRAS, consistent with the dose-dependent inhibition of RAS–RAF interaction in H358 lung cancer cells ($KRAS^{G12C}$) as well as MEK, ERK and AKT phosphorylation in H1299 cells ($NRAS^{Q61K}$) and H1650 cells (wild-type RAS; mutant EGFR) (Upadhyaya et al., 2015). The enhanced membrane permeability of Cyclorasin 9A5 may be due to its more compact, amphipathic structure in DMSO, with converged aromatic amino acids surrounded by arginine residues thereby allowing for more effective membrane penetration (Takeuchi et al., 2021).

$KRAS^{G12D}$ represents the most frequent KRAS mutant in human tumors (Vatansever, Erman, & Gumus, 2019). Although there are no targeted therapeutics against $KRAS^{G12D}$ in the clinic, recent research has provided hope for selective inhibition of $KRAS^{G12D}$. KRpep-2, a cyclic peptide, was the first reported $KRAS^{G12D}$-selective inhibitor (Sakamoto et al., 2017). KRpep-2 was identified by phage display screening against $KRAS^{G12D}$ and found to compete with SOS1 binding. Affinity optimization of KRpep-2 was achieved by the addition of arginine residues to both the N- and C-termini, resulting in KRpep-2d (Sakamoto et al., 2017). Both KRpep-2 and KRpep-2d inhibited $KRAS^{G12D}$ nucleotide exchange in a concentration-dependent manner *in vitro* and KRpep-2d decreased ERK activation in A427 cells ($KRAS^{G12D}$) but not in A549 cells ($KRAS^{G12C}$) (Sakamoto et al., 2017). However, due to poor membrane permeability and decreased inhibitory activity, KRpep-2d was not effective *in vivo* (Sakamoto et al., 2017). To address this concern, KRpep-2d was modified using three approaches: (1) replacing the disulfide bond with an amide bond to increase resistance to reductive cleavage, (2) introducing unnatural amino acids to the binding region to increase its affinity for $KRAS^{G12D}$, and (3) connecting the N- and C-termini via amide bond to introduce protease-resistant, membrane permeable cyclization (Sakamoto et al., 2020). This optimization led to KS-58 (Table 1), the first reported $KRAS^{G12D}$ inhibitory peptide displaying *in vivo* anti-cancer activity (Sakamoto et al., 2020). Further development of this lead compound will be needed prior to moving into clinical trials.

Utilizing a different screening approach, Zhang et al. recently reported another $KRAS^{G12D}$-specific cyclic peptide, KD2 (Table 1). Random non-standard Peptides Integrated Discovery (RaPID) (Ong et al., 2017) was used to screen for cyclic peptides that bound GppNHp-loaded $KRAS^{G12D/T35S}$ to minimize the probability of identifying peptides that bound RAS but did not disrupt effector binding or peptides that stabilized the effector binding-competent state (Zhang et al., 2020). Five total rounds of screening were performed against GppNHp-bound $KRAS^{G12D/T35S}$ and compared to binding of GDP-bound $KRAS^{G12D/T35S}$ and empty streptavidin beads for quantification (Zhang et al., 2020). KD2 inhibited the RAS–RAF interaction and, to a weaker extent, SOS1-mediated nucleotide exchange; however, unlike KRpep-2d, KD2 bound $KRAS^{G12D}$-GTP (Zhang et al., 2020). Structural analysis revealed that KD2 bound and expanded the Switch II groove by shifting the α2 helix and Switch II loop and appeared to make direct contact with Asp12 of $KRAS^{G12D}$ via Thr10 (Fig. 2F) (Zhang et al., 2020). Structure-guided optimization led to a >10-fold increase in potency (IC_{50} of 0.80 μM) at inhibiting the RAS–RAF interaction *in vitro* (Zhang et al., 2020). While these results remain encouraging, KD2 was cell impermeable, thereby requiring further optimization to translate into cell-based and *in vivo* assays (Zhang et al., 2020).

2.3.2 Stapled peptides

Stapled peptides represent yet another group of macromolecules to disrupt RAS function. Stapled peptides usually consist of α-helical segments linked or "stapled" by side chains that can be covalently bonded (Ali, Atmaj, Van Oosterwijk, Groves, & Domling, 2019). This linkage renders the alpha helical structure more rigid, increasing stability and resistance to proteolysis in the cell. Furthermore, the incorporation of non-natural amino acids can further enhance both of these properties (Ali et al., 2019). In addition, the peptide nature of these compounds allows easier optimization to confer affinity and selectivity to the target of interest (Leshchiner et al., 2015).

Patgiri et al. created a SOS1 αH helix mimic, HBS3, using a hydrogen bond surrogate approach. HBS3 bound to nucleotide-free (K_d of ~28 μM) and GDP-loaded RAS (K_d of ~158 μM) near the Switch regions and nucleotide pocket where the αH helix of SOS1 normally binds (Patgiri, Yadav, Arora, & Bar-Sagi, 2011). HBS treatment of cells decreased RAS activation following EGF stimulation of HeLa cells without altering the phosphorylation status of EGFR (Patgiri et al., 2011). Leshchiner et al. used the primary sequence of the RAS-interacting α-helix of SOS1 to generate stabilized alpha helices of SOS1 (SAH-SOS1). Of the stapled peptides screened, one was further optimized by the addition of two arginine residues to the N-terminus to decrease its overall charge and increase its cell permeability (Leshchiner et al., 2015). The resulting stapled peptide, SAH-$SOS1_A$ (Table 1), bound all RAS mutants (both GDP- and GTP-bound) with nanomolar affinity (K_d of 100–175 nM) and competed with RAS–SOS1 binding (Leshchiner et al., 2015). SAH-$SOS1_A$ additionally blocked the association of nucleotide with both WT and $KRAS^{G12D}$ *in vitro*. In cell-based assays, SAH-$SOS1_A$ dose-dependently impaired the viability of various KRAS-mutant cancer cell lines and dose-dependently decreased MEK, ERK, and AKT phosphorylation in EGF-stimulated Panc 10.5 cells (Leshchiner et al., 2015).

3. Targeting RAS using peptide-PROTACs: A new frontier for RAS inhibition

An emerging strategy to selectively degrade proteins utilizes the ubiquitin–proteasome system (UPS). In contrast to technologies such as RNAi and CRISPR, which knock-down or eliminate expression of proteins at a genetic level, proteolysis-targeting chimeras (PROTACs) result in the specific post-translational degradation of a target protein (Chopra, Sadok, & Collins, 2019; Pettersson & Crews, 2019). Traditionally, PROTACs are bifunctional small molecules containing two moieties: a ligand for a specific target of interest coupled to a ligand for a specific E3 ubiquitin ligase complex (Chopra et al., 2019; Pettersson & Crews, 2019). The PROTAC acts to recruit the ubiquitin ligase machinery directly to a target protein for polyubiquitination and subsequent proteasomal degradation (Chopra et al., 2019; Pettersson & Crews, 2019). Thus, PROTACs "hijack" the ubiquitin–proteasome machinery to specifically degrade target proteins. This approach holds advantages over genetic knockdown methods, especially when considering the long-lived half-life of certain proteins (Pettersson & Crews, 2019). An additional benefit of PROTACs is their event-driven mechanism of action (MOA) as opposed to traditional pharmacological occupancy-driven MOA (Pettersson & Crews, 2019). This is particularly beneficial for small molecule PROTACs: as they are not consumed in the reaction, they can carry out multiple rounds proteasomal targeting (Pettersson & Crews, 2019).

Recently, the Crews laboratory developed the first oncogenic RAS-targeting PROTAC, termed LC2, by combining the $KRAS^{G12C}$-specific inhibitor, MRTX849, with a VHL E3 ligase-recruiting ligand (Bond, Chu, Nalawansha, Li, & Crews, 2020). Although this represents an encouraging development in anti-RAS pharmacologics, the current $KRAS^{G12C}$ inhibitors are already covalent, irreversible inhibitors of KRAS. Furthermore, the problem still remains that $KRAS^{G12C}$ is present in only ~10% of KRAS-mutant tumors (Khan, Rhett, & O'Bryan, 2020) with a ~2.2% overall prevalence in cancer (Zuberi et al., 2020). Furthermore, the lack of high affinity small molecule inhibitors of other KRAS alleles limits the translation of these findings to additional KRAS mutants.

Peptide-PROTACs (also referred to as peptide degraders, biodegraders, bioPROTACs, and Affinity-Directed Protein Missiles) provide an alternative approach to small molecule PROTACs (Fig. 3). Peptide-PROTACs contain three main components: an E3 ubiquitin ligase, a high affinity protein binder specific for the target of interest, and a linker joining the two protein domains. The most commonly used E3 ligases thus far are VHL and CRBN (Jin et al., 2020; Lim et al., 2021). Gly/Ser linkers of

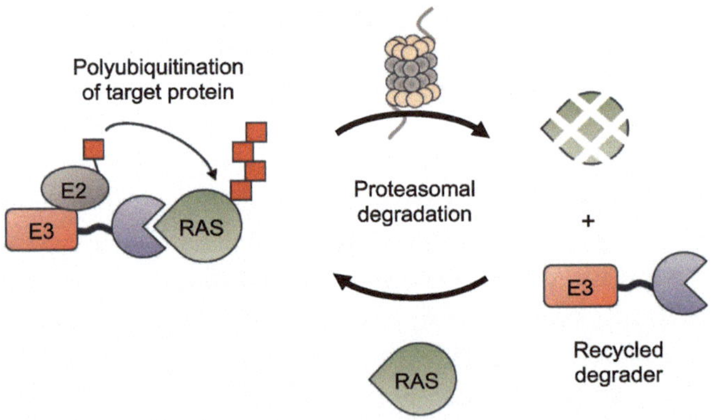

Fig. 3 RAS-targeting peptide-PROTACs. Anti-RAS biologics are genetically fused to an E3 ubiquitin ligase resulting in recruitment of the complex to RAS, polyubiquitination and subsequent proteasomal degradation. The peptide-PROTAC, however, can be recycled to initiate multiple rounds of RAS degradation.

varying lengths are most often used to connect the E3 ligase and protein-targeting ligands (Jin et al., 2020). Peptide-PROTACs are advantageous as they can be genetically encoded (Jin et al., 2020), include an E3 ligase *or* an E3 ligase-recruiting motif (i.e., degrons) (Jin et al., 2020; Qu et al., 2020), and can include high-affinity biologics such as Mbs (Lim et al., 2021; Roth et al., 2020) or DARPins (Bery, Miller, & Rabbitts, 2020) as protein-targeting ligands when small molecule ligands are not an option.

Ma et al. effectively applied the peptide-PROTAC approach to inhibit KRAS (Ma et al., 2013). They coupled the RBD and cysteine-rich domain (CRD) of c-RAF to the U-Box E3 ligase domain of the carboxyl terminus of Hsc70 interacting protein (CHIP) to generate the peptide-PROTAC, RC-U (Ma et al., 2013). RC-U co-immunoprecipitated with FLAG-tagged KRAS in HEK 293T cells and decreased the expression of endogenous KRAS in PANC-1 cells in a polyubiquitination-dependent manner (Ma et al., 2013). Treatment with the proteasome inhibitor, MG-132, rescued KRAS expression in RC-U-transfected PANC-1 cells, demonstrating that the decrease in KRAS expression in RC-U-transfected cells was proteasome-dependent (Ma et al., 2013). Lastly, expression of RC-U decreased tumor volume of PANC-1 xenografts (Ma et al., 2013). Collectively, these results demonstrated that proteasomal targeting of KRAS was indeed possible using a peptide-PROTAC approach.

Another KRAS-targeting peptide-PROTAC, VHL-aHRAS, was recently reported by Roth et al. (Roth et al., 2020). This approach linked the NS1 Mb discussed in Section 2.2.1 (termed aHRAS in the study) to VHL. VHL-aHRAS or VHL-aGFP16 (a control specific for GFP) were retrovirally transduced in A549 cells bearing a homozygous knock-in of GFP-tagged KRAS (Roth et al., 2020). VHL-aHRAS reduced both endogenous KRAS (and HRAS) and GFP-KRAS protein levels whereas the VHL-aGFP16 only reduced GFP-KRAS protein levels (Roth et al., 2020). Furthermore, MG-132 and bortezomib both induced accumulation of polyubiquitinated proteins and rescued RAS protein expression in A549 cells transduced with VHL-aHRAS (Roth et al., 2020). Proteomic analysis revealed that HRAS was decreased >2-fold by VHL-aHRAS compared to VHL alone; however, KRAS was not significantly reduced (Roth et al., 2020). This same trend was observed (via Western blot analysis) in HT-29 and SW620 cell lines (Roth et al., 2020). This selectivity for HRAS is likely due to the higher affinity of NS1 for HRAS vs KRAS (Spencer-Smith et al., 2017). Nonetheless, VHL-aHRAS was effective at targeting HRAS, and KRAS to a lesser extent, for proteasomal degradation.

Rabbitts and colleagues also developed anti-RAS peptide-PROTACs using their RAS targeting iDabs and DARPins as described in Section 2.1.2 and 2.2.2, respectively (Bery et al., 2020). These anti-RAS biologics were genetically fused to either VHL or the U-Box domain of CHIP and examined for their ability to degrade endogenous H, K, and NRAS in HCT116 cells. Fusion of the U-Box domain to the COOH-terminus of the iDab (pan-RAS degrader) was more efficient than N-terminal fusion and was more potent than VHL fusions (Bery et al., 2020). Conversely, fusion of VHL to the N-terminus of the anti-KRAS DARPin (KRAS degrader) was the most effective peptide-PROTAC (Bery et al., 2020). These findings emphasize the importance of optimizing the orientation of the specific ubiquitin ligase domain relative to the specific targeting domain in the design of a particular peptide-PROTAC.

The pan-RAS PROTAC (i.e., iDab) reduced expression of H, K, and NRAS whereas the KRAS-specific PROTAC (i.e., DARPin) selectively reduced KRAS levels in a time- and dose-dependent manner in H358 cells resulting in a commensurate reduction in AKT, MEK, and especially ERK phosphorylation (Bery et al., 2020). RAS expression levels were rescued by inhibition of the proteasome suggesting that the mechanism of targeted knockdown was indeed proteasome-dependent degradation (Bery et al., 2020). Furthermore, the specificities of the pan-RAS and KRAS degraders were consistent in a panel of RAS mutant cell lines and recapitulated *in vivo* using doxycycline-inducible pan-RAS degrader and KRAS degrader, both of which reduced KRAS-mutant tumor growth (Bery et al., 2020). However, only the pan-RAS degrader elicited effects in non-oncogenic KRAS mice xenografts (Bery et al., 2020).

Lim et al. have also explored the use of peptide-PROTACs in targeting RAS degradation. They examined the effectiveness of 10 different E3 ligases and identified the speckle type POZ protein (SPOP) as the most robust E3 ligase for this format. SPOP was then fused to various anti-RAS biologics including NS1 Mb, K27 and K55 DARPins, the R11.1.6 Sso7d, and the c-RAF RBD-CRD (Lim et al., 2021). All peptide-PROTACs with the exception of K55 effectively depleted KRAS levels, with the RBD-CRD-SPOP and K27-SPOP promoting the most robust RAS depletion (Lim et al., 2021). In addition, K27-reduced endogenous KRAS expression in AsPC-1 PDAC tumor cells ($KRAS^{G12D}$) resulting in decreased proliferation and increased apoptosis. Furthermore, these studies provided novel insights into the prevalence of the inactive state of KRAS through the use of GDP-state specific biologics such as K27 DARPin. Consistent with the relative population of the GDP-bound state by each KRAS

(Moore et al., 2020), K27-SPOP degraded $KRAS^{WT} > KRAS^{G12C} > KRAS^{G12D} > KRAS^{G12V}$. To further confirm the specificity of K27-SPOP for the GDP-bound state, treatment of $KRAS^{G12C}$ expressing cells with AMG510, which traps $KRAS^{G12C}$ in the GDP-bound state, enhanced K27-SPOP mediated degradation of KRAS but blocked the effects of the RBD-CRD-SPOP degrader. Interestingly the $KRAS^{Q61H}$ mutant was efficiently degraded by K27-SPOP despite its reported low catalytic activity (Hunter et al., 2015). Finally, as discussed in Section 2.2.1, Teng et al. recently developed a mutation specific degrader (Mb 12VC1) that selectively binds and degrades $KRAS^{G12V}$ and $KRAS^{G12C}$ (Teng et al., 2021). Together, these studies highlight the potential power of using peptide-PROTACs to target oncogenic RAS in cells. Further development of this technology will be needed to allow for effective and efficient delivery of such degraders to tumor cells in a patient.

4. Conclusion and future outlook

RAS remains at the forefront of cancer research given its prominent role in driving tumorigenesis. Despite the recent success in developing anti-RAS pharmacologics, continued efforts are required to expand on this success. RAS biologics have proven effective at inhibiting oncogenic RAS function through multiple mechanisms including disruption of PPI with effectors or activators of RAS as well as allosteric inhibition. However, these biologics require additional optimization to enhance their cellular uptake and tumor specificity in order to become useful clinical agents. With the utilization of delivery mechanisms such as cell-penetrating peptides, this shift appears ever more realistic. Additionally, with the entrance of liposomal-mediated cellular delivery of mRNA in the vaccine arena, many of the aforementioned genetically encodable biologics may yet reach the clinic setting. Nevertheless, biologics have proven powerful tools to further understand the biology of RAS, and these new insights are unveiling additional vulnerabilities of the formerly "undruggable" RAS.

Acknowledgments

The authors wish to apologize in advance for any omissions of references to our colleagues' work. M.W was supported by an NIH T32 (GM132055). J.P.O. was supported in part by a Merit Review Award (1I01BX002095) from the United States (US) Department of Veterans Affairs Biomedical Laboratory Research and Development Service, NIH awards (CA212608, CA138313, and GM103542), and start-up funds from the Hollings Cancer Center at MUSC. The contents of this article do not represent the views of the US. Department of Veterans Affairs or the United States Government.

Disclosures

J.P.O is listed as an inventor on a patent application on Monobodies targeting RAS filed by the Medical University of South Carolina and New York University (No. 62/862,924).

References

Ali, A. M., Atmaj, J., Van Oosterwijk, N., Groves, M. R., & Domling, A. (2019). Stapled peptides inhibitors: A new window for target drug discovery. *Computational and Structural Biotechnology Journal*, *17*, 263–281.

Antic, I., Biancucci, M., Zhu, Y., Gius, D. R., & Satchell, K. J. F. (2015). Site-specific processing of Ras and Rap1 switch I by a MARTX toxin effector domain. *Nature Communications*, *6*, 7396.

Bery, N., Legg, S., Debreczeni, J., Breed, J., Embrey, K., Stubbs, C., et al. (2019). KRAS-specific inhibition using a DARPin binding to a site in the allosteric lobe. *Nature Communications*, *10*, 2607.

Bery, N., Miller, A., & Rabbitts, T. (2020). A potent KRAS macromolecule degrader specifically targeting tumours with mutant KRAS. *Nature Communications*, *11*, 3233.

Biancucci, M., Minasov, G., Banerjee, A., Herrera, A., Woida, P. J., Kieffer, M. B., et al. (2018). The bacterial Ras/Rap1 site-specific endopeptidase RRSP cleaves Ras through an atypical mechanism to disrupt Ras-ERK signaling. *Science Signaling*, *11*, eaat8335.

Biocca, S., Pierandrei-Amaldi, P., Campioni, N., & Cattaneo, A. (1994). Intracellular immunization with cytosolic recombinant antibodies. *Biotechnology (N Y)*, *12*, 396–399.

Biocca, S., Pierandrei-Amaldi, P., & Cattaneo, A. (1993). Intracellular expression of anti-p21ras single chain Fv fragments inhibits meiotic maturation of xenopus oocytes. *Biochemical and Biophysical Research Communications*, *197*, 422–427.

Bond, M. J., Chu, L., Nalawansha, D. A., Li, K., & Crews, C. M. (2020). Targeted degradation of oncogenic KRAS(G12C) by VHL-recruiting PROTACs. *ACS Central Science*, *6*, 1367–1375.

Canon, J., Rex, K., Saiki, A. Y., Mohr, C., Cooke, K., Bagal, D., et al. (2019). The clinical KRAS(G12C) inhibitor AMG 510 drives anti-tumour immunity. *Nature*, *575*, 217–223.

Cardinale, A., Lener, M., Messina, S., Cattaneo, A., & Biocca, S. (1998). The mode of action of Y13-259 scFv fragment intracellularly expressed in mammalian cells. *FEBS Letters*, *439*, 197–202.

Cetin, M., Evenson, W. E., Gross, G. G., Jalali-Yazdi, F., Krieger, D., Arnold, D., et al. (2017). RasIns: Genetically encoded intrabodies of activated Ras proteins. *Journal of Molecular Biology*, *429*, 562–573.

Choi, D. K., Bae, J., Shin, S. M., Shin, J. Y., Kim, S., & Kim, Y. S. (2014). A general strategy for generating intact, full-length IgG antibodies that penetrate into the cytosol of living cells. *MAbs*, *6*, 1402–1414.

Chopra, R., Sadok, A., & Collins, I. (2019). A critical evaluation of the approaches to targeted protein degradation for drug discovery. *Drug Discovery Today: Technologies*, *31*, 5–13.

Clark, R., Wong, G., Arnheim, N., Nitecki, D., & McCormick, F. (1985). Antibodies specific for amino acid 12 of the ras oncogene product inhibit GTP binding. *Proceedings of the National Academy of Sciences of the United States of America*, *82*, 5280–5284.

Cochet, O., Kenigsberg, M., Delumeau, I., Virone-Oddos, A., Multon, M. C., Fridman, W. H., et al. (1998). Intracellular expression of an antibody fragment-neutralizing p21 ras promotes tumor regression. *Cancer Research*, *58*, 1170–1176.

Fell, J. B., Fischer, J. P., Baer, B. R., Blake, J. F., Bouhana, K., Briere, D. M., et al. (2020). Identification of the clinical development candidate MRTX849, a covalent KRAS(G12C) inhibitor for the treatment of cancer. *Journal of Medicinal Chemistry*, *63*, 6679–6693.

Feramisco, J. R., Clark, R., Wong, G., Arnheim, N., Milley, R., & McCormick, F. (1985). Transient reversion of ras oncogene-induced cell transformation by antibodies specific for amino acid 12 of ras protein. *Nature, 314*, 639–642.

Friedman, M., & Stahl, S. (2009). Engineered affinity proteins for tumour-targeting applications. *Biotechnology and Applied Biochemistry, 53*, 1–29.

Furth, M. E., Davis, L. J., Fleurdelys, B., & Scolnick, E. M. (1982). Monoclonal antibodies to the p21 products of the transforming gene of Harvey murine sarcoma virus and of the cellular ras gene family. *Journal of Virology, 43*, 294–304.

Galaine, J., Turco, C., Vauchy, C., Royer, B., Mercier-Letondal, P., Queiroz, L., et al. (2019). CD4 T cells target colorectal cancer antigens upregulated by oxaliplatin. *International Journal of Cancer, 145*, 3112–3125.

Grimm, S., Lundberg, E., Yu, F., Shibasaki, S., Vernet, E., Skogs, M., et al. (2010). Selection and characterisation of affibody molecules inhibiting the interaction between Ras and Raf in vitro. *New Biotechnology, 27*, 766–773.

Guillard, S., Kolasinska-Zwierz, P., Debreczeni, J., Breed, J., Zhang, J., Bery, N., et al. (2017). Structural and functional characterization of a DARPin which inhibits Ras nucleotide exchange. *Nature Communications, 8*, 16111.

Hallin, J., Engstrom, L. D., Hargis, L., Calinisan, A., Aranda, R., Briere, D. M., et al. (2020). The KRAS(G12C) inhibitor MRTX849 provides insight toward therapeutic susceptibility of KRAS-mutant cancers in mouse models and patients. *Cancer Discovery, 10*, 54–71.

Hingorani, S. R., Wang, L., Multani, A. S., Combs, C., Deramaudt, T. B., Hruban, R. H., et al. (2005). Trp53R172H and KrasG12D cooperate to promote chromosomal instability and widely metastatic pancreatic ductal adenocarcinoma in mice. *Cancer Cell, 7*, 469–483.

Hobbs, G. A., Baker, N. M., Miermont, A. M., Thurman, R. D., Pierobon, M., Tran, T. H., et al. (2020). Atypical KRAS(G12R) mutant is impaired in PI3K signaling and macropinocytosis in pancreatic cancer. *Cancer Discovery, 10*, 104–123.

Hobbs, G. A., Der, C. J., & Rossman, K. L. (2016). RAS isoforms and mutations in cancer at a glance. *Journal of Cell Science, 129*, 1287–1292.

Hong, D. S., Fakih, M. G., Strickler, J. H., Desai, J., Durm, G. A., Shapiro, G. I., et al. (2020). KRAS(G12C) inhibition with sotorasib in advanced solid tumors. *The New England Journal of Medicine, 383*, 1207–1217.

Hunter, J. C., Manandhar, A., Carrasco, M. A., Gurbani, D., Gondi, S., & Westover, K. D. (2015). Biochemical and structural analysis of common cancer-associated KRAS mutations. *Molecular Cancer Research, 13*, 1325–1335.

Hurd, C. A., Mott, H. R., & Owen, D. (2020). Therapeutic peptides targeting the Ras superfamily. *Peptide Science, 112*, e24165.

Huston, J. S., Levinson, D., Mudgett-Hunter, M., Tai, M. S., Novotny, J., Margolies, M. N., et al. (1988). Protein engineering of antibody binding sites: Recovery of specific activity in an anti-digoxin single-chain Fv analogue produced in Escherichia coli. *Proceedings of the National Academy of Sciences of the United States of America, 85*, 5879–5883.

Jin, J., Wu, Y., Chen, J., Shen, Y., Zhang, L., Zhang, H., et al. (2020). The peptide PROTAC modality: A novel strategy for targeted protein ubiquitination. *Theranostics, 10*, 10141–10153.

Karnoub, A. E., & Weinberg, R. A. (2008). Ras oncogenes: Split personalities. *Nature Reviews. Molecular Cell Biology, 9*, 517–531.

Kauke, M. J., Traxlmayr, M. W., Parker, J. A., Kiefer, J. D., Knihtila, R., McGee, J., et al. (2017). An engineered protein antagonist of K-Ras/B-Raf interaction. *Scientific Reports, 7*, 5831.

Khan, I., MarElia-Bennet, C., Lefler, J., Zuberi, M., Denbaum, E., Koide, A., et al. (2021). Targeting the KRAS alpha4-alpha5 allosteric interface inhibits pancreatic cancer tumorigenesis. *Small GTPases*, 1–14.

Khan, I., Rhett, J. M., & O'Bryan, J. P. (2020). Therapeutic targeting of RAS: New hope for drugging the "undruggable". *Biochimica et Biophysica Acta, Molecular Cell Research, 1867,* 118570.

Khan, I., Spencer-Smith, R., & O'Bryan, J. P. (2019). Targeting the alpha4-alpha5 dimerization interface of K-RAS inhibits tumor formation in vivo. *Oncogene, 38,* 2984–2993.

Koide, A., Bailey, C. W., Huang, X., & Koide, S. (1998). The fibronectin type III domain as a scaffold for novel binding proteins. *Journal of Molecular Biology, 284,* 1141–1151.

Koide, A., Wojcik, J., Gilbreth, R. N., Hoey, R. J., & Koide, S. (2012). Teaching an old scaffold new tricks: Monobodies constructed using alternative surfaces of the FN3 scaffold. *Journal of Molecular Biology, 415,* 393–405.

Kordiak, J., Szemraj, J., Grabska-Kobylecka, I., Bialasiewicz, P., Braun, M., Kordek, R., et al. (2019). Intratumor heterogeneity and tissue distribution of KRAS mutation in non-small cell lung cancer: Implications for detection of mutated KRAS oncogene in exhaled breath condensate. *Journal of Cancer Research and Clinical Oncology, 145,* 241–251.

Koschubs, T., Dengl, S., Durr, H., Kaluza, K., Georges, G., Hartl, C., et al. (2012). Allosteric antibody inhibition of human hepsin protease. *The Biochemical Journal, 442,* 483–494.

Kreiter, S., Vormehr, M., van de Roemer, N., Diken, M., Lower, M., Diekmann, J., et al. (2015). Mutant MHC class II epitopes drive therapeutic immune responses to cancer. *Nature, 520,* 692–696.

Kummer, L., Parizek, P., Rube, P., Millgramm, B., Prinz, A., Mittl, P. R., et al. (2012). Structural and functional analysis of phosphorylation-specific binders of the kinase ERK from designed ankyrin repeat protein libraries. *Proceedings of the National Academy of Sciences of the United States of America, 109,* E2248–E2257.

Kwak, J. S., Jeong, H. G., & Satchell, K. J. (2011). Vibrio vulnificus rtxA1 gene recombination generates toxin variants with altered potency during intestinal infection. *Proceedings of the National Academy of Sciences of the United States of America, 108,* 1645–1650.

Lacal, J. C., & Aaronson, S. A. (1986). Monoclonal antibody Y13-259 recognizes an epitope of the p21 ras molecule not directly involved in the GTP-binding activity of the protein. *Molecular and Cellular Biology, 6,* 1002–1009.

Lamy, A., Blanchard, F., Le Pessot, F., Sesboue, R., Di Fiore, F., Bossut, J., et al. (2011). Metastatic colorectal cancer KRAS genotyping in routine practice: Results and pitfalls. *Modern Pathology, 24,* 1090–1100.

Leshchiner, E. S., Parkhitko, A., Bird, G. H., Luccarelli, J., Bellairs, J. A., Escudero, S., et al. (2015). Direct inhibition of oncogenic KRAS by hydrocarbon-stapled SOS1 helices. *Proceedings of the National Academy of Sciences of the United States of America, 112,* 1761–1766.

Liao, W., Overman, M. J., Boutin, A. T., Shang, X., Zhao, D., Dey, P., et al. (2019). KRAS-IRF2 axis drives immune suppression and immune therapy resistance in colorectal cancer. *Cancer Cell, 35,* 559–572 e557.

Lim, S., Khoo, R., Juang, Y. C., Gopal, P., Zhang, H., Yeo, C., et al. (2021). Exquisitely specific anti-KRAS biodegraders inform on the cellular prevalence of nucleotide-loaded states. *ACS Central Science,* 7, 274–291.

Lin, Y., Chen, Z., Hu, C., Chen, Z. S., & Zhang, L. (2020). Recent progress in antitumor functions of the intracellular antibodies. *Drug Discovery Today, 25,* 1109–1120.

Linnemann, C., van Buuren, M. M., Bies, L., Verdegaal, E. M., Schotte, R., Calis, J. J., et al. (2015). High-throughput epitope discovery reveals frequent recognition of neo-antigens by CD4+ T cells in human melanoma. *Nature Medicine, 21,* 81–85.

Lu, R. M., Hwang, Y. C., Liu, I. J., Lee, C. C., Tsai, H. Z., Li, H. J., et al. (2020). Development of therapeutic antibodies for the treatment of diseases. *Journal of Biomedical Science, 27,* 1.

Ma, Y., Gu, Y., Zhang, Q., Han, Y., Yu, S., Lu, Z., et al. (2013). Targeted degradation of KRAS by an engineered ubiquitin ligase suppresses pancreatic cancer cell growth in vitro and in vivo. *Molecular Cancer Therapeutics*, *12*, 286–294.

Marschall, A. L., & Dubel, S. (2016). Antibodies inside of a cell can change its outside: Can intrabodies provide a new therapeutic paradigm? *Computational and Structural Biotechnology Journal*, *14*, 304–308.

McGee, J. H., Shim, S. Y., Lee, S. J., Swanson, P. K., Jiang, S. Y., Durney, M. A., et al. (2018). Exceptionally high-affinity Ras binders that remodel its effector domain. *The Journal of Biological Chemistry*, *293*, 3265–3280.

Montano, X., & Jimenez, A. (1995). Intracellular expression of the monoclonal anti-ras antibody Y13-259 blocks the transforming activity of ras oncogenes. *Cell Growth & Differentiation*, *6*, 597–605.

Moore, A. R., Rosenberg, S. C., McCormick, F., & Malek, S. (2020). RAS-targeted therapies: Is the undruggable drugged? *Nature Reviews. Drug Discovery*, *19*, 533–552.

Mulcahy, L. S., Smith, M. R., & Stacey, D. W. (1985). Requirement for ras proto-oncogene function during serum-stimulated growth of NIH 3T3 cells. *Nature*, *313*, 241–243.

Nielsen, D. S., Shepherd, N. E., Xu, W., Lucke, A. J., Stoermer, M. J., & Fairlie, D. P. (2017). Orally absorbed cyclic peptides. *Chemical Reviews*, *117*, 8094–8128.

Nilsson, F. Y., & Tolmachev, V. (2007). Affibody molecules: New protein domains for molecular imaging and targeted tumor therapy. *Current Opinion in Drug Discovery & Development*, *10*, 167–175.

Nygren, P. A. (2008). Alternative binding proteins: Affibody binding proteins developed from a small three-helix bundle scaffold. *The FEBS Journal*, *275*, 2668–2676.

Oltedal, S., Aasprong, O. G., Moller, J. H., Korner, H., Gilje, B., Tjensvoll, K., et al. (2011). Heterogeneous distribution of K-ras mutations in primary colon carcinomas: Implications for EGFR-directed therapy. *International Journal of Colorectal Disease*, *26*, 1271–1277.

Ong, Y. S., Gao, L., Kalesh, K. A., Yu, Z., Wang, J., Liu, C., et al. (2017). Recent advances in synthesis and identification of cyclic peptides for bioapplications. *Current Topics in Medicinal Chemistry*, *17*, 2302–2318.

Ostrem, J. M., Peters, U., Sos, M. L., Wells, J. A., & Shokat, K. M. (2013). K-Ras(G12C) inhibitors allosterically control GTP affinity and effector interactions. *Nature*, *503*, 548–551.

Ott, P. A., Hu, Z., Keskin, D. B., Shukla, S. A., Sun, J., Bozym, D. J., et al. (2017). An immunogenic personal neoantigen vaccine for patients with melanoma. *Nature*, *547*, 217–221.

Pan, X. Y., Liu, X. J., Li, J., Zhen, S. J., Liu, D. X., Feng, Q., et al. (2017). The antitumor efficacy of anti-p21Ras scFv mediated by the dual-promoter-regulated recombinant adenovirus KGHV300. *Gene Therapy*, *24*, 40–48.

Parizek, P., Kummer, L., Rube, P., Prinz, A., Herberg, F. W., & Pluckthun, A. (2012). Designed ankyrin repeat proteins (DARPins) as novel isoform-specific intracellular inhibitors of c-Jun N-terminal kinases. *ACS Chemical Biology*, *7*, 1356–1366.

Patgiri, A., Yadav, K. K., Arora, P. S., & Bar-Sagi, D. (2011). An orthosteric inhibitor of the Ras-Sos interaction. *Nature Chemical Biology*, *7*, 585–587.

Pei, D., Chen, K., & Liao, H. (2018). Targeting Ras with macromolecules. *Cold Spring Harbor Perspectives in Medicine*, *8*, a031476.

Pettersson, M., & Crews, C. M. (2019). PROteolysis TArgeting chimeras (PROTACs)—Past, present and future. *Drug Discovery Today: Technologies*, *31*, 15–27.

Philippe, G. J. B., Craik, D. J., & Henriques, S. T. (2021). Converting peptides into drugs targeting intracellular protein-protein interactions. *Drug Discovery Today*, *26*, 1521–1531.

Poulin, E. J., Bera, A. K., Lu, J., Lin, Y. J., Strasser, S. D., Paulo, J. A., et al. (2019). Tissue-specific oncogenic activity of KRAS(A146T). *Cancer Discovery*, *9*, 738–755.
Qu, J., Ren, X., Xue, F., He, Y., Zhang, R., Zheng, Y., et al. (2020). Specific knockdown of alpha-synuclein by peptide-directed proteasome degradation rescued its associated neurotoxicity. *Cell Chemical Biology*, *27*, 751–762 e754.
Richman, S. D., Chambers, P., Seymour, M. T., Daly, C., Grant, S., Hemmings, G., et al. (2011). Intra-tumoral heterogeneity of KRAS and BRAF mutation status in patients with advanced colorectal cancer (aCRC) and cost-effectiveness of multiple sample testing. *Analytical Cellular Pathology (Amsterdam)*, *34*, 61–66.
Roth, S., Macartney, T. J., Konopacka, A., Chan, K. H., Zhou, H., Queisser, M. A., et al. (2020). Targeting endogenous K-RAS for degradation through the affinity-directed protein missile system. *Cell Chemical Biology*, *27*, 1151–1163 e1156.
Ryan, M. B., & Corcoran, R. B. (2018). Therapeutic strategies to target RAS-mutant cancers. *Nature Reviews. Clinical Oncology*, *15*, 709–720.
Sahin, U., Derhovanessian, E., Miller, M., Kloke, B. P., Simon, P., Lower, M., et al. (2017). Personalized RNA mutanome vaccines mobilize poly-specific therapeutic immunity against cancer. *Nature*, *547*, 222–226.
Sakamoto, K., Kamada, Y., Sameshima, T., Yaguchi, M., Niida, A., Sasaki, S., et al. (2017). K-Ras(G12D)-selective inhibitory peptides generated by random peptide T7 phage display technology. *Biochemical and Biophysical Research Communications*, *484*, 605–611.
Sakamoto, K., Masutani, T., & Hirokawa, T. (2020). Generation of KS-58 as the first K-Ras(G12D)-inhibitory peptide presenting anti-cancer activity in vivo. *Scientific Reports*, *10*, 21671.
Santini, B. L., & Zacharias, M. (2020). Rapid in silico design of potential cyclic peptide binders targeting protein-protein interfaces. *Frontiers in Chemistry*, *8*, 573259.
Sha, F., Salzman, G., Gupta, A., & Koide, S. (2017). Monobodies and other synthetic binding proteins for expanding protein science. *Protein Science*, *26*, 910–924.
Shibasaki, S., Karasaki, M., Graslund, T., Nygren, P. A., Sano, H., & Iwasaki, T. (2014). Inhibitory effects of H-Ras/Raf-1-binding affibody molecules on synovial cell function. *AMB Express*, *4*, 82.
Shin, S. M., Choi, D. K., Jung, K., Bae, J., Kim, J. S., Park, S. W., et al. (2017). Antibody targeting intracellular oncogenic Ras mutants exerts anti-tumour effects after systemic administration. *Nature Communications*, *8*, 15090.
Shin, S. M., Kim, J. S., Park, S. W., Jun, S. Y., Kweon, H. J., Choi, D. K., et al. (2020). Direct targeting of oncogenic RAS mutants with a tumor-specific cytosol-penetrating antibody inhibits RAS mutant-driven tumor growth. *Science Advances*, *6*, eaay2174.
Skoulidis, F., Li, B. T., Dy, G. K., Price, T. J., Falchook, G. S., Wolf, J., et al. (2021). Sotorasib for lung cancers with KRAS p.G12C mutation. *The New England Journal of Medicine*, *384*, 2371–2381.
Spencer-Smith, R., Koide, A., Zhou, Y., Eguchi, R. R., Sha, F., Gajwani, P., et al. (2017). Inhibition of RAS function through targeting an allosteric regulatory site. *Nature Chemical Biology*, *13*, 62–68.
Spencer-Smith, R., Li, L., Prasad, S., Koide, A., Koide, S., & O'Bryan, J. P. (2019). Targeting the alpha4-alpha5 interface of RAS results in multiple levels of inhibition. *Small GTPases*, *10*, 378–387.
Stocks, M. R. (2004). Intrabodies: Production and promise. *Drug Discovery Today*, *9*, 960–966.
Stumpp, M. T., Binz, H. K., & Amstutz, P. (2008). DARPins: A new generation of protein therapeutics. *Drug Discovery Today*, *13*, 695–701.
Takeuchi, K., Misaki, I., Tokunaga, Y., Fujisaki, M., Kamoshida, H., Takizawa, T., et al. (2021). Conformational plasticity of cyclic ras-inhibitor peptides defines cell permeabilization activity. *Angewandte Chemie (International Ed. in English)*, *60*, 6567–6572.

Tanaka, T., Lobato, M. N., & Rabbitts, T. H. (2003). Single domain intracellular antibodies: A minimal fragment for direct in vivo selection of antigen-specific intrabodies. *Journal of Molecular Biology*, *331*, 1109–1120.

Tanaka, T., & Rabbitts, T. H. (2003). Intrabodies based on intracellular capture frameworks that bind the RAS protein with high affinity and impair oncogenic transformation. *The EMBO Journal*, *22*, 1025–1035.

Tanaka, T., & Rabbitts, T. H. (2010). Interfering with RAS-effector protein interactions prevent RAS-dependent tumour initiation and causes stop-start control of cancer growth. *Oncogene*, *29*, 6064–6070.

Tanaka, T., Williams, R. L., & Rabbitts, T. H. (2007). Tumour prevention by a single antibody domain targeting the interaction of signal transduction proteins with RAS. *The EMBO Journal*, *26*, 3250–3259.

Tay, R. E., Richardson, E. K., & Toh, H. C. (2021). Revisiting the role of CD4(+) T cells in cancer immunotherapy-new insights into old paradigms. *Cancer Gene Therapy*, *28*, 5–17.

Teng, K. W., Tsai, S. T., Hattori, T., Fedele, C., Koide, A., Yang, C., et al. (2021). Selective and noncovalent targeting of RAS mutants for inhibition and degradation. *Nature Communications*, *12*, 2656.

Traut, T. W. (1994). Physiological concentrations of purines and pyrimidines. *Molecular and Cellular Biochemistry*, *140*, 1–22.

Traxlmayr, M. W., Kiefer, J. D., Srinivas, R. R., Lobner, E., Tisdale, A. W., Mehta, N. K., et al. (2016). Strong enrichment of aromatic residues in binding sites from a charge-neutralized hyperthermostable Sso7d scaffold library. *The Journal of Biological Chemistry*, *291*, 22496–22508.

Tse, E., Lobato, M. N., Forster, A., Tanaka, T., Chung, G. T., & Rabbitts, T. H. (2002). Intracellular antibody capture technology: Application to selection of intracellular antibodies recognising the BCR-ABL oncogenic protein. *Journal of Molecular Biology*, *317*, 85–94.

Upadhyaya, P., Qian, Z., Habir, N. A., & Pei, D. (2014). Direct Ras inhibitors identified from a structurally rigidified bicyclic peptide library. *Tetrahedron*, *70*, 7714–7720.

Upadhyaya, P., Qian, Z., Selner, N. G., Clippinger, S. R., Wu, Z., Briesewitz, R., et al. (2015). Inhibition of Ras signaling by blocking Ras-effector interactions with cyclic peptides. *Angewandte Chemie (International Ed. in English)*, *54*, 7602–7606.

Vatansever, S., Erman, B., & Gumus, Z. H. (2019). Oncogenic G12D mutation alters local conformations and dynamics of K-Ras. *Scientific Reports*, *9*, 11730.

Vidimar, V., Beilhartz, G. L., Park, M., Biancucci, M., Kieffer, M. B., Gius, D. R., et al. (2020). An engineered chimeric toxin that cleaves activated mutant and wild-type RAS inhibits tumor growth. *Proceedings of the National Academy of Sciences of the United States of America*, *117*, 16938–16948.

Visintin, M., Settanni, G., Maritan, A., Graziosi, S., Marks, J. D., & Cattaneo, A. (2002). The intracellular antibody capture technology (IACT): Towards a consensus sequence for intracellular antibodies. *Journal of Molecular Biology*, *317*, 73–83.

Wang, M., Hong, Y., Feng, Q., Pan, X., Song, S., Cui, J., et al. (2018). Recombinant adenovirus KGHV500 and CIK cells codeliver anti-p21-Ras scFv for the treatment of gastric Cancer with wild-type Ras overexpression. *Molecular Therapy Oncolytics*, *11*, 90–101.

Waters, A. M., & Der, C. J. (2018). KRAS: The critical driver and therapeutic target for pancreatic cancer. *Cold Spring Harbor Perspectives in Medicine*, *8*, a031435.

Weiner, G. J. (2007). Monoclonal antibody mechanisms of action in cancer. *Immunologic Research*, *39*, 271–278.

Werge, T. M., Baldari, C. T., & Telford, J. L. (1994). Intracellular single chain Fv antibody inhibits Ras activity in T-cell antigen receptor stimulated Jurkat cells. *FEBS Letters*, *351*, 393–396.

Wu, X., Upadhyaya, P., Villalona-Calero, M. A., Briesewitz, R., & Pei, D. (2013). Inhibition of Ras-effector interaction by cyclic peptides. *MedChemComm*, *4*, 378–382.
Yang, J. L., Liu, D. X., Zhen, S. J., Zhou, Y. G., Zhang, D. J., Yang, L. Y., et al. (2016). A novel anti-p21Ras scFv antibody reacting specifically with human tumour cell lines and primary tumour tissues. *BMC Cancer*, *16*, 131.
Zhang, Z., Gao, R., Hu, Q., Peacock, H., Peacock, D. M., Dai, S., et al. (2020). GTP-state-selective cyclic peptide ligands of K-Ras(G12D) block its interaction with Raf. *ACS Central Science*, *6*, 1753–1761.
Zhang, L., Yu, X., Zheng, L., Zhang, Y., Li, Y., Fang, Q., et al. (2018). Lineage tracking reveals dynamic relationships of T cells in colorectal cancer. *Nature*, *564*, 268–272.
Zuberi, M., Khan, I., & O'Bryan, J. P. (2020). Inhibition of RAS: Proven and potential vulnerabilities. *Biochemical Society Transactions*, *48*, 1831–1841.

CHAPTER NINE

Unraveling and targeting RAS-driven metabolic signaling for therapeutic gain

Jonathan M. DeLiberty[a], Ryan Robb[a], Claire E. Gates[a], and Kirsten L. Bryant[a,b,*]

[a]Department of Pharmacology, University of North Carolina at Chapel Hill, Chapel Hill, NC, United States
[b]Lineberger Comprehensive Cancer Center, University of North Carolina at Chapel Hill, Chapel Hill, NC, United States
*Corresponding author: e-mail address: bryantkl@email.unc.edu

Contents

1. Introduction 268
   1.1 RAS basics 268
   1.2 Mutant RAS drives pancreatic cancer 270
   1.3 Direct and indirect strategies for targeting mutant RAS 271
2. Glycolysis 274
   2.1 Enhanced glycolytic flux in KRAS mutant cancer 274
   2.2 Pentose phosphate pathway 276
   2.3 Hexosamine biosynthesis pathway 276
3. Mitochondrial signaling and dynamics 277
   3.1 Glutamine metabolism and bioenergetics 278
   3.2 Mitochondrial morphological dynamics 279
4. Autophagy 280
   4.1 An overview of the autophagic pathway and machinery 280
   4.2 AMPK and mTOR 283
   4.3 ULK and the initiation complex 284
   4.4 VPS34, Beclin-1, nucleation complex 285
   4.5 Maturation and degradation 285
   4.6 Mitophagy 288
5. Lipid scavenging and metabolism 289
6. Macropinocytosis 290
7. Concluding remarks 291
Acknowledgments 292
References 292

Abstract

RAS mutations are among the most frequent oncogenic drivers observed in human cancers. With a lack of available treatment options, RAS-mutant cancers account for many of the deadliest cancers in the United States. Recent studies established that

Advances in Cancer Research, Volume 153
ISSN 0065-230X
https://doi.org/10.1016/bs.acr.2021.07.010

altered metabolic requirements are a hallmark of cancer, and many of these alterations are driven by aberrant RAS signaling. Specifically, RAS-driven cancers are characterized by upregulated glycolysis, the differential channeling of glycolytic intermediates, upregulated nutrient scavenging pathways such as autophagy and macropinocytosis, and altered glutamine utilization and mitochondrial function. This unique metabolic landscape promotes tumorigenesis, proliferation, survival in nutrient deficient environments and confers resistance to conventional cytotoxic and targeted therapies. Emerging work demonstrates how these dependencies can be therapeutically exploited in vitro and in vivo with many metabolic inhibitors currently in clinical trials. This review aims to outline the unique metabolic requirements induced by aberrant RAS signaling and how these altered dependencies present opportunities for therapeutic intervention.

1. Introduction

1.1 RAS basics

The discovery that mutationally activated *RAS* genes are capable of cellular transformation initiated nearly 40 years of intense study (Cox & Der, 2010). Understanding RAS protein biology became imperative to the cancer biology field due to the high frequency of *RAS* mutations in human disease, including the three deadliest cancers (lung adenocarcinoma (LAC), colorectal cancer (CRC), and pancreatic ductal adenocarcinoma (PDAC)) in the United States (Siegel, Miller, Fuchs, & Jemal, 2021). The three human *RAS* genes (*HRAS, NRAS*, and *KRAS*) encode four highly similar RAS proteins (HRAS, NRAS, KRAS4A, and KRAS4B) that function as intracellular GTPases (Cox & Der, 2010). These small GTPases function as molecular "on/off" switches, cycling between a GTP-bound "on-state" and a GDP-bound "off-state" (Fig. 1). Two classes of proteins facilitate this cycle: Guanine nucleotide exchange factors (GEFs), such as SOS1, function to exchange GDP for GTP, thereby activating RAS. GTPase-activating proteins (GAPs) inactivate RAS by increasing RAS-mediated GTP hydrolysis. Interestingly, nearly all *RAS* mutations occur at one of three residue hotspots: G12, G13, or Q61. These single nucleotide point mutations cause constitutive RAS protein activation either by preventing GAPs from inactivating RAS (G12 and G13) or through interference of the coordination of a water molecule required for the hydrolysis of GTP (Q61). Mutationally activated RAS proteins continuously signal to downstream effectors, thus resulting in accelerated proliferation, metabolic rewiring, evasion of the immune response, metastasis, and many other hallmarks of cancer. While single nucleotide point mutations are a common mechanism of

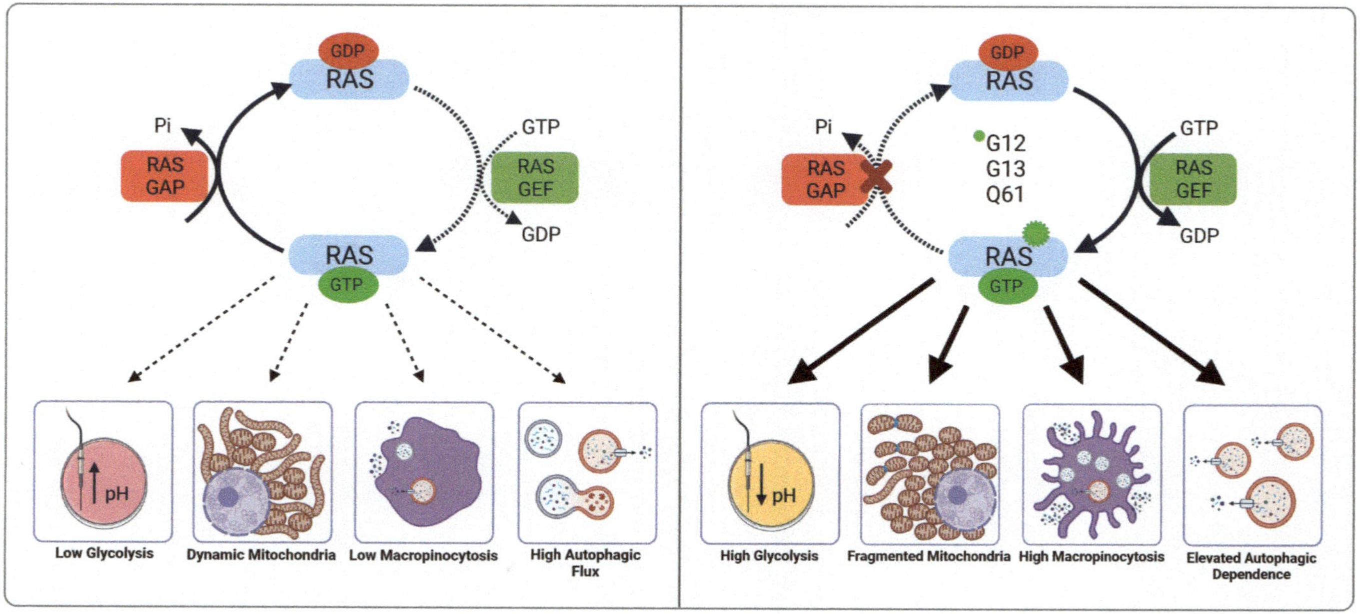

Fig. 1 RAS proteins are small GTPases that regulate critical metabolic phenotypes. Wild-type (WT) RAS proteins function as molecular "on/off" switches that cycle between the inactive GDP-bound (red) and the active GTP-bound (green) state. Active RAS proteins transduce signals to pathways involved in proliferation, metabolism, cell death, and other critical cellular pathways. Single nucleotide point mutations, typically observed at residues G12, G13, and Q61, result in persistent GTP-bound RAS proteins and constitutive activation. Oncogenic RAS over activates downstream effectors such as the RAF-MEK-ERK MAPK cascade and the PI3K-AKT-mTOR pathway to drive proliferation, survival, anti-apoptotic factors, and metastasis. Aberrant RAS signaling also metabolically rewires the cell, upregulating macropinocytosis, increasing the dependence on autophagy and glycolysis, and altering mitochondrial dynamics in order to support accelerated growth.

oncogenic RAS signaling, there are other processes by which RAS can be inappropriately activated. Loss of the RasGAP neurofibromatosis type 1 (NF-1) impairs the negative regulation of RAS activity and promotes the development of a variety of malignancies (Philpott, Tovell, Frayling, Cooper, & Upadhyaya, 2017). RAS also functions downstream of receptor tyrosine kinases. Hence, when receptor tyrosine kinases such as EGFR and HER2 are amplified or mutated in cancers, RAS is aberrantly activated (Du & Lovly, 2018). RAS signaling can also be overactivated by copy number amplification (Sasaki et al., 2011).

While activated RAS proteins share similarities, there is an unequal distribution in RAS mutational frequencies and an increasing body of evidence that not all RAS mutants govern the exact same cellular phenotypes. The distribution of *RAS* missense mutations across all cancers display a significant bias toward *KRAS* (85%), with mutations in *NRAS* being less frequent (12%) and rarer in *HRAS* (3%) (Cox & Der, 2010). Furthermore, the vast majority of KRAS mutations in PDAC occur at the G12 residue. Among this population, the predominant G12 substitutions are G12D (51%), G12V (30%), and G12R (12%), while G12C, G12S, and G12A occur less frequently. Notably, this mutation frequency is distributed differently in other tissue types harboring *KRAS*-mutations (Cook, Melloni, Gulhan, Park, & Haigis, 2021). A more comprehensive detailing on mutation frequency and biological differences of RAS mutants is reviewed elsewhere (Cook et al., 2021; Haigis, 2017). This review will focus on KRAS-mutant PDAC and will make note of other RAS-driven cancers when relevant.

1.2 Mutant RAS drives pancreatic cancer

PDAC is a particularly deadly cancer with a 5-year survival rate of 10% (Siegel et al., 2021). It is an exceptional model for studying KRAS biology as over 95% of PDACs harbor a *KRAS* missense mutation, which encodes mutationally activated KRAS proteins capable of driving tumorigenesis and tumor maintenance (Waters & Der, 2018). In fact, *RAS* mutation is the initiating, critical genetic alteration for the cellular transformation to pancreatic intraepithelial neoplasia (PanIN) lesions, which are precursors to PDAC (Yanagisawa et al., 1993). Following oncogenic, gain-of-function *KRAS* mutations, p16/*CDKN2A* loss precedes PanIN-2 development; and later, *TP53* and *SMAD4* inactivating mutations allow for development of fully invasive PDAC (Maitra & Hruban, 2008).

Canonically, KRAS signals to a wide array of important effectors and cellular pathways. Specifically, the most essential pathway responsible for transducing mutant KRAS signaling is the RAF-MEK-ERK mitogen activated protein kinase (MAPK) pathway. Support for this concept developed from studies where genetically engineered mouse models (GEMMs) with pancreas-specific expression of $BRAF^{V600E}$, but not $PIK3CA^{H1047R}$, formed PanIN lesions (Collisson et al., 2012). Expression of both $BRAF^{V600E}$ and $TP53^{R270H}$, led to the robust development of PDAC in mice. The ability of an oncogenic driver, such as $BRAF^{V600E}$, to phenocopy $KRAS^{G12D}$ strongly implicates that the ERK MAPK cascade encompasses the key effectors of aberrant RAS signaling. Notably, the phosphoinositide-3 kinase (PI3K) pathway is also an important effector for RAS signaling (Castellano et al., 2013), as inhibition of this pathway is known to reduce RAS-mutant tumor growth (Engelman et al., 2008).

1.3 Direct and indirect strategies for targeting mutant RAS

There are five general strategies for targeting aberrant RAS signaling in human cancers: (1) direct RAS inhibition, (2) inhibition of RAS membrane association, (3) targeting key RAS effectors, (4) pursuing synthetic lethal interactors of mutant RAS, and (5) targeting the unique metabolic requirements of *RAS*-mutant cancers (Papke & Der, 2017). Direct RAS inhibition recently saw success with the first ever FDA-approved $KRAS^{G12C}$ inhibitor for NSCLC. While this is a massive step forward for the field, $KRAS^{G12C}$ mutations are present in only around half of mutant KRAS-driven LAC (~46%), more infrequently in CRC (~8%), and rarely in PDAC (~2%) (Cox, Fesik, Kimmelman, Luo, & Der, 2014), Therefore, additional mutation-specific or pan-KRAS inhibitors are needed to reach a larger subset of patients. RAS plasma membrane association is required for proper RAS signaling (Ahearn, Haigis, Bar-Sagi, & Philips, 2012). In the early 1990s farnesyl transferase inhibitors (FTIs) appeared to be a promising strategy to inhibit this association (Berndt, Hamilton, & Sebti, 2011). Farnesylation is the first step in the process of RAS posttranslational modifications whereby a farnesyl isoprenoid is added to the C-termini to allow for membrane association (Cox & Der, 2010). However, FTIs are only effective at preventing HRAS plasma membrane association, as both NRAS and KRAS proteins can be geranylgeranylated, effectively compensating for the loss of farnesylation (Cox et al., 2014). While recent work has reinvigorated an interest in FTIs for the treatment of HRAS-driven cancers, such as head and neck squamous cell carcinoma (Ho et al., 2021), it is likely alternative strategies will be necessary for NRAS- and KRAS-driven cancers.

Targeting key RAS effectors, specifically the RAF-MEK-ERK MAPK cascade and the PI3K-AKT-mTOR pathway, is an intense focus of anti-KRAS research (Papke & Der, 2017). It is well established that when used as monotherapy, inhibitors of either of these two pathways are not effective in RAS mutant cancers. For example, inhibitors of the MAPK pathway (i.e., RAF and MEK inhibitors) have a limited efficacy in RAS-mutant cancers, due to ERK reactivation following treatment. Hayes et al. studied the effects of an ERK inhibitor (ERKi), SCH772984, in PDAC cell lines and demonstrated that MEKi-resistant PDAC cells are responsive to ERK inhibition through a MYC degradation-dependent mechanism (Hayes et al., 2016). MYC is a transcription factor functioning as a critical terminal effector transducing oncogenic KRAS signaling (Soucek et al., 2008, 2013). Previous studies show that ERK1/2 and ERK5 integrate KRAS signals and exert direct effects on MYC, underpinning ERK as a critical effector of KRAS signaling and a potentially efficacious therapeutic target (Vaseva et al., 2018). These targeted therapies appear to be limited by acquired resistance (Duncan et al., 2012; Heidorn et al., 2010) and combination therapies targeting the pathways responsible for identified resistance mechanisms are actively being explored. For instance, dual targeting of RAF and ERK (vertical inhibition) caused increased cell death, prevented compensatory ERK reactivation, and synergistically decreased PDAC proliferation at low doses (Ozkan-Dagliyan et al., 2020).

Altered metabolism and metabolic rewiring has emerged as a hallmark of cancer (Hanahan & Weinberg, 2011). KRAS-driven cancers display an increased dependence on glycolysis, altered mitochondrial signaling and dynamics, and an increased reliance on nutrient scavenging pathways such as autophagy and macropinocytosis, as well as many other key metabolic processes (Bryant, Mancias, Kimmelman, & Der, 2014). Herein, we will provide an update on the most recent literature related to the metabolic consequences of oncogenic RAS signaling with a particular emphasis on KRAS-mutant PDAC. (Fig. 2). We will particularly highlight the multitude of recent advances related to different types of nutrient scavenging pathways employed by RAS-mutant cancers, including autophagy and macropinocytosis. While nutrient scavenging can be induced via nutrient deprivation, hypoxia, and loss of growth factor stimulation, it is also upregulated as a resistance mechanism to many cancer therapies, and the pathways that regulate these cellular processes are targets for anti-cancer drug discovery.

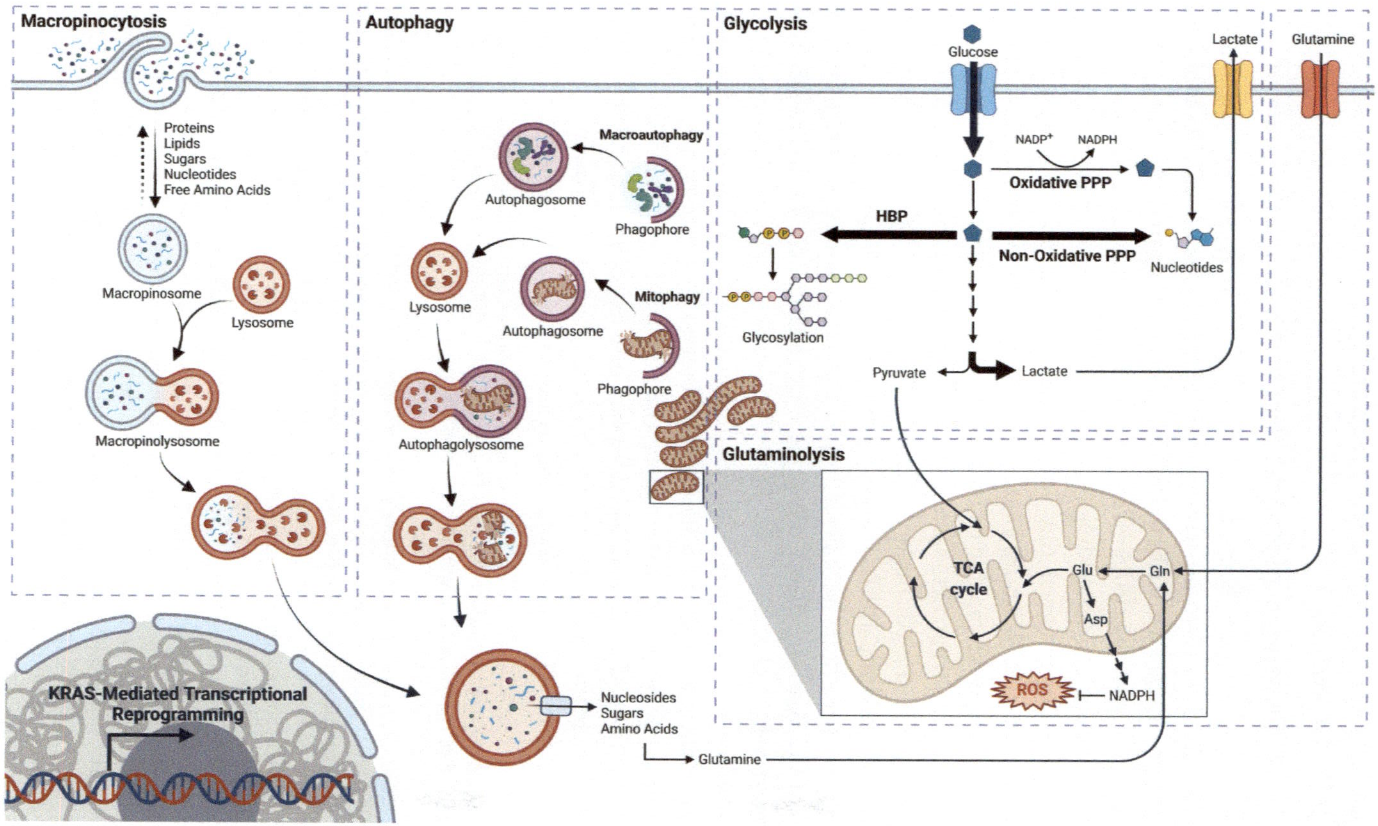

Fig. 2 See figure legend on next page.

2. Glycolysis

2.1 Enhanced glycolytic flux in KRAS mutant cancer

An early and well-appreciated study by Otto Warburg concluded that under aerobic conditions, tumor tissues metabolize approximately 10-fold more glucose to lactate than their normal cell counterparts (Warburg, 1925). Warburg originally misinterpreted his findings as evidence that cancer cells have defective mitochondria (Warburg, 1956). However, in the 1970s, Efraim Racker revisited this work, coined the term "Warburg Effect," and reinvigorated Warburg's earlier findings that supported the concept that tumor tissues were capable of respiration. Furthermore, Racker and others demonstrated that growth factor signaling was responsible for regulating glycolysis in tumors (Birnbaum, Haspel, & Rosen, 1987; Flier, Mueckler, Usher, & Lodish, 1987; Racker, Resnick, & Feldman, 1985). While a variety of hypotheses have been proposed in an attempt to describe the function of the Warburg Effect, a complete explanation has yet to be established (Liberti & Locasale, 2016). However, it remains clear that RAS signaling is responsible for upregulated glycolytic signaling in several different tumor types. High glycolytic rates have been observed both in model cell systems transformed with the *RAS* oncogene (Chiaradonna et al., 2005, 2006; Vizan et al., 2005) as well as established human cancers (Yun et al., 2009), particularly pancreatic cancer (Yan, Raj, Yao, & Ying, 2019; Zhou et al., 2012). According to TCGA analysis, pancreatic cancer patients with *KRAS*- or

Fig. 2 Metabolic rewiring is a hallmark of RAS-mutant cancers. Oncogenic RAS upregulates many key metabolic pathways to sustain the aggressive growth observed with many RAS-mutant cancers. Macropinocytosis is a bulk nutrient uptake process whereby cells nonselectively endocytose extracellular fluid to supply the cell with amino acids, lipids, proteins, and sugars to fuel growth. RAS-mutant cancers are highly addicted to autophagy, a "self-eating" process by which the cell degrades and recycles damaged proteins and macromolecules via a lysosome-dependent mechanism to sustain growth and survival in nutrient poor conditions. Selective autophagic processes such as mitophagy, a process by which damaged or impaired mitochondria are degraded to prevent inefficient energy production and the buildup of reactive oxygen species are also upregulated. Glycolysis is a critical metabolic pathway for RAS-driven cancer cell growth. and RAS-driven cancers upregulate the nonoxidative arm of the pentose phosphate pathway (PPP) and hexosamine biosynthesis pathway (HBP) but remain dependent on oxidative phosphorylation. Mitochondrial functions and dynamics are significantly altered following RAS-transformation. RAS-transformed cells demonstrate a highly fragmented morphology and a dependency on glutamine metabolism to drive the TCA cycle.

MYC-amplified tumors have a higher expression of glycolytic genes, which was in turn associated with poor survival (Karasinska et al., 2020).

Glucose transport across the membrane is accomplished through two different classes of transporters: ATP-dependent glucose transporters (GLUTs) and sodium dependent glucose cotransporters (SGLTs). While there are 14 different GLUT family members, only GLUT1 is associated with PDAC progression (Basturk et al., 2011). Multiple GLUT1 inhibitors are currently under investigation as possible anti-cancer therapies (Meng et al., 2019). The discovery of the alternative glucose transporter SGLT2 is more recent (Scafoglio et al., 2015); however, it appears that targeting glucose transport through this mechanism may be a tractable treatment option. The SGLT2 inhibitor canagliflorzin was demonstrated to inhibit PDAC xenograft tumor growth alone and also sensitized tumors to PI3K inhibition (Hopkins et al., 2018; Scafoglio et al., 2015). The effects of mutant KRAS on glycolysis have been attributed to the transcriptional upregulation of glucose transporters and glycolytic genes (Kerr, Gaude, Turrell, Frezza, & Martins, 2016; Ying et al., 2012). Interestingly, it has recently been established that a direct interaction between KRAS4A and hexokinase 1 (HK1) alters the kinase activity of HK1, providing evidence that it is a direct substrate of KRAS (Amendola et al., 2019). It should be noted that additionally, HK2 is required for tumor initiation and maintenance of mutant KRAS-driven models of nonsmall cell lung carcinoma (NSCLC) (Patra et al., 2013). Hence, the glycolytic rewiring triggered by mutant KRAS is induced by multiple mechanisms. Another mechanism for targeting glycolysis is with the synthetic glucose derivative, 2-deoxy-D-glucose (2-DG), which has a hydrogen replacing the 2-hydoxyl group present in glucose. Consequently, it competitively inhibits the production of glucose-6-phosphate from glucose. There have been preclinical demonstrations that 2-DG treatment enhanced PDAC susceptibility to gemcitabine (Dai et al., 2020) as well as inhibitors of proteins downstream of RAS, such as MEK (Yan et al., 2021).

In a landmark study published in 2012, Ying and colleagues developed an inducible $Kras^{G12D}$-driven mouse model of PDAC (*iKras*) and performed comprehensive transcriptomic and metabolomic analyses in the presence and absence of mutant Kras signaling (Ying et al., 2012). This study not only confirmed previous findings regarding the link between KRAS and glucose import in PDAC, but also demonstrated how mutant KRAS signaling results in the funneling of glycolytic intermediates into particular branches of the glycolytic pathway. The biology associated with these alterations as well as the therapeutic opportunities these changes suggest, will be described in more detail below.

2.2 Pentose phosphate pathway

The pentose phosphate pathway (PPP) branches from glucose-6-phosphate and is composed of both an oxidative and a nonoxidative arm. The product of the oxidative arm is ribose-5-phosphate, which is utilized to generate nucleic acids, and the product of the nonoxidative arm is NADPH, which provides the reducing power necessary for the production of fatty acids, nucleotides, and nonessential amino acids (Ge et al., 2020). It had previously been suggested that the oxidative and nonoxidative arm of the PPP could decouple (Boros et al., 1998). Indeed, Ying et al. demonstrated that oncogenic Kras signaling through Mek and Myc leads to a downregulation of flux through the oxidative arm of the PPP, and increased flux through the nonoxidative arm of the PPP to sustain the nucleotide pool (Ying et al., 2012). Studies performed in human PDAC supported these findings and further demonstrated that human PDAC cell lines were sensitive to targeting of the PPP independent of their reliance on KRAS for proliferation (Santana-Codina et al., 2018). This was demonstrated both through genetic depletion of ribose-5-phosphate isomerase or by inhibiting pyrimidine biosynthesis through targeting dihydroorotate dehydrogenase (DHODH) with leflunomide (Santana-Codina et al., 2018). Oncogenic KRAS-driven upregulation of flux through the PPP has also been shown to contribute to resistance to standard of care treatment in PDAC. For example, increased pyrimidine biosynthesis leads to high intrinsic levels of deoxycytidine (dCTP) in PDAC, which can in turn out-compete the pyrimidine antagonist, gemcitabine (Shukla et al., 2017). Interestingly, a recent study also proposed that metabolite transfer of pyrimidines from tumor-associated macrophages contributes to the high pool of dCTP in PDAC (Halbrook et al., 2019). It should be noted that the oxidative arm of the PPP has been implicated in KRAS-driven transformation (Weinberg et al., 2010) and that RAS signaling has been proposed to promote resistance to reactive oxygen species (ROS) stress through the production of NADPH generated through the oxidative arm of the PPP (Cheung et al., 2020; Denicola et al., 2011). Thus, the regulation of the PPP by oncogenic KRAS is still not fully elucidated and may be context dependent.

2.3 Hexosamine biosynthesis pathway

The hexosamine biosynthesis pathway (HBP) branches from the glycolytic pathway when fructose-6-phosphate is converted to glucosamine-6-phosphate (GlcN-6P) by glutamine-fructose-6-phosphate transaminase 1

(GFPT1), whereas the end product is uridine diphosphate *N*-acetyl glucosamine (UDP-GlcNAc). The HBP and UDP-GlcNAc are important regulators of cell signaling and their upregulation is associated with tumorigenesis (Akella, Ciraku, & Reginato, 2019). Kras silencing in the *iKras* model resulted in decrease in GlcN-6P levels as well as transcriptional down-regulation of *Gfpt1*. Furthermore, western blotting confirmed a decrease in cellular levels of the total O-linked *N*-acetylglucosamine (O-GlcNAc) posttranslational modification (Ying et al., 2012). Subsequently, another study demonstrated that glucose starvation in KRAS-transformed cells promotes cell death through interference with the HBP, while the addition of *N*-acetyl-D-glucosamine, a substrate of the HBP pathway, can rescue glucose-depletion related cell death (Palorini et al., 2013). Targeting the HBP is emerging as a potential strategy for KRAS-mutant cancers. Recent studies show that inhibition of the HBP can overcome gemcitabine resistance in PDAC (Ricciardiello et al., 2020), as well as sensitized PDAC cells to the pan-RAS inhibitor BI-2852 (Ricciardiello et al., 2021). This vulnerability does not seem to be restricted to KRAS-mutant PDAC, as inhibition of the HBP pathway through inhibition of glutamine-fructose-6-phosphate transaminase (GFPT2) negatively affected the growth of *KRAS/LKB1* co-mutant NSCLC in cell culture, xenografts, and genetically modified mice (Kim et al., 2020).

3. Mitochondrial signaling and dynamics

While the differential channeling of glycolytic intermediates clearly contributes to oncogenic KRAS-derived tumorigenicity, it is also clear that glucose-derived carbon does still enter the TCA-cycle directly (Davidson et al., 2016; Hensley et al., 2016) and indirectly (Hui et al., 2017). Hence, the mitochondria of cancer cells still function to promote energy utilization. Strikingly, using the *iKras* mouse model of PDAC described above, Viale and colleagues demonstrated that a subpopulation of tumor cells that survive oncogene ablation are particularly reliant on oxidative phosphorylation for survival (Viale et al., 2014). It is well appreciated that the TCA cycle is fueled by other nutrients in cancer cells (i.e., glutamine) (Altman, Stine, & Dang, 2016), and targeting this unique dependence is an area of intense research. Furthermore, mitochondrial membrane dynamics are altered by oncogenic RAS signaling in RAS-mutant cancers (Giacomello, Pyakurel, Glytsou, & Scorrano, 2020).

3.1 Glutamine metabolism and bioenergetics

Glutamine is particularly important for driving TCA cycle flux, especially when cells are transformed with oncogenic RAS (Fan et al., 2013; Gaglio et al., 2011). Additionally, PDAC cells utilize glutamine in a non-canonical pathway through which glutamine-derived aspartate is transported into the cytoplasm where it is converted into oxaloacetate by the aspartate transaminase (GOT1) (Son et al., 2013). This pathway ultimately culminates with an increase in the NADPH/NADP^{+} ratio, which in turn contributes to the maintenance of the cellular redox state and perhaps compensates for the loss of glycolytic flux through the oxidative arm of the PPP (Ying et al., 2012). The mitochondrial transporter responsible for aspartate efflux was previously unknown, until a recent study by Raho et al. suggested that it might be uncoupling protein 2 (UCP2) (Raho et al., 2020). Their study shows that UCP2 silencing in KRAS-mutant PDAC results in decreased glutaminolysis, lower NADPH/NADP^{+} ratios, and higher levels of ROS (Raho et al., 2020).

GOT1 expression and the reprogramming of glutamine utilization through this pathway is reliant on oncogenic KRAS signaling (Son et al., 2013). Son et al. demonstrated the potential utility of a GOT1-specific inhibitor; however, the development of selective and potent GOT1 inhibitors is challenging, as none of the inhibitors identified in screening efforts performed well in cell-based assays (Anglin et al., 2018; Holt et al., 2018; Sun et al., 2019). Current efforts are aimed at targeting proteins necessary for GOT1 function, such as the transaminase cofactor pyridoxal-5′-phosphate (PLP) (Yoshida et al., 2020). While current molecules are not specific for PLP's regulation of GOT1 only, they are active in PDAC cells, and their effects can be rescued by the addition of GOT1 reaction products (Yoshida et al., 2020). Interestingly, it appears that GOT1 signaling is not universally essential for all KRAS-driven cancers. Nelson et al. discovered that while GOT1 loss sensitizes PDAC to radiotherapy, this effect is not observed in KRAS-mutant CRC (Nelson et al., 2020). This furthers the concept that RAS signaling is isoform and cancer-type specific.

In addition to GOT1 inhibition, there are other proposed methods of targeting glutamine utilization in RAS-mutant cancers. One potential target is glutaminase (GLS), which mediates an early step in glutamine metabolism. Earlier GLS inhibitors displayed low potency and modest bioavailability (Robinson et al., 2007), but the compound CB-839 is potent and orally bioavailable (Gross et al., 2014). However, when KRAS-mutant PDAC cells were treated with CB-839, they displayed sustained proliferation despite an initial sensitivity observed in vitro (Biancur et al., 2017). Thus, in order to attain durable responses to inhibition of glutamine metabolism combinatorial approaches may be necessary. In fact, dual MEK and GLS inhibition has recently been demonstrated to enhance anti-tumor activity in KRAS-mutant NSCLC both in vitro and in vivo (Xia, Li, Diao, Du, & Li, 2021).

3.2 Mitochondrial morphological dynamics

The relationship between mitochondrial morphological dynamics and mitochondrial activities is a topic of intense research (Cogliati, Enriquez, & Scorrano, 2016). KRAS transformation is associated with a highly fragmented mitochondrial matrix (Kashatus et al., 2015; Serasinghe et al., 2015). Dynamin-related protein 1 (DRP1) is a GTPase that promotes mitochondrial fission and is a direct substrate of ERK2 (Kashatus et al., 2015). Hence, loss of ERK signaling leads to loss of phosphorylation of DRP1 and the subsequent inactivation of the protein, resulting in mitochondrial fusion. Additionally, ERK inhibition has been shown to increase autophagy in KRAS-mutant PDAC (Bryant et al., 2019) and activation of autophagy results in increased mitochondrial fusion (Gomes, Di Benedetto, & Scorrano, 2011; Rambold, Kostelecky, Elia, & Lippincott-Schwartz, 2011). Using MEFs derived from a novel KPD ($Kras^{LSL\text{-}G12D/+}$; $Trp53^{flox/flox}$; $Drp1^{flox/flo}$) mouse, Nagdas and colleagues demonstrated that Drp1 is required for oncogenic Kras-driven transformation and loss of Drp1 inhibits pancreatic tumorigenesis and impairs mitochondrial metabolism (Nagdas et al., 2019). There are two available inhibitors for DRP1: Mdivi-1 and P110. Mdivi-1 is reported to act as an allosteric inhibitor of GTPase assembly and prevent GTPase hydrolysis (Cassidy-Stone et al., 2008); however, it has also been shown to reversibly inhibit mitochondrial complex 1 and thus its cellular effects may be more attributable to this target (Bordt et al., 2017). P110 is a DRP1-specific inhibitor that alters DRP1 multimer assembly (Qi, Qvit, Su, & Mochly-Rosen, 2013) that is understudied as compared to Mdivi-1. While the importance of DRP1 as a therapeutic target has been established, potent and specific pharmacological inhibitors for this target are still needed (Breitzig, Alleyn, Lockey, & Kolliputi, 2018). Importantly, targets other than DRP1 may exist. The mitofusins (MFN1 and MFN2) are transmembrane GTPases that are responsible for fusion of the outer mitochondrial membrane; whereas, the GTPase optic atrophy 1 (OPA1) is responsible for fusion of the inner mitochondrial membrane (Cogliati et al., 2016). It was recently demonstrated that oral leflunomide, an FDA-approved arthritis drug, promoted increased MFN2 expression and increased mitochondrial fusion, which in turn suppressed tumor growth.

4. Autophagy

4.1 An overview of the autophagic pathway and machinery

Autophagy is a lysosome-mediated catabolic process whereby cells degrade and recycle macromolecules to fuel and sustain growth (Kimmelman & White, 2017; Rabinowitz & White, 2010). The autophagic pathway can be divided into four distinct phases: initiation, nucleation, maturation, and degradation (Fig. 3). While autophagy is constitutively maintained at a basal level in cells, it can be upregulated in response to stress, nutrient deprivation, as well as growth factor signaling. Autophagy is initiated by activation of unc-51 like activating kinase 1 (ULK1), a critical initiator of autophagosome formation (Chan, Kir, & Tooze, 2007; Lee & Tournier, 2011). When cellular energy levels decrease during times of stress, AMP-activated protein kinase (AMPK) inhibits mammalian target of rapamycin complex 1 (mTORC1) activity causing a release of the inhibitory effect mTORC1 exerts on ULK1 (Ganley et al., 2009). AMPK also activates the ULK1 complex via direct phosphorylation of ULK1 in order to initiate autophagy (Egan et al., 2011; Gwinn et al., 2008; Inoki, Zhu, & Guan, 2003). The serine/threonine kinase ULK1 forms an initiation complex with three other proteins: autophagy-related protein 13 (ATG13), ATG101, and the focal adhesion kinase family-interacting protein of 200 kDa (FIP200) scaffold protein (Ganley et al., 2009). These binding partners are required for correct localization and inducing the maximal kinase activity of ULK. The ULK1 complex translocates to autophagy initiation sites within subdomains of the endoplasmic reticulum where it regulates the activity of the nucleation complex: Beclin-1, VPS34, ATG14, and VPS15.

The nucleation complex promotes phagophore formation by producing phosphatidylinositol 3-phosphate (PI3P), a phospholipid that recruits PI3P-binding proteins such as WD repeat domain phosphoinositide-interacting protein 2 (WIPI2B) and double FYVE containing protein 1 (DFCP1) (Axe et al., 2008; Pyo, Nah, & Jung, 2012). Recruitment of these PI3P-binding proteins, among others, facilitate the nucleation of the phagophore. As the forming phagophore expands, the conjugation complex, made up of ATG5, ATG12, and ATG16L (formation promoted by ATG7 and ATG10), is recruited to further facilitate elongation. This complex functions

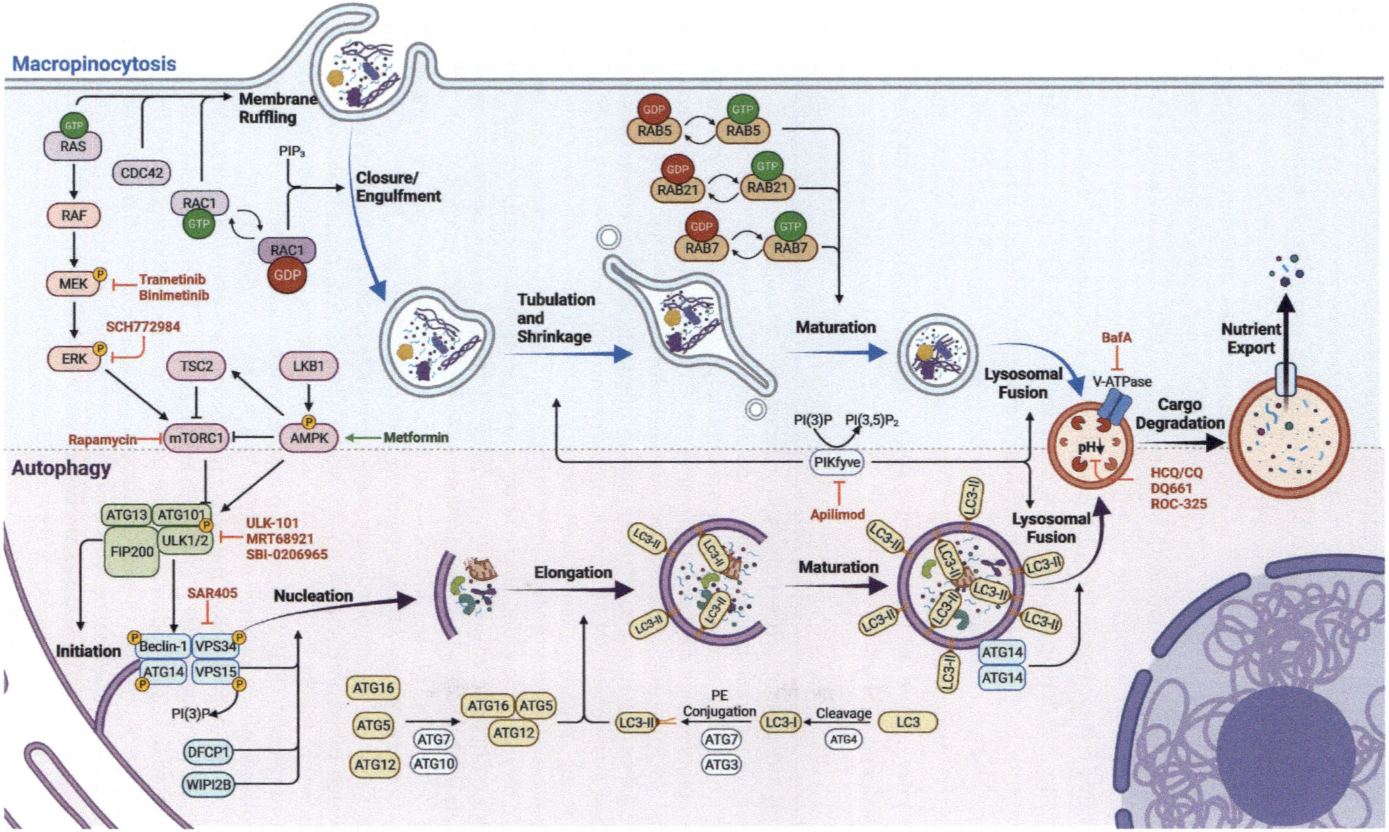

Fig. 3 See figure legend on next page.

to catalyze the conjugation of the microtubule-associated protein 1 light chain 3 (LC3A/B/C), GABA type A receptor associated protein (GABARAP), and GABARAP-like 1/2 (GABARAPL1/2) to the phosphatidylethanolamine (PE) in the elongating phagophore (Pyo et al., 2012). Simultaneously, LC3 (ATG8) is cleaved by the ATG4 cysteine protease and then conjugated with the phospholipid PE by ATG7 and ATG3, a necessary process for mature autophagosome formation (Ichimura et al., 2000). Lipidation of LC3B allows for association within the membrane of the growing autophagosome. This budding phagophore engulfs cytoplasmic components such as misfolded proteins and damaged organelles, ultimately fusing with the lysosome and allowing for cargo degradation within the autolysosome via lysosomal hydrolases. The products generated are then released into the cytoplasm and used to fuel critical metabolic and biosynthetic pathways. More specifically, autophagy provides the cell with sugars, lipids, amino acids, and nucleosides in order to fuel glycolysis, the TCA cycle, and rapid proliferation.

Fig. 3 RAS-mutant cancers are highly dependent on nutrient scavenging pathways. Mutant RAS signaling upregulates macropinocytosis in order to transport extracellular fluid containing amino acids, phospholipids, proteins, and other macromolecules. Macropinocytosis is dependent on many small GTPases including RAS, RAC and multiple RABs. Macropinosomes mature and eventually fuse with the lysosome where the contents are degraded, transported out, and shuttled into their respective pathways to fuel carbon metabolism, DNA synthesis, protein generation, and other key functions in the cell. Autophagy is a nutrient scavenging process that occurs within the cell to selectively degrade misfolded proteins, damaged organelles, and other defective cellular machinery to recycle nucleosides, amino acids, and sugars to sustain cancer growth in nutrient poor and hypoxic environments. Inhibition of the ERK-MAPK pathway and or mTOR results in upregulated autophagic flux in RAS-mutant cells. Autophagy can be divided into four distinct phases and inhibited with a variety of strategies. First, the initiation step is controlled by AMPK-dependent phosphorylation of ULK1, a serine/threonine kinase that can be inhibited with many chemically distinct and potent inhibitors. The activated initiation complex signals to the second step, nucleation, whereby many members of the Beclin-1-VPS34-ATG14-VPS15 complex are phosphorylated and function to generate PI3P in order to promote phagophore formation. This step can be pharmacologically inhibited directly by VPS34 inhibition (SAR405). The third step, elongation/maturation involves the proteolytic processing of pro-LC3 into LC3B-I that can then be conjugated to phosphatidylethanolamine. Following lipidation, LC3B-I is converted to LC3B-II and can associate with the forming autophagosome. The final step is degradation, in which the mature autophagosome fuses with the lysosome and cargo is degraded by acidification. This final step can be inhibited using lysosomal-pH inhibitors such as HCQ/CQ, ROC-325, or Bafilomycin A1 as well as inhibitors against vesicular dynamics such as apilimod, a compound that inhibits the lipid kinase PIKfyve that regulates endosomal dynamics.

4.2 AMPK and mTOR

Two essential nutrient sensors in cells, AMPK and mTOR, facilitate coordination of the cellular response to metabolic stressors. AMPK is activated in response to oxidative stress and increased AMP:ADP and ADP:ATP ratios, which can be induced by glucose deprivation. The binding of AMP or ADP activates AMPK, as does phosphorylation by the serine/threonine liver-kinase-B1 (LKB1) kinase (Shaw et al., 2004). AMPK ultimately functions to restore ATP levels by inhibiting anabolic processes and promoting catabolic pathways in order to generate ATP (Garcia & Shaw, 2017). Activated AMPK is able to phosphorylate many downstream effectors involved in autophagy such as ULK1 (Egan et al., 2011), VPS34 (Kim et al., 2013), and Beclin-1 (Kim et al., 2013). The ability of AMPK to induce autophagy is also mediated by direct inhibition of mTORC1 activity. More specifically, AMPK is able to phosphorylate and activate TSC2 (Inoki et al., 2003) (a negative regulator of mTORC1) as well as phosphorylate and inhibit Raptor (Gwinn et al., 2008) (a component of the mTOR complex). In contrast to AMPK, mTOR suppresses autophagy by direct inhibitory phosphorylation of ULK1 as well as disrupting the ULK1 and AMPK interaction, during nutrient abundance and in response to growth factor signaling (Kim, Kundu, Viollet, & Guan, 2011).

As AMPK lies at the nexus of multiple nutrient sensing and stress pathways, it has been suggested to have multifaceted roles in both tumor progression and tumor suppression (Faubert, Vincent, Poffenberger, & Jones, 2015). Indirect AMPK agonists, such as metformin, were shown to suppress the growth of a p53 null, RAS-mutant, HCT116 CRC cell line (Buzzai et al., 2007). This observation was extended to RAS-mutant human LAC cell lines in which metformin treatment induced apoptosis and inhibited growth (Wu et al., 2011) as well as other tissue types (Kuznetsov, Leclerc, Leclerc, & Barredo, 2011). Supporting these findings, a direct agonist of AMPK (MT 63–78) exerts an anti-proliferative effect on prostate cancer through a lipogenesis-inhibitory mechanism (Zadra et al., 2014). Loss of AMPK activity can also have pro-tumorigenic consequences such as increased glycolytic flux (Faubert et al., 2015) and loss of mTORC1 regulation (Shaw et al., 2004). The role of AMPK as a tumor suppressor is likely due in part to the ability of this metabolic sensor to negatively regulate the Warburg effect (Faubert et al., 2013). However, the capability of AMPK to mitigate cellular stress can also aid in cancer cell growth and survival. In fact, AMPK has been shown to maintain NADPH levels as a stress response to the loss of acetyl CoA carboxylases (ACC1 or ACC2) ultimately resulting in cancer cell survival (Jeon, Chandel, & Hay, 2012). Similarly, it was demonstrated that AMPK protects leukemia-initiating cells from metabolic stress promoting development of more advanced disease, which could be abrogated with AMPK inhibition (Saito, Chapple, Lin, Kitano, & Nakada, 2015). The dual role of AMPK in tumor progression and tumor suppression, as well as a lack of specific and selective drugs, makes targeting AMPK a challenging task. There are clearly multiple roles AMPK can play in tumorigenesis, tumor maintenance, and cell survival following energetic stress. Further studies will be required to fully elucidate the tissue-specific role of AMPK and the potential therapeutic interventions (Eichner, Brun, Ross, Shaw, & Svensson, 2019).

4.3 ULK and the initiation complex

In mammals, the initiation complex is comprised of ULK1, ATG13, FIP200, and ATG101, and is constitutively formed in the cell independent of nutrient levels (Ganley et al., 2009). Upon ULK activation (via AMPK activation and release of mTOR inhibition), ULK1 drives phagophore formation via direct phosphorylation of Beclin-1, VPS34, and ATG14. ULK is an attractive drug target as it is potentially the only serine/threonine kinase specific to the autophagic pathway. A number of groups have identified and characterized efficacious pharmacologic inhibitors of ULK1 and ULK2 (Egan et al., 2015; Martin et al., 2018; Petherick et al., 2015). Additionally, combined pharmacologic inhibition of ULK, with MRT68921 (MRT) or SBI-0206965 (SBI), and ERK or $KRAS^{G12C}$ synergistically reduced the proliferation of human PDAC cell lines (Bryant et al., 2019). Furthermore, Hobbs and colleagues showed that the proliferation of $KRAS^{G12R}$-mutant PDAC cell lines was more effectively reduced by MRT and SBI than by chloroquine, and that dual ULK and ERK inhibition synergistically reduced the proliferation of the majority of the cell lines tested (Hobbs et al., 2020). These findings extend to other tissue types such as KRAS-mutant NSCLCs harboring a variety of oncogenic KRAS mutations that showed sensitivity to ULK inhibition by ULK-101 as well as a sensitization to nutrient deprivation (Martin et al., 2018). It has also been shown that ULK inhibition by SBI induces apoptosis in combination with mTOR inhibition in KRAS-mutant LAC (Egan et al., 2015). These results are promising as mTOR inhibition typically results in a cytostatic growth arrest; however, it appears that loss of ULK activity can trigger a cytotoxic response when paired with mTOR inhibition. Furthermore, a recent study demonstrated that high ULK1 expression in NSCLC is correlated with poor prognosis and patient outcomes, and further demonstrated that inhibition of ULK reduced RAS-mutant NSCLC cell line proliferation and sensitized cells to cisplatin treatment (Tang et al., 2017). Very recently, the first clinical trial to assess the efficacy of ULK inhibition in patients was opened to test DCC-3116 from Deciphera Pharmaceuticals (NCT04892017).

4.4 VPS34, Beclin-1, nucleation complex

The nucleation complex comprises VPS34, Beclin-1, ATG14, and VPS15. Initially, Beclin-1, a core subunit of the nucleation complex binds VPS34. The class III phosphatidylinositol 3-kinase (PI3K) vacuolar protein sorting 34 (VPS34) functions to produce PI3P, a molecule important for membrane elongation. The complex also contains ATG14, a binding partner of Beclin-1 that promotes early autophagosome biogenesis (Matsunaga et al., 2009) and fusion of autophagosomes with lysosomes (Diao et al., 2015). Finally, phosphoinositide 3-kinase regulatory subunit 4 (VPS15) functions to regulate the kinase activity of VPS34 (Lindmo et al., 2008).

The nucleation step can be pharmacologically inhibited by targeting the lipid kinase VPS34 directly with SAR405 (Ronan et al., 2014), as well as indirectly with Spautin-1 via inhibition of the deubiquitinating activity of USP10 and USP13 (Liu et al., 2011). While these compounds have not been widely tested in RAS-mutant cancers, they have demonstrated efficacy in other cancers, such as renal cancer and chronic myeloid leukemia. (Ronan et al., 2014; Shao et al., 2014), and will likely be characterized more in autophagy-dependent RAS-mutant cancers. Treatment with SAR405 resulted in a reduction in autophagic flux in $KRAS^{G12C}$-mutant PDAC cell lines (Kinsey et al., 2019). Additionally, spautin-1 treatment reduced the proliferation of human, KRAS-mutant PDAC cell lines, as well as synergistically reduced PDAC proliferation when combined with an ERK inhibitor (Bryant et al., 2019).

4.5 Maturation and degradation

The budding phagophore engulfs cytoplasmic components, and as the phagophore expands, the ATG-mediated conjugation complex is recruited to further facilitate elongation. Full sequestration of cargo forms the autophagosome, a double-membraned organelle. Following maturation, the mature autophagosome fuses with the lysosome. The acidic environment degrades the contents of the resulting autophagolysosome that are then transported into the cytosol and used to fuel and sustain cancer growth. It is clear that autophagy is an adaptive response to loss of KRAS signaling, as demonstrated in response to both pharmacologic inhibition and genetic ablation of KRAS (Santana-Codina et al., 2020; Viale et al., 2014). Ablation of autophagy as a therapeutic mechanism against RAS-mutant cancers has been extensively studied. The deletion of autophagy-related genes (ATGs) typically results in reduced tumor growth, highlighting the necessity of this pathway for tumor maintenance. For example, *ATG5* deleted PDAC tumors display elevated levels of apoptosis and reduced proliferation when grown as xenografts (Yang & Kimmelman, 2014). Furthermore, using an autochthonous mouse model of PDAC, driven by oncogenic *Kras* and stochastic loss of heterozygosity of *Trp53*, Yang and colleagues demonstrated in that premalignant lesions were unable to progress to invasive cancer (Yang et al., 2014). This study also reported on a mouse preclinical trial using genetically characterized patient-derived xenografts that showed consistent responses to hydroxychloroquine (HCQ) across the group of tumors (Yang et al., 2014). Similarly, genetic deletion of *Atg7* in a Kras-driven mouse model of NSCLC resulted in altered tumor fate from adenomas and carcinomas to a benign oncocytoma (Guo et al., 2013). One potential mechanism for these findings is that autophagy, among other roles, functions to maintain nucleotide pools to promote and sustain survival (Guo et al., 2016).

RAS-mutant cancers display high levels of basal autophagy, and this metabolic process can be upregulated following general cytotoxic chemotherapies or targeted therapies as a resistance mechanism. In fact, lysosomal inhibitors have been tested in combination with current standard of care in PDAC patients (Karasic et al., 2019; Zeh et al., 2020). Multiple independent studies recently demonstrated that autophagy is robustly upregulated following inhibition of each distinct node of the RAF-MEK-ERK MAPK pathway (Bryant et al., 2019; Kinsey et al., 2019; Lee et al., 2019). An initial surprising result was that *KRAS* suppression, both genetically and pharmacologically, resulted in increased autophagic flux (Bryant et al., 2019). This phenotype was recapitulated upon targeting of RAF (Lee et al., 2019), MEK (Kinsey et al., 2019), and ERK (Bryant et al., 2019). Similar metabolic adaptations were elucidated following direct targeting of $KRAS^{G12C}$ (Santana-Codina et al., 2020). The only clinically approved autophagy inhibitors to date are chloroquine (CQ) and the analog hydroxychloroquine (HCQ), which indirectly inhibit autophagy by preventing the acidification of the lysosome (Onorati, Dyczynski, Ojha, & Amaravadi, 2018; Schrezenmeier & Dörner, 2020). Not only did CQ/HCQ reduce the growth of PDAC alone

(Yang et al., 2011, 2014), they also sensitized cells to ERK MAPK inhibition and synergistically reduced KRAS-mutant PDAC growth in multiple preclinical models (Bryant et al., 2019; Kinsey et al., 2019; Lee et al., 2019). These results are likely not a mutation-specific response and are hypothesized to be applicable to a large subset of PDAC patients (Hobbs et al., 2020) and other RAS-driven cancers such as CRC, NRAS-driven melanoma, and *GNAQ/11*-mutated uveal melanoma (Kinsey et al., 2019; Truong et al., 2020). These findings have led to a resurgence in the interest in targeting autophagy for cancer treatment (Dolgin, 2019). In fact, a number of clinical trials have been initiated to assess the safety and efficacy of dual MAPK and autophagy inhibition in RAS-mutant cancers (NCT04386057, NCT03825289, NCT04132505, NCT04214418, NCT03979651, NCT04145297).

HCQ exerts inhibitory effects on processes outside of autophagy and has limited pharmacodynamics in vivo, creating a demand for better more specific inhibitors of autophagy (Shi, Yu, Yan, & Xiao, 2017). One direction is HCQ/CQ derivatives (McAfee et al., 2012; Rebecca et al., 2017). Indeed, the dimeric quinacrine DQ661 exerts a significantly strong anti-proliferative effect on $BRAF^{V600E}$-mutant melanoma tumors that display resistance to HCQ treatment (Rebecca et al., 2017). Similar to HCQ, ROC-325, inhibits autophagy by preventing lysosome acidification and appears to induce apoptosis in renal cell carcinoma (Carew et al., 2017), but has yet to be characterized extensively in RAS-mutant cancer models. Another mechanism to inhibit lysosome acidification is through inhibiting the vacuolar type H^+-ATPase, that pumps protons into the lysosome, with bafilomycin A1 (BafA) (Ohkuma et al., 1993). BafA has been shown to reduce the viability of RAS-mutant cell lines to varying degrees, but studies did not find a correlation between RAS- or p53-mutation status and BafA response (Morgan et al., 2014). Additionally, late endosomal and lysosome dynamics are highly regulated through the phosphoinositide kinase, 5-type zinc finger containing protein (PIKfyve), which generates $PI(3,5)P_2$ from PI(3)P (Ikonomov, Sbrissa, & Shisheva, 2001; Nicot et al., 2006). PIKfyve activity can be pharmacologically inhibited via apilimod treatment (Cai et al., 2013), which interferes with vesicular trafficking, thus inhibiting processes such as autophagy and macropinocytosis. Apilimod treatment prevented $Kras^{G12D}$-expressing mouse embryo fibroblasts (MEFs) from proliferating and abolished mTOR inhibition-dependent increase in proliferation (Krishna et al., 2016). These results were recapitulated in KRAS-mutant PDAC as well as HRAS-mutant bladder cancer (Krishna et al., 2016).

Many KRAS-mutant dependencies have been identified that have yet to be targeted pharmacologically. For example, autophagy can function to supply the cell with important amino acids such as cysteine, from the extracellular environment (Mukhopadhyay et al., 2021), and alanine, secreted from pancreatic stellate cells (Sousa et al., 2016) for growth. In fact, PDAC cells failed to overcome the loss of the amino acid uptake transporter SLC38A2, the loss of which significantly reduced tumor size in multiple PDAC mouse models (Parker et al., 2020). However, more work is necessary in order to pharmacologically target these metabolic vulnerabilities.

4.6 Mitophagy

Macroautophagy is a bulk recycling process that degrades cellular contents such as damaged organelles and misfolded proteins. Additionally, there are many types of selective autophagy to specifically degrade particular intracellular components. For example, the selective degradation of mitochondria, mitophagy, is a highly important housekeeping function to ensure damaged, defective, or excessive mitochondria have a quality control mechanism (Ding & Yin, 2012). The signaling governing how mitochondria are selectively targeted for degradation is not fully understood. It appears that mitophagy is preceded by fission and selective fusion of healthy mitochondria to separate the dysfunctional mitochondria and encourage their selective degradation (Twig et al., 2008). Mitochondria can be spared from this selective autophagy through a PKA-DRP1-dependent mechanism that induces mitochondria elongation, resulting in viable organelles that maintain the energetic ATP pool (Gomes et al., 2011). Mitochondrial depolarization can also signal the organelle to be selectively taken up by the autophagosome. Loss of membrane potential results in stabilization of PTEN-induced putative kinase 1 (PINK1), allowing for recruitment of Parkin and subsequent initiation of mitophagy (Matsuda et al., 2010).

As the mitochondria acts as a critical hub for energy production to fuel cancer, it is also integral that there is a balance between mitochondrial biogenesis and mitophagy. Due to their altered metabolic demands, RAS-mutant cancers are highly dependent on mitochondrial function to fuel aberrant growth and proliferation (Weinberg et al., 2010). However, it is critical that a healthy pool of mitochondria are maintained to prevent build-up of reactive oxygen species, less efficient oxygen consumption, and overall energy exhaustion (Guo et al., 2011). It was recently shown that KRAS drives PDAC growth via increased mitochondrial function through DRP1 (Nagdas et al., 2019). As DRP1 is a direct substrate of ERK, a key effector of KRAS signaling, these findings highlight connections between RAS signaling, mitochondrial function and glycolytic flux. While oncogenic RAS drives increased mitochondrial function, it was recently reported that mutationally activated KRAS drives mitophagy via a NIX-mediated mechanism, ultimately supporting PDAC tumorigenesis and metabolic transformation (Humpton et al., 2019). Selectively targeting mitophagy could be an effective therapeutic strategy for RAS-mutant cancers that are uniquely highly dependent on this process for maintaining efficient mitochondrial function and respiration. In summary, there is clearly a delicate balance between mitochondrial function and selective degradation by mitophagy to maintain energy homeostasis and fuel aberrant cancer growth during times of nutrient depletion.

5. Lipid scavenging and metabolism

Lipid scavenging has recently emerged as another important mechanism for RAS-driven metabolic reprogramming and ATP production (Swierczynski, Hebanowska, & Sledzinski, 2014). Interestingly, it appears that mutant RAS has an elevated dependence on the extracellular uptake of lipids such that serum lipid depletion sensitizes cells to mTOR inhibition by rapamycin (Salloum, Mukhopadhyay, Tung, Polonetskaya, & Foster, 2014). It appears that exogenous lipids can effectively support pancreatic cancer growth (Clerc et al., 1991), but there is a clear variation on the overall impact individual lipids may have (Yu et al., 2015). Indeed, RAS-driven cells phenocopy hypoxic reprogramming that resulted in increased extracellular uptake of phospholipids to overcome metabolic stress (Kamphorst et al., 2013). In fact, it has been suggested that lipid metabolism presents a therapeutic vulnerability in KRAS-mutant NSCLC (Singh et al., 2018). As with many critical metabolic processes, it was demonstrated that mutant KRAS activity upregulates lipid synthesis to support PDAC growth (Bian, Yu, Wang, & Li, 2015). Recent reports on the metabolic vulnerabilities presented by a reliance on lipid scavenging has prompted the development and characterization of inhibitors of these pathways, some of which have advanced to the clinic (Falchook et al., 2021).

6. Macropinocytosis

Macropinocytosis is a nonselective nutrient uptake process by which extracellular fluid and macromolecules are taken up through endocytic vesicles termed macropinosomes (Fig. 3) (Finicle, Jayashankar, & Edinger, 2018). It is widely accepted that activated RAS and RAC proteins result in enhanced membrane ruffling and pinocytosis or "cell drinking" (Lewis, 1937) and macropinocytosis induction (Bar-Sagi & Feramisco, 1986). In fact, macropinocytosis can be initiated with the activation of the small GTPase RAC1 (Fujii, Kawai, Egami, & Araki, 2013; West, Prescott, Eskelinen, Ridley, & Watts, 2000), cell division control protein 42 homolog (CDC42) (Garrett et al., 2000), as well as RAB and ARF GTPases (Lanzetti, Palamidessi, Areces, Scita, & Di Fiore, 2004; Machin, Tsonou, Hornigold, & Welch, 2021). Activation of these proteins induces membrane protrusions that form circular cups that close into macropinosomes (Swanson, 2008). Macropinocytosis is dependent on both production of phosphatidylinositol (3,4,5)-triphosphate (PIP_3) (Yoshida, Hoppe, Araki, & Swanson, 2009) as well as RAC1 deactivation (Fujii et al., 2013). Following formation, macropinosomes undergo maturation first by tubulation and shrinking (Kerr et al., 2006), regulated by several proteins, including PIKfyve (Krishna et al., 2016). Following this, sequential activation and inactivation of RAB5, RAB21, and RAB7 are critical for maturation of macropinosomes (Song, Zhang, Ding, Ji, & Zhao, 2021). Upon maturation, macropinosomes fuse with the lysosome, contents are degraded, and transported out for the cell to utilize.

RAS-transformed cancers utilize macropinocytosis to supply macromolecules to fuel the cell's unique metabolic demands. Specifically, KRAS-mutant PDAC displays higher levels of macropinocytosis relative to cells expressing wild-type KRAS (Commisso et al., 2013). Macropinocytosis is crucial for RAS-driven cancer growth because it can increase the overall biomass of the cell, unlike autophagy that can only utilize resources already present in the cell. It has become increasingly clear that KRAS-mutant PDAC relies heavily on macropinocytosis as an essential amino acid source from extracellular proteins to fuel cell growth (Davidson et al., 2017; Kamphorst et al., 2015). In fact, loss of *PTEN*, a negative regulator of the PI3K-AKT-mTOR pathway, enhances macropinocytosis through an AKT-mTORC2-dependent mechanism (Michalopoulou et al., 2020). Macropinocytosis conferred resistance to mTOR inhibition in PDAC, a phenotype that was abrogated upon dual mTOR and CQ treatment (Michalopoulou et al., 2020). These results further establish the importance of the PI3K-AKT-mTOR pathway in regulating macropinocytosis as a critical source of amino acids to fuel growth of RAS-mutant cancers (Palm, Araki, King, DeMatteo, & Thompson, 2017). It has also been shown that mTORC1 activity suppresses the catabolism of proteins that have been acquired via macropinocytosis (Palm et al., 2015). Furthermore, mTOR inhibition alleviates this suppression, allowing these proteins to be degraded as an amino acid source to fuel growth, proliferation, and likely confer resistance to mTOR inhibitors in RAS-mutant cells (Palm et al., 2015). New players involved in regulating this scavenging pathway are constantly being elucidated. For example, it was recently demonstrated that the transcriptional co-activators YAP and TAZ act as important transcriptional regulators of macropinocytotic uptake in PDAC cells (King, Araki, Palm, & Thompson, 2020). Interestingly, cancers driven by different RAS mutations result in varied dependence on macropinocytosis. For instance, $KRAS^{G12R}$ is impaired in driving macropinocytosis because it is defective in interacting with the key effector, p110α PI3K, (PI3Kα) (Hobbs et al., 2020). However, $KRAS^{G12R}$-mutant PDAC cell lines are still very sensitive to the ULK inhibitors SBI and MRT, as well as the lysosomal inhibitor, CQ (Hobbs et al., 2020). By inhibiting the final stage of nutrient scavenging, lysosome degradation, with CQ/HCQ it may be possible to target both autophagy and macropinocytosis, producing a more toxic result (Onorati et al., 2018; Schrezenmeier & Dörner, 2020).

7. Concluding remarks

A substantial body of literature supports the function altered metabolism performs in driving RAS-mutant cancer development, maintenance, and proliferation. This review summarizes the central role oncogenic RAS signaling has in driving metabolic functions to ensure survival of these aggressive cancers and highlights a variety of strategies to therapeutically target potential vulnerabilities given the unique metabolic demand of these cancers. More specifically, it is well accepted that RAS-driven cancers have increased reliance on glycolysis, scavenging pathways such as autophagy and macropinocytosis, and display markedly altered mitochondrial signaling and dynamics. More work remains to be done to further evaluate the specific role the four different RAS isoforms have on individual processes, how the different hotspot mutations drive metabolic pathways, and the likely tissue-specific function these metabolic vulnerabilities present. The studies outlined above, among others, have led to development and characterization of a number of metabolic inhibitors that are currently being tested in the clinic and with the hope that these metabolic vulnerabilities will prove to be effective therapeutic targets.

Acknowledgments

We apologize to colleagues whose studies we did not cite due to space limitations. K.L.B. is supported by grants from the National Institutes of Health (NIH; R37 CA251877), the Pancreatic Cancer Action Network/American Association for Cancer Research (15-70-25-BRYA), and the Sky Pancreatic Cancer Foundation Inc. All figures were created with BioRender.com.

References

Ahearn, I. M., Haigis, K., Bar-Sagi, D., & Philips, M. R. (2012). Regulating the regulator: Post-translational modification of RAS. *Nature Reviews Molecular Cell Biology*, *13*, 39–51. https://doi.org/10.1038/nrm3255.

Akella, N. M., Ciraku, L., & Reginato, M. J. (2019). Fueling the fire: Emerging role of the hexosamine biosynthetic pathway in cancer. *BMC Biology*, *17*, 1–14. https://doi.org/10.1186/s12915-019-0671-3.

Altman, B. J., Stine, Z. E., & Dang, C. V. (2016). From Krebs to clinic: Glutamine metabolism to cancer therapy. *Nature Reviews Cancer*, *16*, 619–634. https://doi.org/10.1038/nrc.2016.71.

Amendola, C. R., Mahaffey, J. P., Parker, S. J., Ahearn, I. M., Chen, W. C., Zhou, M., et al. (2019). KRAS4A directly regulates hexokinase 1. *Nature*, *576*(7787), 482–486. https://doi.org/10.1038/s41586-019-1832-9.

Anglin, J., Zavareh, R. B., Sander, P. N., Haldar, D., Mullarky, E., Cantley, L. C., et al. (2018). Discovery and optimization of aspartate aminotransferase 1 inhibitors to target redox balance in pancreatic ductal adenocarcinoma. *Bioorganic and Medicinal Chemistry Letters*, *28*(16), 2675–2678. https://doi.org/10.1016/j.bmcl.2018.04.061.

Axe, E. L., Walker, S. A., Manifava, M., Chandra, P., Roderick, H. L., Habermann, A., et al. (2008). Autophagosome formation from membrane compartments enriched in phosphatidylinositol 3-phosphate and dynamically connected to the endoplasmic reticulum. *Journal of Cell Biology*, *182*(4), 685–701. https://doi.org/10.1083/jcb.200803137.

Bar-Sagi, D., & Feramisco, J. R. (1986). Induction of membrane ruffling and fluid-phase pinocytosis in quiescent fibroblasts by ras proteins. *Science*, *233*(4768), 1061–1068. https://doi.org/10.1126/science.3090687.

Basturk, O., Singh, R., Kaygusuz, E., Balci, S., Dursun, N., Culhaci, N., et al. (2011). GLUT-1 expression in pancreatic neoplasia: Implications in pathogenesis, diagnosis, and prognosis. *Pancreas*, *40*(2), 187–192. https://doi.org/10.1097/MPA.0b013e318201c935.

Berndt, N., Hamilton, A. D., & Sebti, S. M. (2011). Targeting protein prenylation for cancer therapy. *Nature Reviews Cancer*, *11*, 775–791. https://doi.org/10.1038/nrc3151.

Bian, Y., Yu, Y., Wang, S., & Li, L. (2015). Up-regulation of fatty acid synthase induced by EGFR/ERK activation promotes tumor growth in pancreatic cancer. *Biochemical and Biophysical Research Communications*, *463*(4), 612–617. https://doi.org/10.1016/j.bbrc.2015.05.108.

Biancur, D. E., Paulo, J. A., Małachowska, B., Del Rey, M. Q., Sousa, C. M., Wang, X., et al. (2017). Compensatory metabolic networks in pancreatic cancers upon perturbation of glutamine metabolism. *Nature Communications*, *8*(1), 1–15. https://doi.org/10.1038/ncomms15965.

Birnbaum, M. J., Haspel, H. C., & Rosen, O. M. (1987). Transformation of rat fibroblasts by FSV rapidly increases glucose transporter gene transcription. *Science*, *235*(4795), 1495–1498. https://doi.org/10.1126/science.3029870.

Bordt, E. A., Clerc, P., Roelofs, B. A., Saladino, A. J., Tretter, L., Adam-Vizi, V., et al. (2017). The putative Drp1 inhibitor mdivi-1 is a reversible mitochondrial complex I inhibitor that modulates reactive oxygen species. *Developmental Cell*, *40*(6), 583–594.e6. https://doi.org/10.1016/j.devcel.2017.02.020.

Boros, L. G., Lee, P. W. N., Brandes, J. L., Cascante, M., Muscarella, P., Schirmer, W. J., et al. (1998). Nonoxidative pentose phosphate pathways and their direct role in ribose synthesis in tumors: Is cancer a disease of cellular glucose metabolism? *Medical Hypotheses*, *50*(1), 55–59. https://doi.org/10.1016/S0306-9877(98)90178-5.

Breitzig, M. T., Alleyn, M. D., Lockey, R. F., & Kolliputi, N. (2018). A mitochondrial delicacy: Dynamin-related protein 1 and mitochondrial dynamics. *American Journal of Physiology. Cell Physiology*, *315*, 80–90. https://doi.org/10.1152/ajpcell.00042.2018.

Bryant, K. L., Mancias, J. D., Kimmelman, A. C., & Der, C. J. (2014). KRAS: Feeding pancreatic cancer proliferation. *Trends in Biochemical Sciences*, *39*, 91–100. https://doi.org/10.1016/j.tibs.2013.12.004.

Bryant, K. L., Stalnecker, C. A., Zeitouni, D., Klomp, J. E., Peng, S., Tikunov, A. P., et al. (2019). Combination of ERK and autophagy inhibition as a treatment approach for pancreatic cancer. *Nature Medicine*, *25*(4), 628–640. https://doi.org/10.1038/s41591-019-0368-8.

Buzzai, M., Jones, R. G., Amaravadi, R. K., Lum, J. J., DeBerardinis, R. J., Zhao, F., et al. (2007). Systemic treatment with the antidiabetic drug metformin selectively impairs p53-deficient tumor cell growth. *Cancer Research*, *67*(14), 6745–6752. https://doi.org/10.1158/0008-5472.CAN-06-4447.

Cai, X., Xu, Y., Cheung, A. K., Tomlinson, R. C., Alcázar-Román, A., Murphy, L., et al. (2013). PIKfyve, a class III PI kinase, is the target of the small molecular IL-12/IL-23 inhibitor apilimod and a player in toll-like receptor signaling. *Chemistry and Biology*, *20*(7), 912–921. https://doi.org/10.1016/j.chembiol.2013.05.010.

Carew, J. S., Espitia, C. M., Zhao, W., Han, Y., Visconte, V., Phillips, J., et al. (2017). Disruption of autophagic degradation with ROC-325 antagonizes renal cell carcinoma pathogenesis. *Clinical Cancer Research*, *23*(11), 2869–2879. https://doi.org/10.1158/1078-0432.CCR-16-1742.

Cassidy-Stone, A., Chipuk, J. E., Ingerman, E., Song, C., Yoo, C., Kuwana, T., et al. (2008). Chemical inhibition of the mitochondrial division dynamin reveals its role in Bax/Bak-dependent mitochondrial outer membrane Permeabilization. *Developmental Cell*, *14*(2), 193–204. https://doi.org/10.1016/j.devcel.2007.11.019.

Castellano, E., Sheridan, C., Thin, M. Z., Nye, E., Spencer-Dene, B., Diefenbacher, M. E., et al. (2013). Requirement for interaction of PI3-kinase p110α with RAS in lung tumor maintenance. *Cancer Cell*, *24*(5), 617–630. https://doi.org/10.1016/j.ccr.2013.09.012.

Chan, E. Y. W., Kir, S., & Tooze, S. A. (2007). siRNA screening of the kinome identifies ULK1 as a multidomain modulator of autophagy. *Journal of Biological Chemistry*, *282*(35), 25464–25474. https://doi.org/10.1074/jbc.M703663200.

Cheung, E. C., DeNicola, G. M., Nixon, C., Blyth, K., Labuschagne, C. F., Tuveson, D. A., et al. (2020). Dynamic ROS control by TIGAR regulates the initiation and progression of pancreatic cancer. *Cancer Cell*, *37*(2), 168–182.e4. https://doi.org/10.1016/j.ccell.2019.12.012.

Chiaradonna, F., Magnani, C., Sacco, E., Manzoni, R., Alberghina, L., & Vanoni, M. (2005). Acquired glucose sensitivity of k-ras transformed fibroblasts. *Biochemical Society Transactions*, *33*(1), 297–299. https://doi.org/10.1042/BST0330297.

Chiaradonna, F., Sacco, E., Manzoni, R., Giorgio, M., Vanoni, M., & Alberghina, L. (2006). Ras-dependent carbon metabolism and transformation in mouse fibroblasts. *Oncogene*, *25*(39), 5391–5404. https://doi.org/10.1038/sj.onc.1209528.

Clerc, P., Bensaadi, N., Pradel, P., Estival, A., Clemente, F., & Vaysse, N. (1991). Lipid-dependent proliferation of pancreatic cancer cell lines. *Cancer Research*, *51*(14), 3633–3638.

Cogliati, S., Enriquez, J. A., & Scorrano, L. (2016). Mitochondrial cristae: Where beauty meets functionality. *Trends in Biochemical Sciences*, *41*, 261–273. https://doi.org/10.1016/j.tibs.2016.01.001.

Collisson, E. A., Trejo, C. L., Silva, J. M., Gu, S., Korkola, J. E., Heiser, L. M., et al. (2012). A central role for RAF→MEK→ERK signaling in the genesis of pancreatic ductal adenocarcinoma. *Cancer Discovery*, *2*(8), 685–693. https://doi.org/10.1158/2159-8290.CD-11-0347.

Commisso, C., Davidson, S. M., Soydaner-Azeloglu, R. G., Parker, S. J., Kamphorst, J. J., Hackett, S., et al. (2013). Macropinocytosis of protein is an amino acid supply route in Ras-transformed cells. *Nature*, *497*(7451), 633–637. https://doi.org/10.1038/nature12138.

Cook, J. H., Melloni, G. E. M., Gulhan, D. C., Park, P. J., & Haigis, K. M. (2021). The origins and genetic interactions of KRAS mutations are allele- and tissue-specific. *Nature Communications*, *12*(1), 1–14. https://doi.org/10.1038/s41467-021-22125-z.

Cox, A. D., & Der, C. J. (2010). Ras history: The saga continues. *Small GTPases*, *1*(1), 2–27. https://doi.org/10.4161/sgtp.1.1.12178.

Cox, A. D., Fesik, S. W., Kimmelman, A. C., Luo, J., & Der, C. J. (2014). Drugging the undruggable RAS: Mission possible? *Nature Reviews Drug Discovery*, *13*, 828–851. https://doi.org/10.1038/nrd4389.

Dai, S., Peng, Y., Zhu, Y., Xu, D., Zhu, F., Xu, W., et al. (2020). Glycolysis promotes the progression of pancreatic cancer and reduces cancer cell sensitivity to gemcitabine. *Biomedicine and Pharmacotherapy*, *121*, 109521. https://doi.org/10.1016/j.biopha.2019.109521.

Davidson, S. M., Jonas, O., Keibler, M. A., Hou, H. W., Luengo, A., Mayers, J. R., et al. (2017). Direct evidence for cancer-cell-autonomous extracellular protein catabolism in pancreatic tumors. *Nature Medicine*, *23*(2), 235–241. https://doi.org/10.1038/nm.4256.

Davidson, S. M., Papagiannakopoulos, T., Olenchock, B. A., Heyman, J. E., Keibler, M. A., Luengo, A., et al. (2016). Environment impacts the metabolic dependencies of ras-driven non-small cell lung cancer. *Cell Metabolism*, *23*(3), 517–528. https://doi.org/10.1016/j.cmet.2016.01.007.

Denicola, G. M., Karreth, F. A., Humpton, T. J., Gopinathan, A., Wei, C., Frese, K., et al. (2011). Oncogene-induced Nrf2 transcription promotes ROS detoxification and tumorigenesis. *Nature*, *475*(7354), 106–110. https://doi.org/10.1038/nature10189.

Diao, J., Liu, R., Rong, Y., Zhao, M., Zhang, J., Lai, Y., et al. (2015). ATG14 promotes membrane tethering and fusion of autophagosomes to endolysosomes. *Nature*, *520*(7548), 563–566. https://doi.org/10.1038/nature14147.

Ding, W. X., & Yin, X. M. (2012). Mitophagy: Mechanisms, pathophysiological roles, and analysis. *Biological Chemistry*, *393*, 547–564. https://doi.org/10.1515/hsz-2012-0119.

Dolgin, E. (2019). Anticancer autophagy inhibitors attract "resurgent" interest. *Nature Reviews. Drug Discovery*, *18*, 408–410. https://doi.org/10.1038/d41573-019-00072-1.

Du, Z., & Lovly, C. M. (2018). Mechanisms of receptor tyrosine kinase activation in cancer. *Molecular Cancer*, *17*, 1–13. https://doi.org/10.1186/s12943-018-0782-4.

Duncan, J. S., Whittle, M. C., Nakamura, K., Abell, A. N., Midland, A. A., Zawistowski, J. S., et al. (2012). Dynamic reprogramming of the kinome in response to targeted MEK inhibition in triple-negative breast cancer. *Cell*, *149*(2), 307–321. https://doi.org/10.1016/j.cell.2012.02.053.

Egan, D. F., Chun, M. G. H., Vamos, M., Zou, H., Rong, J., Miller, C. J., et al. (2015). Small molecule inhibition of the autophagy kinase ULK1 and identification of ULK1 substrates. *Molecular Cell*, *59*(2), 285–297. https://doi.org/10.1016/j.molcel.2015.05.031.

Egan, D. F., Shackelford, D. B., Mihaylova, M. M., Gelino, S., Kohnz, R. A., Mair, W., et al. (2011). Phosphorylation of ULK1 (hATG1) by AMP-activated protein kinase connects energy sensing to mitophagy. *Science*, *331*(6016), 456–461. https://doi.org/10.1126/science.1196371.

Eichner, L. J., Brun, S. N., Ross, D. S., Shaw, R. J., & Svensson, R. U. (2019). Genetic analysis reveals AMPK is required to support tumor growth in murine Kras-dependent lung cancer models. *Cell Metabolism*, *29*, 285–302. https://doi.org/10.1016/j.cmet.2018.10.005.

Engelman, J. A., Chen, L., Tan, X., Crosby, K., Guimaraes, A. R., Upadhyay, R., et al. (2008). Effective use of PI3K and MEK inhibitors to treat mutant Kras G12D and PIK3CA H1047R murine lung cancers. *Nature Medicine*, *14*(12), 1351–1356. https://doi.org/10.1038/nm.1890.

Falchook, G., Infante, J., Arkenau, H. T., Patel, M. R., Dean, E., Borazanci, E., et al. (2021). First-in-human study of the safety, pharmacokinetics, and pharmacodynamics of first-in-class fatty acid synthase inhibitor TVB-2640 alone and with a taxane in advanced tumors. *EClinicalMedicine*, *34*, 100797. https://doi.org/10.1016/j.eclinm.2021.100797.

Fan, J., Kamphorst, J. J., Mathew, R., Chung, M. K., White, E., Shlomi, T., et al. (2013). Glutamine-driven oxidative phosphorylation is a major ATP source in transformed mammalian cells in both normoxia and hypoxia. *Molecular Systems Biology*, *9*, 712. https://doi.org/10.1038/msb.2013.65.

Faubert, B., Boily, G., Izreig, S., Griss, T., Samborska, B., Dong, Z., et al. (2013). AMPK is a negative regulator of the Warburg effect and suppresses tumor growth in vivo. *Cell Metabolism*, *17*(1), 113–124. https://doi.org/10.1016/j.cmet.2012.12.001.

Faubert, B., Vincent, E. E., Poffenberger, M. C., & Jones, R. G. (2015). The AMP-activated protein kinase (AMPK) and cancer: Many faces of a metabolic regulator. *Cancer Letters*, *356*, 165–170. https://doi.org/10.1016/j.canlet.2014.01.018.

Finicle, B. T., Jayashankar, V., & Edinger, A. L. (2018). Nutrient scavenging in cancer. *Nature Reviews Cancer*, *18*, 619–633. https://doi.org/10.1038/s41568-018-0048-x.

Flier, J. S., Mueckler, M. M., Usher, P., & Lodish, H. F. (1987). Elevated levels of glucose transport and transporter messenger RNA are induced by ras or src oncogenes. *Science*, *235*(4795), 1492–1495. https://doi.org/10.1126/science.3103217.

Fujii, M., Kawai, K., Egami, Y., & Araki, N. (2013). Dissecting the roles of Rac1 activation and deactivation in macropinocytosis using microscopic photo-manipulation. *Scientific Reports*, *3*, 2385. https://doi.org/10.1038/srep02385.

Gaglio, D., Metallo, C. M., Gameiro, P. A., Hiller, K., Danna, L. S., Balestrieri, C., et al. (2011). Oncogenic K-Ras decouples glucose and glutamine metabolism to support cancer cell growth. *Molecular Systems Biology*, 7, 523. https://doi.org/10.1038/msb.2011.56.

Ganley, I. G., Lam, D. H., Wang, J., Ding, X., Chen, S., & Jiang, X. (2009). ULK1·ATG13·FIP200 complex mediates mTOR signaling and is essential for autophagy. *Journal of Biological Chemistry*, *284*(18), 12297–12305. https://doi.org/10.1074/jbc.M900573200.

Garcia, D., & Shaw, R. J. (2017). AMPK: Mechanisms of cellular energy sensing and restoration of metabolic balance. *Molecular Cell*, *66*, 789–800. https://doi.org/10.1016/j.molcel.2017.05.032.

Garrett, W. S., Chen, L. M., Kroschewski, R., Ebersold, M., Turley, S., Trombetta, S., et al. (2000). Developmental control of endocytosis in dendritic cells by Cdc42. *Cell*, *102*(3), 325–334. https://doi.org/10.1016/S0092-8674(00)00038-6.

Ge, T., Yang, J., Zhou, S., Wang, Y., Li, Y., & Tong, X. (2020). The role of the pentose phosphate pathway in diabetes and cancer. *Frontiers in Endocrinology*, *11*, 365. https://doi.org/10.3389/fendo.2020.00365.

Giacomello, M., Pyakurel, A., Glytsou, C., & Scorrano, L. (2020). The cell biology of mitochondrial membrane dynamics. *Nature Reviews Molecular Cell Biology*, *21*, 204–224. https://doi.org/10.1038/s41580-020-0210-7.

Gomes, L. C., Di Benedetto, G., & Scorrano, L. (2011). During autophagy mitochondria elongate, are spared from degradation and sustain cell viability. *Nature Cell Biology*, *13*(5), 589–598. https://doi.org/10.1038/ncb2220.

Gross, M. I., Demo, S. D., Dennison, J. B., Chen, L., Chernov-Rogan, T., Goyal, B., et al. (2014). Antitumor activity of the glutaminase inhibitor CB-839 in triple-negative breast cancer. *Molecular Cancer Therapeutics*, *13*(4), 890–901. https://doi.org/10.1158/1535-7163.MCT-13-0870.

Guo, J. Y., Chen, H. Y., Mathew, R., Fan, J., Strohecker, A. M., Karsli-Uzunbas, G., et al. (2011). Activated Ras requires autophagy to maintain oxidative metabolism and tumorigenesis. *Genes and Development*, *25*(5), 460–470. https://doi.org/10.1101/gad.2016311.

Guo, J. Y., Karsli-Uzunbas, G., Mathew, R., Aisner, S. C., Kamphorst, J. J., Strohecker, A. M., et al. (2013). Autophagy suppresses progression of K-ras-induced lung tumors to oncocytomas and maintains lipid homeostasis. *Genes and Development*, *27*(13), 1447–1461. https://doi.org/10.1101/gad.219642.113.

Guo, J. Y., Teng, X., Laddha, S. V., Ma, S., Van Nostrand, S. C., Yang, Y., et al. (2016). Autophagy provides metabolic substrates to maintain energy charge and nucleotide pools in Ras-driven lung cancer cells. *Genes and Development*, *30*(15), 1704–1717. https://doi.org/10.1101/gad.283416.116.

Gwinn, D. M., Shackelford, D. B., Egan, D. F., Mihaylova, M. M., Mery, A., Vasquez, D. S., et al. (2008). AMPK phosphorylation of raptor mediates a metabolic checkpoint. *Molecular Cell*, *30*(2), 214–226. https://doi.org/10.1016/j.molcel.2008.03.003.

Haigis, K. M. (2017). KRAS alleles: The devil is in the detail. *Trends in Cancer*, *3*, 686–697. https://doi.org/10.1016/j.trecan.2017.08.006.

Halbrook, C. J., Pontious, C., Kovalenko, I., Lapienyte, L., Dreyer, S., Lee, H. J., et al. (2019). Macrophage-released pyrimidines inhibit gemcitabine therapy in pancreatic cancer. *Cell Metabolism*, *29*(6), 1390–1399.e6. https://doi.org/10.1016/j.cmet.2019.02.001.

Hanahan, D., & Weinberg, R. A. (2011). Hallmarks of cancer: The next generation. *Cell*, *144*, 646–674. https://doi.org/10.1016/j.cell.2011.02.013.

Hayes, T. K., Neel, N. F., Hu, C., Gautam, P., Chenard, M., Long, B., et al. (2016). Long-term ERK inhibition in KRAS-mutant pancreatic cancer is associated with MYC degradation and senescence-like growth suppression. *Cancer Cell*, *29*(1), 75–89. https://doi.org/10.1016/j.ccell.2015.11.011.

Heidorn, S. J., Milagre, C., Whittaker, S., Nourry, A., Niculescu-Duvas, I., Dhomen, N., et al. (2010). Kinase-dead BRAF and oncogenic RAS cooperate to drive tumor progression through CRAF. *Cell*, *140*(2), 209–221. https://doi.org/10.1016/j.cell.2009.12.040.

Hensley, C. T., Faubert, B., Yuan, Q., Lev-Cohain, N., Jin, E., Kim, J., et al. (2016). Metabolic heterogeneity in human lung Tumors. *Cell*, *164*(4), 681–694. https://doi.org/10.1016/j.cell.2015.12.034.

Ho, A. L., Brana, I., Haddad, R., Bauman, J., Bible, K., Oosting, S., et al. (2021). Tipifarnib in head and neck squamous cell carcinoma with HRAS mutations. *Journal of Clinical Oncology*, *39*(17), 1856–1864. https://doi.org/10.1200/jco.20.02903.

Hobbs, G. A., Baker, N. M., Miermont, A. M., Thurman, R. D., Pierobon, M., Tran, T. H., et al. (2020). Atypical KRASG12R mutant is impaired in PI3K signaling and macropinocytosis in pancreatic cancer. *Cancer Discovery*, *10*(1), 104–123. https://doi.org/10.1158/2159-8290.CD-19-1006.

Holt, M. C., Assar, Z., Beheshti Zavareh, R., Lin, L., Anglin, J., Mashadova, O., et al. (2018). Biochemical characterization and structure-based mutational analysis provide insight into the binding and mechanism of action of novel aspartate aminotransferase inhibitors. *Biochemistry*, *57*(47), 6604–6614. https://doi.org/10.1021/acs.biochem.8b00914.

Hopkins, B. D., Pauli, C., Xing, D., Wang, D. G., Li, X., Wu, D., et al. (2018). Suppression of insulin feedback enhances the efficacy of PI3K inhibitors. *Nature*, *560*(7719), 499–503. https://doi.org/10.1038/s41586-018-0343-4.

Hui, S., Ghergurovich, J. M., Morscher, R. J., Jang, C., Teng, X., Lu, W., et al. (2017). Glucose feeds the TCA cycle via circulating lactate. *Nature*, *551*(7678), 115–118. https://doi.org/10.1038/nature24057.

Humpton, T. J., Alagesan, B., Denicola, G. M., Lu, D., Yordanov, G. N., Leonhardt, C. S., et al. (2019). Oncogenic KRAS induces NIX-mediated mitophagy to promote pancreatic cancer. *Cancer Discovery*, *9*(9), 1268–1287. https://doi.org/10.1158/2159-8290.CD-18-1409.

Ichimura, Y., Kirisako, T., Takao, T., Satomi, Y., Shimonishi, Y., Ishihara, N., et al. (2000). A ubiquitin-like system mediates protein lipidation. *Nature*, *408*(6811), 488–492. https://doi.org/10.1038/35044114.

Ikonomov, O. C., Sbrissa, D., & Shisheva, A. (2001). Mammalian cell morphology and endocytic membrane homeostasis require enzymatically active phosphoinositide 5-kinase PIKfyve. *Journal of Biological Chemistry*, *276*(28), 26141–26147. https://doi.org/10.1074/jbc.M101722200.

Inoki, K., Zhu, T., & Guan, K. L. (2003). TSC2 mediates cellular energy response to control cell growth and survival. *Cell*, *115*(5), 577–590. https://doi.org/10.1016/S0092-8674(03)00929-2.

Jeon, S. M., Chandel, N. S., & Hay, N. (2012). AMPK regulates NADPH homeostasis to promote tumour cell survival during energy stress. *Nature*, *485*(7400), 661–665. https://doi.org/10.1038/nature11066.

Kamphorst, J. J., Cross, J. R., Fan, J., De Stanchina, E., Mathew, R., White, E. P., et al. (2013). Hypoxic and Ras-transformed cells support growth by scavenging unsaturated fatty acids from lysophospholipids. *Proceedings of the National Academy of Sciences of the United States of America*, *110*(22), 8882–8887. https://doi.org/10.1073/pnas.1307237110.

Kamphorst, J. J., Nofal, M., Commisso, C., Hackett, S. R., Lu, W., Grabocka, E., et al. (2015). Human pancreatic cancer tumors are nutrient poor and tumor cells actively scavenge extracellular protein. *Cancer Research*, *75*(3), 544–553. https://doi.org/10.1158/0008-5472.CAN-14-2211.

Karasic, T. B., O'Hara, M. H., Loaiza-Bonilla, A., Reiss, K. A., Teitelbaum, U. R., Borazanci, E., et al. (2019). Effect of gemcitabine and nab-paclitaxel with or without hydroxychloroquine on patients with advanced pancreatic cancer: A phase 2 randomized clinical trial. *JAMA Oncology*, *5*(7), 993–998. https://doi.org/10.1001/jamaoncol.2019.0684.

Karasinska, J. M., Topham, J. T., Kalloger, S. E., Jang, G. H., Denroche, R. E., Culibrk, L., et al. (2020). Altered gene expression along the glycolysis–cholesterol synthesis axis is associated with outcome in pancreatic cancer. *Clinical Cancer Research*, *26*(1), 135–146. https://doi.org/10.1158/1078-0432.CCR-19-1543.

Kashatus, J. A., Nascimento, A., Myers, L. J., Sher, A., Byrne, F. L., Hoehn, K. L., et al. (2015). Erk2 phosphorylation of Drp1 promotes mitochondrial fission and MAPK-driven tumor growth. *Molecular Cell*, *57*(3), 537–551. https://doi.org/10.1016/j.molcel.2015.01.002.

Kerr, E. M., Gaude, E., Turrell, F. K., Frezza, C., & Martins, C. P. (2016). Mutant Kras copy number defines metabolic reprogramming and therapeutic susceptibilities. *Nature, 531*(7592), 110–113. https://doi.org/10.1038/nature16967.

Kerr, M. C., Lindsay, M. R., Luetterforst, R., Hamilton, N., Simpson, F., Parton, R. G., et al. (2006). Visualisation of macropinosome maturation by the recruitment of sorting nexins. *Journal of Cell Science, 119*(19), 3967–3980. https://doi.org/10.1242/jcs.03167.

Kim, J., Kim, Y. C., Fang, C., Russell, R. C., Kim, J. H., Fan, W., et al. (2013). Differential regulation of distinct Vps34 complexes by AMPK in nutrient stress and autophagy. *Cell, 152*(1–2), 290–303. https://doi.org/10.1016/j.cell.2012.12.016.

Kim, J., Kundu, M., Viollet, B., & Guan, K. L. (2011). AMPK and mTOR regulate autophagy through direct phosphorylation of Ulk1. *Nature Cell Biology, 13*(2), 132–141. https://doi.org/10.1038/ncb2152.

Kim, J., Lee, H. M., Cai, F., Ko, B., Yang, C., Lieu, E. L., et al. (2020). The hexosamine biosynthesis pathway is a targetable liability in KRAS/LKB1 mutant lung cancer. *Nature Metabolism, 2*(12), 1401–1412. https://doi.org/10.1038/s42255-020-00316-0.

Kimmelman, A. C., & White, E. (2017). Autophagy and tumor metabolism. *Cell Metabolism, 25*, 1037–1043. https://doi.org/10.1016/j.cmet.2017.04.004.

King, B., Araki, J., Palm, W., & Thompson, C. B. (2020). Yap/Taz promote the scavenging of extracellular nutrients through macropinocytosis. *Genes and Development, 34*(19–20), 1345–1358. https://doi.org/10.1101/gad.340661.120.

Kinsey, C. G., Camolotto, S. A., Boespflug, A. M., Guillen, K. P., Foth, M., Truong, A., et al. (2019). Protective autophagy elicited by RAF→MEK→ERK inhibition suggests a treatment strategy for RAS-driven cancers. *Nature Medicine, 25*(4), 620–627. https://doi.org/10.1038/s41591-019-0367-9.

Krishna, S., Palm, W., Lee, Y., Yang, W., Bandyopadhyay, U., Xu, H., et al. (2016). PIKfyve regulates vacuole maturation and nutrient recovery following engulfment. *Developmental Cell, 38*(5), 536–547. https://doi.org/10.1016/j.devcel.2016.08.001.

Kuznetsov, J. N., Leclerc, G. J., Leclerc, G. M., & Barredo, J. C. (2011). AMPK and Akt determine apoptotic cell death following perturbations of one-carbon metabolism by regulating ER stress in acute lymphoblastic leukemia. *Molecular Cancer Therapeutics, 10*(3), 437–447. https://doi.org/10.1158/1535-7163.MCT-10-0777.

Lanzetti, L., Palamidessi, A., Areces, L., Scita, G., & Di Fiore, P. P. (2004). Rab5 is a signalling GTPase involved in actin remodelling by receptor tyrosine kinases. *Nature, 429*(6989), 309–314. https://doi.org/10.1038/nature02542.

Lee, C. S., Lee, L. C., Yuan, T. L., Chakka, S., Fellmann, C., Lowe, S. W., et al. (2019). MAP kinase and autophagy pathways cooperate to maintain RAS mutant cancer cell survival. *Proceedings of the National Academy of Sciences of the United States of America, 116*(10), 4508–4517. https://doi.org/10.1073/pnas.1817494116.

Lee, E. J., & Tournier, C. (2011). The requirement of uncoordinated 51-like kinase 1 (ULK1) and ULK2 in the regulation of autophagy. *Autophagy, 7*(7), 689–695. https://doi.org/10.4161/auto.7.7.15450.

Lewis, W. H. (1937). Pinocytosis by malignant cells. *American Journal of Cancer, 29*(4), 666–679. https://doi.org/10.1158/ajc.1937.666.

Liberti, M. V., & Locasale, J. W. (2016). The Warburg effect: How does it benefit cancer cells? *Trends in Biochemical Sciences, 41*, 211–218. https://doi.org/10.1016/j.tibs.2015.12.001.

Lindmo, K., Brech, A., Finley, K. D., Gaumer, S., Contamine, D., Rusten, T. E., et al. (2008). The PI 3-kinase regulator Vps15 is required for autophagic clearance of protein aggregates. *Autophagy, 4*(4), 500–506. https://doi.org/10.4161/auto.5829.

Liu, J., Xia, H., Kim, M., Xu, L., Li, Y., Zhang, L., et al. (2011). Beclin1 controls the levels of p53 by regulating the deubiquitination activity of USP10 and USP13. *Cell, 147*(1), 223–234. https://doi.org/10.1016/j.cell.2011.08.037.

Machin, P. A., Tsonou, E., Hornigold, D. C., & Welch, H. C. E. (2021). Rho family GTPases and rho GEFs in glucose homeostasis. *Cell*, *10*(4), 915. https://doi.org/10.3390/cells10040915.

Maitra, A., & Hruban, R. H. (2008). Pancreatic cancer. *Annual Review of Pathology: Mechanisms of Disease*, *3*, 157–188. https://doi.org/10.1146/annurev.pathmechdis.3.121806.154305.

Martin, K. R., Celano, S. L., Solitro, A. R., Gunaydin, H., Scott, M., O'Hagan, R. C., et al. (2018). A potent and selective ULK1 inhibitor suppresses autophagy and sensitizes cancer cells to nutrient stress. *IScience*, *8*, 74–84. https://doi.org/10.1016/j.isci.2018.09.012.

Matsuda, N., Sato, S., Shiba, K., Okatsu, K., Saisho, K., Gautier, C. A., et al. (2010). PINK1 stabilized by mitochondrial depolarization recruits parkin to damaged mitochondria and activates latent parkin for mitophagy. *Journal of Cell Biology*, *189*(2), 211–221. https://doi.org/10.1083/jcb.200910140.

Matsunaga, K., Saitoh, T., Tabata, K., Omori, H., Satoh, T., Kurotori, N., et al. (2009). Two Beclin 1-binding proteins, Atg14L and Rubicon, reciprocally regulate autophagy at different stages. *Nature Cell Biology*, *11*(4), 385–396. https://doi.org/10.1038/ncb1846.

McAfee, Q., Zhang, Z., Samanta, A., Levi, S. M., Ma, X. H., Piao, S., et al. (2012). Autophagy inhibitor Lys05 has single-agent antitumor activity and reproduces the phenotype of a genetic autophagy deficiency. *Proceedings of the National Academy of Sciences of the United States of America*, *109*(21), 8253–8258. https://doi.org/10.1073/pnas.1118193109.

Meng, Y., Xu, X., Luan, H., Li, L., Dai, W., Li, Z., et al. (2019). The progress and development of GLUT1 inhibitors targeting cancer energy metabolism. *Future Medicinal Chemistry*, *11*, 2333–2352. https://doi.org/10.4155/fmc-2019-0052.

Michalopoulou, E., Auciello, F. R., Bulusu, V., Strachan, D., Campbell, A. D., Tait-Mulder, J., et al. (2020). Macropinocytosis renders a subset of pancreatic tumor cells resistant to mTOR inhibition. *Cell Reports*, *30*(8), 2729–2742.e4. https://doi.org/10.1016/j.celrep.2020.01.080.

Morgan, M. J., Gamez, G., Menke, C., Hernandez, A., Thorburn, J., Gidan, F., et al. (2014). Regulation of autophagy and chloroquine sensitivity by oncogenic RAS in vitro is context-dependent. *Autophagy*, *10*(10), 1814–1826. https://doi.org/10.4161/auto.32135.

Mukhopadhyay, S., Biancur, D. E., Parker, S. J., Yamamoto, K., Banh, R. S., Paulo, J. A., et al. (2021). Autophagy is required for proper cysteine homeostasis in pancreatic cancer through regulation of SLC7A11. *Proceedings of the National Academy of Sciences of the United States of America*, *118*(6). e2021475118. https://doi.org/10.1073/pnas.2021475118.

Nagdas, S., Kashatus, J. A., Nascimento, A., Hussain, S. S., Trainor, R. E., Pollock, S. R., et al. (2019). Drp1 promotes KRas-driven metabolic changes to drive pancreatic tumor growth. *Cell Reports*, *28*(7), 1845–1859.e5. https://doi.org/10.1016/j.celrep.2019.07.031.

Nelson, B. S., Lin, L., Kremer, D. M., Sousa, C. M., Cotta-Ramusino, C., Myers, A., et al. (2020). Tissue of origin dictates GOT1 dependence and confers synthetic lethality to radiotherapy. *Cancer & Metabolism*, *8*(1), 1–16. https://doi.org/10.1186/s40170-019-0202-2.

Nicot, A. S., Fares, H., Payrastre, B., Chisholm, A. D., Labouesse, M., & Laporte, J. (2006). The phosphoinositide kinase PIKfyve/Fab1p regulates terminal lysosome maturation in Caenorhabditis elegans. *Molecular Biology of the Cell*, *17*(7), 3062–3074. https://doi.org/10.1091/mbc.e05-12-1120.

Ohkuma, S., Shimizu, S., Noto, M., Sai, Y., Kinoshita, K., & Tamura, H. O. (1993). Inhibition of cell growth by bafilomycin A1, a selective inhibitor of vacuolar H+-ATPase. *In Vitro Cellular & Developmental Biology - Animal: Journal of the Society for In Vitro Biology*, *29*(11), 862–866. https://doi.org/10.1007/BF02631364.

Onorati, A. V., Dyczynski, M., Ojha, R., & Amaravadi, R. K. (2018). Targeting autophagy in cancer. *Cancer, 124*, 3307–3318. https://doi.org/10.1002/cncr.31335.

Ozkan-Dagliyan, I., Diehl, J. N., George, S. D., Schaefer, A., Papke, B., Klotz-Noack, K., et al. (2020). Low-dose vertical inhibition of the RAF-MEK-ERK Cascade causes apoptotic death of KRAS mutant cancers. *Cell Reports, 31*(11), 107764. https://doi.org/10.1016/j.celrep.2020.107764.

Palm, W., Araki, J., King, B., DeMatteo, R. G., & Thompson, C. B. (2017). Critical role for PI3-kinase in regulating the use of proteins as an amino acid source. *Proceedings of the National Academy of Sciences of the United States of America, 114*(41), E8628–E8636. https://doi.org/10.1073/pnas.1712726114.

Palm, W., Park, Y., Wright, K., Pavlova, N. N., Tuveson, D. A., & Thompson, C. B. (2015). The utilization of extracellular proteins as nutrients is suppressed by mTORC1. *Cell, 162*(2), 259–270. https://doi.org/10.1016/j.cell.2015.06.017.

Palorini, R., Cammarata, F., Balestrieri, C., Monestiroli, A., Vasso, M., Gelfi, C., et al. (2013). Glucose starvation induces cell death in K-ras-transformed cells by interfering with the hexosamine biosynthesis pathway and activating the unfolded protein response. *Cell Death & Disease, 4*(7), 732. https://doi.org/10.1038/cddis.2013.257.

Papke, B., & Der, C. J. (2017). Drugging RAS: Know the enemy. *Science, 355*, 1158–1163. https://doi.org/10.1126/science.aam7622.

Parker, S. J., Amendola, C. R., Hollinshead, K. E. R., Yu, Q., Yamamoto, K., Encarnación-Rosado, J., et al. (2020). Selective alanine transporter utilization creates a targetable metabolic niche in pancreatic cancer. *Cancer Discovery, 10*(7), 1018–1037. https://doi.org/10.1158/2159-8290.CD-19-0959.

Patra, K. C., Wang, Q., Bhaskar, P. T., Miller, L., Wang, Z., Wheaton, W., et al. (2013). Hexokinase 2 is required for tumor initiation and maintenance and its systemic deletion is therapeutic in mouse models of cancer. *Cancer Cell, 24*(2), 213–228. https://doi.org/10.1016/j.ccr.2013.06.014.

Petherick, K. J., Conway, O. J. L., Mpamhanga, C., Osborne, S. A., Kamal, A., Saxty, B., et al. (2015). Pharmacological inhibition of ULK1 kinase blocks mammalian target of rapamycin (mTOR)-dependent autophagy. *Journal of Biological Chemistry, 290*(18), 11376–11383. https://doi.org/10.1074/jbc.C114.627778.

Philpott, C., Tovell, H., Frayling, I. M., Cooper, D. N., & Upadhyaya, M. (2017). The NF1 somatic mutational landscape in sporadic human cancers. *Human Genomics, 11*(1), 13. https://doi.org/10.1186/s40246-017-0109-3.

Pyo, J. O., Nah, J., & Jung, Y. K. (2012). Molecules and their functions in autophagy. *Experimental and Molecular Medicine, 44*(2), 73–80. https://doi.org/10.3858/emm.2012.44.2.029.

Qi, X., Qvit, N., Su, Y. C., & Mochly-Rosen, D. (2013). A novel Drp1 inhibitor diminishes aberrant mitochondrial fission and neurotoxicity. *Journal of Cell Science, 126*(3), 789–802. https://doi.org/10.1242/jcs.114439.

Rabinowitz, J. D., & White, E. (2010). Autophagy and metabolism. *Science, 330*, 1344–1348. https://doi.org/10.1126/science.1193497.

Racker, E., Resnick, R. J., & Feldman, R. (1985). Glycolysis and methylaminoisobutyrate uptake in rat-1 cells transfected with ras or myc oncogenes. *Proceedings of the National Academy of Sciences of the United States of America, 82*(11), 3535–3538. https://doi.org/10.1073/pnas.82.11.3535.

Raho, S., Capobianco, L., Malivindi, R., Vozza, A., Piazzolla, C., De Leonardis, F., et al. (2020). KRAS-regulated glutamine metabolism requires UCP2-mediated aspartate transport to support pancreatic cancer growth. *Nature Metabolism, 2*(12), 1373–1381. https://doi.org/10.1038/s42255-020-00315-1.

Rambold, A. S., Kostelecky, B., Elia, N., & Lippincott-Schwartz, J. (2011). Tubular network formation protects mitochondria from autophagosomal degradation during nutrient starvation. *Proceedings of the National Academy of Sciences of the United States of America, 108*(25), 10190–10195. https://doi.org/10.1073/pnas.1107402108.

Rebecca, V. W., Nicastri, M. C., McLaughlin, N., Fennelly, C., McAfee, Q., Ronghe, A., et al. (2017). A unified approach to targeting the lysosome's degradative and growth signaling roles. *Cancer Discovery*, 7(11), 1266–1283. https://doi.org/10.1158/2159-8290.CD-17-0741.

Ricciardiello, F., Bergamaschi, L., De Vitto, H., Gang, Y., Zhang, T., Palorini, R., et al. (2021). Suppression of the HBP function increases pancreatic cancer cell sensitivity to a pan-RAS inhibitor. *Cell, 10*(2), 431. https://doi.org/10.3390/cells10020431.

Ricciardiello, F., Gang, Y., Palorini, R., Li, Q., Giampà, M., Zhao, F., et al. (2020). Hexosamine pathway inhibition overcomes pancreatic cancer resistance to gemcitabine through unfolded protein response and EGFR-Akt pathway modulation. *Oncogene, 39*(20), 4103–4117. https://doi.org/10.1038/s41388-020-1260-1.

Robinson, M. M., McBryant, S. J., Tsukamoto, T., Rojas, C., Ferraris, D. V., Hamilton, S. K., et al. (2007). Novel mechanism of inhibition of rat kidney-type glutaminase by bis-2-(5-phenylacetamido-1,2,4-thiadiazol-2-yl)ethyl sulfide (BPTES). *Biochemical Journal, 406*(3), 407–414. https://doi.org/10.1042/BJ20070039.

Ronan, B., Flamand, O., Vescovi, L., Dureuil, C., Durand, L., Fassy, F., et al. (2014). A highly potent and selective Vps34 inhibitor alters vesicle trafficking and autophagy. *Nature Chemical Biology, 10*(12), 1013–1019. https://doi.org/10.1038/nchembio.1681.

Saito, Y., Chapple, R. H., Lin, A., Kitano, A., & Nakada, D. (2015). AMPK protects Leukemia-initiating cells in myeloid leukemias from metabolic stress in the bone marrow. *Cell Stem Cell, 17*(5), 585–596. https://doi.org/10.1016/j.stem.2015.08.019.

Salloum, D., Mukhopadhyay, S., Tung, K., Polonetskaya, A., & Foster, D. A. (2014). Mutant Ras elevates dependence on serum lipids and creates a synthetic lethality for rapamycin. *Molecular Cancer Therapeutics, 13*(3), 733–741. https://doi.org/10.1158/1535-7163.MCT-13-0762.

Santana-Codina, N., Chandhoke, A. S., Yu, Q., Małachowska, B., Kuljanin, M., Gikandi, A., et al. (2020). Defining and targeting adaptations to oncogenic KRASG12C inhibition using quantitative temporal proteomics. *Cell Reports, 30*(13), 4584–4599.e4. https://doi.org/10.1016/j.celrep.2020.03.021.

Santana-Codina, N., Roeth, A. A., Zhang, Y., Yang, A., Mashadova, O., Asara, J. M., et al. (2018). Oncogenic KRAS supports pancreatic cancer through regulation of nucleotide synthesis. *Nature Communications, 9*(1), 1–13. https://doi.org/10.1038/s41467-018-07472-8.

Sasaki, H., Hikosaka, Y., Kawano, O., Moriyama, S., Yano, M., & Fujii, Y. (2011). Evaluation of Kras gene mutation and copy number gain in non-small cell lung cancer. *Journal of Thoracic Oncology, 6*(1), 15–20. https://doi.org/10.1097/JTO.0b013e31820594f0.

Scafoglio, C., Hirayama, B. A., Kepe, V., Liu, J., Ghezzi, C., Satyamurthy, N., et al. (2015). Functional expression of sodium-glucose transporters in cancer. *Proceedings of the National Academy of Sciences of the United States of America, 112*(30), E4111–E4119. https://doi.org/10.1073/pnas.1511698112.

Schrezenmeier, E., & Dörner, T. (2020). Mechanisms of action of hydroxychloroquine and chloroquine: Implications for rheumatology. *Nature Reviews Rheumatology, 16*, 155–166. https://doi.org/10.1038/s41584-020-0372-x.

Serasinghe, M. N., Wieder, S. Y., Renault, T. T., Elkholi, R., Asciolla, J. J., Yao, J. L., et al. (2015). Mitochondrial division is requisite to RAS-induced transformation and targeted by oncogenic MAPK pathway inhibitors. *Molecular Cell, 57*(3), 521–536. https://doi.org/10.1016/j.molcel.2015.01.003.

Shao, S., Li, S., Qin, Y., Wang, X., Yang, Y., Bai, H., et al. (2014). Spautin-1, a novel autophagy inhibitor, enhances imatinib-induced apoptosis in chronic myeloid leukemia. *International Journal of Oncology, 44*(5), 1661–1668. https://doi.org/10.3892/ijo.2014.2313.

Shaw, R. J., Bardeesy, N., Manning, B. D., Lopez, L., Kosmatka, M., DePinho, R. A., et al. (2004). The LKB1 tumor suppressor negatively regulates mTOR signaling. *Cancer Cell, 6*(1), 91–99. https://doi.org/10.1016/j.ccr.2004.06.007.

Shaw, R. J., Kosmatka, M., Bardeesy, N., Hurley, R. L., Witters, L. A., DePinho, R. A., et al. (2004). The tumor suppressor LKB1 kinase directly activates AMP-activated kinase and regulates apoptosis in response to energy stress. *Proceedings of the National Academy of Sciences of the United States of America, 101*(10), 3329–3335. https://doi.org/10.1073/pnas.0308061100.

Shi, T. T., Yu, X. X., Yan, L. J., & Xiao, H. T. (2017). Research progress of hydroxychloroquine and autophagy inhibitors on cancer. *Cancer Chemotherapy and Pharmacology, 79*, 287–294. https://doi.org/10.1007/s00280-016-3197-1.

Shukla, S. K., Purohit, V., Mehla, K., Gunda, V., Chaika, N. V., Vernucci, E., et al. (2017). MUC1 and HIF-1alpha Signaling crosstalk induces anabolic glucose metabolism to impart gemcitabine resistance to pancreatic cancer. *Cancer Cell, 32*(1), 71–87.e7. https://doi.org/10.1016/j.ccell.2017.06.004.

Siegel, R. L., Miller, K. D., Fuchs, H. E., & Jemal, A. (2021). Cancer statistics, 2021. *CA: a Cancer Journal for Clinicians, 71*(1), 7–33. https://doi.org/10.3322/caac.21654.

Singh, A., Ruiz, C., Bhalla, K., Haley, J. A., Li, Q. K., Acquaah-Mensah, G., et al. (2018). De novo lipogenesis represents a therapeutic target in mutant Kras non-small cell lung cancer. *FASEB Journal, 32*(12), 7018–7027. https://doi.org/10.1096/fj.201800204.

Son, J., Lyssiotis, C. A., Ying, H., Wang, X., Hua, S., Ligorio, M., et al. (2013). Glutamine supports pancreatic cancer growth through a KRAS-regulated metabolic pathway. *Nature, 496*(7443), 101–105. https://doi.org/10.1038/nature12040.

Song, S., Zhang, Y., Ding, T., Ji, N., & Zhao, H. (2021). The dual role of macropinocytosis in cancers: Promoting growth and inducing Methuosis to participate in anticancer therapies as targets. *Frontiers in Oncology, 10*, 3099. https://doi.org/10.3389/fonc.2020.570108.

Soucek, L., Whitfield, J., Martins, C. P., Finch, A. J., Murphy, D. J., Sodir, N. M., et al. (2008). Modelling Myc inhibition as a cancer therapy. *Nature, 455*(7213), 679–683. https://doi.org/10.1038/nature07260.

Soucek, L., Whitfield, J. R., Sodir, N. M., Massó-Vallés, D., Serrano, E., Karnezis, A. N., et al. (2013). Inhibition of Myc family proteins eradicates KRas-driven lung cancer in mice. *Genes and Development, 27*(5), 504–513. https://doi.org/10.1101/gad.205542.112.

Sousa, C. M., Biancur, D. E., Wang, X., Halbrook, C. J., Sherman, M. H., Zhang, L., et al. (2016). Pancreatic stellate cells support tumour metabolism through autophagic alanine secretion. *Nature, 536*(7617), 479–483. https://doi.org/10.1038/nature19084.

Sun, W., Luan, S., Qi, C., Tong, Q., Yan, S., Li, H., et al. (2019). Aspulvinone O, a natural inhibitor of GOT1 suppresses pancreatic ductal adenocarcinoma cells growth by interfering glutamine metabolism. *Cell Communication and Signaling: CCS, 17*(1), 1–11. https://doi.org/10.1186/s12964-019-0425-4.

Swanson, J. A. (2008). Shaping cups into phagosomes and macropinosomes. *Nature Reviews Molecular Cell Biology, 9*, 639–649. https://doi.org/10.1038/nrm2447.

Swierczynski, J., Hebanowska, A., & Sledzinski, T. (2014). Role of abnormal lipid metabolism in development, progression, diagnosis and therapy of pancreatic cancer. *World Journal of Gastroenterology, 20*(9), 2279–2303. https://doi.org/10.3748/wjg.v20.i9.2279.

Tang, F., Hu, P., Yang, Z., Xue, C., Gong, J., Sun, S., et al. (2017). SBI0206965, a novel inhibitor of Ulk1, suppresses non-small cell lung cancer cell growth by modulating both autophagy and apoptosis pathways. *Oncology Reports, 37*(6), 3449–3458. https://doi.org/10.3892/or.2017.5635.

Truong, A., Yoo, J. H., Scherzer, M. T., Michael, J., Sanchez, S., Dale, K. J., et al. (2020). Chloroquine sensitizes GNAQ/11-mutated melanoma to MEK1/2 inhibition. *Clinical Cancer Research, 26*(23), 6374–6386. https://doi.org/10.1158/1078-0432.CCR-20-1675.

Twig, G., Elorza, A., Molina, A. J. A., Mohamed, H., Wikstrom, J. D., Walzer, G., et al. (2008). Fission and selective fusion govern mitochondrial segregation and elimination by autophagy. *EMBO Journal*, *27*(2), 433–446. https://doi.org/10.1038/sj.emboj.7601963.

Vaseva, A. V., Blake, D. R., Gilbert, T. S. K., Ng, S., Hostetter, G., Azam, S. H., et al. (2018). KRAS suppression-induced degradation of MYC is antagonized by a MEK5-ERK5 compensatory mechanism. *Cancer Cell*, *34*(5), 807–822.e7. https://doi.org/10.1016/j.ccell.2018.10.001.

Viale, A., Pettazzoni, P., Lyssiotis, C. A., Ying, H., Sánchez, N., Marchesini, M., et al. (2014). Oncogene ablation-resistant pancreatic cancer cells depend on mitochondrial function. *Nature*, *514*(7524), 628–632. https://doi.org/10.1038/nature13611.

Vizan, P., Boros, L. G., Figueras, A., Capella, G., Mangues, R., Bassilian, S., et al. (2005). K-ras codon-specific mutations produce distinctive metabolic phenotypes in human fibroblasts. *Cancer Research*, *65*(13), 5512–5515. https://doi.org/10.1158/0008-5472.CAN-05-0074.

Warburg, O. (1925). The metabolism of carcinoma cells 1. *The Journal of Cancer Research*, *9*(1), 148–163. https://doi.org/10.1158/jcr.1925.148.

Warburg, O. (1956). On the origin of cancer cells. *Science*, *123*(3191), 309–314. https://doi.org/10.1126/science.123.3191.309.

Waters, A. M., & Der, C. J. (2018). KRAS: The critical driver and therapeutic target for pancreatic cancer. *Cold Spring Harbor Perspectives in Medicine*, *8*(9), a031435. https://doi.org/10.1101/cshperspect.a031435.

Weinberg, F., Hamanaka, R., Wheaton, W. W., Weinberg, S., Joseph, J., Lopez, M., et al. (2010). Mitochondrial metabolism and ROS generation are essential for Kras-mediated tumorigenicity. *Proceedings of the National Academy of Sciences of the United States of America*, *107*(19), 8788–8793. https://doi.org/10.1073/pnas.1003428107.

West, M. A., Prescott, A. R., Eskelinen, E. L., Ridley, A. J., & Watts, C. (2000). Rac is required for constitutive macropinocytosis by dendritic cells but does not control its downregulation. *Current Biology*, *10*(14), 839–848. https://doi.org/10.1016/S0960-9822(00)00595-9.

Wu, N., Gu, C., Gu, H., Hu, H., Han, Y., & Li, Q. (2011). Metformin induces apoptosis of lung cancer cells through activating JNK/p38 MAPK pathway and GADD153. *Neoplasma*, *58*(6), 482–490. https://doi.org/10.4149/neo_2011_06_482.

Xia, M., Li, X., Diao, Y., Du, B., & Li, Y. (2021). Targeted inhibition of glutamine metabolism enhances the antitumor effect of selumetinib in KRAS-mutant NSCLC. *Translational Oncology*, *14*(1), 100920. https://doi.org/10.1016/j.tranon.2020.100920.

Yan, L., Raj, P., Yao, W., & Ying, H. (2019). Glucose metabolism in pancreatic cancer. *Cancers*, *11*(10), 1460. https://doi.org/10.3390/cancers11101460.

Yan, L., Tu, B., Yao, J., Gong, J., Carugo, A., Bristow, C. A., et al. (2021). Targeting glucose metabolism sensitizes pancreatic cancer to MEK inhibition. *Cancer Research*, *81*(15), 4054–4065. canres.3792.2020. https://doi.org/10.1158/0008-5472.CAN-20-3792.

Yanagisawa, A., Ohtake, K., Ohashi, K., Hori, M., Kitagawa, T., Sugano, H., et al. (1993). Frequent c-Ki-ras oncogene activation in mucous cell Hyperplasias of pancreas suffering from chronic inflammation. *Cancer Research*, *53*(5), 953–956.

Yang, A., & Kimmelman, A. C. (2014). Inhibition of autophagy attenuates pancreatic cancer growth independent of TP53/TRP53 status. *Autophagy*, *10*(9), 1683–1684. https://doi.org/10.1158/2159-8290.CD-14-0362.

Yang, A., Rajeshkumar, N. V., Wang, X., Yabuuchi, S., Alexander, B. M., Chu, G. C., et al. (2014). Autophagy is critical for pancreatic tumor growth and progression in tumors with p53 alterations. *Cancer Discovery*, *4*(8), 905–913. https://doi.org/10.1158/2159-8290.CD-14-0362.

Yang, S., Wang, X., Contino, G., Liesa, M., Sahin, E., Ying, H., et al. (2011). Pancreatic cancers require autophagy for tumor growth. *Genes and Development*, *25*, 717–729. http://www.genesdev.org/cgi/doi/10.1101/gad.2016111.

Ying, H., Kimmelman, A. C., Lyssiotis, C. A., Hua, S., Chu, G. C., Fletcher-Sananikone, E., et al. (2012). Oncogenic kras maintains pancreatic tumors through regulation of anabolic glucose metabolism. *Cell*, *149*(3), 656–670. https://doi.org/10.1016/j.cell.2012.01.058.

Yoshida, S., Hoppe, A. D., Araki, N., & Swanson, J. A. (2009). Sequential signaling in plasma-membrane domains during macropinosome formation in macrophages. *Journal of Cell Science*, *122*(18), 3250–3261. https://doi.org/10.1242/jcs.053207.

Yoshida, T., Yamasaki, S., Kaneko, O., Taoka, N., Tomimoto, Y., Namatame, I., et al. (2020). A covalent small molecule inhibitor of glutamate-oxaloacetate transaminase 1 impairs pancreatic cancer growth. *Biochemical and Biophysical Research Communications*, *522*(3), 633–638. https://doi.org/10.1016/j.bbrc.2019.11.130.

Yu, M., Liu, H., Duan, Y., Zhang, D., Li, S., & Wang, F. (2015). Four types of fatty acids exert differential impact on pancreatic cancer growth. *Cancer Letters*, *360*(2), 187–194. https://doi.org/10.1016/j.canlet.2015.02.002.

Yun, J., Rago, C., Cheong, I., Pagliarini, R., Angenendt, P., Rajagopalan, H., et al. (2009). Glucose deprivation contributes to the development of KRAS pathway mutations in tumor cells. *Science*, *325*(5947), 1555–1559. https://doi.org/10.1126/science.1174229.

Zadra, G., Photopoulos, C., Tyekucheva, S., Heidari, P., Weng, Q. P., Fedele, G., et al. (2014). A novel direct activator of AMPK inhibits prostate cancer growth by blocking lipogenesis. *EMBO Molecular Medicine*, *6*(4), 519–538. https://doi.org/10.1002/emmm.201302734.

Zeh, H. J., Bahary, N., Boone, B. A., Singhi, A. D., Miller-Ocuin, J. L., Normolle, D. P., et al. (2020). A randomized phase II preoperative study of autophagy inhibition with high-dose hydroxychloroquine and gemcitabine/nab-paclitaxel in pancreatic cancer patients. *Clinical Cancer Research*, *26*(13), 3126–3134. https://doi.org/10.1158/1078-0432.CCR-19-4042.

Zhou, W., Capello, M., Fredolini, C., Racanicchi, L., Piemonti, L., Liotta, L. A., et al. (2012). Proteomic analysis reveals Warburg effect and anomalous metabolism of glutamine in pancreatic cancer cells. *Journal of Proteome Research*, *11*(2), 554–563. https://doi.org/10.1021/pr2009274.

CHAPTER TEN

The RASopathies: Biology, genetics and therapeutic options

Jody Fromm Longo and Steven L. Carroll*
Department of Pathology and Laboratory Medicine, Medical University of South Carolina, Charleston, SC, United States
*Corresponding author: e-mail address: carrolst@musc.edu

Contents

1. Introduction 306
2. Neurofibromatosis type 1 (NF1) and related syndromes 306
3. Legius syndrome 317
4. Noonan syndrome and related syndromes 319
5. Noonan-related syndromes 322
6. Costello syndrome 324
7. Cardiofaciocutaneous syndrome 325
8. Gingival fibromatosis 326
9. Capillary malformation-arteriovenous malformation syndrome 326
10. Future directions 327
References 328

Abstract

The RASopathies are a group of genetic diseases in which the Ras/MAPK signaling pathway is inappropriately activated as a result of mutations in genes encoding proteins within this pathway. As their causative mutations have been identified, this group of diseases has expanded to include neurofibromatosis type 1 (NF1), Legius syndrome, Noonan syndrome, CBL syndrome, Noonan syndrome-like disorder with loose anagen hair, Noonan syndrome with multiple lentigines, Costello syndrome, cardiofaciocutaneous syndrome, gingival fibromatosis and capillary malformation-arteriovenous malformation syndrome. Many of these genetic disorders share clinical features in common such as abnormal facies, short stature, varying degrees of cognitive impairment, cardiovascular abnormalities, skeletal abnormalities and a predisposition to develop benign and malignant neoplasms. Others are more dissimilar, even though their mutations are in the same gene that is mutated in a different RASopathy. Here, we describe the clinical features of each RASopathy and contrast them with the other RASopathies. We discuss the genetics of these disorders, including the causative mutations for each RASopathy, the impact that these mutations have on the function of an individual protein and how this dysregulates the Ras/MAPK signaling pathway. As several of these individual disorders are genetically heterogeneous, we also consider the different genes that can be mutated to produce disease with the same phenotype. We also discuss how our

Advances in Cancer Research, Volume 153
ISSN 0065-230X
https://doi.org/10.1016/bs.acr.2021.07.007

growing understanding of dysregulated Ras/MAPK signaling had led to the development of new therapeutic agents and what work will be critically important in the future to improve the lives of patients with RASopathies.

1. Introduction

For decades, the group of genetic diseases that we now know as the RASopathies have presented a daunting diagnostic challenge for clinicians. In large part, this is because the clinical features seen in diseases such as neurofibromatosis type 1 (NF1), Legius syndrome and Noonan syndrome overlap. It was not until the genes that were mutated in these conditions were identified that clinicians and geneticists began to be able to reliably distinguish these diseases from one another. The identification of the genes that were mutated in these diseases also brought to light a common feature that they share—they all result from mutations in genes encoding individual Ras proteins, proteins that regulate Ras activity or key upstream and downstream effectors of Ras action. The subsequent identification of the causative genetic mutations in other conditions such as Costello syndrome, cardiofaciocutaneous syndrome and capillary malformation-arteriovenous malformation syndrome showed that these diseases are also caused by mutations of genes encoding proteins in the Ras signaling cascade; indeed, some of these conditions result from different mutations in the same genes that are mutated in other RASopathies. Here, we will explain these relationships by discussing the clinical presentation of each of these diseases, the genetic mutation(s) associated with each RASopathy, how these mutations impact the Ras signaling cascade and how this understanding is shaping attempts to treat these diseases. We will first consider NF1, the longest known and most thoroughly studied RASopathy, in some detail as it illustrates key principles about these diseases. We will then discuss the other RASopathies, starting with the ones that are most similar to each other clinically before turning to more clinically distinct conditions that have been recently identified as RASopathies.

2. Neurofibromatosis type 1 (NF1) and related syndromes

NF1 (OMIM #162200), also known as von Recklinghausen disease or peripheral neurofibromatosis, is estimated to occur in 1 in every 2500–3500 newborn infants. NF1, which was the first RASopathy described, is the most common genetic disease affecting the human nervous system (Carroll, 2012; Carroll & Ratner, 2008; Shen, Harper, & Upadhyaya, 1996; Williams et al., 2009) and the second most common RASopathy. NF1 is caused by a mutation in the *NF1* tumor suppressor gene located on the long arm of chromosome 17 (17q11.2). In families with NF1, the disease is inherited in an autosomal dominant fashion and is fully penetrant, affecting men and women equally (see Table 1, for a comparison of the inheritance, mutated gene and chromosomal location of the mutated gene in all of the RASopathies discussed here). That does not, however, mean that the manifestations of the disease will be identical in all members of the family; intriguingly, even closely related family members often present with different clinical features, an observation that potentially is due to the presence of as yet unidentified modifier genes that alter the disease phenotype (Easton, Ponder, Huson, & Ponder, 1993). The *NF1* gene has one of highest *de novo* mutation rates of any gene in the human genome, with about 50% of cases occurring in individuals without a family history of NF1. These *de novo* mutations occur in the copy of chromosome 17 inherited from the father and arise during spermatogenesis (Jadayel et al., 1990).

NF1 patients are at increased risk for the development of several types of benign and malignant neoplasms that occur when a "second hit" mutation inactivates the remaining functional *NF1* allele in the tumor cell of origin. As a result of both this and the vascular diseases to which they are prone (see below), individuals with NF1 have a lifespan that is approximately 15 years shorter than the general population (Rasmussen, Yang, & Friedman, 2001). NF1-associated tumors arise in the nervous system as well as at other sites including the endocrine system, breast, bone marrow, soft tissues and the gastrointestinal tract. Tumors of the peripheral nervous system are the neoplasms most commonly encountered in NF1. NF1 patients develop large numbers of neurofibromas, which are benign tumors that arise in the skin (dermal neurofibromas) as well as in large nerves (*e.g.*, spinal nerve roots) and nerve plexuses (plexiform neurofibromas). Dermal neurofibromas and plexiform neurofibromas are identical histologically, being composed of a complex mixture of cell types including Schwann cells (the neoplastic component of these tumors), mast cells, fibroblasts, macrophages, perineurial-like cells and vascular elements. However, the natural history of dermal neurofibromas and plexiform neurofibromas is quite different. Dermal neurofibromas are not present at birth (Longo, Weber, Turner-Ivey, & Carroll, 2018), typically beginning to appear when a child with NF1 enters puberty. Further, if a woman with NF1 becomes pregnant, she develops new dermal

Table 1 Genetic features of RASopathies.

Syndrome	Inheritance	Chromosomal location	Gene mutated
Neurofibromatosis type 1	Autosomal dominant	17q11.2	*NF1*
Neurofibromatosis-Noonan	Autosomal dominant	17q11.2	*NF1*
Legius syndrome	Autosomal dominant	15q14	*SPRED1*
Noonan syndrome 1	Autosomal dominant	12q24.13	*PTPN11*
Noonan syndrome 2	Autosomal recessive	22q11.21	*LZTR1*
Noonan syndrome 3	Autosomal dominant	12p12.1	*KRAS*
Noonan syndrome 4	Autosomal dominant	2p22.1	*SOS1*
Noonan syndrome 5	Autosomal dominant	3p25.2	*RAF1*
Noonan syndrome 6	Autosomal dominant	1p13.2	*NRAS*
Noonan syndrome 7	Autosomal dominant	7p34	*BRAF*
Noonan syndrome 8	Autosomal dominant	1p22	*RIT1*
Noonan syndrome 9	Autosomal dominant	14q21.3	*SOS2*
Noonan syndrome 10	Autosomal dominant	22q11.21	*LZTR1*
Noonan syndrome 11	Autosomal dominant	3q22.3	*MRAS*
Noonan syndrome 12	Autosomal dominant	11p15.2	*RRAS2*
Noonan syndrome 13	Autosomal dominant	22q11.22	*MAPK1*
Noonan syndrome-like disorder with or without juvenile myelomonocytic leukemia	Autosomal dominant	11q23.3	*CBL*
Noonan syndrome-like disorder with loose anagen hair 1	Autosomal dominant	10q25.2	*SHOC2*

Noonan syndrome-like disorder with loose anagen hair 2	Autosomal dominant	2p23.2	*PPP1CB*
LEOPARD syndrome 1	Autosomal dominant	12q24.13	*PTPN11*
LEOPARD syndrome 2	?	3p25.2	*RAF1*
LEOPARD syndrome 3	Autosomal dominant	7q34	*BRAF*
Costello syndrome	Autosomal dominant	11p15.5	*HRAS*
Cardiofaciocutaneous syndrome 1	Autosomal dominant	7q34	*BRAF*
Cardiofaciocutaneous syndrome 2	Autosomal dominant	12p12.1	*KRAS*
Cardiofaciocutaneous syndrome 3	Autosomal dominant	15q22.31	*MAP2K1*
Cardiofaciocutaneous syndrome 4	Autosomal dominant	19p13.3	*MAP2K2*
Gingival fibromatosis 1	Autosomal dominant	2p22.1	*SOS1*
Gingival fibromatosis 2	?	5q13-q22	?
Gingival fibromatosis 3	?	2p23.3	?
Gingival fibromatosis 4	?	11p15	?
Gingival fibromatosis 5	Autosomal dominant	4q12	*REST*
Capillary malformation-arteriovenous malformation 1	Autosomal dominant	5q14.3	*RASA1*
Capillary malformation-arteriovenous malformation 2	Autosomal dominant	7q22.1	*EPHB4*

neurofibromas and the ones that she already had enlarge. Taken together, these observations suggest that dermal neurofibromas are hormonally responsive; however, the hormone(s) driving dermal neurofibroma development and growth have not yet been identified. It is also notable that dermal neurofibromas have no malignant potential. In contrast, plexiform neurofibromas are typically congenital and are prone to transform into aggressive spindle cell neoplasms derived from the Schwann cell lineage that are known as malignant peripheral nerve sheath tumors (MPNSTs) (Brossier & Carroll, 2012; Carroll, 2012). MPNSTs are the most common malignancy and the leading cause of death in NF1 patients. Two large clinical series have estimated that an NF1 patient's lifetime risk for developing an MPNST is 8–13% (Evans et al., 2002) and 5.9–10.3% (McCaughan, Holloway, Davidson, & Lam, 2007). The prognosis for patients with MPNSTs is grim, with several large studies reporting that the 5-year disease-free survival rate for these individuals is between 34 and 60% (Anghileri et al., 2006; Ducatman, Scheithauer, Piepgras, Reiman, & Ilstrup, 1986; Fan, Yang, & Wang, 2014; Porter et al., 2009; Stucky et al., 2012; Valentin et al., 2016; Watson et al., 2017; Zou et al., 2009). These poor outcomes reflect the fact that complete surgical resection is still the only effective means of treating MPNSTs—currently available radio- and chemotherapeutic regimens are ineffective against these neoplasms.

In the central nervous system (CNS), NF1 patients most commonly develop optic gliomas, a variant of the low grade neoplasms known as pilocytic astrocytomas, during childhood or early adolescence. Optic gliomas, which can be found anywhere along the optic nerves or in the optic chiasm, occur in about 15% of young NF1 patients (Ferner et al., 2007). Although optic gliomas are benign [World Health Organization (WHO) grade I] neoplasms, they produce significant morbidity. About 50% of patients with optic gliomas report moderate to severe visual impairment (Balcer et al., 2001; Thiagalingam, Flaherty, Billson, & North, 2004) and a small subset of children with gliomas in the optic chiasm develop precocious puberty as a result of pressure exerted by the tumor on the hypothalamus (Listernick, Charrow, Greenwald, & Mets, 1994; Listernick, Darling, Greenwald, Strauss, & Charrow, 1995). Although not encountered as frequently, NF1 patients are also at heightened risk for the development of higher grade diffuse astrocytomas (WHO grade II), anaplastic astrocytomas (WHO grade III) and glioblastomas (WHO grade IV) (Rodriguez et al., 2008). This observation is consistent with the fact that The Cancer Genome Atlas Research Network has found that approximately 18% of sporadic glioblastomas encountered in the general population have homozygous deletions or mutations of the *NF1* gene (Cancer Genome Atlas Research, 2008).

NF1 patients are also prone to develop a variety of other tumor types elsewhere in the body. During early childhood, individuals with NF1, especially boys, are at increased risk for the development of juvenile myelomonocytic leukemia (Clark & Hutter, 1982; Stiller, Chessells, & Fitchett, 1994). Tumors of the adrenal medulla known as pheochromocytomas are estimated to occur in 0.1–5.7% of NF1 patients, being most commonly encountered in the 5th decade of life (Bausch et al., 2006; Eisenhofer et al., 2011; Mannelli et al., 2009; Riccardi, 1991; Walther, Herring, Enquist, Keiser, & Linehan, 1999). NF1 patients are also prone to develop gastrointestinal stromal tumors (GISTs), with a lifetime risk of developing these tumors that is 45-fold higher than that of the general population (Barahona-Garrido et al., 2009). For reasons that are unclear, 60% of the GISTs encountered in NF1 patients occur in the small intestine rather than in the stomach as is typically seen in the general population (Miettinen, Fetsch, Sobin, & Lasota, 2006). Although relatively uncommon, individuals with NF1 are also at increased risk for the occurrence of gastrointestinal carcinoids (Griffiths, Williams, & Williams, 1987; Swinburn, Yeong, Lane, Nicholson, & Holdaway, 1988). Glomus tumors, which are small painful but typically benign neoplasms that develop on fingertips or under the nail of the finger, also appear to occur at increased frequency in NF1 patients (Brems et al., 2009). Although they are rare manifestations of NF1, NF1 patients are also at increased risk for the development of several types of sarcomas including embryonal rhabdomyosarcomas (Crucis et al., 2015; McGaughran et al., 1999; Patil & Chamberlain, 2012), leiomyosarcomas and osteosarcomas (Afsar et al., 2013). In addition, women with NF1 are two times more likely to develop breast cancer than the general population and, when they do, their survival is poorer than that of breast cancer patients without NF1 (Uusitalo et al., 2017; Wang et al., 2018).

Although inactivation of both copies of the *NF1* gene is required for the development of NF1-associated neoplasms, *NF1* haploinsufficiency is apparently sufficient to produce a number of non-neoplastic abnormalities that are characteristic of NF1. Melanocytic lesions affecting the eye (Lisch nodules) and skin (café-au-lait macules, axillary freckling) are so common that they are part of the consensus diagnostic criteria for NF1 (see below). Macrocephaly is a common finding in NF1 as is, to a lesser extent, short stature (Clementi et al., 1999; Szudek, Birch, & Friedman, 2000a, 2000b). About half of NF1 patients have skeletal abnormalities such as scoliosis, vertebral scalloping, unilateral sphenoid wing dysplasia and tibial pseudoarthrosis (Elefteriou et al., 2009). Bone mineral density is also often decreased in individuals with NF1. This may reflect the fact that serum 25-hydroxy-vitamin D levels are reduced in patients with large numbers of dermal neurofibromas; indeed, the more dermal neurofibromas an individual has, the lower their 25-hydroxy-vitamin D levels (Lammert et al., 2006). About 10% of NF1 patients have cardiovascular abnormalities such as coarctation of the abdominal aorta, cardiac valvular malformations, and stenosis of cerebral, coronary, renal and/or pulmonary arteries (D'Arco et al., 2014; Lin et al., 2000; Oderich et al., 2007; Rasmussen et al., 2001; Rea et al., 2009). Learning disabilities are present in up to 80% of children with NF1 (Schwetye & Gutmann, 2014), with 40% suffering from attention deficit/hyperactivity disorder and 30% having an autism spectrum disorder (Garg et al., 2013). In addition, many of these children have visuospatial deficits (Cutting & Levine, 2010).These learning disabilities may be related to the presence of "unidentified bright objects" (UBOs) that can be seen radiologically in the subcortical white matter, basal ganglia, thalamus, cerebellum and brainstem of these children (Griffiths et al., 1999; Itoh et al., 1994; Torres Nupan, Velez Van Meerbeke, Lopez Cabra, & Herrera Gomez, 2017; Van Es, North, McHugh, & De Silva, 1996). At present, it is unclear what type of pathology these UBOs represent. Possibilities include regions of abnormal myelination (Smirniotopoulos & Murphy, 1992) or vacuolated myelin (DiPaolo et al., 1995), hamartomas (Braffman, Bilaniuk, & Zimmerman, 1988) and heterotopias (Bognanno et al., 1988).

As noted above, the spectrum of findings that are present in an individual NF1 patient is highly variable and, to make matters worse, diseases such as Legius syndrome have clinical features that overlap with NF1. To address this problem, NIH sponsored a Consensus Development Conference on Neurofibromatosis to define the diagnostic criteria for NF1 (Neurofibromatosis Conference Statement, 1988). The consensus diagnostic criteria this group developed, based on the most common manifestations of NF1, were that at least two of the following had to be present: (1) 6 café-au-lait macules (>1.5 cm in diameter post-puberty; >0.5 cm prepuberty); (2) armpit or inguinal freckling; (3) 1 plexiform neurofibroma or ≥2 cutaneous (dermal) neurofibromas; (4) an optic glioma; (5) ≥2 Lisch nodules; (6) bony dysplasia; or (7) a first-degree relative with NF1. In most cases, these diagnostic criteria, particularly when partnered with more recently developed technology such as NextGen sequencing, allow a diagnosis of NF1 to be confidently made. However, diagnosis is still problematic in a subset of potential NF1 patients whose clinical findings extensively overlap with other RASopathies. Overlap syndromes such as neurofibromatosis-Noonan syndrome (OMIM #601321) are particularly difficult. Patients with neurofibromatosis-Noonan syndrome present with café-au-lait macules, Lisch nodules and axillary freckling in the absence of neurofibromas as well as features of Noonan syndrome including epicanthal folds, a webbed neck, low-set ears, hypertelorism, and pulmonary stenosis (Nystrom et al., 2009; Stevenson, Viskochil, Rope, & Carey, 2006). Neurofibromatosis-Noonan syndrome patients have heterozygous *NF1* mutations, often with in-frame deletions of exons 24 and 25 (De Luca et al., 2005), which encode part of neurofibromin's Ras GTPase-stimulating protein (GAP) domain; in addition, *NF1* missense mutations affecting Arg1809 have a high incidence of Noonan syndrome-like features (Rojnueangnit et al., 2015). Intriguingly, a small subset of neurofibromatosis-Noonan syndrome patients have coexisting mutations in the *PTPN11* gene (Bertola et al., 2005), which is linked to the pathogenesis of a subset of Noonan syndrome patients (see below).

The *NF1* gene contains 60 exons and spans approximately 283,000 bp (Viskochil et al., 1990; Wallace et al., 1990). The germline mutations encountered in NF1 patients are highly variable. The majority of these are point mutations, including nonsense mutations, missense mutations, frameshift mutations and mutations that interfere with RNA splicing (Messiaen et al., 2000; Wimmer et al., 2007). Although point mutations have been identified throughout the *NF1* locus, germline mutations have a tendency to cluster at certain locations, with mutations in exons 10a-c and 37 being notably common (Messiaen et al., 2000). 5% of NF1 patients have whole gene deletion (also known as *NF1* microdeletions) produced by nonallelic homologous recombination between sequences flanking the *NF1* gene (Wimmer et al., 2006). Type I deletions are produced by recombination between the *NF1-REP A* and *NF1-REP C* clusters of paralogous loci and delete 1.4 MB of chromosome 17 (Dorschner, Sybert, Weaver, Pletcher, & Stephens, 2000; Jenne et al., 2001; Lopez-Correa et al., 2001; Raedt et al., 2006). Type II deletions are a result of recombination between the *SUZ12* gene and the *SUZ12P1* pseudogene and remove 1.2 MB of sequence (Kehrer-Sawatzki et al., 2004; Steinmann et al., 2007). Notably, the *OMPG*, *EVI2B* and *EVI2A* genes, which are embedded in an intron of the *NF1* gene, are also lost in these microdeletion events which raises the question of whether this loss contribute to the phenotype of NF1 microdeletion patients. The phenotype of NF1 microdeletion patients is indeed distinct from that of other NF1 patients; *NF1* microdeletion patients have thousands of dermal neurofibromas, substantial cognitive impairment and dysmorphic features and more frequently develop plexiform neurofibromas and MPNSTs (Pasmant et al., 2010).

Neurofibromin, the 220 kDa protein encoded by the *NF1* gene, contains 2818 amino acids. Shortly after the *NF1* gene was identified, it was noted that neurofibromin contains a domain highly similar to the Ras GTPase-activating protein (GAP) domain present in two yeast proteins, IRA1 and IRA2 (Xu, O'Connell, et al., 1990). Ras proteins are highly conserved small GTP-binding proteins (G-proteins) that regulate critical cellular functions such as proliferation, survival, migration and differentiation. Ras proteins become activated when they bind GTP, a process that is facilitated by several proteins known collectively as guanine nucleotide exchange factors (GEFs). Ras inactivation occurs when a GTPase activity intrinsic to the Ras protein cleaves a phosphate from GTP, converting it to GDP. However, in isolation, this GTPase activity is inefficient—IRA1 and IRA2 bind to Ras and enhance Ras's GTPase activity by several orders of magnitude. In keeping with the similarity of neurofibromin's GAP-related domain (GRD) to IRA1 and IRA2, the neurofibromin GRD rescued the phenotype of IRA1- and IRA2-deficient yeast and stimulated the GTPase activity of yeast RAS2 and human H-Ras (Xu, Lin, et al., 1990). Neurofibromin is now recognized as a member of a family of mammalian Ras GAPs that includes RASA1 (p120GAP), RASA2 ($Gap1^{m}$), RASA3 ($GAP1^{IP4BP}$), RASA4 (CAPRI), RASAL1 and SYNGAP (p135SynGAP).

Given the Ras GAP activity of neurofibromin, it was reasonable to predict that loss of neurofibromin in *NF1*-null neoplasms would result in Ras hyperactivation (see Fig. 1, for a schematic of the affected signaling cascades). This prediction was confirmed when it was shown that Ras activation was increased in NF1-associated neurofibromas (Sherman, Atit, Rosenbaum, Cox, & Ratner, 2000) and MPNSTs (Basu et al., 1992). Since these observations suggested that inhibiting Ras hyperactivation in these tumors would be an effective means of treatment, a clinical trial was established in which children with plexiform neurofibromas were treated with tipifarnib, a farnesyltransferase inhibitor that blocks a posttranslational modification that is required for proper maturation and localization of some Ras proteins (*e.g.*, H-Ras). Tipifarnib was well tolerated in phase I (Widemann et al., 2006) but proved ineffective against plexiform neurofibromas in phase II (Widemann et al., 2014). One potential explanation for this failure is the fact that

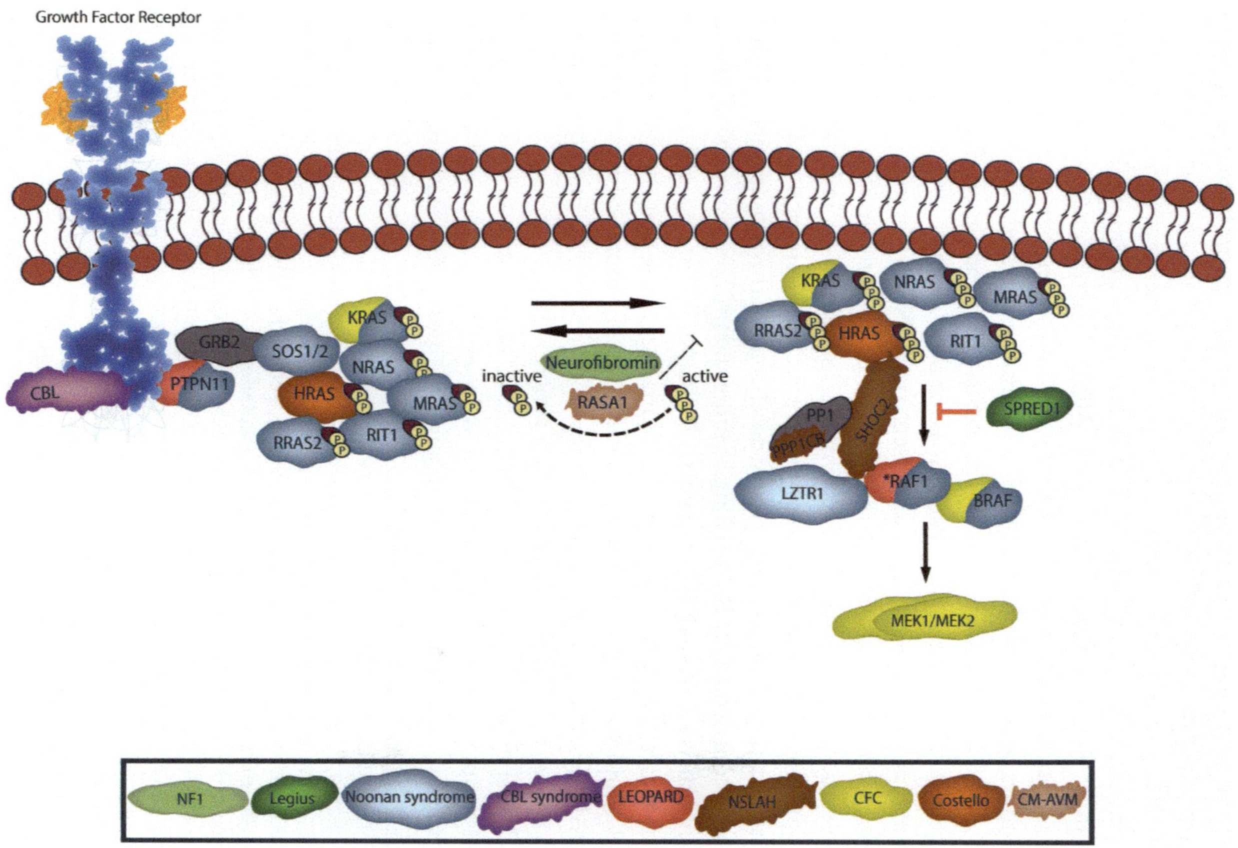

Fig. 1 See figure legend on next page.

neurofibromin serves as a Ras GAP for multiple Ras proteins in the classic Ras (H-Ras, N-Ras, K-Ras) and R-Ras (R-Ras, R-Ras2/TC21, M-Ras) subfamilies (Ohba et al., 2000). Consequently, neurofibromin loss can potentially simultaneously activate multiple Ras proteins and several of these Ras proteins can bypass the effects of farnesyltransferase inhibition by utilizing other modifications such as geranylgeranylation. To determine whether this was the case in *NF1*-null MPNSTs, we assessed the expression and activation of the classic Ras and R-Ras family members in these tumors. We found that *NF1*-null MPNSTs uniformly expressed activated H-Ras, K-Ras, N-Ras (Brossier et al., 2015) and R-Ras2 and that R-Ras was also expressed and activated in a major subset of these tumors . Inhibition of the classic Ras proteins with a dominant negative (DN) H-Ras mutant inhibited the proliferation and survival of MPNST cells, but not their migration. However, when we knocked down the expression of individual classic Ras proteins, MPNST cells simply increased the expression and activation of other family members and we saw no change in the proliferation of these cells. In contrast, a DN R-Ras mutant inhibited proliferation and migration, but not survival. Knockdown of either R-Ras or R-Ras2 in cells expressing both of these proteins inhibited migration, suggesting that the functions of R-Ras and R-Ras2 were not redundant. Phosphoproteomic analyses confirmed the existence of signaling cascades that are differentially activated downstream of classic Ras and R-Ras proteins.

These observations suggested that agents targeting individual Ras proteins would have limited effectiveness in neurofibromas and MPNSTs. This led several laboratories to examine the possibility that targeting key

Fig. 1 The Ras signaling pathway, indicating the proteins that are mutated in each RASopathy. Each protein is color-coded to indicate which RASopathy they correspond to; if a protein is mutated in more than one RASopathy, two colors are indicated. Abbreviations: BRAF, B-Raf proto-oncogene serine/threonine-protien kinase; CBL, the Cbl proto-oncogene; CFC, cardiofaciocutaneous syndrome; CV-AVM, capillary malformation-arteriovenous malformation syndrome; GRB2, growth factor receptor bound protein 2; HRAS, H-Ras; KRAS, K-Ras; LZTR1, leucine zipper like transcription regulator 1; MEK1/MEK2, mitogen-activated protein kinase kinase 1 and 2; MRAS, M-Ras; NF1, neurofibromatosis type 1; NRAS, N-Ras; NSLAH, Noonan syndrome-like disorder with loose anagen hair; PP1, protein phosphatase 1 complex; PPP1CB, protein phosphatase 1 catalytic subunit beta; PTPN11, protein tyrosine phosphatase non-receptor type 11; RAF1, Raf-1 proto-oncogene; RASA1, Ras p21 protein activator 1; RIT1, Ras-Like without CAAX 1; RRAS, R-Ras; RRAS2, R-Ras2; SHOC2, SHOC2 leucine rich repeat scaffold protein; SOS1, SOS Ras/Rac guanine nucleotide exchange factor 1; SOS2, SOS Ras/Rho guanine nucleotide exchange factor 2; SPRED1, sprouty-related EVH1 domain containing 1.

downstream effectors of Ras signaling such as the Raf/MEK/ERK and PI3K/AKT/mTOR pathways would be a better means of treating *NF1*-null neoplasms. This approach has had some success. In particular, MEK inhibitors have proven to be effective against neurofibromas (Jessen et al., 2013) and the MEK inhibitor selumetinib has now been approved by the U.S. Food and Drug Administration for the treatment of symptomatic, inoperable plexiform neurofibromas in pediatric NF1 patients 3 years of age and older. In contrast, agents targeting other proteins in these signaling cascades have not performed as well. The Raf inhibitor sorafenib demonstrated no objective responses in plexiform neurofibromas (Kim et al., 2013) or MPNSTs (Maki et al., 2009), while the mTORC1 inhibitor rapamycin, which showed some effectiveness against cultured tumor cells, had only a transient effect against MPNST xenografts (Johansson et al., 2008). Surprisingly, the MEK inhibitors that were effective against neurofibromas had only moderate, variable effects on MPNSTs (Jessen et al., 2013). This disparity may reflect the fact that the only mutations evident in neurofibromas are in *NF1*, while the pathogenesis of MPNSTs is driven by mutations in both *NF1* and multiple other signaling pathways (Carroll, 2012; Longo et al., 2018).

3. Legius syndrome

Legius syndrome (OMIM #611431), also known as neurofibromatosis type 1-like syndrome, is an autosomal dominant disorder that shares many clinical findings with NF1 (Table 1). Unlike NF1, Legius syndrome is rare, with approximately 200 cases reported in the literature (Tajan, Paccoud, Branka, Edouard, & Yart, 2018); its estimated birth prevalence is estimated at 1 in 46,000–75,000 (Pasmant et al., 2015). Individuals with Legius syndrome typically have café-au-lait macules and, to a lesser extent, skin fold freckling. These pigmentary abnormalities are relatively mild and are difficult to distinguish from those seen in NF1 patients, especially in young children (Spurlock et al., 2009). Affected individuals often have mild learning disabilities (Pasmant et al., 2009), hyperactivity (Brems et al., 2007) or decreased IQs (Laycock-van Spyk et al., 2011). Dysmorphic features including macrocephaly and hypertelorism are variably present (Brems et al., 2007; Spurlock et al., 2009). Legius syndrome differs from NF1 in that these patients do not have dermal neurofibromas, plexiform neurofibromas, optic gliomas or Lisch nodules and are not at increased risk for the other malignant neoplasms that arise in NF1 patients. However, there have been reports of some other tumor types in Legius syndrome patients. The most commonly reported tumors are benign lipomas (Brems et al., 2007, 2012). In addition, Brems et al. reported that three of the 37 patients they studied had lung cancer, childhood renal cancer and a colonic adenoma (Brems et al., 2007). However, given the rarity of Legius syndrome and the fact that these tumor types are common in the general population, it is difficult to ascertain whether the occurrence of these tumors is part of the syndrome or simply a coincidence.

Because of the overlap in clinical findings between NF1 and Legius syndrome, an international panel recently revised the diagnostic criteria for NF1 and established new diagnostic criteria for Legius syndrome (Legius et al., 2021). Per these criteria, Legius syndrome can be diagnosed in an individual who does not have a parent diagnosed with Legius syndrome if the following are established: (1) the presence of six or more café-au-lait macules that are bilaterally distributed and no other NF1-related diagnostic criteria except for axillary or inguinal freckling; and (2) a heterozygous pathogenic variant in *SPRED1* with a variant allele fraction of 50% in apparently normal tissue such as white blood cells.

Early genome-wide linkage analyses mapped the causative gene for Legius syndrome to chromosome 15 in two families with this disorder (Brems et al., 2007). It was subsequently demonstrated that >90% of cases of Legius syndrome have mutations in the *SPRED1* gene, with the majority of these mutations resulting in the production of truncated non-functional proteins (Brems et al., 2007, 2012; Laycock-van Spyk et al., 2011; Pasmant et al., 2009; Spencer et al., 2011; Spurlock et al., 2009). SPRED1 (Sprouty-related EVH1 domain-containing protein (1) is a member of the Sprouty/SPRED family of proteins (Fig. 1). This protein, which has a molecular mass of 50 kDa and contains 444 amino acids, suppresses the phosphorylation and activation of Raf by forming a complex with Ras proteins (Wakioka et al., 2001). Increased activation of the Ras/MAPK signaling cascade is evident in Legius syndrome patients (Brems et al., 2012), indicating that loss-of-function of one copy of the *SPRED1* gene is sufficient to produce the clinical findings evident in these patients. Due to the difficulties inherent to enrolling enough patients with this rare disease, clinical trials testing potential treatments for the clinical manifestations of Legius syndrome have not yet been conducted. In principle, though, the agents targeting MEK signaling that have proved effective against some manifestations of NF1 may be effective in Legius syndrome patients.

4. Noonan syndrome and related syndromes

Noonan syndrome, which was first described by Jacqueline Noonan in 1968, is the most common RASopathy, with an estimated prevalence of 1 in every 2000–2500 live births (Noonan, 1968). Noonan syndrome is a genetically heterogeneous disorder that is inherited (with the exception of Noonan syndrome 2; Johnston et al., 2018) in an autosomal dominant fashion (Table 1). Despite this genetic heterogeneity, individuals with Noonan syndrome characteristically demonstrate facial dysmorphism, a variety of congenital heart defects and short stature. The distinctive facial features seen in Noonan syndrome include a broad forehead, hypertelorism, ptosis, a high arched palate, downslanting palpebral fissues and posteriorly rotated low-set ears; intriguingly, these features are striking from birth through middle childhood and become less obvious in adults with Noonan syndrome (Allanson et al., 2010). Although congenital heart defects are present in 80% of Noonan syndrome patients, the type of defect present varies from individual to individual. In a series of 293 patients, pulmonary valvular stenosis was most common, being present in 57% of individuals with Noonan syndrome, followed by secundum atrial septal defects (32%) and hypertrophic cardiomyopathy (16%). However other valvular defects, aortic coarctation and persistently patent ductus arteriosus are also encountered at increased frequency in Noonan syndrome (Prendiville et al., 2014). Hypertrophic cardiomyopathy is particularly concerning during the first year of life as these patients are at increased risk for death from heart failure (Shaw, Kalidas, Crosby, Jeffery, & Patton, 2007).

The birth weight and length of newborns with Noonan syndrome is usually normal. However, during early childhood, growth failure that may be due to either feeding difficulties or growth hormone insensitivity (Noonan & Kappelgaard, 2015) sets in and up to 80% of children with Noonan syndrome have mean heights that are 2 standard deviations below the mean prior to puberty. The onset of puberty is also delayed (13.4 years for boys and 13.0 years in girls) in Noonan syndrome (Romano et al., 2009). As a result of this delayed puberty and attenuated pubertal growth spurts, the mean height of men and women with Noonan syndrome is also 2 standard deviations below the mean (Cessans et al., 2016; Malaquias et al., 2012; Ranke et al., 1988; Romano et al., 2010; Witt, Keena, Hall, & Allanson, 1986; Yart & Edouard, 2018). Patients with Noonan syndrome are less prone to obesity, which may be due to their early feeding difficulties or changes in adipose and muscle tissues that occur despite a normal caloric intake (da Silva et al., 2016).

Although severe intellectual disability is uncommon in Noonan syndrome patients (Pierpont, Tworog-Dube, & Roberts, 2015; Roelofs, Janssen, Wingbermuhle, Kessels, & Egger, 2016), half or more of them will exhibit attention deficits, language impairment and impaired social and communication skills (Cesarini et al., 2009). As in NF1, decreased bone mineral density is often evident (Baldassarre, Mussa, Carli, Molinatto, & Ferrero, 2017; Choudhry et al., 2012) and skeletal abnormalities such as scoliosis, pectus carinatum (Robicsek & Watts, 2010) and craniosynostosis (Addissie et al., 2015) are frequently encountered in these patients. Cryptorchidism occurs in up to 80% of male Noonan syndrome patients (Ankarberg-Lindgren, Westphal, & Dahlgren, 2011); in keeping with this, deficient spermatogenesis and diminished fertility is seen in men with Noonan syndrome (Sharland, Burch, McKenna, & Paton, 1992). Hemostatic abnormalities secondary to coagulation factor deficiencies and platelet defects also occur in individuals with Noonan syndrome (Artoni et al., 2014). Finally, Noonan syndrome patients carrying a mutation in the most commonly affected gene in this disease, *PTPN11*, have a 3.5-fold elevated risk of developing cancer compared to the general population (Jongmans et al., 2011). Most of these malignancies are hematologic, with juvenile myelomonocytic leukemia (JMML, the same malignancy encountered in NF1) or JMML-like myeloproliferative disorders predominating (Kratz et al., 2015; Strullu et al., 2014).

Mutations in multiple genes have been linked to Noonan syndrome including *PTPN11* (Noonan Syndrome 1, also known as Male Turner Syndrome; OMIM #163950), *LZTR1* (Noonan Syndrome 2; OMIM #605275), *KRAS* (Noonan Syndrome 3; OMIM #609942), *SOS1* (Noonan Syndrome 4; OMIM #610733), *RAF1* (Noonan Syndrome 5; OMIM #611553), *NRAS* (Noonan Syndrome 6; OMIM #613224), *BRAF* (Noonan Syndrome 7; OMIM #613706), *RIT1* (Noonan Syndrome 8; OMIM #615355), *SOS2* (Noonan Syndrome 9; OMIM #616559), *LZTR1* (Noonan Syndrome 10; OMIM #616564), *MRAS* (Noonan Syndrome 11; OMIM #618499), *RRAS2* (Noonan Syndrome 12; OMIM #618624) and *MAPK1* (Noonan Syndrome 13; OMIM #619087); Noonan Syndrome 2 differs from Noonan Syndrome 10 in that it has homozygous or compound heterozygous mutations in *LZTR1*, while Noonan Syndrome 10 is caused by heterozygous mutations in the same locus.

Mutations in the *PTPN11* (protein tyrosine non-receptor type 11) gene, which encodes SHP2 (Fig. 1), are most commonly encountered in Noonan syndrome, occurring in 50–60% of these individuals (Tajan et al., 2018). *PTPN11* mutations in Noonan Syndrome 1 cluster in the amino terminal Src homology 2 (SH2) and phosphotyrosine phosphatase domains, which interact to switch the proteins between its active and inactive configurations (Tartaglia et al., 2001); these mutations release auto-inhibitory restraints and promote SHP2 activation which in turn hyperactivates ERK1 (MAPK3) and ERK2 (MAPK1) (Araki et al., 2004; Bonetti et al., 2014; Eminaga & Bennett, 2008; Fragale, Tartaglia, Wu, & Gelb, 2004; Krenz, Yutzey, & Robbins, 2005; Oishi et al., 2006). Mutations in *SOS1* (Noonan Syndrome 4) and *RAF1* (Noonan Syndrome 5) both occur in about 10% of individuals with Noonan syndrome (Tajan et al., 2018). As with *PTPN11*, *SOS1* mutations destabilize autoinhibition, resulting in increased activity of this GEF and subsequent activation of several Ras proteins (Lepri et al., 2011); the *SOS2* mutations in Noonan syndrome 9 have a similar effect on SOS2 activity (Cordeddu et al., 2015). Many of the *RAF1* and *BRAF* mutations encountered in Noonan syndrome occur at sites encoding the protein's kinase domain and result in hyperactivation of these proteins (Tajan et al., 2018). However, this is not uniformly true—some Noonan syndrome-associated RAF1 and BRAF mutants are kinase-impaired. In these instances, the disease phenotype may be explained by the finding that such RAF1 mutants heterodimerize more avidly with BRAF which enhances ERK/MEK activation (Wu et al., 2012). Mutations affecting genes encoding Ras proteins such as *NRAS*, *RRAS*, *MRAS* and *RIT*, which are found in a rare subset of Noonan syndrome patients (Weber & Carroll, 2021) typically result in hyperactivation of the corresponding protein (Aoki et al., 2013; Chen et al., 2014; Cirstea et al., 2010; Flex et al., 2014; Higgins et al., 2017).

How *LZTR1* (Leucine Zipper-Like Transcriptional Regulator 1) mutations produce a Noonan syndrome phenotype is less obvious. This protein, which belongs to the BTB/POZ (broad complex, tramtrack and bric-a-brac/poxvirus and zinc finger) superfamily is normally thought of as a transcription factor. However, it has recently been demonstrated that LZTR1 protein binds to the RAF1-SHOC2-PPP1CB complex; mutations in the genes encoding all three of these latter proteins are associated with Noonan syndrome variants or other Noonan syndrome-like conditions (see below). In keeping with this association, knockdown of LZTR1 with siRNAs decreases RAF1 phosphorylation (Umeki et al., 2019). It thus appears that LZTR1 also plays a role in regulating the activity of the Ras/MAPK pathway.

5. Noonan-related syndromes

Three syndromes that have both clinical features overlapping with Noonan syndrome and other findings not typically observed in Noonan syndrome are known (Table 1). The first of these is Noonan syndrome-like disorder with or without juvenile myelomonocytic leukemia (OMIM #613563), which is increasingly known as CBL syndrome (Martinelli et al., 2010). Individuals with CBL syndrome have facial dysmorphism, cardiac abnormalities such as bicuspid aortic valves with stenosis, mitral valve insufficiency and cardiomyopathy, abnormalities of the lymphatics, short stature, developmental delays, cryptorchidism, cognitive deficits, ectodermal and musculoskeletal abnormalities and increased risk for malignancies including juvenile myelomonocytic leukemia (Bulow et al., 2015; Niemeyer et al., 2010; Perez et al., 2010). There is extensive phenotypic heterogeneity among individuals with this disorder and they typically do not meet the full diagnostic criteria for Noonan syndrome.

Individuals with Noonan syndrome-like disorder with loose anagen hair have facial features similar to those seen in Noonan syndrome (Cordeddu et al., 2009; Gripp et al., 2013; Mazzanti et al., 2003; Tosti et al., 1991, 1997). These patients also have short statures associated with severe growth hormone deficiency, relative macrocephaly, cognitive deficits, hyperactivity, cardiac defects (particularly mitral valve dysplasia and septal defects) and hypernasal voices. Their skulls have small posterior fossae associated with a Chiari type I malformation in which the tonsils of the cerebellum extend beyond the foramen magnum. Their defining characteristic, however, is their sparse, thin, slow-growing and easily pluckable hair; "anagen" refers to the active phase of the hair growth cycle. Approximately 100 cases of this condition have been reported in the literature (Tajan et al., 2018).

Noonan syndrome with multiple lentigines (NS-ML) was formerly referred to as LEOPARD syndrome, an acronym derived from the findings characteristic of this disorder: multiple lentigines, ECG conduction abnormalities, ocular hypertelorism, pulmonic stenosis, abnormal genitalia, retardation of growth and sensorineural deafness (Coppin & Temple, 1997). The lentigines that were the initial inspiration for the LEOPARD acronym are brown macular lesions that are scattered on the skin of afflicted individuals. This is also a rare condition that is estimated to occur in approximately 1 in every 100,000 newborns (Tajan et al., 2018).

As implied by its alternative name, patients with Noonan-like disorder with or without juvenile myelomonocytic leukemia have germline missense or splice site mutations in the *CBL* proto-oncogene (Fig. 1), which encodes CBL (Casitas B-lineage lymphoma), a RING finger E3 ubiquitin ligase that negatively regulates receptor tyrosine kinases by targeting them for internalization and recycling or degradation (Swaminathan & Tsygankov, 2006). These mutations predominantly occur in exons 8 and 9 of the CBL gene, which encode the RING finger domain and/or the adjacent linker helix region which connects the RING finger domain to an N-terminal tyrosine kinase binding domain. About 50% of these mutations affect Tyr371, which is critically important for CBL's E3 ligase activity, while others destabilize the CBL protein (Martinelli et al., 2015).

Noonan syndrome-like disorder with loose anagen hair is genetically heterogeneous, being caused by mutations in either SHOC2 (Noonan syndrome-like disorder with loose anagen hair 1 (NSLH1), also known as Mazzanti syndrome; OMIM #607721) or PPP1CB (Noonan syndrome-like disorder with loose anagen hair 2 (NSLH2); OMIM #617506). SHOC2 is a scaffolding protein that interacts with M-Ras and protein phosphatase-1 (PP1) to form a holoenzyme that dephosphorylates an inhibitory site (Ser259) on RAF kinases, which activates downstream signaling (Young et al., 2018). NSLH1 patients consistently have the same mutation (S2G) which introduces an N-myristoylation site into the protein that aberrantly targets the mutant SHOC2 protein to the cell membrane and prevents it from translocating into the nucleus following growth factor stimulation. When expressed in cultured cells, this SHOC2 mutant enhances MAPK activation (Cordeddu et al., 2009). Individuals with NSLH2 instead have missense mutations in the gene encoding protein phosphatase 1 catalytic subunit beta (PPP1CB). As noted above, PPP1CB interacts with MRAS and SHOC2. As with SHOC2, only a limited repertoire of *PPP1CB* mutations are found in NSLH2 patients and these mutations target highly conserved amino acids at sites predicted to disrupt PP1 subunit interactions and phosphorylation (Gripp et al., 2016; Ma et al., 2016); the investigators identifying these mutations have suggested that they enhance MAPK signaling but the mechanism by which this occurs remains elusive.

NS-ML is also genetically heterogeneous, being caused by mutations in *PTPN11* (LEOPARD syndrome 1; OMIM #151100), *RAF1* (LEOPARD syndrome 2; OMIM #611554) or *BRAF* (LEOPARD syndrome 3; OMIM #613707). The *PTPN11* mutations in LEOPARD syndrome 1 differ from those in Noonan syndrome 1 in that they target conserved residues in the protein tyrosine phosphatase domain of the protein (Tajan et al., 2018). Consistent with this, enzymatic analyses demonstrated that these mutants had lost phosphatase activity (Tartaglia et al., 2006). The effect that this has on the activity of the Ras/MAPK pathway is controversial. Some investigators have presented evidence that overexpression of these mutants in model systems reduces ERK1/2 activation (Kontaridis, Swanson, David, Barford, & Neel, 2006; Stewart et al., 2010) and argue that they function as dominant negatives rather than null mutations. Other investigators, however, have presented evidence that these *PTPN11* mutants are gain-of-function mutants that exert a stimulatory effect on the Ras/MAPK pathway (Bonetti et al., 2014; Oishi et al., 2006, 2009). Given the phenotype of LEOPARD syndrome 1 patients, the latter possibility seems more likely. The situation in LEOPARD syndrome 2 is more straightforward as the *RAF1* mutations found in these patients have been shown to be activating mutations (Pandit et al., 2007). In LEOPARD syndrome 3, T241P and L245F *BRAF* mutations have been identified, but the function of these mutants has not been assessed (Koudova, Seemanova, & Zenker, 2009; Sarkozy et al., 2009).

6. Costello syndrome

Costello syndrome (OMIM #218040), also known as faciocutaneoskeletal syndrome, is characterized by a coarse facial appearance that sometimes is described as "aged," short stature, thick redundant skin on the neck, palms, soles and fingers, a distinctive hand posture, severe feeding difficulty with failure to thrive, papillomata around the mouth, nares and anus and severe intellectual disability (Costello, 1977; Der Kaloustian, Moroz, McIntosh, Watters, & Blaichman, 1991; Rauen et al., 2011). Congenital heart defects, primarily tachyarrhythmias and hypertrophic cardiomyopathies, are also common in Costello syndrome. Individuals with Costello syndrome are at very high risk for developing cancers such as rhabdomyosarcomas, bladder carcinoma and neuroblastoma (Kratz et al., 2015). Costello syndrome is inherited in an autosomal dominant fashion and is quite rare, occurring in 1 in every 400,000 newborns (Tajan et al., 2018).

Heterozygous mutations in the *HRAS* gene are the cause of Costello syndrome. These mutations typically occur in codons 12 and 13, with the majority (slightly over 90%) being a G12S missense mutation

(Estep, Tidyman, Teitell, Cotter, & Rauen, 2006; Gripp, Stabley, Nicholson, Hoffman, & Sol-Church, 2006; Kerr et al., 2006). These mutations result in the generation of a constitutively active H-Ras protein.

7. Cardiofaciocutaneous syndrome

Individuals with cardiofaciocutaneous (CFC) syndrome, as implied by the name, have a distinctive facial appearance with hair and skin abnormalities, cardiac defects and intellectual disability (Roberts et al., 2006). Cardiac abnormalities in CFC syndrome are variable and include hypertrophic cardiomyopathies, pulmonary stenosis and atrial septal defects. The facial features characteristic of CFC syndrome are a high forehead with bitemporal constriction, downslanting palpebral fissues, a depressed nasal bridge, hypoplastic supraorbital ridges and posteriorly angled ears with prominent helices. The skin of CFC syndrome patients often demonstrates a generalized ichthyosis-like condition with hyperkeratotic skin lesions and sparse, curly, friable hair. CFC syndrome is autosomal dominant and quite rare, occurring in 1 in every 200,000 newborns (Tajan et al., 2018).

CFC syndrome is genetically heterogeneous, being caused by mutations in *BRAF* (CFC1; OMIM #115150), *KRAS* (CFC2; OMIM #615278), *MAP2K1* (CFC3; OMIM #615279) or *MAP2K2* (OMIM #615280). *BRAF* mutations are the most common cause of CFC syndrome, having been reported in 78% and 52% of CFC syndrome cases in two series (Rodriguez-Viciana et al., 2006; Sarkozy et al., 2009). These mutations clustered in exons 11, 12, 14 and 15 of the *BRAF* gene, which encodes conserved region 1 and the protein kinase domain of B-Raf. Heterozygous knockin mice carrying *BRAF* genes with the most common mutation encountered in CFC syndrome, Q241R, had embryonic lethality with craniofacial defects and cardiac abnormalities that could be partially rescued by prenatal treatment with a MEK inhibitor (Inoue et al., 2014), suggesting that at least some of these mutants enhance MEK activity. Missense mutations in *MAP2K1* (MEK1) and *MAP2K2* (MEK2) are the next most common mutations, each occurring in about 10% of CFC syndrome patients. The *MEK1* (F53S) and *MEK2* (F57C) mutants that have been identified in these cases have been demonstrated to be more active than their wild-type MEK counterparts at stimulating ERK phosphorylation (Rodriguez-Viciana et al., 2006). Several *KRAS* mutations have been identified in CFC syndrome 2 patients including D153V, G60R, Y71H and K147E (Niihori et al., 2006; Stark et al., 2012). Functional assessment of the activity of the *KRAS* D153V mutant indicated that it has a strong gain-of-function effect (Schubbert et al., 2007).

8. Gingival fibromatosis

The phenotype of gingival fibromatosis is much more restricted that of the other RASopathies and is characterized simply by a benign, slowly progressive enlargement of the maxillary and mandibular keratinized gingiva (Hart et al., 1998, 2002). Gingival fibromatosis is also genetically heterogeneous, although the mutated gene responsible for the pathology in individual afflicted with this condition has only been identified in two of the five types of gingival fibromatosis. In the autosomal dominant disorder gingival fibromatosis type 1 (OMIM #135300), an insertion mutation in *SOS1* produces a frameshift and premature termination, resulting in the production of a chimeric 1105 amino acid protein composed of 1083 amino acids followed by 22 novel amino acids (Jang et al., 2007). The portion of the SOS1 protein that is lost consists of four C-terminal proline-rich motifs that are required for Grb2 binding. This truncated mutant protein localizes to the plasma membrane in the absence of growth factor signaling, where it drives prolonged activation of the Ras/MAPK signaling pathway. Gingival fibromatosis type 5 is associated with heterozygous truncating or frameshift mutations in the *REST* gene located on chromosome 4q12. *REST* encodes the RE1-silencing transcription factor (REST) which functions as a transcriptional repressor. At present, the relationship of REST to the Ras/MAPK signaling pathway is poorly understood. However, it is notable that ERK-dependent phosphorylation in combination with Pin1 activity promotes REST degradation in at least some settings (Nesti, Corson, McCleskey, Oyer, & Mandel, 2014).

9. Capillary malformation-arteriovenous malformation syndrome

Capillary malformation-arteriovenous malformation (CMAVM) syndrome is an autosomal dominant disorder characterized by capillary malformations that are often found in association with arteriovenous malformations and arteriovenous fistulas. The arteriovenous malformations in these patients are found in the brain, face or extremities. These latter malformations are at times found in association with Parkes Weber syndrome, a condition evident as a cutaneous blush produced by multiple micro-arteriovenous malformations that occur in association with soft tissue and skeletal hypertrophy in an affected limb. CVAVM is also genetically heterogeneous. CMAVM 1 (OMIM #608354) is caused by heterozygous germline inactivating *RASA1* mutations (Eerola et al., 2003). Interestingly, capillary malformation lesion tissue shows evidence of a second somatic *RASA1* mutation in endothelial cells (Lapinski et al., 2018), suggesting that these lesions have features analogous to benign neoplasms. *Rasa1*^{R780Q} knockin mice that express a form of RASA1 lacking Ras GAP activity show vascular abnormalities reminiscent of CMAVM syndrome (Lubeck et al., 2014), suggesting that Ras hyperactivation drives the pathology in this disease. CMAVM 2 (OMIM #618196) is caused by heterozygous mutations in the *EPHB4* gene. EPHB4 encodes a receptor tyrosine kinase that binds ephrin B2, a transmembrane protein expressed in endothelial cells. In a group of 54 patients with *EPHB4* mutations, 27 had nonsense or splice-site mutations while the remainder had scattered at multiple locations in the gene (Amyere et al., 2017). These mutations are thought to result in a loss of *EPHB4* function; since EPHB4 inhibits the Ras/MAPK pathway by promoting the action of p120GAP (Kawasaki et al., 2014; Xiao et al., 2012), the Ras GAP encoded by *RASA1*, a loss of EPHB4 expression would be predicted to enhance Ras/MAPK signaling.

10. Future directions

Although great strides have been made in understanding how the mutations in the RASopathies cause the phenotype seen in these patients, our understanding of these pathways are still not completely understood and more work is needed to understand the complicated interactions between these proteins. However, there is an even more pressing need to translate the knowledge that we have gained thus far into effective treatments for these diseases. Most of the therapeutic development efforts thus far have focused on developing effective new treatments for the neoplasms that develop in patients with RASopathies. As new therapeutics are developed that target proteins such as SHP2, individual Ras proteins and MEK, it will be important to establish clinical trials that focus on developing treatments for other problems that these patients have such as their intellectual disabilities and behavioral abnormalities. It is also important to remember that other abnormalities seen in individuals with RASopathies do not occur in isolation. Going forward, we need to understand how the different features seen in these patients interact and worsen their disease.

References

Addissie, Y. A., Kotecha, U., Hart, R. A., Martinez, A. F., Kruszka, P., & Muenke, M. (2015). Craniosynostosis and Noonan syndrome with KRAS mutations: Expanding the phenotype with a case report and review of the literature. *American Journal of Medical Genetics. Part A*, *167A*(11), 2657–2663. https://doi.org/10.1002/ajmg.a.37259.

Afsar, C. U., Kara, I. O., Kozat, B. K., Demiryurek, H., Duman, B. B., & Doran, F. (2013). Neurofibromatosis type 1, gastrointestinal stromal tumor, leiomyosarcoma and osteosarcoma: Four cases of rare tumors and a review of the literature. *Critical Reviews in Oncology/Hematology*, *86*(2), 191–199. https://doi.org/10.1016/j.critrevonc.2012.11.001.

Allanson, J. E., Bohring, A., Dorr, H. G., Dufke, A., Gillessen-Kaesbach, G., Horn, D., et al. (2010). The face of Noonan syndrome: Does phenotype predict genotype. *American Journal of Medical Genetics. Part A*, *152A*(8), 1960–1966. https://doi.org/10.1002/ajmg.a.33518.

Amyere, M., Revencu, N., Helaers, R., Pairet, E., Baselga, E., Cordisco, M., et al. (2017). Germline loss-of-function mutations in EPHB4 cause a second form of capillary malformation-arteriovenous malformation (CM-AVM2) deregulating RAS-MAPK signaling. *Circulation*, *136*(11), 1037–1048. https://doi.org/10.1161/CIRCULATIONAHA.116.026886.

Anghileri, M., Miceli, R., Fiore, M., Mariani, L., Ferrari, A., Mussi, C., et al. (2006). Malignant peripheral nerve sheath tumors: Prognostic factors and survival in a series of patients treated at a single institution. *Cancer*, *107*(5), 1065–1074. https://doi.org/10.1002/cncr.22098.

Ankarberg-Lindgren, C., Westphal, O., & Dahlgren, J. (2011). Testicular size development and reproductive hormones in boys and adult males with Noonan syndrome: A longitudinal study. *European Journal of Endocrinology*, *165*(1), 137–144. https://doi.org/10.1530/EJE-11-0092.

Aoki, Y., Niihori, T., Banjo, T., Okamoto, N., Mizuno, S., Kurosawa, K., et al. (2013). Gain-of-function mutations in RIT1 cause Noonan syndrome, a RAS/MAPK pathway syndrome. *American Journal of Human Genetics*, *93*(1), 173–180. https://doi.org/10.1016/j.ajhg.2013.05.021.

Araki, T., Mohi, M. G., Ismat, F. A., Bronson, R. T., Williams, I. R., Kutok, J. L., et al. (2004). Mouse model of Noonan syndrome reveals cell type- and gene dosage-dependent effects of Ptpn11 mutation. *Nature Medicine*, *10*(8), 849–857. https://doi.org/10.1038/nm1084.

Artoni, A., Selicorni, A., Passamonti, S. M., Lecchi, A., Bucciarelli, P., Cerutti, M., et al. (2014). Hemostatic abnormalities in Noonan syndrome. *Pediatrics*, *133*(5), e1299–e1304. https://doi.org/10.1542/peds.2013-3251.

Balcer, L. J., Liu, G. T., Heller, G., Bilaniuk, L., Volpe, N. J., Galetta, S. L., et al. (2001). Visual loss in children with neurofibromatosis type 1 and optic pathway gliomas: Relation to tumor location by magnetic resonance imaging. *American Journal of Ophthalmology*, *131*(4), 442–445.

Baldassarre, G., Mussa, A., Carli, D., Molinatto, C., & Ferrero, G. B. (2017). Constitutional bone impairment in Noonan syndrome. *American Journal of Medical Genetics. Part A*, *173*(3), 692–698. https://doi.org/10.1002/ajmg.a.38086.

Barahona-Garrido, J., Aguirre-Gutierrez, R., Gutierrez-Manjarrez, J. I., Tellez-Avila, F. I., Lopez-Arce, G., Fomperoza-Torres, A., et al. (2009). Association of GIST and somatostatinoma in a patient with type-1 neurofibromatosis: Is there a common pathway? *The American Journal of Gastroenterology*, *104*(3), 797–799. https://doi.org/10.1038/ajg.2008.133.

Basu, T. N., Gutmann, D. H., Fletcher, J. A., Glover, T. W., Collins, F. S., & Downward, J. (1992). Aberrant regulation of ras proteins in malignant tumour cells from type 1 neurofibromatosis patients. *Nature, 356*(6371), 713–715. https://doi.org/10.1038/356713a0.

Bausch, B., Koschker, A. C., Fassnacht, M., Stoevesandt, J., Hoffmann, M. M., Eng, C., et al. (2006). Comprehensive mutation scanning of NF1 in apparently sporadic cases of pheochromocytoma. *The Journal of Clinical Endocrinology and Metabolism, 91*(9), 3478–3481. https://doi.org/10.1210/jc.2006-0780.

Bertola, D. R., Pereira, A. C., Passetti, F., de Oliveira, P. S., Messiaen, L., Gelb, B. D., et al. (2005). Neurofibromatosis-Noonan syndrome: Molecular evidence of the concurrence of both disorders in a patient. *American Journal of Medical Genetics. Part A, 136*(3), 242–245. https://doi.org/10.1002/ajmg.a.30813.

Bognanno, J. R., Edwards, M. K., Lee, T. A., Dunn, D. W., Roos, K. L., & Klatte, E. C. (1988). Cranial MR imaging in neurofibromatosis. *AJR. American Journal of Roentgenology, 151*(2), 381–388. https://doi.org/10.2214/ajr.151.2.381.

Bonetti, M., Paardekooper Overman, J., Tessadori, F., Noel, E., Bakkers, J., & den Hertog, J. (2014). Noonan and LEOPARD syndrome Shp2 variants induce heart displacement defects in zebrafish. *Development, 141*(9), 1961–1970. https://doi.org/10.1242/dev.106310.

Braffman, B. H., Bilaniuk, L. T., & Zimmerman, R. A. (1988). The central nervous system manifestations of the phakomatoses on MR. *Radiologic Clinics of North America, 26*(4), 773–800.

Brems, H., Chmara, M., Sahbatou, M., Denayer, E., Taniguchi, K., Kato, R., et al. (2007). Germline loss-of-function mutations in SPRED1 cause a neurofibromatosis 1-like phenotype. *Nature Genetics, 39*(9), 1120–1126. https://doi.org/10.1038/ng2113.

Brems, H., Park, C., Maertens, O., Pemov, A., Messiaen, L., Upadhyaya, M., et al. (2009). Glomus tumors in neurofibromatosis type 1: Genetic, functional, and clinical evidence of a novel association. *Cancer Research, 69*(18), 7393–7401. https://doi.org/10.1158/0008-5472.CAN-09-1752.

Brems, H., Pasmant, E., Van Minkelen, R., Wimmer, K., Upadhyaya, M., Legius, E., et al. (2012). Review and update of SPRED1 mutations causing Legius syndrome. *Human Mutation, 33*(11), 1538–1546. https://doi.org/10.1002/humu.22152.

Brossier, N. M., & Carroll, S. L. (2012). Genetically engineered mouse models shed new light on the pathogenesis of neurofibromatosis type I-related neoplasms of the peripheral nervous system. *Brain Research Bulletin, 88*(1), 58–71. https://doi.org/10.1016/j.brainresbull.2011.08.005.

Brossier, N. M., Prechtl, A. M., Longo, J. F., Barnes, S., Wilson, L. S., Byer, S. J., et al. (2015). Classic ras proteins promote proliferation and survival via distinct phosphoproteome alterations in neurofibromin-null malignant peripheral nerve sheath tumor cells. *Journal of Neuropathology and Experimental Neurology, 74*(6), 568–586. https://doi.org/10.1097/NEN.0000000000000201.

Bulow, L., Lissewski, C., Bressel, R., Rauch, A., Stark, Z., Zenker, M., et al. (2015). Hydrops, fetal pleural effusions and chylothorax in three patients with CBL mutations. *American Journal of Medical Genetics. Part A, 167A*(2), 394–399. https://doi.org/10.1002/ajmg.a.36838.

Cancer Genome Atlas Research, N. (2008). Comprehensive genomic characterization defines human glioblastoma genes and core pathways. *Nature, 455*(7216), 1061–1068. https://doi.org/10.1038/nature07385.

Carroll, S. L. (2012). Molecular mechanisms promoting the pathogenesis of Schwann cell neoplasms. *Acta Neuropathologica, 123*(3), 321–348. https://doi.org/10.1007/s00401-011-0928-6.

Carroll, S. L., & Ratner, N. (2008). How does the Schwann cell lineage form tumors in NF1? *Glia, 56*(14), 1590–1605. https://doi.org/10.1002/glia.20776.
Cesarini, L., Alfieri, P., Pantaleoni, F., Vasta, I., Cerutti, M., Petrangeli, V., et al. (2009). Cognitive profile of disorders associated with dysregulation of the RAS/MAPK signaling cascade. *American Journal of Medical Genetics. Part A, 149A*(2), 140–146. https://doi.org/10.1002/ajmg.a.32488.
Cessans, C., Ehlinger, V., Arnaud, C., Yart, A., Capri, Y., Barat, P., et al. (2016). Growth patterns of patients with Noonan syndrome: Correlation with age and genotype. *European Journal of Endocrinology, 174*(5), 641–650. https://doi.org/10.1530/EJE-15-0922.
Chen, P. C., Yin, J., Yu, H. W., Yuan, T., Fernandez, M., Yung, C. K., et al. (2014). Next-generation sequencing identifies rare variants associated with Noonan syndrome. *Proceedings of the National Academy of Sciences of the United States of America, 111*(31), 11473–11478. https://doi.org/10.1073/pnas.1324128111.
Choudhry, K. S., Grover, M., Tran, A. A., O'Brian Smith, E., Ellis, K. J., & Lee, B. H. (2012). Decreased bone mineralization in children with Noonan syndrome: Another consequence of dysregulated RAS MAPKinase pathway? *Molecular Genetics and Metabolism, 106*(2), 237–240. https://doi.org/10.1016/j.ymgme.2012.04.003.
Cirstea, I. C., Kutsche, K., Dvorsky, R., Gremer, L., Carta, C., Horn, D., et al. (2010). A restricted spectrum of NRAS mutations causes Noonan syndrome. *Nature Genetics, 42*(1), 27–29. https://doi.org/10.1038/ng.497.
Clark, R. D., & Hutter, J. J., Jr. (1982). Familial neurofibromatosis and juvenile chronic myelogenous leukemia. *Human Genetics, 60*(3), 230–232. https://doi.org/10.1007/BF00303009.
Clementi, M., Milani, S., Mammi, I., Boni, S., Monciotti, C., & Tenconi, R. (1999). Neurofibromatosis type 1 growth charts. *American Journal of Medical Genetics, 87*(4), 317–323.
Coppin, B. D., & Temple, I. K. (1997). Multiple lentigines syndrome (LEOPARD syndrome or progressive cardiomyopathic lentiginosis). *Journal of Medical Genetics, 34*(7), 582–586. https://doi.org/10.1136/jmg.34.7.582.
Cordeddu, V., Di Schiavi, E., Pennacchio, L. A., Ma'ayan, A., Sarkozy, A., Fodale, V., et al. (2009). Mutation of SHOC2 promotes aberrant protein N-myristoylation and causes Noonan-like syndrome with loose anagen hair. *Nature Genetics, 41*(9), 1022–1026. https://doi.org/10.1038/ng.425.
Cordeddu, V., Yin, J. C., Gunnarsson, C., Virtanen, C., Drunat, S., Lepri, F., et al. (2015). Activating mutations affecting the Dbl homology domain of SOS2 cause Noonan syndrome. *Human Mutation, 36*(11), 1080–1087. https://doi.org/10.1002/humu.22834.
Costello, J. M. (1977). A new syndrome: Mental subnormality and nasal papillomata. *Australian Paediatric Journal, 13*(2), 114–118. https://doi.org/10.1111/j.1440-1754.1977.tb01135.x.
Crucis, A., Richer, W., Brugieres, L., Bergeron, C., Marie-Cardine, A., Stephan, J. L., et al. (2015). Rhabdomyosarcomas in children with neurofibromatosis type I: A national historical cohort. *Pediatric Blood & Cancer, 62*(10), 1733–1738. https://doi.org/10.1002/pbc.25556.
Cutting, L. E., & Levine, T. M. (2010). Cognitive profile of children with neurofibromatosis and reading disabilities. *Child Neuropsychology, 16*(5), 417–432. https://doi.org/10.1080/09297041003761985.
da Silva, F. M., Jorge, A. A., Malaquias, A., da Costa Pereira, A., Yamamoto, G. L., Kim, C. A., et al. (2016). Nutritional aspects of Noonan syndrome and Noonan-related disorders. *American Journal of Medical Genetics. Part A, 170*(6), 1525–1531. https://doi.org/10.1002/ajmg.a.37639.

D'Arco, F., D'Amico, A., Caranci, F., Di Paolo, N., Melis, D., & Brunetti, A. (2014). Cerebrovascular stenosis in neurofibromatosis type 1 and utility of magnetic resonance angiography: Our experience and literature review. *La Radiologia Medica, 119*(6), 415–421. https://doi.org/10.1007/s11547-013-0358-8.

De Luca, A., Bottillo, I., Sarkozy, A., Carta, C., Neri, C., Bellacchio, E., et al. (2005). NF1 gene mutations represent the major molecular event underlying neurofibromatosis-Noonan syndrome. *American Journal of Human Genetics*, 77(6), 1092–1101. https://doi.org/10.1086/498454.

Der Kaloustian, V. M., Moroz, B., McIntosh, N., Watters, A. K., & Blaichman, S. (1991). Costello syndrome. *American Journal of Medical Genetics, 41*(1), 69–73. https://doi.org/10.1002/ajmg.1320410118.

DiPaolo, D. P., Zimmerman, R. A., Rorke, L. B., Zackai, E. H., Bilaniuk, L. T., & Yachnis, A. T. (1995). Neurofibromatosis type 1: Pathologic substrate of high-signal-intensity foci in the brain. *Radiology, 195*(3), 721–724. https://doi.org/10.1148/radiology.195.3.7754001.

Dorschner, M. O., Sybert, V. P., Weaver, M., Pletcher, B. A., & Stephens, K. (2000). NF1 microdeletion breakpoints are clustered at flanking repetitive sequences. *Human Molecular Genetics, 9*(1), 35–46.

Ducatman, B. S., Scheithauer, B. W., Piepgras, D. G., Reiman, H. M., & Ilstrup, D. M. (1986). Malignant peripheral nerve sheath tumors. A clinicopathologic study of 120 cases. *Cancer, 57*(10), 2006–2021.

Easton, D. F., Ponder, M. A., Huson, S. M., & Ponder, B. A. (1993). An analysis of variation in expression of neurofibromatosis (NF) type 1 (NF1): Evidence for modifying genes. *American Journal of Human Genetics, 53*(2), 305–313.

Eerola, I., Boon, L. M., Mulliken, J. B., Burrows, P. E., Dompmartin, A., Watanabe, S., et al. (2003). Capillary malformation-arteriovenous malformation, a new clinical and genetic disorder caused by RASA1 mutations. *American Journal of Human Genetics, 73*(6), 1240–1249. https://doi.org/10.1086/379793.

Eisenhofer, G., Lenders, J. W., Timmers, H., Mannelli, M., Grebe, S. K., Hofbauer, L. C., et al. (2011). Measurements of plasma methoxytyramine, normetanephrine, and metanephrine as discriminators of different hereditary forms of pheochromocytoma. *Clinical Chemistry, 57*(3), 411–420. https://doi.org/10.1373/clinchem.2010.153320.

Elefteriou, F., Kolanczyk, M., Schindeler, A., Viskochil, D. H., Hock, J. M., Schorry, E. K., et al. (2009). Skeletal abnormalities in neurofibromatosis type 1: Approaches to therapeutic options. *American Journal of Medical Genetics. Part A, 149A*(10), 2327–2338. https://doi.org/10.1002/ajmg.a.33045.

Eminaga, S., & Bennett, A. M. (2008). Noonan syndrome-associated SHP-2/Ptpn11 mutants enhance SIRPalpha and PZR tyrosyl phosphorylation and promote adhesion-mediated ERK activation. *The Journal of Biological Chemistry, 283*(22), 15328–15338. https://doi.org/10.1074/jbc.M801382200.

Estep, A. L., Tidyman, W. E., Teitell, M. A., Cotter, P. D., & Rauen, K. A. (2006). HRAS mutations in Costello syndrome: Detection of constitutional activating mutations in codon 12 and 13 and loss of wild-type allele in malignancy. *American Journal of Medical Genetics. Part A, 140*(1), 8–16. https://doi.org/10.1002/ajmg.a.31078.

Evans, D. G., Baser, M. E., McGaughran, J., Sharif, S., Howard, E., & Moran, A. (2002). Malignant peripheral nerve sheath tumours in neurofibromatosis 1. *Journal of Medical Genetics, 39*(5), 311–314.

Fan, Q., Yang, J., & Wang, G. (2014). Clinical and molecular prognostic predictors of malignant peripheral nerve sheath tumor. *Clinical & Translational Oncology, 16*(2), 191–199. https://doi.org/10.1007/s12094-013-1061-x.

Ferner, R. E., Huson, S. M., Thomas, N., Moss, C., Willshaw, H., Evans, D. G., et al. (2007). Guidelines for the diagnosis and management of individuals with neurofibromatosis 1. *Journal of Medical Genetics*, *44*(2), 81–88. https://doi.org/10.1136/jmg.2006.045906.

Flex, E., Jaiswal, M., Pantaleoni, F., Martinelli, S., Strullu, M., Fansa, E. K., et al. (2014). Activating mutations in RRAS underlie a phenotype within the RASopathy spectrum and contribute to leukaemogenesis. *Human Molecular Genetics*, *23*(16), 4315–4327. https://doi.org/10.1093/hmg/ddu148.

Fragale, A., Tartaglia, M., Wu, J., & Gelb, B. D. (2004). Noonan syndrome-associated SHP2/PTPN11 mutants cause EGF-dependent prolonged GAB1 binding and sustained ERK2/MAPK1 activation. *Human Mutation*, *23*(3), 267–277. https://doi.org/10.1002/humu.20005.

Garg, S., Green, J., Leadbitter, K., Emsley, R., Lehtonen, A., Evans, D. G., et al. (2013). Neurofibromatosis type 1 and autism spectrum disorder. *Pediatrics*, *132*(6), e1642–e1648. https://doi.org/10.1542/peds.2013-1868.

Griffiths, P. D., Blaser, S., Mukonoweshuro, W., Armstrong, D., Milo-Mason, G., & Cheung, S. (1999). Neurofibromatosis bright objects in children with neurofibromatosis type 1: A proliferative potential? *Pediatrics*, *104*(4), e49.

Griffiths, D. F., Williams, G. T., & Williams, E. D. (1987). Duodenal carcinoid tumours, phaeochromocytoma and neurofibromatosis: Islet cell tumour, phaeochromocytoma and the von Hippel-Lindau complex: Two distinctive neuroendocrine syndromes. *The Quarterly Journal of Medicine*, *64*(245), 769–782.

Gripp, K. W., Aldinger, K. A., Bennett, J. T., Baker, L., Tusi, J., Powell-Hamilton, N., et al. (2016). A novel rasopathy caused by recurrent de novo missense mutations in PPP1CB closely resembles Noonan syndrome with loose anagen hair. *American Journal of Medical Genetics. Part A*, *170*(9), 2237–2247. https://doi.org/10.1002/ajmg.a.37781.

Gripp, K. W., Stabley, D. L., Nicholson, L., Hoffman, J. D., & Sol-Church, K. (2006). Somatic mosaicism for an HRAS mutation causes Costello syndrome. *American Journal of Medical Genetics. Part A*, *140*(20), 2163–2169. https://doi.org/10.1002/ajmg.a.31456.

Gripp, K. W., Zand, D. J., Demmer, L., Anderson, C. E., Dobyns, W. B., Zackai, E. H., et al. (2013). Expanding the SHOC2 mutation associated phenotype of Noonan syndrome with loose anagen hair: Structural brain anomalies and myelofibrosis. *American Journal of Medical Genetics. Part A*, *161A*(10), 2420–2430. https://doi.org/10.1002/ajmg.a.36098.

Hart, T. C., Pallos, D., Bowden, D. W., Bolyard, J., Pettenati, M. J., & Cortelli, J. R. (1998). Genetic linkage of hereditary gingival fibromatosis to chromosome 2p21. *American Journal of Human Genetics*, *62*(4), 876–883. https://doi.org/10.1086/301797.

Hart, T. C., Zhang, Y., Gorry, M. C., Hart, P. S., Cooper, M., Marazita, M. L., et al. (2002). A mutation in the SOS1 gene causes hereditary gingival fibromatosis type 1. *American Journal of Human Genetics*, *70*(4), 943–954. https://doi.org/10.1086/339689.

Higgins, E. M., Bos, J. M., Mason-Suares, H., Tester, D. J., Ackerman, J. P., MacRae, C. A., et al. (2017). Elucidation of MRAS-mediated Noonan syndrome with cardiac hypertrophy. *JCI Insight*, *2*(5), e91225. https://doi.org/10.1172/jci.insight.91225.

Inoue, S., Moriya, M., Watanabe, Y., Miyagawa-Tomita, S., Niihori, T., Oba, D., et al. (2014). New BRAF knockin mice provide a pathogenetic mechanism of developmental defects and a therapeutic approach in cardio-facio-cutaneous syndrome. *Human Molecular Genetics*, *23*(24), 6553–6566. https://doi.org/10.1093/hmg/ddu376.

Itoh, T., Magnaldi, S., White, R. M., Denckla, M. B., Hofman, K., Naidu, S., et al. (1994). Neurofibromatosis type 1: The evolution of deep gray and white matter MR abnormalities. *AJNR. American Journal of Neuroradiology*, *15*(8), 1513–1519.

Jadayel, D., Fain, P., Upadhyaya, M., Ponder, M. A., Huson, S. M., Carey, J., et al. (1990). Paternal origin of new mutations in von Recklinghausen neurofibromatosis. *Nature, 343*(6258), 558–559. https://doi.org/10.1038/343558a0.

Jang, S. I., Lee, E. J., Hart, P. S., Ramaswami, M., Pallos, D., & Hart, T. C. (2007). Germ line gain of function with SOS1 mutation in hereditary gingival fibromatosis. *The Journal of Biological Chemistry, 282*(28), 20245–20255. https://doi.org/10.1074/jbc.M701609200.

Jenne, D. E., Tinschert, S., Reimann, H., Lasinger, W., Thiel, G., Hameister, H., et al. (2001). Molecular characterization and gene content of breakpoint boundaries in patients with neurofibromatosis type 1 with 17q11.2 microdeletions. *American Journal of Human Genetics, 69*(3), 516–527. https://doi.org/10.1086/323043.

Jessen, W. J., Miller, S. J., Jousma, E., Wu, J., Rizvi, T. A., Brundage, M. E., et al. (2013). MEK inhibition exhibits efficacy in human and mouse neurofibromatosis tumors. *The Journal of Clinical Investigation, 123*(1), 340–347. https://doi.org/10.1172/JCI60578.

Johansson, G., Mahller, Y. Y., Collins, M. H., Kim, M. O., Nobukuni, T., Perentesis, J., et al. (2008). Effective in vivo targeting of the mammalian target of rapamycin pathway in malignant peripheral nerve sheath tumors. *Molecular Cancer Therapeutics*, 7(5), 1237–1245. https://doi.org/10.1158/1535-7163.MCT-07-2335.

Johnston, J. J., van der Smagt, J. J., Rosenfeld, J. A., Pagnamenta, A. T., Alswaid, A., Baker, E. H., et al. (2018). Autosomal recessive Noonan syndrome associated with biallelic LZTR1 variants. *Genetics in Medicine, 20*(10), 1175–1185. https://doi.org/10.1038/gim.2017.249.

Jongmans, M. C., van der Burgt, I., Hoogerbrugge, P. M., Noordam, K., Yntema, H. G., Nillesen, W. M., et al. (2011). Cancer risk in patients with Noonan syndrome carrying a PTPN11 mutation. *European Journal of Human Genetics, 19*(8), 870–874. https://doi.org/10.1038/ejhg.2011.37.

Kawasaki, J., Aegerter, S., Fevurly, R. D., Mammoto, A., Mammoto, T., Sahin, M., et al. (2014). RASA1 functions in EPHB4 signaling pathway to suppress endothelial mTORC1 activity. *The Journal of Clinical Investigation, 124*(6), 2774–2784. https://doi.org/10.1172/JCI67084.

Kehrer-Sawatzki, H., Kluwe, L., Sandig, C., Kohn, M., Wimmer, K., Krammer, U., et al. (2004). High frequency of mosaicism among patients with neurofibromatosis type 1 (NF1) with microdeletions caused by somatic recombination of the JJAZ1 gene. *American Journal of Human Genetics, 75*(3), 410–423. https://doi.org/10.1086/423624.

Kerr, B., Delrue, M. A., Sigaudy, S., Perveen, R., Marche, M., Burgelin, I., et al. (2006). Genotype-phenotype correlation in Costello syndrome: HRAS mutation analysis in 43 cases. *Journal of Medical Genetics, 43*(5), 401–405. https://doi.org/10.1136/jmg.2005.040352.

Kim, A., Dombi, E., Tepas, K., Fox, E., Martin, S., Wolters, P., et al. (2013). Phase I trial and pharmacokinetic study of sorafenib in children with neurofibromatosis type I and plexiform neurofibromas. *Pediatric Blood & Cancer, 60*(3), 396–401. https://doi.org/10.1002/pbc.24281.

Kontaridis, M. I., Swanson, K. D., David, F. S., Barford, D., & Neel, B. G. (2006). PTPN11 (Shp2) mutations in LEOPARD syndrome have dominant negative, not activating, effects. *The Journal of Biological Chemistry, 281*(10), 6785–6792. https://doi.org/10.1074/jbc.M513068200.

Koudova, M., Seemanova, E., & Zenker, M. (2009). Novel BRAF mutation in a patient with LEOPARD syndrome and normal intelligence. *European Journal of Medical Genetics, 52*(5), 337–340. https://doi.org/10.1016/j.ejmg.2009.04.006.

Kratz, C. P., Franke, L., Peters, H., Kohlschmidt, N., Kazmierczak, B., Finckh, U., et al. (2015). Cancer spectrum and frequency among children with Noonan, Costello, and cardio-facio-cutaneous syndromes. *British Journal of Cancer, 112*(8), 1392–1397. https://doi.org/10.1038/bjc.2015.75.

Krenz, M., Yutzey, K. E., & Robbins, J. (2005). Noonan syndrome mutation Q79R in Shp2 increases proliferation of valve primordia mesenchymal cells via extracellular signal-regulated kinase 1/2 signaling. *Circulation Research*, *97*(8), 813–820. https://doi.org/10.1161/01.RES.0000186194.06514.b0.

Lammert, M., Friedman, J. M., Roth, H. J., Friedrich, R. E., Kluwe, L., Atkins, D., et al. (2006). Vitamin D deficiency associated with number of neurofibromas in neurofibromatosis 1. *Journal of Medical Genetics*, *43*(10), 810–813. https://doi.org/10.1136/jmg.2006.041095.

Lapinski, P. E., Doosti, A., Salato, V., North, P., Burrows, P. E., & King, P. D. (2018). Somatic second hit mutation of RASA1 in vascular endothelial cells in capillary malformation-arteriovenous malformation. *European Journal of Medical Genetics*, *61*(1), 11–16. https://doi.org/10.1016/j.ejmg.2017.10.004.

Laycock-van Spyk, S., Jim, H. P., Thomas, L., Spurlock, G., Fares, L., Palmer-Smith, S., et al. (2011). Identification of five novel SPRED1 germline mutations in Legius syndrome. *Clinical Genetics*, *80*(1), 93–96. https://doi.org/10.1111/j.1399-0004.2010.01618.x.

Legius, E., Messiaen, L., Wolkenstein, P., Pancza, P., Avery, R. A., Berman, Y., et al. (2021). Revised diagnostic criteria for neurofibromatosis type 1 and Legius syndrome: An international consensus recommendation. *Genetics in Medicine*. https://doi.org/10.1038/s41436-021-01170-5.

Lepri, F., De Luca, A., Stella, L., Rossi, C., Baldassarre, G., Pantaleoni, F., et al. (2011). SOS1 mutations in Noonan syndrome: Molecular spectrum, structural insights on pathogenic effects, and genotype-phenotype correlations. *Human Mutation*, *32*(7), 760–772. https://doi.org/10.1002/humu.21492.

Lin, A. E., Birch, P. H., Korf, B. R., Tenconi, R., Niimura, M., Poyhonen, M., et al. (2000). Cardiovascular malformations and other cardiovascular abnormalities in neurofibromatosis 1. *American Journal of Medical Genetics*, *95*(2), 108–117.

Listernick, R., Charrow, J., Greenwald, M., & Mets, M. (1994). Natural history of optic pathway tumors in children with neurofibromatosis type 1: A longitudinal study. *The Journal of Pediatrics*, *125*(1), 63–66.

Listernick, R., Darling, C., Greenwald, M., Strauss, L., & Charrow, J. (1995). Optic pathway tumors in children: The effect of neurofibromatosis type 1 on clinical manifestations and natural history. *The Journal of Pediatrics*, *127*(5), 718–722.

Longo, J. F., Weber, S. M., Turner-Ivey, B. P., & Carroll, S. L. (2018). Recent advances in the diagnosis and pathogenesis of neurofibromatosis type 1 (NF1)-associated peripheral nervous system neoplasms. *Advances in Anatomic Pathology*, *25*(5), 353–368. https://doi.org/10.1097/PAP.0000000000000197.

Lopez-Correa, C., Dorschner, M., Brems, H., Lazaro, C., Clementi, M., Upadhyaya, M., et al. (2001). Recombination hotspot in NF1 microdeletion patients. *Human Molecular Genetics*, *10*(13), 1387–1392.

Lubeck, B. A., Lapinski, P. E., Bauler, T. J., Oliver, J. A., Hughes, E. D., Saunders, T. L., et al. (2014). Blood vascular abnormalities in Rasa1(R780Q) knockin mice: Implications for the pathogenesis of capillary malformation-arteriovenous malformation. *The American Journal of Pathology*, *184*(12), 3163–3169. https://doi.org/10.1016/j.ajpath.2014.08.018.

Ma, L., Bayram, Y., McLaughlin, H. M., Cho, M. T., Krokosky, A., Turner, C. E., et al. (2016). De novo missense variants in PPP1CB are associated with intellectual disability and congenital heart disease. *Human Genetics*, *135*(12), 1399–1409. https://doi.org/10.1007/s00439-016-1731-1.

Maki, R. G., D'Adamo, D. R., Keohan, M. L., Saulle, M., Schuetze, S. M., Undevia, S. D., et al. (2009). Phase II study of sorafenib in patients with metastatic or recurrent sarcomas. *Journal of Clinical Oncology*, *27*(19), 3133–3140. https://doi.org/10.1200/JCO.2008.20.4495.

Malaquias, A. C., Brasil, A. S., Pereira, A. C., Arnhold, I. J., Mendonca, B. B., Bertola, D. R., et al. (2012). Growth standards of patients with Noonan and Noonan-like syndromes with mutations in the RAS/MAPK pathway. *American Journal of Medical Genetics. Part A, 158A*(11), 2700–2706. https://doi.org/10.1002/ajmg.a.35519.

Mannelli, M., Castellano, M., Schiavi, F., Filetti, S., Giacche, M., Mori, L., et al. (2009). Clinically guided genetic screening in a large cohort of italian patients with pheochromocytomas and/or functional or nonfunctional paragangliomas. *The Journal of Clinical Endocrinology and Metabolism, 94*(5), 1541–1547. https://doi.org/10.1210/jc.2008-2419.

Martinelli, S., De Luca, A., Stellacci, E., Rossi, C., Checquolo, S., Lepri, F., et al. (2010). Heterozygous germline mutations in the CBL tumor-suppressor gene cause a Noonan syndrome-like phenotype. *American Journal of Human Genetics, 87*(2), 250–257. https://doi.org/10.1016/j.ajhg.2010.06.015.

Martinelli, S., Stellacci, E., Pannone, L., D'Agostino, D., Consoli, F., Lissewski, C., et al. (2015). Molecular diversity and associated phenotypic spectrum of germline CBL mutations. *Human Mutation, 36*(8), 787–796. https://doi.org/10.1002/humu.22809.

Mazzanti, L., Cacciari, E., Cicognani, A., Bergamaschi, R., Scarano, E., & Forabosco, A. (2003). Noonan-like syndrome with loose anagen hair: A new syndrome? *American Journal of Medical Genetics. Part A, 118A*(3), 279–286. https://doi.org/10.1002/ajmg.a.10923.

McCaughan, J. A., Holloway, S. M., Davidson, R., & Lam, W. W. (2007). Further evidence of the increased risk for malignant peripheral nerve sheath tumour from a Scottish cohort of patients with neurofibromatosis type 1. *Journal of Medical Genetics, 44*(7), 463–466. https://doi.org/10.1136/jmg.2006.048140.

McGaughran, J. M., Harris, D. I., Donnai, D., Teare, D., MacLeod, R., Westerbeek, R., et al. (1999). A clinical study of type 1 neurofibromatosis in north west England. *Journal of Medical Genetics, 36*(3), 197–203.

Messiaen, L. M., Callens, T., Mortier, G., Beysen, D., Vandenbroucke, I., Van Roy, N., et al. (2000). Exhaustive mutation analysis of the NF1 gene allows identification of 95% of mutations and reveals a high frequency of unusual splicing defects. *Human Mutation, 15*(6), 541–555. https://doi.org/10.1002/1098-1004(200006)15:6<541::AID-HUMU6>3.0.CO;2-N.

Miettinen, M., Fetsch, J. F., Sobin, L. H., & Lasota, J. (2006). Gastrointestinal stromal tumors in patients with neurofibromatosis 1: A clinicopathologic and molecular genetic study of 45 cases. *The American Journal of Surgical Pathology, 30*(1), 90–96.

Nesti, E., Corson, G. M., McCleskey, M., Oyer, J. A., & Mandel, G. (2014). C-terminal domain small phosphatase 1 and MAP kinase reciprocally control REST stability and neuronal differentiation. *Proceedings of the National Academy of Sciences of the United States of America, 111*(37), E3929–E3936. https://doi.org/10.1073/pnas.1414770111.

Neurofibromatosis Conference Statement. (1988). National Institutes of Health consensus development conference. *Archives of Neurology, 45*(5), 575–578.

Niemeyer, C. M., Kang, M. W., Shin, D. H., Furlan, I., Erlacher, M., Bunin, N. J., et al. (2010). Germline CBL mutations cause developmental abnormalities and predispose to juvenile myelomonocytic leukemia. *Nature Genetics, 42*(9), 794–800. https://doi.org/10.1038/ng.641.

Niihori, T., Aoki, Y., Narumi, Y., Neri, G., Cave, H., Verloes, A., et al. (2006). Germline KRAS and BRAF mutations in cardio-facio-cutaneous syndrome. *Nature Genetics, 38*(3), 294–296. https://doi.org/10.1038/ng1749.

Noonan, J. A. (1968). Hypertelorism with Turner phenotype. A new syndrome with associated congenital heart disease. *American Journal of Diseases of Children, 116*(4), 373–380. https://doi.org/10.1001/archpedi.1968.02100020377005.

Noonan, J. A., & Kappelgaard, A. M. (2015). The efficacy and safety of growth hormone therapy in children with noonan syndrome: A review of the evidence. *Hormone Research in Pædiatrics*, *83*(3), 157–166. https://doi.org/10.1159/000369012.

Nystrom, A. M., Ekvall, S., Allanson, J., Edeby, C., Elinder, M., Holmstrom, G., et al. (2009). Noonan syndrome and neurofibromatosis type I in a family with a novel mutation in NF1. *Clinical Genetics*, *76*(6), 524–534. https://doi.org/10.1111/j.1399-0004.2009.01233.x.

Oderich, G. S., Sullivan, T. M., Bower, T. C., Gloviczki, P., Miller, D. V., Babovic-Vuksanovic, D., et al. (2007). Vascular abnormalities in patients with neurofibromatosis syndrome type I: Clinical spectrum, management, and results. *Journal of Vascular Surgery*, *46*(3), 475–484. https://doi.org/10.1016/j.jvs.2007.03.055.

Ohba, Y., Mochizuki, N., Yamashita, S., Chan, A. M., Schrader, J. W., Hattori, S., et al. (2000). Regulatory proteins of R-Ras, TC21/R-Ras2, and M-Ras/R-Ras3. *The Journal of Biological Chemistry*, *275*(26), 20020–20026. https://doi.org/10.1074/jbc.M000981200.

Oishi, K., Gaengel, K., Krishnamoorthy, S., Kamiya, K., Kim, I. K., Ying, H., et al. (2006). Transgenic Drosophila models of Noonan syndrome causing PTPN11 gain-of-function mutations. *Human Molecular Genetics*, *15*(4), 543–553. https://doi.org/10.1093/hmg/ddi471.

Oishi, K., Zhang, H., Gault, W. J., Wang, C. J., Tan, C. C., Kim, I. K., et al. (2009). Phosphatase-defective LEOPARD syndrome mutations in PTPN11 gene have gain-of-function effects during Drosophila development. *Human Molecular Genetics*, *18*(1), 193–201. https://doi.org/10.1093/hmg/ddn336.

Pandit, B., Sarkozy, A., Pennacchio, L. A., Carta, C., Oishi, K., Martinelli, S., et al. (2007). Gain-of-function RAF1 mutations cause Noonan and LEOPARD syndromes with hypertrophic cardiomyopathy. *Nature Genetics*, *39*(8), 1007–1012. https://doi.org/10.1038/ng2073.

Pasmant, E., Parfait, B., Luscan, A., Goussard, P., Briand-Suleau, A., Laurendeau, I., et al. (2015). Neurofibromatosis type 1 molecular diagnosis: What can NGS do for you when you have a large gene with loss of function mutations? *European Journal of Human Genetics*, *23*(5), 596–601. https://doi.org/10.1038/ejhg.2014.145.

Pasmant, E., Sabbagh, A., Hanna, N., Masliah-Planchon, J., Jolly, E., Goussard, P., et al. (2009). SPRED1 germline mutations caused a neurofibromatosis type 1 overlapping phenotype. *Journal of Medical Genetics*, *46*(7), 425–430. https://doi.org/10.1136/jmg.2008.065243.

Pasmant, E., Sabbagh, A., Spurlock, G., Laurendeau, I., Grillo, E., Hamel, M. J., et al. (2010). NF1 microdeletions in neurofibromatosis type 1: From genotype to phenotype. *Human Mutation*, *31*(6), E1506–E1518. https://doi.org/10.1002/humu.21271.

Patil, S., & Chamberlain, R. S. (2012). Neoplasms associated with germline and somatic NF1 gene mutations. *The Oncologist*, *17*(1), 101–116. https://doi.org/10.1634/theoncologist.2010-0181.

Perez, B., Mechinaud, F., Galambrun, C., Ben Romdhane, N., Isidor, B., Philip, N., et al. (2010). Germline mutations of the CBL gene define a new genetic syndrome with predisposition to juvenile myelomonocytic leukaemia. *Journal of Medical Genetics*, *47*(10), 686–691. https://doi.org/10.1136/jmg.2010.076836.

Pierpont, E. I., Tworog-Dube, E., & Roberts, A. E. (2015). Attention skills and executive functioning in children with Noonan syndrome and their unaffected siblings. *Developmental Medicine and Child Neurology*, *57*(4), 385–392. https://doi.org/10.1111/dmcn.12621.

Porter, D. E., Prasad, V., Foster, L., Dall, G. F., Birch, R., & Grimer, R. J. (2009). Survival in malignant peripheral nerve sheath tumours: A comparison between sporadic and neurofibromatosis type 1-associated tumours. *Sarcoma*, *2009*, 756395. https://doi.org/10.1155/2009/756395.

Prendiville, T. W., Gauvreau, K., Tworog-Dube, E., Patkin, L., Kucherlapati, R. S., Roberts, A. E., et al. (2014). Cardiovascular disease in Noonan syndrome. *Archives of Disease in Childhood, 99*(7), 629–634. https://doi.org/10.1136/archdischild-2013-305047.

Raedt, T. D., Stephens, M., Heyns, I., Brems, H., Thijs, D., Messiaen, L., et al. (2006). Conservation of hotspots for recombination in low-copy repeats associated with the NF1 microdeletion. *Nature Genetics, 38*(12), 1419–1423. https://doi.org/10.1038/ng1920.

Ranke, M. B., Heidemann, P., Knupfer, C., Enders, H., Schmaltz, A. A., & Bierich, J. R. (1988). Noonan syndrome: Growth and clinical manifestations in 144 cases. *European Journal of Pediatrics, 148*(3), 220–227. https://doi.org/10.1007/BF00441408.

Rasmussen, S. A., Yang, Q., & Friedman, J. M. (2001). Mortality in neurofibromatosis 1: An analysis using U.S. death certificates. *American Journal of Human Genetics, 68*(5), 1110–1118. https://doi.org/10.1086/320121.

Rauen, K. A., Banerjee, A., Bishop, W. R., Lauchle, J. O., McCormick, F., McMahon, M., et al. (2011). Costello and cardio-facio-cutaneous syndromes: Moving toward clinical trials in RASopathies. *American Journal of Medical Genetics. Part C, Seminars in Medical Genetics, 157C*(2), 136–146. https://doi.org/10.1002/ajmg.c.30294.

Rea, D., Brandsema, J. F., Armstrong, D., Parkin, P. C., deVeber, G., MacGregor, D., et al. (2009). Cerebral arteriopathy in children with neurofibromatosis type 1. *Pediatrics, 124*(3), e476–e483. https://doi.org/10.1542/peds.2009-0152.

Riccardi, V. M. (1991). Neurofibromatosis: Past, present, and future. *The New England Journal of Medicine, 324*(18), 1283–1285. https://doi.org/10.1056/NEJM199105023241812.

Roberts, A., Allanson, J., Jadico, S. K., Kavamura, M. I., Noonan, J., Opitz, J. M., et al. (2006). The cardiofaciocutaneous syndrome. *Journal of Medical Genetics, 43*(11), 833–842. https://doi.org/10.1136/jmg.2006.042796.

Robicsek, F., & Watts, L. T. (2010). Pectus carinatum. *Thoracic Surgery Clinics, 20*(4), 563–574. https://doi.org/10.1016/j.thorsurg.2010.07.007.

Rodriguez, F. J., Perry, A., Gutmann, D. H., O'Neill, B. P., Leonard, J., Bryant, S., et al. (2008). Gliomas in neurofibromatosis type 1: A clinicopathologic study of 100 patients. *Journal of Neuropathology and Experimental Neurology, 67*(3), 240–249. https://doi.org/10.1097/NEN.0b013e318165eb75.

Rodriguez-Viciana, P., Tetsu, O., Tidyman, W. E., Estep, A. L., Conger, B. A., Cruz, M. S., et al. (2006). Germline mutations in genes within the MAPK pathway cause cardio-facio-cutaneous syndrome. *Science, 311*(5765), 1287–1290. https://doi.org/10.1126/science.1124642.

Roelofs, R. L., Janssen, N., Wingbermuhle, E., Kessels, R. P., & Egger, J. I. (2016). Intellectual development in Noonan syndrome: A longitudinal study. *Brain and Behavior: A Cognitive Neuroscience Perspective, 6*(7), e00479. https://doi.org/10.1002/brb3.479.

Rojnueangnit, K., Xie, J., Gomes, A., Sharp, A., Callens, T., Chen, Y., et al. (2015). High incidence of Noonan syndrome features including short stature and pulmonic stenosis in patients carrying NF1 missense mutations affecting p.Arg1809: Genotype-phenotype correlation. *Human Mutation, 36*(11), 1052–1063. https://doi.org/10.1002/humu.22832.

Romano, A. A., Allanson, J. E., Dahlgren, J., Gelb, B. D., Hall, B., Pierpont, M. E., et al. (2010). Noonan syndrome: Clinical features, diagnosis, and management guidelines. *Pediatrics, 126*(4), 746–759. https://doi.org/10.1542/peds.2009-3207.

Romano, A. A., Dana, K., Bakker, B., Davis, D. A., Hunold, J. J., Jacobs, J., et al. (2009). Growth response, near-adult height, and patterns of growth and puberty in patients with noonan syndrome treated with growth hormone. *The Journal of Clinical Endocrinology and Metabolism, 94*(7), 2338–2344. https://doi.org/10.1210/jc.2008-2094.

Sarkozy, A., Carta, C., Moretti, S., Zampino, G., Digilio, M. C., Pantaleoni, F., et al. (2009). Germline BRAF mutations in Noonan, LEOPARD, and cardiofaciocutaneous syndromes: Molecular diversity and associated phenotypic spectrum. *Human Mutation*, *30*(4), 695–702. https://doi.org/10.1002/humu.20955.

Schubbert, S., Bollag, G., Lyubynska, N., Nguyen, H., Kratz, C. P., Zenker, M., et al. (2007). Biochemical and functional characterization of germ line KRAS mutations. *Molecular and Cellular Biology*, *27*(22), 7765–7770. https://doi.org/10.1128/MCB.00965-07.

Schwetye, K. E., & Gutmann, D. H. (2014). Cognitive and behavioral problems in children with neurofibromatosis type 1: Challenges and future directions. *Expert Review of Neurotherapeutics*, *14*(10), 1139–1152. https://doi.org/10.1586/14737175.2014.953931.

Sharland, M., Burch, M., McKenna, W. M., & Paton, M. A. (1992). A clinical study of Noonan syndrome. *Archives of Disease in Childhood*, *67*(2), 178–183. https://doi.org/10.1136/adc.67.2.178.

Shaw, A. C., Kalidas, K., Crosby, A. H., Jeffery, S., & Patton, M. A. (2007). The natural history of Noonan syndrome: A long-term follow-up study. *Archives of Disease in Childhood*, *92*(2), 128–132. https://doi.org/10.1136/adc.2006.104547.

Shen, M. H., Harper, P. S., & Upadhyaya, M. (1996). Molecular genetics of neurofibromatosis type 1 (NF1). *Journal of Medical Genetics*, *33*(1), 2–17.

Sherman, L. S., Atit, R., Rosenbaum, T., Cox, A. D., & Ratner, N. (2000). Single cell Ras-GTP analysis reveals altered Ras activity in a subpopulation of neurofibroma Schwann cells but not fibroblasts. *The Journal of Biological Chemistry*, *275*(39), 30740–30745. https://doi.org/10.1074/jbc.M001702200.

Smirniotopoulos, J. G., & Murphy, F. M. (1992). The phakomatoses. *AJNR. American Journal of Neuroradiology*, *13*(2), 725–746.

Spencer, E., Davis, J., Mikhail, F., Fu, C., Vijzelaar, R., Zackai, E. H., et al. (2011). Identification of SPRED1 deletions using RT-PCR, multiplex ligation-dependent probe amplification and quantitative PCR. *American Journal of Medical Genetics. Part A*, *155A*(6), 1352–1359. https://doi.org/10.1002/ajmg.a.33894.

Spurlock, G., Bennett, E., Chuzhanova, N., Thomas, N., Jim, H. P., Side, L., et al. (2009). SPRED1 mutations (Legius syndrome): Another clinically useful genotype for dissecting the neurofibromatosis type 1 phenotype. *Journal of Medical Genetics*, *46*(7), 431–437. https://doi.org/10.1136/jmg.2008.065474.

Stark, Z., Gillessen-Kaesbach, G., Ryan, M. M., Cirstea, I. C., Gremer, L., Ahmadian, M. R., et al. (2012). Two novel germline KRAS mutations: Expanding the molecular and clinical phenotype. *Clinical Genetics*, *81*(6), 590–594. https://doi.org/10.1111/j.1399-0004.2011.01754.x.

Steinmann, K., Cooper, D. N., Kluwe, L., Chuzhanova, N. A., Senger, C., Serra, E., et al. (2007). Type 2 NF1 deletions are highly unusual by virtue of the absence of nonallelic homologous recombination hotspots and an apparent preference for female mitotic recombination. *American Journal of Human Genetics*, *81*(6), 1201–1220. https://doi.org/10.1086/522089.

Stevenson, D. A., Viskochil, D. H., Rope, A. F., & Carey, J. C. (2006). Clinical and molecular aspects of an informative family with neurofibromatosis type 1 and Noonan phenotype. *Clinical Genetics*, *69*(3), 246–253. https://doi.org/10.1111/j.1399-0004.2006.00576.x.

Stewart, R. A., Sanda, T., Widlund, H. R., Zhu, S., Swanson, K. D., Hurley, A. D., et al. (2010). Phosphatase-dependent and -independent functions of Shp2 in neural crest cells underlie LEOPARD syndrome pathogenesis. *Developmental Cell*, *18*(5), 750–762. https://doi.org/10.1016/j.devcel.2010.03.009.

Stiller, C. A., Chessells, J. M., & Fitchett, M. (1994). Neurofibromatosis and childhood leukaemia/lymphoma: A population-based UKCCSG study. *British Journal of Cancer, 70*(5), 969–972.

Strullu, M., Caye, A., Lachenaud, J., Cassinat, B., Gazal, S., Fenneteau, O., et al. (2014). Juvenile myelomonocytic leukaemia and Noonan syndrome. *Journal of Medical Genetics, 51*(10), 689–697. https://doi.org/10.1136/jmedgenet-2014-102611.

Stucky, C. C., Johnson, K. N., Gray, R. J., Pockaj, B. A., Ocal, I. T., Rose, P. S., et al. (2012). Malignant peripheral nerve sheath tumors (MPNST): The Mayo Clinic experience. *Annals of Surgical Oncology, 19*(3), 878–885. https://doi.org/10.1245/s10434-011-1978-7.

Swaminathan, G., & Tsygankov, A. Y. (2006). The Cbl family proteins: Ring leaders in regulation of cell signaling. *Journal of Cellular Physiology, 209*(1), 21–43. https://doi.org/10.1002/jcp.20694.

Swinburn, B. A., Yeong, M. L., Lane, M. R., Nicholson, G. I., & Holdaway, I. M. (1988). Neurofibromatosis associated with somatostatinoma: A report of two patients. *Clinical Endocrinology, 28*(4), 353–359. https://doi.org/10.1111/j.1365-2265.1988.tb03666.x.

Szudek, J., Birch, P., & Friedman, J. M. (2000a). Growth charts for young children with neurofibromatosis 1 (NF1). *American Journal of Medical Genetics, 92*(3), 224–228.

Szudek, J., Birch, P., & Friedman, J. M. (2000b). Growth in North American white children with neurofibromatosis 1 (NF1). *Journal of Medical Genetics, 37*(12), 933–938.

Tajan, M., Paccoud, R., Branka, S., Edouard, T., & Yart, A. (2018). The RASopathy family: Consequences of germline activation of the RAS/MAPK pathway. *Endocrine Reviews, 39*(5), 676–700. https://doi.org/10.1210/er.2017-00232.

Tartaglia, M., Martinelli, S., Stella, L., Bocchinfuso, G., Flex, E., Cordeddu, V., et al. (2006). Diversity and functional consequences of germline and somatic PTPN11 mutations in human disease. *American Journal of Human Genetics, 78*(2), 279–290. https://doi.org/10.1086/499925.

Tartaglia, M., Mehler, E. L., Goldberg, R., Zampino, G., Brunner, H. G., Kremer, H., et al. (2001). Mutations in PTPN11, encoding the protein tyrosine phosphatase SHP-2, cause Noonan syndrome. *Nature Genetics, 29*(4), 465–468. https://doi.org/10.1038/ng772.

Thiagalingam, S., Flaherty, M., Billson, F., & North, K. (2004). Neurofibromatosis type 1 and optic pathway gliomas: Follow-up of 54 patients. *Ophthalmology, 111*(3), 568–577. https://doi.org/10.1016/j.ophtha.2003.06.008.

Torres Nupan, M. M., Velez Van Meerbeke, A., Lopez Cabra, C. A., & Herrera Gomez, P. M. (2017). Cognitive and behavioral disorders in children with neurofibromatosis type 1. *Frontiers in Pediatrics, 5*, 227. https://doi.org/10.3389/fped.2017.00227.

Tosti, A., Misciali, C., Borrello, P., Fanti, P. A., Bardazzi, F., & Patrizi, A. (1991). Loose anagen hair in a child with Noonan's syndrome. *Dermatologica, 182*(4), 247–249. https://doi.org/10.1159/000247806.

Tosti, A., Peluso, A. M., Misciali, C., Venturo, N., Patrizi, A., & Fanti, P. A. (1997). Loose anagen hair. *Archives of Dermatology, 133*(9), 1089–1093.

Umeki, I., Niihori, T., Abe, T., Kanno, S. I., Okamoto, N., Mizuno, S., et al. (2019). Delineation of LZTR1 mutation-positive patients with Noonan syndrome and identification of LZTR1 binding to RAF1-PPP1CB complexes. *Human Genetics, 138*(1), 21–35. https://doi.org/10.1007/s00439-018-1951-7.

Uusitalo, E., Kallionpaa, R. A., Kurki, S., Rantanen, M., Pitkaniemi, J., Kronqvist, P., et al. (2017). Breast cancer in neurofibromatosis type 1: Overrepresentation of unfavourable prognostic factors. *British Journal of Cancer, 116*(2), 211–217. https://doi.org/10.1038/bjc.2016.403.

Valentin, T., Le Cesne, A., Ray-Coquard, I., Italiano, A., Decanter, G., Bompas, E., et al. (2016). Management and prognosis of malignant peripheral nerve sheath tumors: The experience of the French Sarcoma Group (GSF-GETO). *European Journal of Cancer, 56*, 77–84. https://doi.org/10.1016/j.ejca.2015.12.015.

Van Es, S., North, K. N., McHugh, K., & De Silva, M. (1996). MRI findings in children with neurofibromatosis type 1: A prospective study. *Pediatric Radiology*, *26*(7), 478–487.

Viskochil, D., Buchberg, A. M., Xu, G., Cawthon, R. M., Stevens, J., Wolff, R. K., et al. (1990). Deletions and a translocation interrupt a cloned gene at the neurofibromatosis type 1 locus. *Cell*, *62*(1), 187–192.

Wakioka, T., Sasaki, A., Kato, R., Shouda, T., Matsumoto, A., Miyoshi, K., et al. (2001). Spred is a Sprouty-related suppressor of Ras signalling. *Nature*, *412*(6847), 647–651. https://doi.org/10.1038/35088082.

Wallace, M. R., Marchuk, D. A., Andersen, L. B., Letcher, R., Odeh, H. M., Saulino, A. M., et al. (1990). Type 1 neurofibromatosis gene: Identification of a large transcript disrupted in three NF1 patients. *Science*, *249*(4965), 181–186.

Walther, M. M., Herring, J., Enquist, E., Keiser, H. R., & Linehan, W. M. (1999). von Recklinghausen's disease and pheochromocytomas. *The Journal of Urology*, *162*(5), 1582–1586.

Wang, X., Teer, J. K., Tousignant, R. N., Levin, A. M., Boulware, D., Chitale, D. A., et al. (2018). Breast cancer risk and germline genomic profiling of women with neurofibromatosis type 1 who developed breast cancer. *Genes, Chromosomes & Cancer*, *57*(1), 19–27. https://doi.org/10.1002/gcc.22503.

Watson, K. L., Al Sannaa, G. A., Kivlin, C. M., Ingram, D. R., Landers, S. M., Roland, C. L., et al. (2017). Patterns of recurrence and survival in sporadic, neurofibromatosis Type 1-associated, and radiation-associated malignant peripheral nerve sheath tumors. *Journal of Neurosurgery*, *126*(1), 319–329. https://doi.org/10.3171/2015.12.JNS152443.

Weber, S. M., & Carroll, S. L. (2021). The role of R-Ras proteins in normal and pathologic migration and morphologic change. *The American Journal of Pathology*. https://doi.org/10.1016/j.ajpath.2021.05.008.

Widemann, B. C., Dombi, E., Gillespie, A., Wolters, P. L., Belasco, J., Goldman, S., et al. (2014). Phase 2 randomized, flexible crossover, double-blinded, placebo-controlled trial of the farnesyltransferase inhibitor tipifarnib in children and young adults with neurofibromatosis type 1 and progressive plexiform neurofibromas. *Neuro-Oncology*, *16*(5), 707–718. https://doi.org/10.1093/neuonc/nou004.

Widemann, B. C., Salzer, W. L., Arceci, R. J., Blaney, S. M., Fox, E., End, D., et al. (2006). Phase I trial and pharmacokinetic study of the farnesyltransferase inhibitor tipifarnib in children with refractory solid tumors or neurofibromatosis type I and plexiform neurofibromas. *Journal of Clinical Oncology*, *24*(3), 507–516. https://doi.org/10.1200/JCO.2005.03.8638.

Williams, V. C., Lucas, J., Babcock, M. A., Gutmann, D. H., Korf, B., & Maria, B. L. (2009). Neurofibromatosis type 1 revisited. *Pediatrics*, *123*(1), 124–133. https://doi.org/10.1542/peds.2007-3204.

Wimmer, K., Roca, X., Beiglbock, H., Callens, T., Etzler, J., Rao, A. R., et al. (2007). Extensive in silico analysis of NF1 splicing defects uncovers determinants for splicing outcome upon 5' splice-site disruption. *Human Mutation*, *28*(6), 599–612. https://doi.org/10.1002/humu.20493.

Wimmer, K., Yao, S., Claes, K., Kehrer-Sawatzki, H., Tinschert, S., De Raedt, T., et al. (2006). Spectrum of single- and multiexon NF1 copy number changes in a cohort of 1,100 unselected NF1 patients. *Genes, Chromosomes & Cancer*, *45*(3), 265–276. https://doi.org/10.1002/gcc.20289.

Witt, D. R., Keena, B. A., Hall, J. G., & Allanson, J. E. (1986). Growth curves for height in Noonan syndrome. *Clinical Genetics*, *30*(3), 150–153. https://doi.org/10.1111/j.1399-0004.1986.tb00587.x.

Wu, X., Yin, J., Simpson, J., Kim, K. H., Gu, S., Hong, J. H., et al. (2012). Increased BRAF heterodimerization is the common pathogenic mechanism for noonan syndrome-associated RAF1 mutants. *Molecular and Cellular Biology*, *32*(19), 3872–3890. https://doi.org/10.1128/MCB.00751-12.

Xiao, Z., Carrasco, R., Kinneer, K., Sabol, D., Jallal, B., Coats, S., et al. (2012). EphB4 promotes or suppresses Ras/MEK/ERK pathway in a context-dependent manner: Implications for EphB4 as a cancer target. *Cancer Biology & Therapy*, *13*(8), 630–637. https://doi.org/10.4161/cbt.20080.

Xu, G. F., Lin, B., Tanaka, K., Dunn, D., Wood, D., Gesteland, R., et al. (1990). The catalytic domain of the neurofibromatosis type 1 gene product stimulates ras GTPase and complements ira mutants of S. cerevisiae. *Cell*, *63*(4), 835–841.

Xu, G. F., O'Connell, P., Viskochil, D., Cawthon, R., Robertson, M., Culver, M., et al. (1990). The neurofibromatosis type 1 gene encodes a protein related to GAP. *Cell*, *62*(3), 599–608.

Yart, A., & Edouard, T. (2018). Noonan syndrome: An update on growth and development. *Current Opinion in Endocrinology, Diabetes, and Obesity*, *25*(1), 67–73. https://doi.org/10.1097/MED.0000000000000380.

Young, L. C., Hartig, N., Boned Del Rio, I., Sari, S., Ringham-Terry, B., Wainwright, J. R., et al. (2018). SHOC2-MRAS-PP1 complex positively regulates RAF activity and contributes to Noonan syndrome pathogenesis. *Proceedings of the National Academy of Sciences of the United States of America*, *115*(45), E10576–E10585. https://doi.org/10.1073/pnas.1720352115.

Zou, C., Smith, K. D., Liu, J., Lahat, G., Myers, S., Wang, W. L., et al. (2009). Clinical, pathological, and molecular variables predictive of malignant peripheral nerve sheath tumor outcome. *Annals of Surgery*, *249*(6), 1014–1022. https://doi.org/10.1097/SLA.0b013e3181a77e9a.

FSC
www.fsc.org
MIX
Papier | Fördert gute Waldnutzung
FSC® C083411

Kolibri 360 GmbH
Ferdinand-Jühlke-Straße 7
99095 Erfurt, Deutschland
produktsicherheit@kolibri360.de